Natural Products and Neuroprotection

Natural Products and Neuroprotection

Special Issue Editors

Cristina Angeloni
David Vauzour

MDPI • Basel • Beijing • Wuhan • Barcelona • Belgrade • Manchester • Tokyo • Cluj • Tianjin

Special Issue Editors
Cristina Angeloni
University of Camerino
Italy

David Vauzour
University of East Anglia
UK

Editorial Office
MDPI
St. Alban-Anlage 66
4052 Basel, Switzerland

This is a reprint of articles from the Special Issue published online in the open access journal *International Journal of Molecular Sciences* (ISSN 1422-0067) (available at: https://www.mdpi.com/journal/ijms/special_issues/NP_Neuroprotection).

For citation purposes, cite each article independently as indicated on the article page online and as indicated below:

LastName, A.A.; LastName, B.B.; LastName, C.C. Article Title. *Journal Name* **Year**, *Article Number*, Page Range.

ISBN 978-3-03936-216-5 (Hbk)
ISBN 978-3-03936-217-2 (PDF)

Contents

About the Special Issue Editors

Cristina Angeloni is Professor of Biochemistry at the School of Pharmacy of the University of Camerino, Italy. She received an MS degree in Computer Science in 1992, an MS degree in Food Science and Technology in 2000, a Ph.D degree in Biochemistry and Physiopathology of Aging in 2005, and a Master's degree in Bioinformatics in 2005 from the University of Bologna, Italy. She is author of more than seventy peer reviewed articles and of four book chapters. She serves on the editorial board of *Oxidative Medicine and Cellular Longevity* and has been the editor of eight special issues. The main focus of her research is the study of the protective/preventive role of nutraceutical bioactive components of the diet in the prevention/counteraction of chronic degenerative diseases such as cardiovascular and neurodegenerative diseases. In particular, she has investigated the protective mechanisms of nutraceutical compounds, studying the radical scavenging activity, the induction of phase II enzymes, the inhibition of apoptosis, and the modulation of signal transduction pathways in in-vitro and in-vivo models.

David Vauzour received his PhD from the Faculty of Pharmacy, University of Montpellier (France) in 2004. His research over the last 15 years, at the University of Reading (2005–2011), and the Norwich Medical School, University of East Anglia, UK (2011–present) has focused on investigating the molecular mechanisms that underlie the positive correlation between the consumption of diets rich in fruits and vegetables and a decreased risk of (neuro)degenerative disorders, and on ways to develop novel dietary strategies to delay brain ageing, cognitive decline and cardiovascular disease. In this context, his initial work has provided considerable insight into the potential for natural products to promote human vascular function, decrease (neuro)inflammation, enhance memory, learning and neuro-cognitive performance and slow the progression of Alzheimer's and Parkinson's diseases. His recent interests concern how food bioactives modulate APOE-genotype-induced cardiovascular risk and neurodegenerative disorders and their underlying mechanisms. To date, Dr Vauzour has published over 80 peer-reviewed articles, and he currently serves as Associate Editor for the journal Nutrition and Healthy Aging. In addition, he is a member of the editorial boards of *Nature Scientific Reports (Neuroscience), PharmaNutrition* and *Peer J (Pharmacology)*. He is currently the co-the Chair of the ILSI Europe Nutrition and Mental Performance Task Force.

International Journal of
Molecular Sciences

MDPI

Editorial

Natural Products and Neuroprotection

Cristina Angeloni [1,*] and David Vauzour [2,*]

1 School of Pharmacy, University of Camerino, 62032 Camerino, Italy
2 Norwich Medical School, University of East Anglia, Norwich NR4 7UQ, UK
* Correspondence: cristina.angeloni@unicam.it (C.A.); D.Vauzour@uea.ac.uk (D.V.)

Received: 1 November 2019; Accepted: 5 November 2019; Published: 7 November 2019

Neurodegenerative diseases are among the most serious health problems affecting millions of people worldwide, and their incidence is dramatically growing together with increased lifespan [1]. These diseases are a heterogeneous group of chronic, progressive disorders characterized by the gradual loss of neurons in the central nervous system, which leads to deficits in specific brain functions. The most common neurodegenerative diseases are Alzheimer's disease (AD), Parkinson's disease (PD), amyotrophic lateral sclerosis, multiple sclerosis, and Huntington's disease. While the etiology of most neurodegenerative diseases is mainly unknown, it is largely recognized that these disorders share common molecular and cellular characteristics that contribute to their progression. These include oxidative stress, mitochondrial dysfunction, protein misfolding, excitotoxicity, dysregulation of calcium homeostasis, and inflammation [2–5]. There are currently no therapeutic approaches to cure or even halt the progression of these disorders, and existing treatments remain largely palliative. In this context, natural products, because of their broad spectrum of pharmacological and biological activities, are considered promising alternatives for the treatment of neurodegeneration as they might play a role in drug development and discovery. A number of studies showed health-promoting properties in the use of natural products as potential therapeutics for neurodegeneration [6–8]. Natural compounds have been reported to possess different biological activities, including antioxidant, anti-inflammatory, and antiapoptotic effects [9,10]. Moreover, natural compounds have been recently shown to counteract protein misfolding and to modulate autophagy and proteasome activity [11,12].

The papers published as part of this Special Issue deal with two different forms of natural products: extracts and isolated compounds. The study of the bioactivity of the extracts is extremely important as in vivo natural compounds are usually obtained through the diet as a complex mixture. The importance of extracts is further supported by the fact that many studies have demonstrated the synergistic effect of the combination of different natural products [13]. On the other hand, the investigation of the activity of specifically isolated natural products can be also important to understand their cellular and molecular mechanisms and to define what are the specific bioactive components in extracts or foods.

Research conducted by Sabti M. and colleagues [14] elucidated the molecular mechanisms underlying the relaxant and anxiolytic properties of *Lippia citriodora* (VEE) and verbascoside (Vs), a phenypropanoid glycoside. *Lippia citriodora* is a plant from the Verbenaceae family and is cultivated in North Africa, Southern Europe and the Middle East. In this study both an in vivo mouse model of anxiety and depression and the in vitro SH-SY5Y cell line were employed. In particular the authors evidenced a relaxation effect of high doses of VEE associated with the regulation of genes playing key roles in calcium homeostasis (calcium channels), cyclic AMP (cAMP) production and energy metabolism. Low doses of VEE and Vs showed an antidepressant-like effect by enhancing brain-derived neurotrophic factor (BDNF), noradrenalin, serotonin and dopamine expressions. These results were further confirmed in vitro as both VEE and Vs enhanced cell viability, mitochondrial activity and calcium uptake in SH-SY5Y cells.

In their manuscript, Lee Y.G. et al. [15] isolated four flavonols, three flavones, four flavanonols, and one flavanone from a *Chionanthus retusus* extract, a deciduous tree of the Oleaceae family mainly cultivated in Korea, Japan and China. Eight of these flavonoids demonstrated to be effective in counteracting inflammation by inhibiting nitric oxide (NO) production in RAW 264.7 cells activated by lipopolysaccharide. In addition, these flavonoids showed a neuroprotective activity counteracting glutamate-induced cell toxicity increasing heme oxygenase 1 (HO-1) protein expression in mouse hippocampal HT22 cells.

Similarly, Jang Y. et al. [16] demonstrated that auraptene (AUR), a 7-geranyloxylated coumarin isolated from citrus fruit, is able to counteract neurotoxin-induced reduction of mitochondrial respiration and to inhibit reactive oxygen species (ROS) generation in SN4741 mouse embryonic substantia nigra dopaminergic neuronal cell line. Moreover, they observed, in a MPTP-induced PD mouse model, that AUR treatment improved movement deficits in association with an increase in the number of dopaminergic neurons in the substantia nigra.

Chiroma S.M. et al. [17] investigated the neuroprotective effect of *Centella asiatica* (CA), a plant from the family of Apiaceae, in a rat model of neurodegeneration induced by d-galactose/aluminum chloride (d-gal/AlC13). These authors previously observed that CA extract can attenuate cognitive deficits in this model of neurodegeneration and can also prevent morphological aberrations in the CA1 region of hippocampus [18]. In the paper published in this Special Issue, they demonstrated that CA significantly increased the levels of protein phosphatase 2 and decreased the levels of glycogen synthase kinase-3 beta. Moreover, CA extract also counteracted apoptosis as it increased the expression of the Bcl-2 mRNA level.

Finally, Javed H. et al. [19] demonstrated the neuroprotective effect of thymol, a dietary monoterpene phenol, in a rat model of PD. In particular, neurodegeneration was induced by rotenone at a dose of 2.5 mg/kg body weight for four weeks. Thymol, co-administered to rotenone for four weeks at a dose of 50 mg/kg body weight, significantly attenuated dopaminergic neuronal loss, oxidative stress and inflammation suggesting a protective effect of thymol in rotenone-induced PD.

Along with research papers, different reviews are also presented in this Special Issue.

As previously underlined, proteostasis failure plays a crucial role in the context of ageing and neurodegeneration. Therefore, natural products targeting the proteostasis elements emerge as a promising neuroprotective therapeutic approach to prevent or ameliorate the progression of these disorders. Cuanalo-Contreras K. et al. [20] focused on this aspect and revised the current knowledge regarding the use of natural products as modulators of different components of the proteostasis machinery to counteract neurodegeneration. The majority of natural modulators of the proteostasis network are of plant-origin, however some compounds of marine-animal-origin are also emerging. They concluded that further studies are required to understand the precise mechanism of action of the natural proteostasis activators, their off-target effects and their in vivo bioavailability. In their review, Cho B. et al. [21] focused on the effect on natural products on the proteostasis elements such as ubiquitin-proteasome system and autophagy (mitophagy) in experimental PD models. Moreover, in the same experimental models, they also revised the neuroprotective effects of natural products on mitochondrial dysfunction, oxidative stress, and hormesis. They summarized the efforts to use natural extracts as lead compounds for the design of novel pharmacological candidates for the treatment of age-related PD. Finally, they addressed two main limitations in the use of natural compounds in counteracting neurodegeneration: the differences of experimental design, such as the quality of the extracts and the forms of dosage, of the studies and the unclear therapeutic mechanism of natural compounds.

Taking into account these two limitations Di Paolo M. et al. [22] analyzed the ethical framework of the potential clinical use of natural products to counteract neurodegeneration, with particular attention paid to the principles of biomedical ethics. They concluded that natural products could represent a great promise for the treatment of neurodegeneration, where traditional therapies, via synthetic drugs, only act to alleviate symptoms. However, lack of knowledge on the efficacy and safety of many natural products underscores the urgent need for further investigation to better characterize the therapeutic mechanism of natural products in order to promote patient safety and ethical care.

Park J.Y. et al. [23] revised the current research on the structural diversity, biosynthesis, and pleiotropic neuronal functions of ascaroside (ascr) pheromones and their implications in animal physiology. Pheromones are neuronal signals that stimulate conspecific individuals to react to environmental stressors or stimuli. The authors also discuss the concentration and stage-dependent pleiotropic neuronal functions of ascr pheromones. They suggest that in the future, translation of the knowledge of nematode ascr pheromones to higher animals might be beneficial, as it has been observed that ascr has some anti-inflammatory effects in mice.

Pervin M. et al. [24] discuss the function of (−)-epigallocatechin gallate (EGCG) and its microbial ring-fission metabolites in the brain as neuroprotective agent. EGCG, the main green tea catechin, is an ester of (−)-epigallocatechin (EGC) and gallic acid (GA). Despite the great number of studies on the neuroprotective effects of green tea catechins against neurological disorders, it should take into account that the concentration of EGCG in systemic circulation is very low and EGCG disappears within several hours. EGCG undergoes extensive metabolism and recent studies suggest that metabolites of EGCG may play an important role, alongside the beneficial activities of EGCG, in reducing neurodegenerative diseases.

Barbalace M.C. et al. [25] focused on the effect of marine algae on neuroinflammation, one of the main contributors to the onset and progression of neurodegenerative diseases. As pointed out by Cuanalo-Contreras K. et al., marine organisms represent a vast source of natural compounds, and among them, algae are an appreciated source of important bioactive components. Barbalace et al. revised the numerous anti-inflammatory compounds that have been recently isolated from marine algae with potential protective efficacy against neuroinflammation.

Polyphenols are among the most studied dietary molecules probably for their multiple and often overlapping reported modes of action. Epidemiological studies suggest a strong association between polyphenol consumption and reduced prevalence of various neurodegenerative diseases; however, ambiguity still exists as to the significance of their influence on human health. Renaud J. and Martinoli M.G. [26] analyzed the characteristics and functions of polyphenols that determine their potential therapeutic actions in neurodegenerative disorders. In particular, they discuss the properties that may influence the functionality and bioavailability of dietary polyphenols in the central nervous system (CNS) with a particular focus on therapeutic applications and limitations.

Among polyphenols, curcumin, a component of *Curcuma longa*, is currently considered one of the most effective nutritional antioxidants due to its activity in multiple antioxidant and anti-inflammatory pathways involved in neurodegeneration. Mhillaj E. et al. [27] provides a summary of the main findings involving the heme oxygenase/biliverdin reductase system as a valid target in mediating the potential neuroprotective properties of curcumin. Moreover, they address the pharmacokinetic properties and concerns about curcumin's safety profile.

Maher P. [27] focused on a wide class of polyphenols, flavonoids. Among the huge number of polyphenols, several epidemiological studies have specifically highlighted the potential beneficial role of flavonoids to counteract neurodegeneration. In particular the author discusses the beneficial effects of multiple flavonoids in different models of neurodegenerative diseases and identified common mechanisms of action. As outlined by other authors of this Special Issue, the conclusions state that further investigations should be carried out in order to use flavonoids in the treatment of neurodegenerative diseases.

Infante-Garcia C. and Garcia-Alloza M. [28] reviewed natural compounds with a protective activity against brain neurodegeneration in animal models of diabetes mellitus, taking into account several therapeutic targets: inflammation and oxidative stress, vascular damage, neuronal loss or cognitive impairment. Diabetic brain is characterized by micro and macrostructural changes, such as neurovascular deterioration or neuroinflammation that lead to neurodegeneration and progressive cognition dysfunction. The authors evidenced that natural compounds and extracts show antioxidant and anti-inflammatory activities at a central level, as well as a relevant capacity to reduce vascular damage, contributing altogether to limit neurodegeneration and cognitive derived alterations. In their conclusion the authors highlighted that natural products could contribute to expand therapeutic options to treat or reduce central complications associated with diabetes mellitus.

Andrade S. et al. [29] focus their attention on a specific neurodegenerative disease, AD, and discuss both the natural compounds already in clinical trial phase and other natural compounds with known potentially beneficial effects in AD in a preclinical development stage. Regarding the preclinical studies, only the most recent reported works have been considered. Clinical trials have demonstrated that different compounds appear to be effective for AD therapy, on the contrary others have failed in human trials. Natural compounds in earlier phases of research need further studies to uncover their therapeutic potential for AD.

Berezowska M. et al. [21] reviewed the effects of vitamin D in multiple sclerosis on pathology and symptoms. Based on specific criteria, they selected ten studies with a size ranging from 40 to 94 people and with a duration of the intervention from 12 to 96 weeks; all the studies compared the use of vitamin D with a placebo or low dose vitamin D. One trial found a significant effect on Expanded Disability Status Scale (EDSS) score, three demonstrated a significant change in serum cytokines level, one found benefits in enhancing lesions and, interestingly, three studies reported no serious adverse events in the use of vitamin D.

In conclusion, the papers published in this Special Issue, despite addressing different topics, can be considered an important contribution to the knowledge of the neuroprotective effect of natural products, and present a great deal of information related to both the benefits but also the limitations of their use in counteracting neurodegeneration.

Conflicts of Interest: The authors declare no conflicts of interest.

References

1. Erkkinen, M.G.; Kim, M.O.; Geschwind, M.D. Clinical Neurology and Epidemiology of the Major Neurodegenerative Diseases. *Cold Spring Harb. Perspect. Biol.* **2018**, *10*. [CrossRef] [PubMed]
2. Ilieva, H.; Polymenidou, M.; Cleveland, D.W. Non-cell autonomous toxicity in neurodegenerative disorders: ALS and beyond. *J. Cell Biol.* **2009**, *187*, 761–772. [CrossRef] [PubMed]
3. Taylor, J.P.; Brown, R.H.; Cleveland, D.W. Decoding ALS: From genes to mechanism. *Nature* **2016**, *539*, 197–206. [CrossRef]
4. Magalingam, K.B.; Radhakrishnan, A.; Ping, N.S.; Haleagrahara, N. Current Concepts of Neurodegenerative Mechanisms in Alzheimer's Disease. *Biomed. Res. Int.* **2018**, *2018*, 3740461. [CrossRef] [PubMed]
5. Zeng, X.S.; Geng, W.S.; Jia, J.J.; Chen, L.; Zhang, P.P. Cellular and Molecular Basis of Neurodegeneration in Parkinson Disease. *Front. Aging Neurosci.* **2018**, *10*, 109. [CrossRef]
6. Bui, T.T.; Nguyen, T.H. Natural product for the treatment of Alzheimer's disease. *J. Basic Clin. Physiol. Pharmacol.* **2017**, *28*, 413–423. [CrossRef] [PubMed]
7. Calis, Z.; Mogulkoc, R.; Baltaci, A.K. The roles of Flavonoles/Flavonoids in Neurodegeneration and Neuroinflammation. *Mini Rev. Med. Chem.* **2019**. [CrossRef]
8. Tarozzi, A.; Angeloni, C.; Malaguti, M.; Morroni, F.; Hrelia, S.; Hrelia, P. Sulforaphane as a potential protective phytochemical against neurodegenerative diseases. *Oxid. Med. Cell Longev.* **2013**, *2013*, 415078. [CrossRef]
9. Flanagan, E.; Müller, M.; Hornberger, M.; Vauzour, D. Impact of Flavonoids on Cellular and Molecular Mechanisms Underlying Age-Related Cognitive Decline and Neurodegeneration. *Curr. Nutr. Rep.* **2018**, *7*, 49–57. [CrossRef]
10. Angeloni, C.; Giusti, L.; Hrelia, S. New neuroprotective perspectives in fighting oxidative stress and improving cellular energy metabolism by oleocanthal. *Neural Regen. Res.* **2019**, *14*, 1217–1218. [CrossRef]
11. Perrone, L.; Squillaro, T.; Napolitano, F.; Terracciano, C.; Sampaolo, S.; Melone, M.A.B. The Autophagy Signaling Pathway: A Potential Multifunctional Therapeutic Target of Curcumin in Neurological and Neuromuscular Diseases. *Nutrients* **2019**, *11*. [CrossRef] [PubMed]
12. Gan, N.; Wu, Y.C.; Brunet, M.; Garrido, C.; Chung, F.L.; Dai, C.; Mi, L. Sulforaphane activates heat shock response and enhances proteasome activity through up-regulation of Hsp27. *J. Biol. Chem.* **2010**, *285*, 35528–35536. [CrossRef] [PubMed]

13. Marrazzo, P.; Angeloni, C.; Hrelia, S. Combined Treatment with Three Natural Antioxidants Enhances Neuroprotection in a SH-SY5Y 3D Culture Model. *Antioxidants* **2019**, *8*, 420. [CrossRef] [PubMed]
14. Sabti, M.; Sasaki, K.; Gadhi, C.; Isoda, H. Elucidation of the Molecular Mechanism Underlying. *Int. J. Mol. Sci.* **2019**, *20*, 3556. [CrossRef]
15. Lee, Y.G.; Lee, H.; Jung, J.W.; Seo, K.H.; Lee, D.Y.; Kim, H.G.; Ko, J.H.; Lee, D.S.; Baek, N.I. Flavonoids from Chionanthus retusus (Oleaceae) Flowers and Their Protective Effects against Glutamate-Induced Cell Toxicity in HT22 Cells. *Int. J. Mol. Sci.* **2019**, *20*, 3517. [CrossRef]
16. Jang, Y.; Choo, H.; Lee, M.J.; Han, J.; Kim, S.J.; Ju, X.; Cui, J.; Lee, Y.L.; Ryu, M.J.; Oh, E.S.; et al. Auraptene Mitigates Parkinson's Disease-Like Behavior by Protecting Inhibition of Mitochondrial Respiration and Scavenging Reactive Oxygen Species. *Int. J. Mol. Sci.* **2019**, *20*, e3409. [CrossRef]
17. Chiroma, S.M.; Baharuldin, M.T.H.; Mat Taib, C.N.; Amom, Z.; Jagadeesan, S.; Ilham Adenan, M.; Mahdi, O.; Moklas, M.A.M. Centella asiatica Protects d -Galactose/$AlCl_3$ Mediated Alzheimer's Disease-Like Rats via PP2A/GSK-3β Signaling Pathway in Their Hippocampus. *Int. J. Mol. Sci.* **2019**, *20*, 1871. [CrossRef]
18. Chiroma, S.M.; Hidayat Baharuldin, M.T.; Mat Taib, C.N.; Amom, Z.; Jagadeesan, S.; Adenan, M.I.; Mohd Moklas, M.A. Protective effect of Centella asiatica against. *Biomed. Pharmacother.* **2019**, *109*, 853–864. [CrossRef]
19. Javed, H.; Azimullah, S.; Meeran, M.F.N.; Ansari, S.A.; Ojha, S. Neuroprotective Effects of Thymol, a Dietary Monoterpene Against Dopaminergic Neurodegeneration in Rotenone-Induced Rat Model of Parkinson's Disease. *Int. J. Mol. Sci.* **2019**, *20*, 1538. [CrossRef]
20. Cuanalo-Contreras, K.; Moreno-Gonzalez, I. Natural Products as Modulators of the Proteostasis Machinery: Implications in Neurodegenerative Diseases. *Int. J. Mol. Sci.* **2019**, *20*, 4666. [CrossRef]
21. Berezowska, M.; Coe, S.; Dawes, H. Effectiveness of Vitamin D Supplementation in the Management of Multiple Sclerosis: A Systematic Review. *Int. J. Mol. Sci.* **2019**, *20*, e1301. [CrossRef] [PubMed]
22. Di Paolo, M.; Papi, L.; Gori, F.; Turillazzi, E. Natural Products in Neurodegenerative Diseases: A Great Promise but an Ethical Challenge. *Int. J. Mol. Sci.* **2019**, *20*, e5170. [CrossRef] [PubMed]
23. Park, J.Y.; Joo, H.J.; Park, S.; Paik, Y.K. Ascaroside Pheromones: Chemical Biology and Pleiotropic Neuronal Functions. *Int. J. Mol. Sci.* **2019**, *20*, 3898. [CrossRef] [PubMed]
24. Pervin, M.; Unno, K.; Takagaki, A.; Isemura, M.; Nakamura, Y. Function of Green Tea Catechins in the Brain: Epigallocatechin Gallate and its Metabolites. *Int. J. Mol. Sci.* **2019**, *20*, 3630. [CrossRef] [PubMed]
25. Barbalace, M.C.; Malaguti, M.; Giusti, L.; Lucacchini, A.; Hrelia, S.; Angeloni, C. Anti-Inflammatory Activities of Marine Algae in Neurodegenerative Diseases. *Int. J. Mol. Sci.* **2019**, *20*, 3061. [CrossRef]
26. Renaud, J.; Martinoli, M.G. Considerations for the Use of Polyphenols as Therapies in Neurodegenerative Diseases. *Int. J. Mol. Sci.* **2019**, *20*, 1883. [CrossRef]
27. Maher, P. The Potential of Flavonoids for the Treatment of Neurodegenerative Diseases. *Int. J. Mol. Sci.* **2019**, *20*, 3056. [CrossRef]
28. Infante-Garcia, C.; Garcia-Alloza, M. Review of the Effect of Natural Compounds and Extracts on Neurodegeneration in Animal Models of Diabetes Mellitus. *Int. J. Mol. Sci.* **2019**, *20*, 2533. [CrossRef]
29. Andrade, S.; Ramalho, M.J.; Loureiro, J.A.; Pereira, M.D.C. Natural Compounds for Alzheimer's Disease Therapy: A Systematic Review of Preclinical and Clinical Studies. *Int. J. Mol. Sci.* **2019**, *20*, 2313. [CrossRef]

International Journal of
Molecular Sciences

MDPI

Article

Elucidation of the Molecular Mechanism Underlying *Lippia citriodora*(Lim.)-Induced Relaxation and Anti-Depression

Mouad Sabti [1,2], Kazunori Sasaki [1,3], Chemseddoha Gadhi [4] and Hiroko Isoda [1,2,3,*]

1 Alliance for Research on the Mediterranean and North Africa (ARENA), University of Tsukuba, 1-1-1 Tennodai, Tsukuba City 305-8572, Ibaraki, Japan
2 Tsukuba Life Science Innovation Program (T-LSI), University of Tsukuba, Tennodai 1-1-1, Tsukuba City 305-8577, Ibaraki, Japan
3 Interdisciplinary Research Center for Catalytic Chemistry, National Institute of Advanced Industrial Science and Technology (AIST), Tsukuba 305-8560, Japan
4 Faculty of Sciences Semlalia, Cadi Ayyad University, Avenue Prince MoulayAbdellah, BP 2390, 40000 Marrakesh, Morocco
* Correspondence: isoda.hiroko.ga@u.tsukuba.ac.jp; Tel.: +81-29-853-5775

Received: 24 June 2019; Accepted: 18 July 2019; Published: 20 July 2019

Abstract: *Lippia citriodora* ethanolic extract (VEE) and verbascoside (Vs), a phenypropanoid glycoside, have been demonstrated to exert relaxant and anxiolytic properties. However, the molecular mechanisms behind their effects are still unclear. In this work, we studied the effects and action mechanisms of VEE and Vs *in vivo* and *in vitro*, on human neurotypic SH-SY5Y cells.TST was conducted on mice treated orally with VEE (25, 50 and 100 mg/Kg), Vs (2.5 and 5 mg/Kg), Bupropion (20 mg/Kg) and Milli-Q water. Higher dose of VEE-treated mice showed an increase of immobility time compared to control groups, indicating an induction of relaxation. This effect was found to be induced by regulation of genes playing key roles in calcium homeostasis (calcium channels), cyclic AMP (cAMP) production and energy metabolism. On the other hand, low doses of VEE and Vs showed an antidepressant-like effect and was confirmed by serotonin, noradrenalin, dopamine and BDNF expressions. Finally, VEE and Vsenhancedcell viability, mitochondrial activity and calcium uptake *in vitro* confirming *in vivo* findings. Our results showed induction of relaxation and antidepressant-like effects depending on the administered dose of VEE and Vs, through modulation of cAMP and calcium.

Keywords: *Lippia citriodora*; VEE; Vs; relaxation; depression; mitochondria; cyclic AMP; calcium

1. Introduction

The Verbenaceae, commonly known as the verbena or vervain family, is composed of 35 genera containing around 1200 species [1]. They have been used for centuries as medicinal plants due to their beneficial effects to cure several ailments. One of the most important genera is *Lippia*, consisting of200 species exerting interesting biological activities [2]. *Lippia citriodora* K., also referred to as *Aloysiatriphylla*(L'Herit.), is commonly named lemon verbena, vervain or Louisa (Arabic). This species is native to South America and has been cultivated in Europe and North Africa mainly in Morocco [3]. All over Morocco, the plant is used as relaxant and sedative [4]. The herbal tea is traditionally used to alleviate insomnia and restlessness in adults as well as babies [5]. Furthermore, it has been used for its anti-inflammatory, antioxidant, antispasmodic effects and also used as a remedy for gastrointestinal disorders [2]. Recent studies have confirmed the antioxidant and spasmolytic activities of the infusion prepared of lemon verbena [6,7]. Verbena aqueous extract given to rats has proven the hypnotic effect of the plant by promoting sleep [8]. Polyphenols extracted from lemon verbena reduced the obesity burden and restored the mitochondrial activity through AMPK-dependent pathways [9].

Verbascoside (Vs), a major phenypropanoid glycoside, is the most abundant polyphenol in lemon verbena tea and its yield is reported to be around 3.94% (*w/w* dry weight of leaves) [10]. Vs contained in *Buddlejia davidii* and *Lippia multiflora* has already been proven to possess an antioxidant activity [11,12]. Vs has also shown an anti-inflammatory effect *in vitro* on macrophages and THP-1 cells [13,14]. Furthermore, Vs has been reported to exert an antimicrobial activity against *Staphylococcus aureus* and a neuroprotective effect, in vitro, on 1-methyl-4-phenylpyridinum ion-induced toxicity using PC12 cells [15,16]. Interestingly, intraperitoneal administration of Vs and lemon verbena aqueous and ethanolic extracts to mice promoted sleep and induced muscle relaxation, alongside alleviation of anxiety [17]. In addition to Vs, hastatoside (Hs) and verbenalin (Vn) are two abundant iridoids in verbena extract and have been proved to possess sleep-promoting effect [18]. To date, very little is known about the molecular mechanism by which lemon verbena or its compounds induce relaxation and act as anti-anxiety remedies.

In the present study, we investigated the effect of lemon verbena and Vs in mice and elucidated the molecular mechanisms underlying their effects in brain. Interestingly, the transcriptomic analysis in vivo showed regulation of genes implicated in activation of the mitochondrial function. Therefore, to confirm this finding we evaluated, in vitro, the effect of VEE and Vs on cells' ATP production using SH-SY5Y, a Human neurotypic cell line. Also, we assessed the toxicity of VEE, Vs, Hs, and Vn, in addition to neuroprotective effect on dexamethasone (Dex) neurotoxicity.

2. Results

2.1. Effect of VEE and Its Compounds on SH-SY5Y Cells'Viability

We performed the MTT assay to assess the effect of VEE on cell viability. We treated the cells with different concentrations of the extract which were 0.5, 1, 2.5 and 5 μg/mL of VEE. As shown in Figure 1A, all VEE concentrations increased cell viability significantly in a dose-dependent manner, with a higher value of 126.68 ± 7.81% at 2.5 μg/mL. The chemical analysis of various Verbenaceae plants, including *Lippia citriodora* and *Verbena officinalis*, showed a high abundance in Vs, also called acteoside, which is a phenylpropanoid glycoside [19–24]. In our study, we evaluated the cell viability of SH-SY5Y cells treated with 5, 50 and 100 μM of Vs, Hs and Vn. The results in Figure 1C show an increase of viable cells in a dose-dependent manner attaining 134.8 ± 3.8% at 100 μM in case of Vs. On the other hand, Hs and Vn decreased the cell viability significantly (Figure 1C). From these results, we selected Vs to be evaluated for its neuroprotective and energy metabolism effects.

In order to evaluate the neuroprotective activity, we used dexamethasone (Dex) as neurotoxic agent. VEE treatment protected SH-SY5Y cells fromDex toxicity with higher increase at 5 μg/mL (42.82% cell viability) (Figure 1B). Interestingly, cells co-treated with Vs and Dex showed an enhancement of cell viability by more than 30% compared to Dex-treated cells (Figure 1D). These data indicate neuroprotective effect exerted by VEE and Vs.

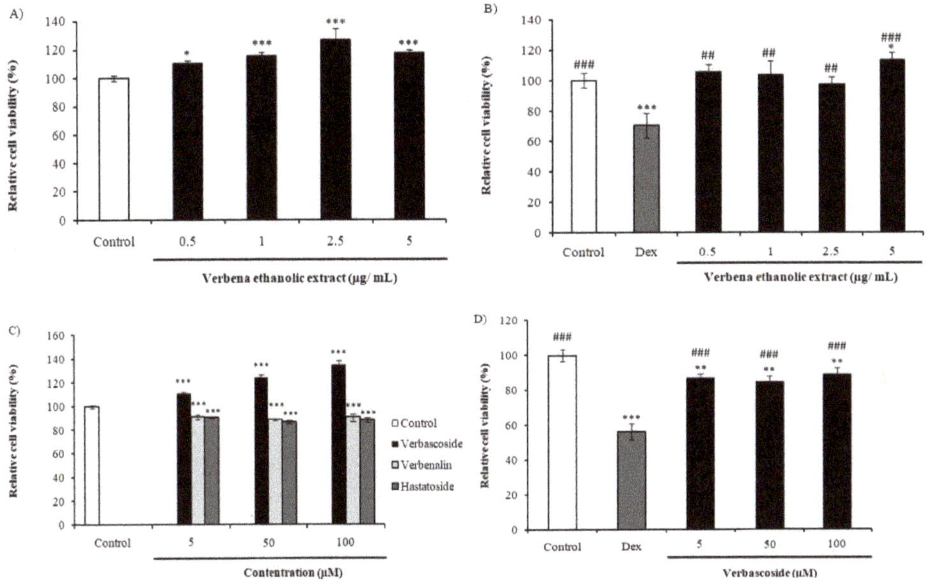

Figure 1. Relative cell viability of SH-SY5Y cells (**A**) treated with *Lippiacitriodora* ethanolic extract (VEE) at doses of 0.5, 1, 2.5 and 5 µg/mL, (**B**) co-treated with VEE and dexamethasone(Dex) (50 µM), (**C**) treated with verbascoside(Vs), hastatoside(Hs), and verbenalin(Vn) (5, 50 and 100 µM) and (**D**) co-treated with Vs and Dex (50 µM). Results were expressed in mean of cell viability ± SD. * $P < 0.05$; ** $P < 0.001$; *** $P < 0.0001$ compared with negative control group. # $P < 0.05$; ## $P < 0.001$; ### $P < 0.0001$ compared to Dex-treated group.

2.2. *Effect of VEE on the Immobility Time of Mice*

The tail suspension test (TST) was used to assess the antidepressant-like effect of VEE 100 mg/Kg compared to the control groups. Normally, drugs having an antidepressant effect decrease the immobility time of mice. In the present study, bupropion was used as a positive control, known for its antidepressant property. Bupropion-treated mice showed a decrease of immobility time on the 4th day of TST to 39.37 s compared to the initial test performed on the 1st day with a value of 42.52 s, resulting of the drug's effect (Figure 2). As for the negative control group, the mice were fed with Milli-Q water and showed a gradual increase of immobility time to day 7 with 114.4 s compared to the initial test with a time of 35.48 s, proving an induction of depression on mice by TST, leading the animals to lack the desire to rectify themselves (Figure 2).

Interestingly, 100 mg/Kg body weight VEE-treated mice showed a highly significant increase of immobility time compared to negative and positive controls starting from day 4 of the test with 202.64 s, which gradually decreased to attain 177.63 s on the 7th day (Figure 2).The low immobility time of the depressant mice receiving only water compared the VEE-treated mice suggested that the effect observed was not a result of the stress induced by TST, but because of the induction of relaxation by VEE, which is a unique effect of VEE.

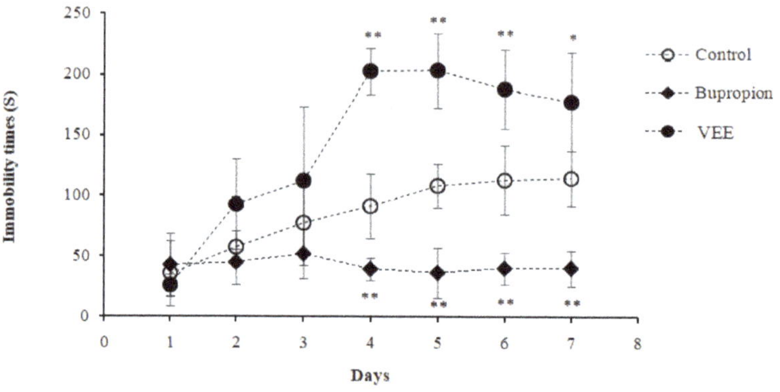

Figure 2. Effect of the oral administration of VEE (100 mg/Kg) and bupropion (20 mg/Kg) on mice immobility times in tail suspension test compared to the control (water 10 mL/Kg, p.o.). Results were expressed in mean of immobility time ± SD. * $P < 0.05$; ** $P < 0.001$ compared with Control group.

2.3. Elucidation of the Genes Regulated by VEE Treatment

To determine the molecular mechanism underlying the effect of VEE on immobility time, we analyzed the mice brains using DNA microarray to detect the transcriptomic changes. The analysis of the data revealed the up-regulation of 62 genes with a fold-change higher than 1.2, while 256 others were down-regulated below 0.65 fold-change. After annotating the genes, they were clustered in order to study their interactions and the pathways they are implicated in. Bupropion and VEE affected interesting pathways controlling the neuronal proliferation, spatial learning and memory, long-term potentiation and depression, inflammation and reactive oxygen species (ROS) production (Table 1). Interestingly, VEE treatment regulated genes such as *Adenylate cyclase (Ac)* implicated in the production of cyclic-Adenosine monophosphate (cAMP). It up-regulated the expression of genes implicated in calcium signaling including *Inositol 1,4,5-trisphosphate receptor type 2 (Itpr2), Protein kinase C (Pkc)* and *Calcium channel voltage-dependent L type alpha 1C subunit (Cacna1c)* [25,26]. VEE treatment increased the expression of *Calcium/calmodulin dependent protein kinase IV (CamkIV)*, one of the genes stimulating mitochondrial biogenesis [27]. The expression of *cGMP-dependent protein kinase (Prkg1)* was affected by verbena treatment, which results in the induction of muscle relaxation [28]. Also, *5 hydroxytryptamine (serotonin) receptor 4 (Htr4)* involved in neurotransmitters production was enhanced, alongside with *AdenosineA2a receptor (Adora2)*, responsible of the development of several neurodegenerative diseases [29–31]. VEE enhanced the expression of *Dopamine receptor D1 (Drd1)*, implicated in activation of *Ac* [32].

As shown in Table 1, out of the all sets of genes, three were highly expressed in the case of VEE-treated mice, which are *Gelsolin (Gsn), Transthyretin (Ttr)* and *Calcium/calmodulin-dependent protein kinase 2 inhibitor 1 (Camk2n1)*. Their expressions were increased 5.26, 3.72 and 2.19 fold, respectively. Recent studies showed a positive correlation between mitochondrial activity and expression of *Ttr* and *Gsn* [33,34]. As for *Camk2n1*, it has been shown to possess a role in controlling cell proliferation [35].

VEE treatment decreased the expression of *melanin-concentrating hormone receptor 1 (Mchr1)* to a fold-change equal to 0.55, while bupropion did not affect its transcription level. The down-regulation of this gene was found to enhance the metabolism [36], which implicates an activation of mitochondria. Also, *Mchr1* antagonist exerted an anti-depressant effect [37].

The *pro-melatonin-concentrating hormone (Pmch)* was drastically down-regulated (Table 1). It has been previously shown to exert a role in energy metabolism [38].

Table 1. Genes regulated by VEE involved in induction of relaxation and the activation of energy metabolism. The ratios were calculated using the data of mice receiving water as reference.

Gene ID	Gene Name	Verbena Ratio	Bupropion Ratio	Function
Gsn	*Gelsolin*	5.26	1.54	Amyloid beta peptides aggregation [33,39]
Ttr	*Transthyretin*	3.72	3.91	
Camk2n1	*Calcium/calmodulin-dependent protein kinase 2 inhibitor 1*	2.19	1.03	Tumor suppressor [35]
CaMK4	*calcium/calmodulin-dependent protein kinase IV*	1.46	1.20	Long-term memory [40]
Cacna1c	*Calcium channel, voltage-dependent, L type, alpha 1C subunit*	1.45	1.07	Cytosolic calcium content [26]
Pkc	*Protein kinase c*	1.45	0.98	*Adenylate cyclase* activation [32,41]
Drd1	*Dopamine receptor 1*	1.43	1.07	
Adora2	*Adenosine A2a receptor*	1.34	1.1	Cyclic-Adenosine monophosphate (cAMP) production [42]
Htr4	*5 hydroxytryptamine (serotonin) receptor 4*	1.34	1.25	Modulation of neurotransmitter release [29]
Itpr2	*Inositol 1,4,5-trisphosphate receptor type 2*	1.30	1.22	Intracellular calcium release [25]
Ac	*Adenylate cyclase*	1.28	0.85	Production of cAMP [43]
Prkg1	*cGMP-dependent protein kinase 1*	1.25	1.32	Induction of relaxation [28]
Mchr1	*melanin-concentrating hormone receptor*	0.55	1.01	Inhibition of cAMP accumulation [44]
Pmch	*pro-melanin-concentrating hormone*	0.12	0.12	Melanin-concentrating hormone activity [45]

2.4. Validation of Expressions of *Gsn*, *Ttr*, *Camk2n1* and *Itpr2*

The microarray analysis of brains collected from mice treated with 100 mg/Kg of VEE showed up-regulation of genes implicated in mitochondrial activity, with fold-changes higher than 2. These genes are *Gsn*, *Ttr*, and *Camk2n1*. Their up-regulations were confirmed and represented in relative gene expression, with the negative control expression as reference. Expressions of *Gsn*, *Ttr*, and *Camk2n1* were increased in the case of VEE-treated mice by 305% (relative gene expression), 115% and 110%, respectively (Figure 3A–C). The *Camk2n1* is an inhibitor that alters the transportation of Ca^{2+}, responsible of the control of the intracellular amount of this ion to avoid its side effects.

Itpr2 is responsible of intracellular calcium release. This gene was up-regulated by VEE treatment. Its expression was confirmed and showed an enhancement of 160% in VEE-treated mice compared to the control group. The effect of bupropion was not significant compared to VEE, with an increase of 19% (Figure 3D).

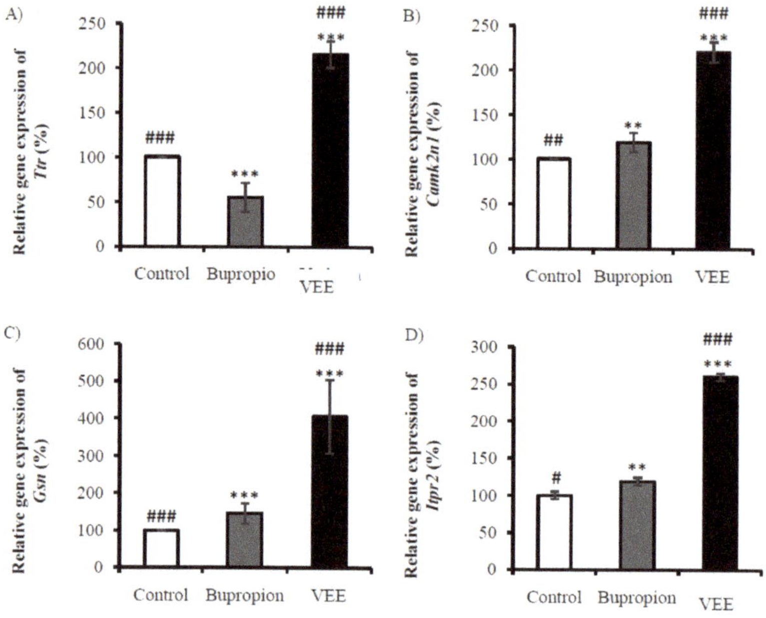

Figure 3. Validation of the expression of genes regulated by VEE treatment (100 mg/Kg) which are (**A**) *Ttr*, (**B**) *Camk2n1*, (**C**) *Gsn*, and (**D**) *Itpr2*. Results were expressed in relative gene expression ± SD. * $P < 0.05$; ** $P < 0.001$; *** $P < 0.0001$ compared with negative control group. # $P < 0.05$; ## $P < 0.001$; ### $P < 0.0001$ compared to bupropion-treated group.

2.5. Antidepressant Effect of Low Doses of VEE and Vs

The control group showed higher immobility time compared to other treatments for 7 days of testing (Figure 4A). The immobility recorded on the first day was 63.81 s for the control, which increased to reach 84.33s on day 7. This increase proved induction of depression in mice. Bupropion treated mice scored an immobility time of 16.96 s on the first day and decreased to 1.56 s on the last day of the test, proving the antidepressant effect of bupropion. Results obtained on first day showed a significant difference between the control group and Vs and VEE at a dose of 25 mg/Kg. On the second day, VEE and Vs treatments decreased the immobility time and the scores were statistically comparable to the bupropion treated group, while the difference was highly significant compared to the control. Similar results were observed for the rest of the test, except on day 3 and 5 where the difference was not significant between the control group and the 25 mg/Kg VEE treated animals.

For decades, depression has been associated with levels of monoamines and catecholamines in the system [46]. Depressive patients have been found to present Sert and NA (norepinephrine) deficiency [47,48]. To confirm the antidepressant effect of the treatments on mice we quantified the amounts of Sert and NA in mice brains. The results showed a low concentration of Sert and NA for control group with an amount of 18 and 171 ng/100 mg total proteins, respectively (Figure 4B,C). Bupropion increased significantly Sert level by 61% compared to control group. A similar effect was observed in case of mice treated with VEE 25 and 50 mg/Kg and Vs 2.5 and 5 mg/Kg showing improvement of 57.90%, 67.05%, 69.19%and 61.04% total proteins, respectively. An enhancement of 19% was observed in NA level in case of bupropion treated mice. Also, the other treatments increased NA concentration with a higher rate of 19.35% for Vs 2.5 mg/Kg treated group.

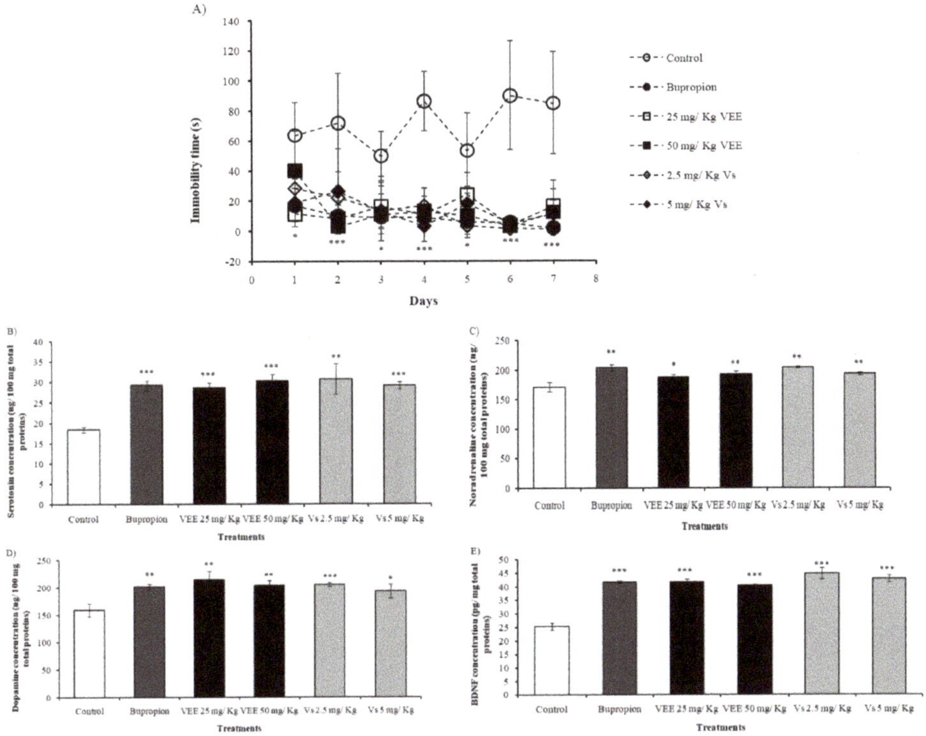

Figure 4. Effect of the oral administration of VEE (25 and 50 mg/Kg), Vs (2.5 and 5 mg/Kg) and bupropion (20 mg/Kg) on (**A**) mice immobility times in tail suspension test compared to the control (water 10 mL/Kg, p.o.) and their respective expression levels of (**B**) serotonin, (**C**) noradrenaline (**D**) dopamine and (**E**) BDNF. Results were expressed in mean of immobility time (s) and protein level ± SD. * $P < 0.05$; ** $P < 0.001$ and *** $P < 0.0001$ compared with Control group.

One of the important targets of antidepressants is the dopaminergic system. We evaluated the effect of our treatments on dopamine levels in mice brains. Bupropion showed an increase of dopamine content by 26% (Figure 4D). The highest dopamine concentration, with an increase of 34.45%, was observed in mice treated with 25 mg/Kg of VEE. The lowest dopamine enhancement (21.21%) was obtained for mice treated with 5 mg/Kg of Vs.

Furthermore, we evaluated the concentration of BDNF, which is one of the markers of depression. Our findings showed an increase of BDNF levels in all treatments. Bupropion enhanced BDNF expression by 64.34% (Figure 4E). Interestingly, VEE at 25 mg/Kg and Vs at 2.5 and 5 mg/Kg were found to exert more substantial effect regarding BDNF level with an enhancement of 64.67%, 76.36% and 69.26%.

2.6. Evaluation of the Mitochondrial Activity of Cells Treated with VEE and Vs

In order to measure the mitochondrial activity, we used the rhodamine 123 that stains the active mitochondria specifically. Both VEE and Vs induced mitochondrial activation of SH-SY5Y cells in a dose-dependent manner, with higher effect at lower concentrations. VEE at 0.5 μg/mL increased mitochondrial activity by 17% and its effect decreased to reach 9.37% for cells treated by 5 μg/mL (Figure 5A). Mitochondrial activity of cells treated with 5 μM of Vs was 115% compared to control, while the higher concentration enhanced the function only by 3% (Figure 5B). These results implicated a stimulation of energy production of VEE and Vs treatments.

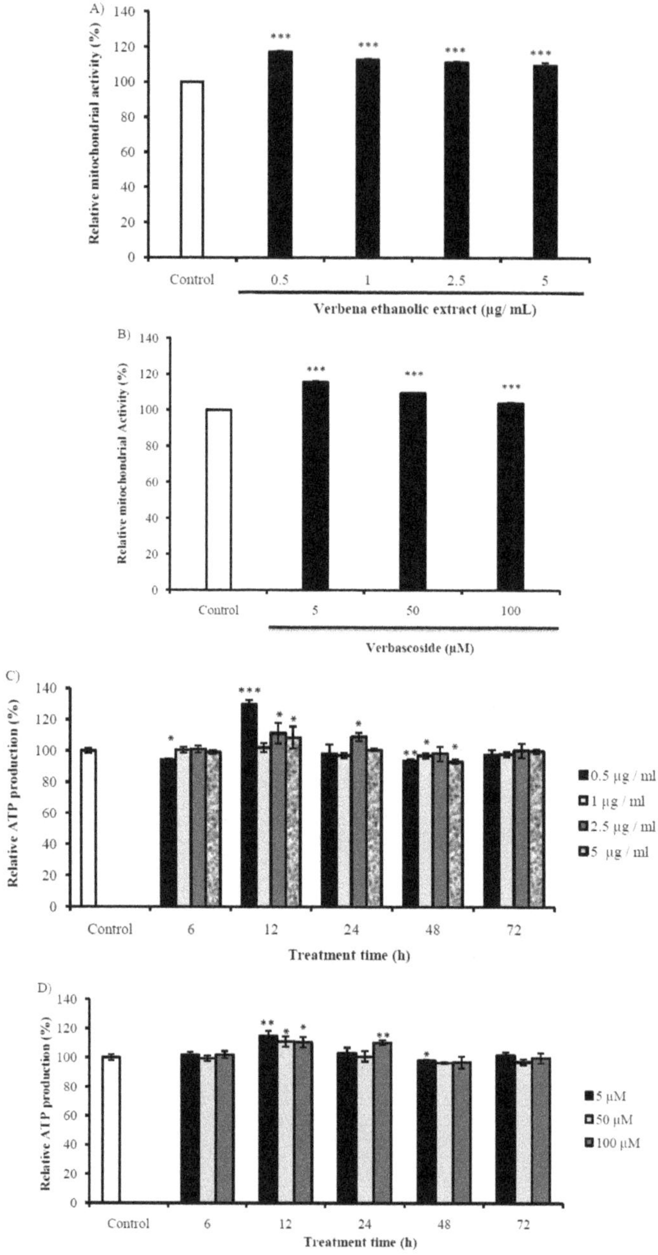

Figure 5. Evaluation of mitochondrial activity of SH-SY5Y cells treated with different concentrations of (**A**) VEE and (**B**) Vs. The intracellular ATP production of SH-SY5Y was assessed *in vitro* using the same concentrations of (**C**) VEE and (**D**) Vs at 6, 12, 24, 48 and 72 h. Results were expressed in mean of relative mitochondrial activity or ATP production (%) ± SD. * $P < 0.05$; ** $P < 0.001$; *** $P < 0.0001$ compared with control cells treated with Opti-MEM.

The same concentrations of VEE and Vs were evaluated for their effect on energy generation by quantifying ATP level. As Figure 5C shows, VEE treatments were not effective on energy metabolism at 6 h, but they show a highly significant increase after 12 h, with a maximum of 129.71 ± 2.73%. The ATP content decreased in a time and dose-dependent manner to reach energy homeostasis after 72 h. Treating the cells with Vs increased ATP production significantly after 12 h (Figure 5D), which decreased gradually to attain the normal status at 72 h.These results proved the stimulation mitochondria by VEE and Vs.

2.7. Effect of VEE and Vs on Intracellular Calcium Levels

Studies have shown a correlation between intracellular calcium uptake and mitochondrial activation. Transcriptomic analysis showed regulation of genes involved in Ca^{2+} in cases of mice treated with VEE. Here, we evaluated the effect of VEE and Vs on Ca^{2+} levels on SH-SY5Y. VEE increased Ca^{2+} uptake after 30 min of treatment in concentration and time-dependent manner, with higher effect at lower concentrations (Figure 6A). Accordingly, Vs showed similar effect on Ca^{2+} with higher activity at lower doses (Figure 6B). These results proved the implication of Ca^{2+} in the observed activities, with Vs being responsible for VEE effects.

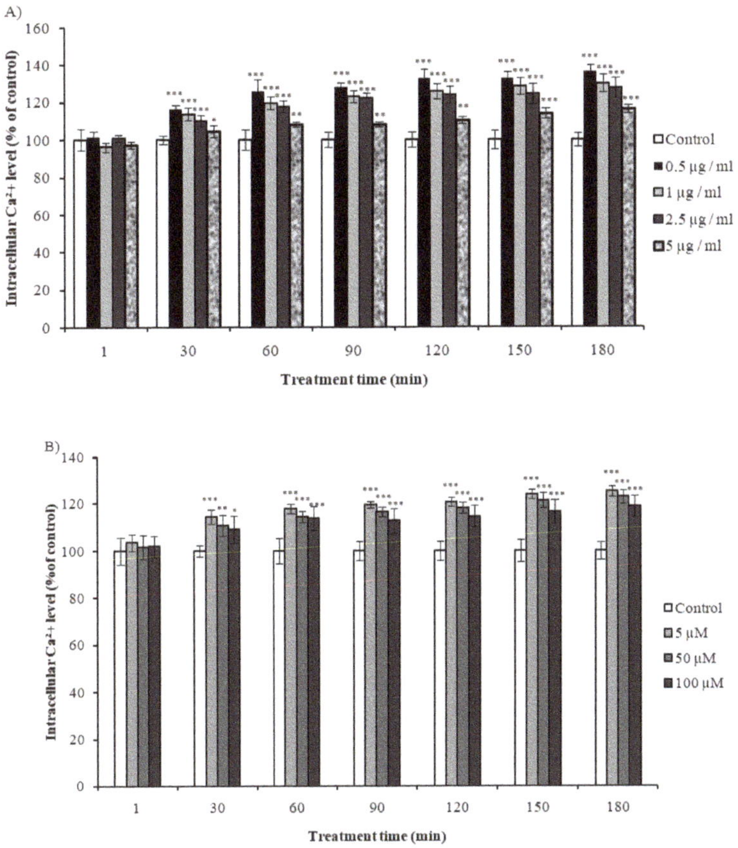

Figure 6. Evaluation of intracellular calcium levels of SH-SY5Y cells treated with different concentrations of (**A**) VEE and (**B**) Vs for 1–180 min. Results were expressed as percentage of control cells treated with Opti-MEM ± SD. * $P < 0.05$; ** $P < 0.001$; *** $P < 0.0001$ compared with control cells.

3. Discussion

Lemon verbena is a medicinal plant exerting important biological activities such as antidepressant, antioxidant, sleep-promoting and analgesic effects [24,49–51]. The molecular mechanisms underlying these effects are still unknown.

The *in vitro* study showed an increase of cell viability of VEE-treated cells compared to the non-treated cells, indicating an activation of cellular functionalities. Co-treatment of VEE and Dex enhanced the cell viability significantly compared to the Dex-treated cells. To determine the compound responsible for the effect observed, we treated the cells with the three most abundant compounds in the extract, Vs, Hs and Vn. The viability was enhanced by Vs in comparison to the control, while Hs and Vn were significantly decreased. The effect observed in the case of the extract is probably due to Vs. Also, Vs was tested for its neuroprotective effect and was found to alleviate Dex toxicity by more than 30%. These findings suggest that VEE and Vs have neuroprotective effects.

In the present work, we studied the effect of VEE on mice at the molecular level by analyzing the expression of all genes. We used the TST to induce psychological stress in mice. The TST results showed increase of immobility time of VEE-treated mice compared to both control groups. In 2017, Razavi et al. reported the anti-anxiety and muscle relaxant effects of VEE and Vs *in vivo* [17]. Another study showed induction of relaxation in mice and rats treated with essential oil extracted from the aerial part of verbena [50]. Accordingly, the aqueous extract of this plant was found to have a sedative effect in rats at high doses (700 and 1000 mg/Kg body weight of extract) [8]. Then, the increase of immobility time observed in this study may suggest the relaxant and sedative effects of VEE.

The evaluation of the transcriptome in the collected brains showed an enhancement of expression of genes implicated in the production of cAMP in the case of mice treated with VEE. *Drd1* expression was increased by VEE, while it remained stable in case of bupropion-treated mice. Previously, the enzymatic activity of *Ac* was found to be tightly regulated by *Drd1* through *Gβα* [32]. Also, VEE increased the expression of *Ac* in mice brains, the enzyme that was down-regulated by Bupropion treatment. Over-expression and activation of *Ac* by VEE implies an increase of cAMP generation, which has been associated with the induction of relaxation effect [52]. Accordingly, the use of apomorphine, a *Drd1* agonist, was proved to induce relaxation [53]. Moreover, treatment with the plant extract increased *Prkg1* expression, a gene that has been associated with induction of relaxation [28].

VEE affected the expression of genes modulating calcium homeostasis. *Itpr2* is one of the intracellular Ca^{2+} release channels, located in the membranes of endoplasmic and sarcoplasmic reticula. These are organelles are rich in Ca^{2+}ion [25]. VEE up-regulated the expression of *Itpr2*, implicating an elevation of the ion in the cytosolic compartment. Ca^{2+}-cytosolic content depends also on channels facilitating the transport of ion from the extracellular compartment [54]. One of these channels is *Cacna1c* which has been over expressed in VEE-treated mice. A previous study evaluated the transcriptomic changes induced by relaxation in humans and *Cacna1c* was found to by over-expressed [54]. Ca^{2+}-induced increase by VEE, up-regulated the expression of *Pkc*, an enzyme found to be dependent to Ca^{2+} concentration in cells, and which activates *Ac* inducing an over-production of cAMP [41,55]. On the other hand, calcium homeostasis has been already proved to play an important role in muscle movement and walking behavior in humans. At the brain level, the calcium signaling regulates different functions, including signal transmission and also the learning and memory [56–60]. When accumulated in cytoplasm, the calcium is transported into mitochondria inducing the activation of enzymes implicated in generation of ATP, including ATP synthase and $NADH^+$ dehydrogenase [61]. The inhibition of the calcium uptake by the mitochondria was found to increase the time needed for relaxation [62]. Accordingly, an increase of Ca^{2+} content has been proved to induce ATP production through cAMP generation [63]. These results suggest that VEE has a relaxant effect on mice through the generation of cAMP, which in addition to high intracellular Ca^{2+}levels, induces activation of mitochondria.

For VEE-treated mice, *Gsn*, *Ttr*, and *Camk2n1* showed the highest expression levels compared to the set of genes analyzed by microarray, and the increase was more than 2 fold-changes, while mice receiving Bupropion showed a decrease of *Ttr* expression, whereas *Gsn* and *Camk2n1* expressions were slightly increased (less than 1.5 fold-change). Mutant mice over-expressing *Gsn* revealed an enhancement of respiratory chain activity [33]. Several studies have demonstrated the neuroprotective role of *Ttr* [64–68], and its positive correlation to mitochondrial function [34]. These findings proved an increase of mitochondrial activity, implying an over-production of ATP. VEE-treated group presented high level of *Camk2n1* expression compared to control group, which implicates a controlled cell proliferation. Previously, a study demonstrated the tumor suppressive effect of *Camk2n1* [35].

Pmch and *Mchr1* were significantly down-regulated by VEE. *Pmch*-deficient mice, as well as *Mchr1*-deficient mice, were found to be more active than wild type mice, and showing an increase in metabolic rate [36,38]. A specific *Mchr1* antagonist has showed antidepressant and anxiolytic effect [37]. The increase of immobility time of VEE-treated mice is due to the relaxant effect of the plant extract, and the molecular analysis proved its antidepressant effect.

In order to evaluate the effect of lower doses of VEE and their respective Vs contents, a second TST was conducted. The treatments used were 25 and 50 mg/Kg of VEE and 2.5 and 5 mg/Kg of Vs. Interestingly, the results showed a decrease in immobility time compared to the control group, and scores were statistically comparable to bupropion treated mice. Our findings suggest low doses have an antidepressant effect. In accordance with the transcriptomic analysis conducted here above, VEE and Vs might be induced mitochondrial activation through accumulation of cAMP and Ca^{2+}, to a lesser extent than high dose of VEE, resulting in agitation of mice rather than their relaxation. To prove the antidepressant effect observed in vivo, we evaluated the levels of different depression markers. Sert and NA implication in depression has been documented and are considered as targets of antidepressants [47,48]. In our study, VEE and Vs were found to enhance Sert and NA levels demonstrating an antidepressant effect of the treatments on mice.

Previous studies found that antidepressants targeting the expressions of Sert and NA only present limitations. Patients might show movement delay, lack of concentration or even persistence of anhedonia [69]. Accordingly, drugs acting on the dopaminergic system have been developed. Dopamine is a catecholamine responsible of expression of emotions such as pleasure and motivation, and stimulates concentration [69]. Hence, we assessed dopamine levels in brains. The results showed a highly significant increase of dopamine expression by VEE and Vs compared to control group. These findings prove the antidepressant activity of VEE and Vs by stimulating the pleasure mechanism.

It has been documented that antidepressants acting on serotonergic and norepinephric mechanisms lead to enhancement of BDNF levels in rodents [70,71]. We evaluated the effect of treatments on BDNF in brains. The results showed a highly significant increase of BDNF by VEE and Vs treatments. Also, it has been documented that Ca^{2+} and cAMP levels regulate BDNF expression through CREB (cAMP response element-binding protein) [72].

The results obtained *in vivo* revealed the activation of mechanisms responsible for the increase of cytosolic Ca^{2+} and cAMP generation, messengers inducing the mitochondrial activity. To confirm this hypothesis, we evaluated the effect of VEE on mitochondrial activity. The results showed enhancement of mitochondrial function in a concentration-dependent manner. Accordingly, Vs increased mitochondrial function in a similar tendency as VEE.ATP production *in vitro* was evaluated to confirm the effect of VEE and Vs mitochondrial activity. Human neurotypicSH-SY5Y cells treated with VEE showed a significant increase of ATP content in a dose-dependent manner after12h treatment. Energy metabolism gradually decreased to regain the initial state. Vs is one of the most important compounds contained in VEE, and has been proven to induce muscle relaxation in mice [17]. Next, we evaluated the potential effect of Vs on mitochondrial activity. We observed that Vs-treatment also showed an increase in ATP production at 12h, which restored to its original condition progressively. In 2013, Bhasin et al. evaluated the transcriptomic changes in humans in response to relaxation condition and showed regulation of genes activating energy metabolism [54]. ATP increase has been found to be regulated positively by activation of mitochondrial calcium uptake, as aresult of different stimuli such as alimentation, hormones and neurotransmitters [61,73–76]. Our *in vitro* study showed that VEE and Vs enhanced intracellular calcium levels in a concentration and time-dependent manner with similar tendency as mitochondrial activation. These results proved the increase of calcium and energy metabolism related genes regulated by the treatments *in vivo*.

4. Materials and Methods

4.1. Plant Material and Extraction Method

The leaves of *Lippia citriodora* were collected in July 2016 from Marrakech Region (Morocco). The species was authenticated by Prof. Ahmed Ouhammou from Cadi Ayyad University, Faculty of Sciences Semlalia, Department of Biology, Marrakech, Morocco. A voucher specimen of plant material (MARK-11186) was deposited in the Herbarium of the same institution. After air drying, the plant material was crushed by a mortar and extracted with ethanol 70%, with a ratio plant material/solvent of 10% (*w/v*). The extraction was carried out in the dark for 2 weeks and vigorously shacked twice a day. The extract was centrifuged and the supernatant filtered through 0.22 μm Millipore (Mark Millipore, Carrigtwohill, Ireland) and solvent evaporated by a rotary evaporator. The yield of VEE was 13.3%.

4.2. Chemicals

Vs, Hs and Vn were purchased from Sigma Aldrich, USA. Dulbecco's Modified Eagle Medium (DMEM)/F-12 and Opti-MEM were obtained from Gibco, USA. Fetal bovine serum was from Gibco, South America. Penicillin - Streptomycin were purchased from Biowhittaker, USA. Non-essential amino acids were from Cosmo Bio Co, LTD, Japan. MTT (3-(4,5-dimethylthiazol-2-yl)-2,5-diphenyltetrazolium bromide) and dexamethasone were from Dojindo, Japan. Bupropion was from Wako, Japan. ATP bioluminescence kit was from TOYO Ink, Japan. ISOGEN kit was purchased from Nippon Gene, Japan. RIPA lysis buffer was from (Santa Cruz Biotechnology, USA). 2-D Quant was purchased from GE Healthcare Life Sciences, USA. Calcium Kit II-Fluo 4 was from Dojindo, Japan.

4.3. Cell Culture

The *in vitro* experiments were conducted on SH-SY5Y cells. This neurotypic cell line was obtained from America Type Culture Collection, Manassas, USA. Cells were maintained in Dulbecco's Modified Eagle Medium (DMEM)/F-12, supplemented with 15% fetal bovine serum, 1% Penicillin (5000 μg/mL) Streptomycin (5000 IU/mL) and 1% of non-essential amino acids. The culture was incubated at 37 °C in a humidified atmosphere of 5% CO_2 incubator. Opti-MEM, a reduced serum medium, was used to culture cells for the evaluation of cell viability and intracellular ATP.

4.4. Determination of Cell Viability

3-(4,5-dimethylthiazol-2-yl)-2,5-diphenyltetrazolium bromide (MTT) assay was used to assess cell viability of SH-SY5Y cells. Briefly, the cells were seeded in a 96-well plate (fibronectin-coated plate) (BD BioCoat, New York City, NY, USA) with a density of 2×10^5 cell/mL. After 24 h, the medium was removed and the cells were VEE (0.5, 1, 2.5 and 5 μg/mL), Vs, Hs or Vn (3.1, 31.2 and 62.4 μg/mL) diluted in Opti-MEM. After 72h incubation period, 10 μL MTT (5 mg/mL) mixed with 100 μL of Opti-MEM was added to each well and the plate was incubated for further 6 h. The formazan crystals formed by the mitochondrial activity were solubilized by adding 100 μL of 10 % SDS (*w/v*). The absorbance was measured at 570 nm using a microtiter plate reader (Dainippon Sumitomo Pharma Co., Ltd., Tokyo, Japan). The results were expressed in percentage of relative cell viability.

To evaluate the neuroprotective effect of VEE, Vs, Hs and Vn, the cells were co-treated with DEX (50 μM), incubated for 72h, then their viability assessed by MTT assay as described above.

4.5. Animals

Male ICR mice, 3 weeks old, weighting between 20 and 30 g were purchased from Charles River laboratories (Tokyo, Japan). Mice were housed individually and had access to food and water *ad libitum*, in a controlled environment (56% humidity, 23 °C temperature, 12/12 h light/dark cycle). Before starting the oral administration and the tail suspension test, the mice were allowed to acclimatize for one week. All experiments were performed in strict accordance with NIH guidelines and were approved by the Animal Ethics Committee of the University of Tsukuba, Japan. The ethical approval code is 16-042 (1/06/2016).

4.6. *Tail Suspension Test*

The animals were divided into three groups. A negative control group receiving Milli-Q water (10 mL/Kg; $n = 6$), a positive control group treated with 20 mg/Kg of Bupropion ($n = 7$) and VEE-treated group ($n = 7$) which received a dose of 100 mg/Kg. The samples were administrated orally every day for 7 days.

The tail suspension test (TST) is a widely used technique to screen the antidepressant effects of drugs. The methodology used in this study is as described by Steru et al., 1985 [77]. Briefly, the TST was performed 60 min after the administration of treatments. The duration of the test was 6 min and the immobility time was measured on the last 4 min of the test. A mouse was considered immobile only when it is hanged passively, showing no resistance to the stress applied by the test. The experiment was recorded using a camera and scored by observing the videos. After completion of the behavioral test, mice were sacrificed by cerebrospinal dislocation, then the whole brains were collected for the subsequent analysis.

A second TST was conducted to evaluate the effect of lower doses of VEE (25 mg/Kg, $n = 6$; 50 mg/Kg, $n = 7$) and their respective Vs content (2.5 mg/Kg, $n = 6$; 5 mg/Kg body weight, $n = 7$) on mice. HPLC analysis showed that VEE contains 10% of Vs (data not shown). Other control groups (Milli-Q, $n = 6$; Bupropion 20 mg/mL, $n = 7$) were used for the second test. TST was performed according to the protocol previously described.

4.7. *DNA Microarray Analysis*

The total RNA was extracted from the brain tissues previously collected using ISOGEN kit and quantified by Nanodrop 2000 spectrophotometer (Thermo Fisher scientific, Wilmington, NC, USA).

To elucidate the molecular mechanism underlying the effect of VEE on neuronal activities, we evaluated the total gene expression of brain tissues by performing microarray on RNAs previously extracted. The experiment was conducted according to the Affymetrix Genechip 3′ IVT PLUS reagent kit user's guide. Briefly, the RNAs were reverse transcripted to generate double stranded DNA. The latter used as a template to synthesize the Biotin-labeled cRNA. After fragmentation of the labeled cRNA, the mixture was hybridized to the Affymetrix mouse 430 PM array strips (Affymetrix) for 16 h at 45 °C in the hybridization station. In Geneatlas Fluidics station, the hybridized arrays were washed and stained, then scanned using Geneatlas imaging station. The total number of genes analyzed by this method is 39,396 genes. All brain samples were analyzed by microarray. The data obtained were analyzed by Expression Console and Pathway Studio software and DAVID and Consensus Path databases.

4.8. *Real Time Polymerase Chain Reaction (qRT-PCR)*

RNA extracts obtained from mice brains were used as templates to validate the microarray results through evaluation of the expression level of some relevant genes regulated by Verbena treatment. First, a reverse transcription was performed, using the Superscript IV reverse transcriptase kit (Invitrogen, Carlsbad, CA, USA) following the manufacturer's protocol. Briefly, we incubated a mixture of RNA samples (0.2 µg/µL) and Oligo(dT)$_{12\text{-}18}$/dNTP (0.5 µg/µL; 10 mM) for 5 min at 65 °C, and then placed for 1 min on ice. The Reverse transcriptase solution was added and incubated the samples at 42 °C for 60 min and then 10 min at 60 °C. The cDNA produced is used to evaluate the expression of 3 genes: *Gelsolin "Gsn"* (Mm00456679_m1), *Transthyretin "Ttr"* (Mm00443267_m1), *Calcium/calmodulin-dependent protein kinase II inhibitor 1 "Camk2n1"* (Mm01718432_s1) and *Inositol 1,4,5-trisphosphate receptor type 2 "Itpr2"*

(Mm00444937_m1). This experiment was conducted using TaqMan Universal PCR mix and TaqMan Probes and the amplifications were performed in a 7500 Fast Real-time PCR (Applied Biosystems, Foster City, CA, USA) with the following conditions: 50 °C for 2 min, followed by 95 °C for 10 min, and 40 cycles of 95 °C for 15 s followed by 60 °C for 1 min.

4.9. Quantification of Neurotransmitters and BDNF

To confirm the antidepressant effect of VEE and Vs at lower doses, we quantified the levels of serotonin (Sert), noradrenaline (NA), dopamine and BDNF in brains. The proteins were measured in the frontal cortex. First we homogenized 100 mg of tissue in 1 mL of RIPA buffer. The homogenate was centrifuged for 5 min at 10,000× *g* and 4 °C. The supernatant was collected and stored at −80 °C. The dopamine, Sert and NA were quantified using ELISA kits (Immusmol SAS, Talence, France). BDNF was measured by colorimetric sandwich ELISA kit (Proteintech, Rosemont, IL, USA). The experiments were conducted following the manufacturer's instructions. The results of each treatment group were corrected by their respective total protein content determined using 2-D Quanti kit.

4.10. Measurement of Mitochondrial Activity

Mitochondrial function was measured using rhodamine 123, a fluorescent dye. The protocol was as described previously by Matsukawa et al., 2017 [78]. Briefly, treated SH-SY5Y were incubated for 20 min at 37 °C after addition of rhodamine 123 (10 μg/mL). Cells were lysed by 1% Triton X-100 and the fluorescence intensity of rhodamine (excitation/emission 485/528 nm) was measured.

4.11. Measurement of the Intracellular ATP Production

The mitochondrial activity was assessed by measuring the intracellular ATP content of cells using ATP bioluminescence kit. Cells were cultured (2×10^5 cell/mL) in a 96-well plate (fibronectin-coated plate) and treated with different concentrations of VEE and Vs for 6, 12, 24, 48 and 72 h. The cells were lysed and the ATP content measured by adding 100 μL of luciferin-luciferase solution. The luminescence was measured using the microtiter plate reader (Dainippon Sumitomo Pharma Co., Ltd., Japan).

4.12. Measurement of Intracellular Calcium Level

Calcium Kit II-Fluo 4 was used to measure intracellular calcium levels of SH-SY5Y. The measurement was conducted according to the manufacturer's protocol. Briefly, SH-SY5Y cells were seeded in black clear-bottom 96 well plates (Corning, NY, USA) and then treated with loading buffer (5% Pluronic F-127, 250-mmol/L Probenecid and 1-μg/μLFluo 4 AM in Hanks'–HEPES Buffer) for 1 h. The supernatant was removed and cells were washed with PBS. Cells were treated with VEE and Vs as described previously. Fluorescence intensity (excitation/emission 485/528 nm) was measured every 30 min using a Powerscan HT plate reader.

4.13. Statistical Analysis

Results are expressed as means ± SD, and statistical analyses were performed using a Student's *t*-test using IBM SPSS Statistics 23 software. Differences were determined statistically significant at a *P*-value of less than 0.05.

5. Conclusions

Taken together, our findings suggest that, depending on the administered dose, VEE and Vs induce either relaxation or anti-depression effects. Higher doses of VEE induced relaxation through regulation of genes, including *Itpr2* and *Ac, responsible* for Ca^{2+} and cAMP generation. Lower doses of VEE and their respective Vs amount treatments was found to induce antidepressant-like effects by enhancing the BDNF, NA, Sert and dopamine expressions, which are cAMP and Ca^{2+} dependent. VEE and Vs increased Ca^{2+} intracellular levels leading to the enhancement of mitochondrial activity and ATP concentration. The effects of VEE observed *in vivo* and *in vitro* are due mostly to Vs.

Author Contributions: Conceptualization, M.S., K.S., C.G. and H.I.; Data curation, M.S. and K.S.; Formal analysis, M.S.; Funding acquisition, H.I.; Investigation, M.S.; Methodology, M.S., K.S. and H.I.; Project administration, H.I.; Resources, H.I.; Software, M.S.; Supervision, K.S. and H.I.; Validation, M.S., K.S. and H.I.; Visualization, M.S.; Writing—original draft, M.S.; Writing—review & editing, K.S., C.G. and H.I.

Funding: This research was funded by SCIENCE AND TECHNOLOGY RESEARCH PARTNERSHIP FOR SUSTAINABLE DEVELOPMENT (SATREPS), grant number JPMJSA1506.

Acknowledgments: Ministry of Education, Culture, Sports, Science and Technology (MEXT) for funding this work partially.

Conflicts of Interest: The authors declare no conflict of interest.

Abbreviations

Ac	Adenylate cyclase
Adora2	Adenosine A2a receptor
Cacna1c	Calcium channel, voltage-dependent, L type, alpha 1C subunit
Camk2n1	Calcium/calmodulin-dependent protein kinase II inhibitor 1
Camk4	Calcium/calmodulin-dependent protein kinase IV
Dex	Dexamethasone
Drd1	Dopamine receptor 1
Gsn	Gelsolin
Hs	Hastatoside
Htr4	5 hydroxytryptamine (serotonin) receptor 4
Itpr2	Inositol 1,4,5-trisphosphate receptor type 2
Mchr1	Melanin-concentrating hormone receptor
MTT	3-(4,5-dimethylthiazol-2-yl)-2,5-diphenyltetrazolium bromide
NA	Noradrenaline
Pkc	Protein kinase c
Pmch	Pro-melanin-concentrating hormone
Prkg1	cGMP-dependent protein kinase 1
Sert	Serotonin
TST	Tail suspension test
Ttr	Transthyretin
VEE	Verbena ethanolic extract
Vn	Verbenalin
Vs	Verbascoside

References

1. Rahmatullah, M.; Jahan, R.; Safiul Azam, F.M.; Hossan, S.; Mollik, M.A.H.; Rahman, T. Folk medicinal uses of verbenaceae family plants in Bangladesh. *Afr. J. Tradit. Complement. Altern. Med.* **2011**, *8*, 53–65. [CrossRef] [PubMed]
2. Pascual, M.E.; Slowing, K.; Carretero, E.; Sánchez Mata, D.; Villar, A. Lippia: Traditional uses, chemistry and pharmacology: A review. *J. Ethnopharmacol.* **2001**, *76*, 201–214. [CrossRef]
3. Carnat, A.; Carnat, A.P.; Fraisse, D.; Lamaison, J.L. The aromatic and polyphenolic composition of lemon verbena tea. *Fitoterapia* **1999**, *70*, 44–49. [CrossRef]
4. Bellakhdar, J. *La Pharmacopée Marocaine Traditionnelle*; Ibis Press: Paris, France, 1997; ISBN 2-910728-0.
5. Montari, B. Aromatic, medicinal plants and vulnerability of traditional herbal knowledge in a berber community of the high atlas mountains of Morocco. *Plant Divers.Resour.* **2014**, *36*, 388–402.
6. Fitsiou, E.; Mitropoulou, G.; Spyridopoulou, K.; Vamvakias, M.; Bardouki, H.; Galanis, A.; Chlichlia, K.; Kourkoutas, Y.; Panayiotidis, M.I.; Pappa, A. Chemical composition and evaluation of the biological properties of the essential oil of the dietary phytochemical Lippia citriodora. *Molecules* **2018**, *23*, 123. [CrossRef] [PubMed]

7. Ragone, M.I.; Sella, M.; Conforti, P.; Volonté, M.G.; Consolini, A.E. The spasmolytic effect of Aloysia citriodora, Palau (South American cedrón) is partially due to its vitexin but not isovitexin on rat duodenums. *J. Ethnopharmacol.* **2007**, *113*, 258–266. [CrossRef] [PubMed]
8. Akanmu, M.A.; Honda, K.; Inoué, S. Hypnotic effects of total aqueous extracts of Vervain hastata (Verbenaceae) in rats. *Psychiatry Clin. Neurosci.* **2002**, *56*, 309–310. [CrossRef]
9. Herranz-López, M.; Barrajón-Catalán, E.; Segura-Carretero, A.; Menéndez, J.A.; Joven, J.; Micol, V. Lemon verbena (Lippia citriodora) polyphenols alleviate obesity-related disturbances in hypertrophic adipocytes through AMPK-dependent mechanisms. *Phytomedicine* **2015**, *22*, 605–614. [CrossRef]
10. Carnat, A.P.; Carnat, A.; Fraisse, D.; Lamaison, J.L. The aromatic and polyphenolic composition of lemon balm (*Melissa officinalis* L. subsp. *officinalis*) tea. *Pharm. Acta Helv.* **1998**, *72*, 301–305. [CrossRef]
11. Arthur, H.; Joubert, E.; De Beer, D.; Malherbe, C.J.; Witthuhn, R.C. Phenylethanoid glycosides as major antioxidants in Lippia multiflora herbal infusion and their stability during steam pasteurisation of plant material. *Food Chem.* **2011**, *127*, 581–588. [CrossRef]
12. Vertuani, S.; Beghelli, E.; Scalambra, E.; Malisardi, G.; Copetti, S.; Toso, R.D.; Baldisserotto, A.; Manfredini, S. Activity and stability studies of verbascoside, a novel antioxidant, in dermo-cosmetic and pharmaceutical topical formulations. *Molecules* **2011**, *16*, 7068–7080. [CrossRef] [PubMed]
13. Speranza, L.; Franceschelli, S.; Pesce, M.; Reale, M.; Menghini, L.; Vinciguerra, I.; De Lutiis, M.A.; Felaco, M.; Grilli, A. Antiinflammatory effects in THP-1 cells treated with verbascoside. *Phytother. Res.* **2010**, *24*, 1398–1404. [CrossRef] [PubMed]
14. Díaz, A.M.; Abad, M.J.; Fernández, L.; Silván, A.M.; De Santos, J.; Bermejo, P. Phenylpropanoid glycosides from Scrophularia scorodonia: In vitro anti-inflammatory activity. *Life Sci.* **2004**, *74*, 2515–2526. [CrossRef] [PubMed]
15. Guillermo Avila, J.; De Liverant, J.G.; Martínez, A.; Martínez, G.; Muñoz, J.L.; Arciniegas, A.; Romo De Vivar, A. Mode of action of Buddleja cordata verbascoside against Staphylococcus aureus. *J. Ethnopharmacol.* **1999**, *66*, 75–78. [CrossRef]
16. Sheng, G.-Q.; Zhang, J.-R.; Pu, X.-P.; Ma, J.; Li, C.-L. Protective effect of verbascoside on 1-methyl-4-phenylpyridinium ion-induced neurotoxicity in PC12 cells. *Eur. J. Pharm.* **2002**, *451*, 119–124. [CrossRef]
17. Razavi, B.M.; Zargarani, N.; Hosseinzadeh, H. Anti-anxiety and hypnotic effects of ethanolic and aqueous extracts of Lippia citriodora leaves and verbascoside in mice. *Avicenna J. Phytomed.* **2017**, *7*, 353–365. [PubMed]
18. Makino, Y.; Kondo, S.; Nishimura, Y.; Tsukamoto, Y.; Huang, Z.L.; Urade, Y. Hastatoside and verbenalin are sleep-promoting components in verbena officinalis. *Sleep Biol. Rhythm.* **2009**, *7*, 211–217. [CrossRef]
19. Nakamura, T.; Okuyama, E.; Tsukada, A.; Yamazaki, M.; Satake, M.; Nishibe, S.; Takeshi, D.; Moriya, A.; Maruno, M.; Nishimura, H. Acteoside as the analgesic priniple of cedron (Lippia triphylla), a Peruvian medicinal plant. *Chem. Pharm. Bull.* **1997**, *45*, 499–504. [CrossRef]
20. Casanova, E.; García-Mina, J.M.; Calvo, M.I. Antioxidant and antifungal activity of *Verbena officinalis* L. leaves. *Plant Food Hum. Nutr.* **2008**, *63*, 93–97. [CrossRef]
21. Deepak, M.; Handa, S.S. Quantitative determination of the major constituents of Verbena officinalis using high performance thin layer chromatography and high pressure liquid chromatography. *Phytochem. Anal.* **2000**, *11*, 351–355. [CrossRef]
22. Funes, L.; Laporta, O.; Cerdán-Calero, M.; Micol, V. Effects of verbascoside, a phenylpropanoid glycoside from lemon verbena, on phospholipid model membranes. *Chem. Phys. Lipids* **2010**, *163*, 190–199. [CrossRef] [PubMed]
23. Funes, L.; Carrera-Quintanar, L.; Cerdán-Calero, M.; Ferrer, M.D.; Drobnic, F.; Pons, A.; Roche, E.; Micol, V. Effect of lemon verbena supplementation on muscular damage markers, proinflammatory cytokines release and neutrophils' oxidative stress in chronic exercise. *Eur. J. Appl. Physiol.* **2011**, *111*, 695–705. [CrossRef] [PubMed]
24. Rehecho, S.; Hidalgo, O.; García-Iñiguez de Cirano, M.; Navarro, I.; Astiasarán, I.; Ansorena, D.; Cavero, R.Y.; Calvo, M.I. Chemical composition, mineral content and antioxidant activity of *Verbena officinalis* L. *LWT Food Sci. Technol.* **2011**, *44*, 875–882. [CrossRef]
25. Santulli, G.; Marks, A.R. Essential roles of intracellular calcium release channels in muscle, brain, metabolism, and aging. *Curr. Mol. Pharm.* **2015**, *8*, 206–222. [CrossRef]

26. Simms, B.A.; Zamponi, G.W. Neuronal voltage-gated calcium channels: Structure, function, and dysfunction. *Neuron* **2014**, *82*, 24–45. [CrossRef] [PubMed]
27. Jornayvaz, F.R.; Shulman, G.I.G. Regulation of mitochondrial biogenesis. *Essays Biochem.* **2010**, *47*, 69–84. [CrossRef]
28. Tang, M.; Wang, G.; Lu, P.; Karas, R.H.; Aronovitz, M.; Heximer, S.P.; Kaltenbronn, K.M.; Blumer, K.J.; Siderovski, D.P.; Zhu, Y.; et al. Regulator of G-protein signaling-2 mediates vascular smooth muscle relaxation and blood pressure. *Nat. Med.* **2003**, *9*, 1506–1512. [CrossRef]
29. Ohtsuki, T.; Ishiguro, H.; Detera-Wadleigh, S.D.; Toyota, T.; Shimizu, H.; Yamada, K.; Yoshitsugu, K.; Hattori, E.; Yoshikawa, T.; Arinami, T. Association between serotonin 4 receptor gene polymorphisms and bipolar disorder in Japanese case-control samples and the NIMH Genetics Initiative Bipolar Pedigrees. *Mol. Psychiatry* **2002**, *7*, 954–961. [CrossRef]
30. Chiu, F.L.; Lin, J.T.; Chuang, C.Y.; Chien, T.; Chen, C.M.; Chen, K.H.; Hsiao, H.Y.; Lin, Y.S.; Chern, Y.; Kuo, H.C. Elucidating the role of the A2Aadenosine receptor in neurodegeneration using neurons derived from Huntington's disease iPSCs. *Hum. Mol. Genet.* **2015**, *24*, 6066–6079. [CrossRef]
31. Horgusluoglu-Moloch, E.; Nho, K.; Risacher, S.L.; Kim, S.; Foroud, T.; Shaw, L.M.; Trojanowski, J.Q.; Aisen, P.S.; Petersen, R.C.; Jack, C.R.; et al. Targeted neurogenesis pathway-based gene analysis identifies ADORA2A associated with hippocampal volume in mild cognitive impairment and Alzheimer's disease. *Neurobiol. Aging* **2017**, *60*, 92–103. [CrossRef]
32. Ehrlich, A.T.; Furuyashiki, T.; Kitaoka, S.; Kakizuka, A.; Narumiya, S. Prostaglandin E receptor EP1 forms a complex with dopamine D1 receptor and directs D1-induced cAMP production to adenylyl cyclase 7 through mobilizing G(βγ) subunits in human embryonic kidney 293T cells. *Mol. Pharm.* **2013**, *84*, 476–486. [CrossRef] [PubMed]
33. Antequera, D.; Vargas, T.; Ugalde, C.; Spuch, C.; Molina, J.A.; Ferrer, I.; Bermejo-Pareja, F.; Carro, E. Cytoplasmic gelsolin increases mitochondrial activity and reduces Aβ burden in a mouse model of Alzheimer's disease. *Neurobiol. Dis.* **2009**, *36*, 42–50. [CrossRef] [PubMed]
34. Oka, S.; Leon, J.; Sakumi, K.; Ide, T.; Kang, D.; LaFerla, F.M.; Nakabeppu, Y. Human mitochondrial transcriptional factor A breaks the mitochondria-mediated vicious cycle in Alzheimer's disease. *Sci. Rep.* **2016**, *6*, 37889. [CrossRef] [PubMed]
35. Wang, T.; Liu, Z.; Guo, S.; Wu, L.; Li, M.; Chen, R.; Xu, H.; Cai, S.; Chen, H.; Li, W.; et al. The tumor suppressive role of CAMK2N1 in castration-resistant prostate cancer. *Oncotarget* **2014**, *5*, 3611. [CrossRef] [PubMed]
36. MacNeil, D.J. The role of melanin-concentrating hormone and its receptors in energy homeostasis. *Front. Endocrinol.* **2013**, *4*, 1–14. [CrossRef] [PubMed]
37. Borowsky, B.; Durkin, M.M.; Ogozalek, K.; Marzabadi, M.R.; DeLeon, J.; Heurich, R.; Lichtblau, H.; Shaposhnik, Z.; Daniewska, I.; Blackburn, T.P.; et al. Antidepressant, anxiolytic and anorectic effects of a melanin-concentrating hormone-1 receptor antagonist. *Nat. Med.* **2002**, *8*, 825–830. [CrossRef] [PubMed]
38. Zhou, D.; Shen, Z.; Strack, A.M.; Marsh, D.J.; Shearman, L.P. Enhanced running wheel activity of both Mch1r- and Pmch-deficient mice. *Regul. Pept.* **2005**, *124*, 53–63. [CrossRef] [PubMed]
39. Li, X.; Buxbaum, J.N. Transthyretin and the brain re-visited: Is neuronal synthesis of transthyretin protective in Alzheimer's disease? *Mol. Neurodegener.* **2011**, *6*, 79. [CrossRef] [PubMed]
40. Kang, H.; Sun, L.D.; Atkins, C.M.; Soderling, T.R.; Wilson, M.A.; Tonegawa, S. An important role of neural activity-dependent CaMKIV signaling in the consolidation of long-term memory. *Cell* **2001**, *106*, 771–783. [CrossRef]
41. Kawabe, J.; Iwami, G.; Ebina, T.; Ohno, S.; Katada, T.; Ueda, Y.; Homcy, C.J.; Ishikawa, Y. Differential activation of adenylyl cyclase by protein kinase C isoenzymes. *J. Biol. Chem.* **1994**, *269*, 16554–16558.
42. Iwamoto, T.; Umemura, S.; Toya, Y.; Uchibori, T.; Kogi, K.; Takagi, N.; Ishii, M. Identification of A2a receptor-cAMP system in Human Aortic Endothelial Cells. *Biochem. Biophys. Res. Commun.* **1994**, *199*, 905–910. [CrossRef] [PubMed]
43. Tresguerres, M.; Levin, L.R.; Buck, J. Intracellular cAMP signaling by soluble adenylyl cyclase. *KidneyInt.* **2011**, *79*, 1277–1288. [CrossRef] [PubMed]
44. Pissios, P.; Maratos-Flier, E. Melanin-concentrating hormone: From fish skin to skinny mammals. *Trends Endocrinol. Metab.* **2003**, *14*, 243–248. [CrossRef]

45. Viale, A.; Ortola, C.; Hervieu, G.; Furuta, M.; Barbero, P.; Steiner, D.F.; Seidah, N.G.; Nahon, J.L. Cellular localization and role of prohormone convertases in the processing of pro-melanin concentrating hormone in mammals. *J. Biol. Chem.* **1999**, *274*, 6536–6545. [CrossRef] [PubMed]
46. Briley, M.; Chantal, M. The importance of norepinephrine in depression. *Neuropsychiatr. Dis. Treat.* **2011**, *7*, 9. [CrossRef] [PubMed]
47. Belmaker, R.; Agam, G. Major depressive disorder. *N. Engl. J. Med.* **2008**, *358*, 55–68. [CrossRef] [PubMed]
48. Krishnan, V.; Nestler, E.J. The molecular neurobiology of depression. *Nature* **2008**, *455*, 894–902. [CrossRef]
49. Jawaid, T.; Imam, S.A.; Kamal, M. Antidepressant activity of methanolic extract of Verbena Officinalis Linn. plant in mice. *Asian J. Pharm. Clin. Res.* **2015**, *8*, 308–310.
50. Makram, S.; Alaoui, K.; Benabboyha, T.; Faridi, B.; Cherrah, Y.; Zellou, A. Extraction et activité psychotrope de l'huile essentielle de la verveine odorante *Lippia citriodora*. *Phytotherapie* **2015**, *13*, 163–167. [CrossRef]
51. Calvo, M.I. Anti-inflammatory and analgesic activity of the topical preparation of *Verbena officinalis* L. *J. Ethnopharmacol.* **2006**, *3*, 380–382. [CrossRef]
52. Kuo, I.Y.; Ehrlich, B.E. Signaling in muscle contraction. *Cold Spring Harb. Perspect. Biol.* **2015**, *7*, 1–15. [CrossRef] [PubMed]
53. Di Villa Bianca, R.D.E.; Sorrentino, R.; Roviezzo, F.; Imbimbo, C.; Palmieri, A.; De Dominicis, G.; Montorsi, F.; Cirino, G.; Mirone, V. Peripheral relaxant activity of apomorphine and of a D1 selective receptor agonist on human corpus cavernosum strips. *Int. J. Impot. Res.* **2005**, *17*, 127–133. [CrossRef] [PubMed]
54. Bhasin, M.K.; Dusek, J.A.; Chang, B.-H.; Joseph, M.G.; Denninger, J.W.; Fricchione, G.L.; Benson, H.; Libermann, T.A. Relaxation response induces temporal transcriptome changes in energy metabolism, insulin secretion and inflammatory pathways. *PLoS ONE* **2013**, *8*, e62817. [CrossRef] [PubMed]
55. Eguchi, S.; Iwasaki, H.; Inagami, T.; Numaguchi, K.; Yamakawa, T.; Motley, E.D.; Owada, K.M.; Fumiaki, M.; Hirata, Y. Vascular smooth muscle cells. *Hypertension* **1999**, *33*, 201–206. [CrossRef] [PubMed]
56. Kovac, J.R.; Preiksaitis, H.G.; Sims, S.M. Functional and molecular analysis of L-type calcium channels in human esophagus and lower esophageal sphincter smooth muscle. *Am. J. Physiol. Gastrointest. Liver Physiol.* **2005**, *289*, G998–G1006. [CrossRef] [PubMed]
57. Arduíno, D.M.; Esteves, A.R.; Cardoso, S.M.; Oliveira, C.R. Endoplasmic reticulum and mitochondria interplay mediates apoptotic cell death: Relevance to Parkinson's disease. *Neurochem. Int.* **2009**, *55*, 341–348. [CrossRef] [PubMed]
58. Schapira, A.H.V.; Olanow, C.W.; Greenamyre, J.T.; Bezard, E. Slowing of neurodegeneration in Parkinson's disease and Huntington's disease: Future therapeutic perspectives. *Lancet* **2014**, *384*, 545–555. [CrossRef]
59. Zatti, G.; Ghidoni, R.; Barbiero, L.; Binetti, G.; Pozzan, T.; Fasolato, C.; Pizzo, P. The presenilin 2 M239I mutation associated with familial Alzheimer's disease reduces Ca2+release from intracellular stores. *Neurobiol. Dis.* **2004**, *15*, 269–278. [CrossRef] [PubMed]
60. Lessard, C.B.; Lussier, M.P.; Cayouette, S.; Bourque, G.; Boulay, G. The overexpression of presenilin2 and Alzheimer's-disease-linked presenilin2 variants influences TRPC6-enhanced Ca2+entry into HEK293 cells. *Cell. Signal.* **2005**, *17*, 437–445. [CrossRef]
61. Tarasov, A.I.; Griffiths, E.J.; Rutter, G.A. Regulation of ATP production by mitochondrial Ca^{2+}. *Cell Calcium* **2012**, *52*, 28–35. [CrossRef]
62. Gillis, J.M. Inhibition of mitochondrial calcium uptake slows down relaxation in mitochondria-rich skeletal muscles. *J. Muscle Res. Cell Motil.* **1997**, *18*, 473–483. [CrossRef] [PubMed]
63. Di Benedetto, G.; Scalzotto, E.; Mongillo, M.; Pozzan, T. Mitochondrial Ca^{2+} uptake induces cyclic AMP generation in the matrix and modulates organelle ATP levels. *Cell Metab.* **2013**, *17*, 965–975. [CrossRef] [PubMed]
64. Sullivan, G.M.; Hatterer, J.A.; Herbert, J.; Chen, X.; Roose, S.P.; Attia, E.; Mann, J.J.; Marangell, L.B.; Goetz, R.R.; Gorman, J.M. Low levels of transthyretin in the CSF of depressed patients. *Am. J. Psychiatry* **1999**, *156*, 710–714. [PubMed]
65. Tsuzuki, K.; Fukatsu, R.; Yamaguchi, H.; Tateno, M.; Imai, K.; Fujii, N.; Yamauchi, T. Transthyretin binds amyloid beta peptides, Abeta1-42 and Abeta1-40 to form complex in the autopsied human kidney—Possible role of transthyretin for abeta sequestration. *Neurosci. Lett.* **2000**, *281*, 171–174. [CrossRef]
66. Buxbaum, J.N.; Ye, Z.; Reixach, N.; Friske, L.; Levy, C.; Das, P.; Golde, T.; Masliah, E.; Roberts, A.R.; Bartfai, T. Transthyretin protects Alzheimer's mice from the behavioral and biochemical effects of A toxicity. *Proc. Natl. Acad. Sci. USA* **2008**, *105*, 2681–2686. [CrossRef] [PubMed]

67. Li, X.; Zhang, X.; Ladiwala, A.R.A.; Du, D.; Yadav, J.K.; Tessier, P.M.; Wright, P.E.; Kelly, J.W.; Buxbaum, J.N. Mechanisms of Transthyretin Inhibition of -Amyloid Aggregation In Vitro. *J. Neurosci.* **2013**, *33*, 19423–19433. [CrossRef] [PubMed]
68. Stein, T.D.; Johnson, J.A. Lack of neurodegeneration in transgenic mice overexpressing mutant amyloid precursor protein is associated with increased levels of transthyretin and the activation of cell survival pathways. *J. Neurosci.* **2002**, *22*, 7380–7388. [CrossRef] [PubMed]
69. Hori, H.; Kunugi, H. Dopamine agonist-responsive depression. *Psychogeriatrics* **2013**, *13*, 189–195. [CrossRef] [PubMed]
70. Lee, B.H.; Kim, Y.K. The roles of BDNF in the pathophysiology of major depression and in antidepressant treatment. *Psychiatry Investig.* **2010**, *7*, 231–235. [CrossRef]
71. Duman, R.S.; Monteggia, L.M. A neurotrophic model for stress-related mood disorders. *Biol. Psychiatry* **2006**, *59*, 1116–1127. [CrossRef]
72. Mizuno, K.; Giese, K.P. Hippocampus-dependent memory formation: Do memory type-specific mechanisms exist? *J. Pharm. Sci.* **2005**, *98*, 191–197. [CrossRef]
73. Denton, R.M.; Mccormack, J.G.; Edgell, N.J. Role of calcium ions in the regulation of intramitochondrial metabolism. *Biochem. J.* **1980**, *190*, 107–117. [CrossRef] [PubMed]
74. Hansford, R.G.; Castro, F. Effect of micromolar concentrations of free calcium ions on the reduction of heart mitochondrial NAD(P) by 2-oxoglutarate. *Biochem. J.* **2015**, *198*, 525–533. [CrossRef]
75. Crompton, M. The regulation of mitochondrial calcium transport in heart. *Curr. Top. Membr. Transp.* **1985**, *25*, 231–276.
76. Jouaville, L.S.; Pinton, P.; Bastianutto, C.; Rutter, G.A.; Rizzuto, R. Regulation of mitochondrial ATP synthesis by calcium: Evidence for a long-term metabolic priming. *Proc. Natl. Acad. Sci. USA* **1999**, *96*, 13807–13812. [CrossRef] [PubMed]
77. Steru, L.; Chermat, R.; Thierry, B.; Simon, P. The tail suspension test: A new method for screening antidepressants in mice. *Psychopharmacology* **1985**, *85*, 367–370. [CrossRef]
78. Matsukawa, T.; Motojima, H.; Sato, Y.; Takahashi, S.; Villareal, M.O.; Isoda, H. Upregulation of skeletal muscle PGC-1α through the elevation of cyclic AMP levels by Cyanidin-3-glucoside enhances exercise performance. *Sci. Rep.* **2017**, *7*, 44799. [CrossRef]

International Journal of
Molecular Sciences

MDPI

Article

Flavonoids from *Chionanthus retusus* (Oleaceae) Flowers and Their Protective Effects against Glutamate-Induced Cell Toxicity in HT22 Cells

Yeong-Geun Lee [1], Hwan Lee [2], Jae-Woo Jung [1], Kyeong-Hwa Seo [3], Dae Young Lee [4], Hyoung-Geun Kim [1], Jung-Hwan Ko [1], Dong-Sung Lee [2] and Nam-In Baek [1,*]

1 Graduate School of Biotechnology and Department of Oriental Medicinal Biotechnology, Kyung Hee University, Yongin 17104, Korea
2 College of Pharmacy, Chosun University, Gwangju 61452, Korea
3 Strategic Planning Division, National Institute of Biological Resources, Incheon 22689, Korea
4 Department of Herbal Crop Research, National Institute of Horticultural and Herbal Science, RDA, Eumseong 27709, Korea
* Correspondence: nibaek@khu.ac.kr; Tel.: +82-31-201-2661

Received: 31 May 2019; Accepted: 15 July 2019; Published: 18 July 2019

Abstract: The dried flowers of *Chionanthus retusus* were extracted with 80% MeOH, and the concentrate was divided into EtOAc, *n*-BuOH, and H_2O fractions. Repeated SiO_2, octadecyl SiO_2 (ODS), and Sephadex LH-20 column chromatography of the EtOAc fraction led to the isolation of four flavonols (**1–4**), three flavones (**5–7**), four flavanonols (**8–11**), and one flavanone (**12**), which were identified based on extensive analysis of various spectroscopic data. Flavonoids **4–6** and **8–11** were isolated from the flowers of *C. retusus* for the first time in this study. Flavonoids **1**, **2**, **5**, **6**, **8**, and **10–12** significantly inhibited NO production in RAW 264.7 cells stimulated by lipopolysaccharide (LPS) and glutamate-induced cell toxicity and effectively increased HO-1 protein expression in mouse hippocampal HT22 cells. Flavonoids with significant neuroprotective activity were also found to recover oxidative-stress-induced cell damage by increasing HO-1 protein expression. This article demonstrates that flavonoids from *C. retusus* flowers have significant potential as therapeutic materials in inflammation and neurodisease.

Keywords: *Chionanthus retusus*; flavonoid; flower; HO-1; neuroprotection; NO

1. Introduction

With the rapid growth of the aging population, the treatment of age-related diseases has become an important global issue, including in Korea [1]. Neurodisease is among the various illnesses induced by aging [2]. Previous studies have revealed the neuroprotective activities of bioactive compounds such as alkaloids, sterols, and flavonoids [3,4]. Flavonoids perform various neuroprotective actions, such as suppressing neuroinflammation; protecting neurons; and promoting memory, cognitive function, and learning [5,6]. Given the many experiments demonstrating their neuroprotective effects, these compounds may have therapeutic potential in neurodisease [3,6–9].

Flavonoids have a phenylchromane (C6-C3-C6) structure and are synthesized from L-phenylalanine and L-tyrosine via the shikimic acid pathway [10]. They comprise one of the most widespread and diverse groups of compounds in nature [11–13]. Among various natural resources, flowers (the reproductive organs of plants) contain diverse secondary metabolites, including volatiles, pigments, and flavonoids, which lure pollinating insects and facilitate pollination [14–16]. Sun et al. previously determined the total flavonoid content of *Chionanthus retusus* flowers to be 10.7% [17]. Thus, in this study, we focused on the isolation, identification, and investigation of the potential therapeutic effects of flavonoids from *C. retusus* flowers.

C. retusus (Oleaceae), a deciduous tree with oval leaves, is widely cultivated and distributed in Korea, China, Taiwan, and Japan, growing to 20–25 m high [18]. This plant has been used as an antipyretic, treatment for palsy and diarrhea in Oriental medicine and is known to contain many kinds of secondary metabolites, including flavonoids, lignans, sterols, and terpenoids [18–20]. These compounds have been reported to exert antioxidant, anti-inflammatory, and neuroprotective effects [6,7,18]. Although numerous active components have been isolated from *C. retusus* leaves and stems, the flowers of *C. retusus* have rarely been studied. This paper describes the isolation of 12 flavonoids from *C. retusus* flowers, determination of their chemical structures through extensive analysis of various spectroscopic data, evaluation of their anti-inflammatory and neuroprotective effects, and the relationship of their structure to their activity.

2. Results and Discussion

2.1. Contents of Total Phenols and Total Flavonoids in C. retusus Flowers

The contents of total phenols and flavonoids in the extract and fractions were determined as gallic acid and catechin equivalent values, respectively. As shown in Table 1, MeOH extract and EtOAc fraction (fr.) showed the highest contents compared to other fr.s. MeOH extract and EtOAc fr. showed a yellowish color on a thin-layer-chromatography (TLC) plate by spraying 10% H_2SO_4 and baking (data not shown), suggesting the extract and EtOAc fr. to include high amounts of flavonoids.

Table 1. Total phenols and flavonoids contents of the extract and fractions from *Chionanthus retusus* flowers.

Samples	Extract	EtOAc fr.	*n*-BuOH fr.	H_2O fr.
Total phenols (mg GA/g DW)	125.4 ± 3.3	245.6 ± 5.2	130.1 ± 2.5	53.1 ± 1.8
Total flavonoids (mg CA/g DW)	119.1 ± 2.7	172.1 ± 2.1	98.2 ± 0.9	18.2 ± 1.2

GA: gallic acid; CA: catechin; fr., fraction.

2.2. Isolation and Identification of Flavonoids from C. retusus Flowers

The dried flowers of *C. retusus* were extracted with MeOH, and the concentrate was divided into EtOAc, *n*-BuOH, and H_2O fr.s. Repeated SiO_2, octadecyl SiO_2 (ODS), and Sephadex LH-20 column chromatography (c.c.). on the EtOAc Fr enabled the isolation of four flavonols (**1**–**4**), three flavones (**5**–**7**), four flavanonols (**8**–**11**), and one flavanone (**12**). These compounds were identified to be quercetin (**1**) [20], kaempferol (**2**) [20], astragalin (**3**) [21], nicotiflorin (**4**) [22], luteolin (**5**) [20], luteolin 4′-*O*-*β*-D-glucopyranoside (**6**) [23], isorhoifolin (**7**) [24], taxifolin (**8**) [25], aromadendrin (**9**) [20], aromadendrin 7-*O*-*β*-D-glucopyranoside (**10**) [26], taxifolin 7-*O*-*β*-D-glucopyranoside (**11**) [27], and eriodictyol 7-*O*-*β*-D-glucopyranoside (**12**) [24] based on extensive analysis of data from various spectroscopic methods, including IR, FAB/MS, 1D-NMR (^{1}H, ^{13}C, DEPT), and 2D-NMR (COSY, HSQC, HMBC) (Figure 1). The identities of the compounds were confirmed by comparing their NMR and MS values with those reported in the literature. We determined the stereochemistry of the chiral centers (C-2 and C-3) in flavonoids **8**–**12** by examining the coupling constants between H-2 and H-3 in the ^{1}H-NMR spectra. They were mostly observed to be 12 Hz, which suggested that the two protons were in a 2,3-*trans* configuration.

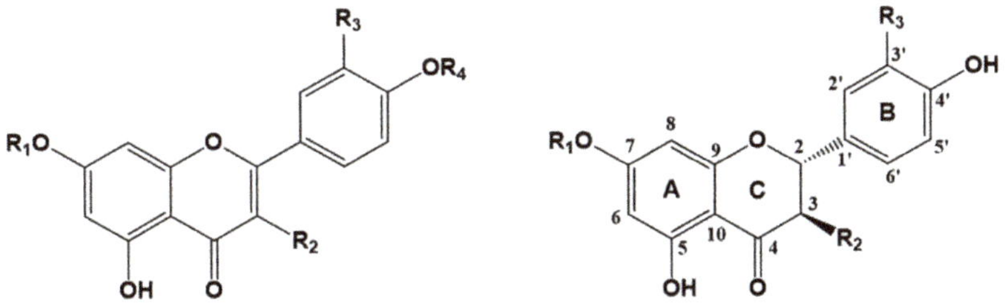

	R1	R2	R3	R4
1	H	OH	OH	H
2	H	OH	H	H
3	H	O-glc	H	H
4	H	O-rut	H	H
5	H	H	OH	H
6	H	H	OH	glc
7	rut	H	H	H

	R1	R2	R3
8	H	OH	OH
9	H	OH	H
10	glc	OH	H
11	glc	OH	OH
12	glc	H	OH

glc: β-D-glucopyranosyl
rut: α-L-rhamnopyranosyl-(1→6)-O-β-D-glucopyranosyl

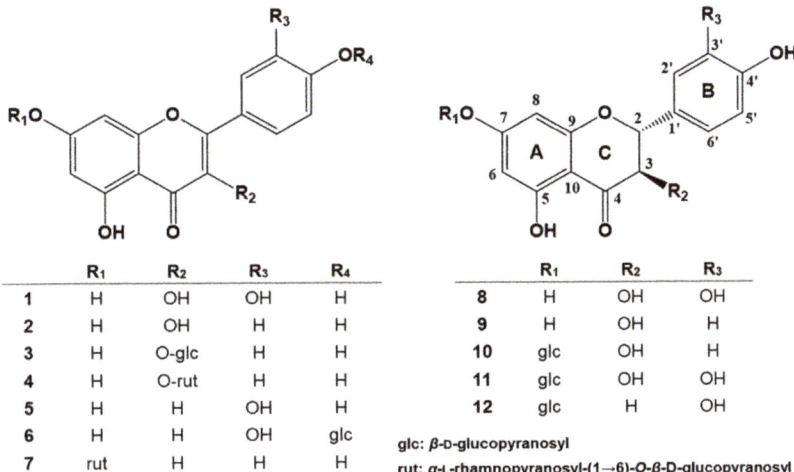

Figure 1. Chemical structures of flavonoids 1-12 isolated from *C. retusus* flowers.

2.3. *Inhibition Effects of Flavonoids* **1–12** *on NO Production in Lipopolysaccharide (LPS)-Induced RAW 264.7 Cells*

Oxidative stress is not only an important feature of several neurodegenerative processes, but also actively triggers intracellular signaling pathways that lead to cell death [28]. We first examined the viability of RAW264.7 cells treated with compounds **1–12** using a 3-(4,5-dimethylthiazol-2-yl)-2,5-diphenyltetrazolium bromide (MTT) assay. It did not show cytotoxicity or cellular proliferation when treated with compounds **1–4** and **6–12** at concentrations of 40 or 80 μM in RAW264.7 cells. However, compound **5** exhibited cytotoxic effects at 80 μM (Figure 2a). To investigate the anti-inflammatory effects of compounds **1–12**, we appreciated their inhibitory effects on NO production in LPS-induced RAW 264.7 cells. These cells were pretreated with flavonoids **1–12** and butein, a positive control, before one day LPS treatment. As shown in Table 2, compounds **1**, **2**, **5**, and **6** highly inhibited NO production, while compounds **8** and **10–12** showed moderate inhibition effect. Flavonoids with a catechol structure in the B ring (**1**, **5**, **6**, **8**, **11**, and **12**) exerted stronger anti-inflammatory effects than those with a phenol structure (**3**, **4**, **7**, **9**, and **10**). In addition, as the number of glucose moieties increased in compounds **1–6**, the NO inhibitory effects of these compounds in RAW 264.7 cells decreased. However, compounds with glucopyranosyl moieties at C-7 (**10** and **11**) exhibited higher activity than aglycones (**8** and **9**). These results indicate that the presence of a catechol structure in the B ring and a glucopyranosyl moiety in the flavonoid structure were key factors of the anti-inflammatory effects of these flavonoids.

Table 2. IC_{50} values of flavonoids **1–12** from *C. retusus* flowers on NO production in lipopolysaccharide (LPS)-induced RAW264.7 cells. The cells were pre-treated with each compound for 12 h, and then stimulated with LPS (1 μg/mL) for 18 h. The production of NO was determined as described in Section 3. Data shown represent the mean ± SD of three experiments.

No.	IC_{50} (μM)	No.	IC_{50} (μM)	No.	IC_{50} (μM)
1	37.93 ± 0.03	**5**	5.99 ± 0.02	**9**	>100
2	21.25 ± 0.03	**6**	30.60 ± 0.05	**10**	71.56 ± 0.08
3	>100	**7**	>100	**11**	57.18 ± 0.03
4	>100	**8**	78.53 ± 0.03	**12**	60.86 ± 0.01

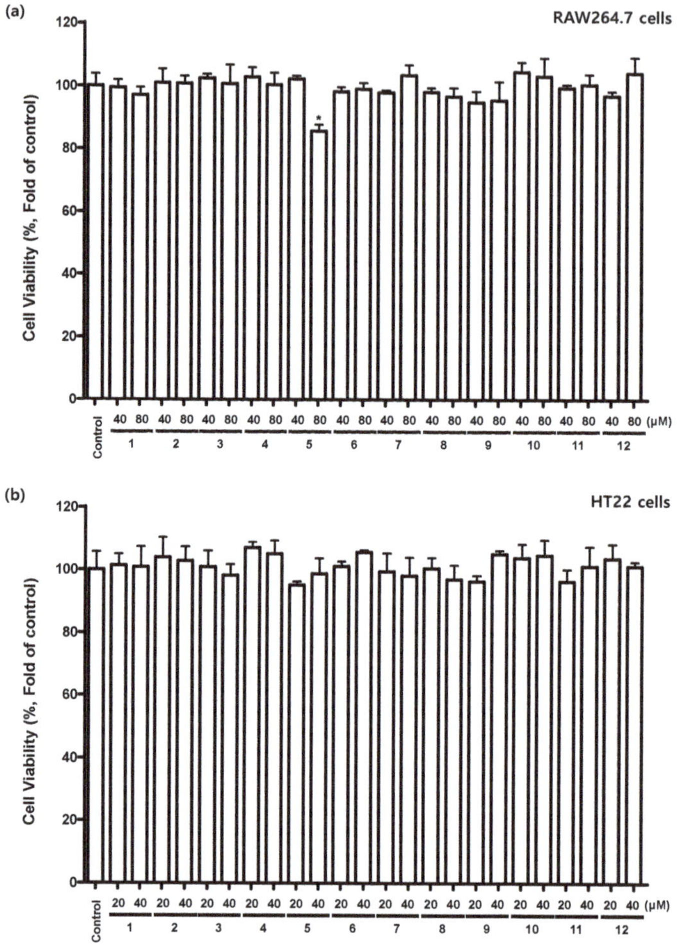

Figure 2. Cytotoxicity of compounds **1-12** on (**a**) RAW264.7 cells and (**b**) mouse hippocampal HT22 cells. (**a**) RAW264.7 cells were treated with 40 or 80 μM of compounds **1–12** for 48 h. (**b**) Mouse hippocampal HT22 cells were treated with 20 or 40 μM of compounds **1–12** for 24 h. Data are presented as the mean ± standard deviation of three independent experiments. * $p < 0.05$ vs. non-treated control.

2.4. *Effects of Flavonoids* 1–12 *on Glutamate-Induced Cell Toxicity in Mouse Hippocampal HT22 Cells*

To investigate the protective effects of compounds **1–12** against glutamate-induced oxidative neuronal cell death, we also examined their effects on the viability of mouse hippocampal HT22 cells. To investigate the potential for cellular proliferation or cytotoxic effects of compounds **1–12**, we first examined the viability of mouse hippocampal HT22 cells treated with compounds **1–12** using an MTT assay. No cytotoxic effects or cellular proliferation by compounds **1–12** were observed at concentrations <40 μM (Figure 2b). These cells were pretreated with compounds **1–12** at concentrations of 20 or 40 μM for 3 h and then were treated with glutamate and reacted for 12 h. Thereafter, cell viability was assessed with an MTT assay. None of the compounds exhibited toxicity at the highest concentration (40 μM). Compounds **1**, **2**, **5**, **6**, **8**, **10**, **11**, and **12** significantly increased cell viability following glutamate treatment (Figure 3). Butein derived from *Rhus verniciflua*, which is known to protect mouse hippocampal HT22 cells from glutamate-induced death [29], was used as a positive control and indeed exhibited cytoprotective effects (Figure 3). Flavonoids with a catechol structure in the B ring (**1**, **5**, **6**, **8**, **11**, and **12**) exerted stronger cytoprotective effects than those with a phenol structure (**3**, **4**, **7**, **9**, and **10**). In addition, as the number of glucose moieties increased in compounds **1–6**, the cytoprotective effects of these compounds in HT22 cells decreased. However, compounds with glucopyranosyl moieties at C-7 (**10** and **11**) exhibited higher activity than aglycones (**8** and **9**). These results indicate that the presence of a catechol structure in the B ring and a glucopyranosyl moiety in the flavonoid structure were key determinants of the effects of these flavonoids on mouse hippocampal HT22 cells.

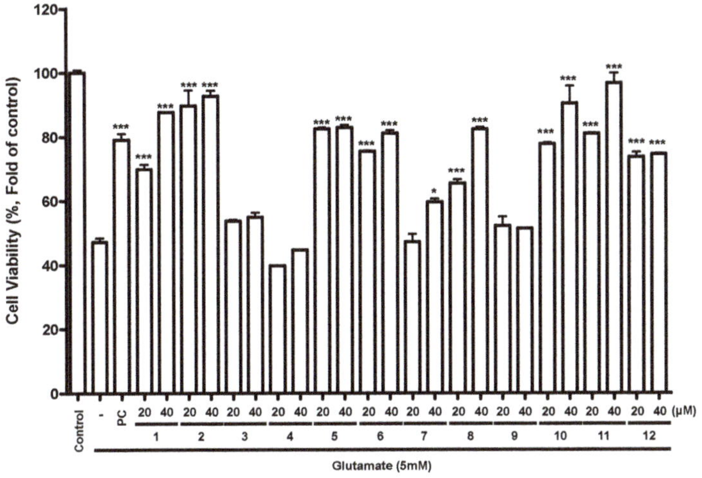

Figure 3. Effects of compounds **1–12** on glutamate-induced oxidative neurotoxicity in mouse hippocampal HT22 cells. Mouse hippocampal HT22 cells were pretreated with 20 or 40 μM of compounds **1–12** and then were treated with glutamate (5 mM) for 12 h. Butein (5 μM) was used as a positive control. Data are presented as the mean ± standard deviation of three independent experiments. * $p < 0.05$, *** $p < 0.001$ vs. glutamate.

2.5. *Effects of Compounds* **1**, **2**, **5**, **6**, **8**, *and* **10–12** *on HO-1 Expression in Mouse Hippocampal HT22 Cells*

Heme oxygenase (HO) is an important enzyme in the antioxidant cell system. HO-1, one of the HO derivatives, decomposes heme in the cell to produce carbon monoxide, iron, and biliverdin [30]. HO-1 expression has been reported to inhibit brain cell damage resulting from oxidative stress [31]. We examined whether compounds **1**, **2**, **5**, **6**, **8**, and **10–12** affected the protein expression of HO-1, given their protection against glutamate-induced toxicity in mouse hippocampal HT22 cells. Mouse hippocampal HT22 cells were treated with compounds **1**, **2**, **5**, **6**, **8**, and **10–12** at three concentrations (10, 20, and 40 μM) and then cultured for 12 h. Cobalt protoporphyrin (CoPP), a well-known HO-1 inducer, was used as a positive control. As shown in Figure 4, compounds **1**, **2**, **5**, **6**, **8**, and **10–12** all increased HO-1 protein expression in a dose-dependent manner in mouse hippocampal HT22 cells. Flavonoid aglycones (**1**, **2**, **5**, and **8**) exhibited higher activity than the glycosides (**10–12**). The flavonol and flavanonol with a catechol structure in the B ring (**1** and **11**) displayed stronger HO-1 expression than those with a phenol structure (**2** and **10**). Flavonoids with a hydroxy group at C-3 (**8** and **11**) exhibited weaker HO-1 expression than those without (**5** and **12**). In addition, a flavonoid with a double bond between C2 and C3 (**1**) was a weaker inhibitor of oxidative-stress-induced brain-cell damage than one with a single bond (**8**). These results indicate that the presence of a hydroxy group at C-3, the structure of the B ring and the type of C2-C3 bond are key determinants of the extent to which these flavonoids protect brain cells from damage due to oxidative stress.

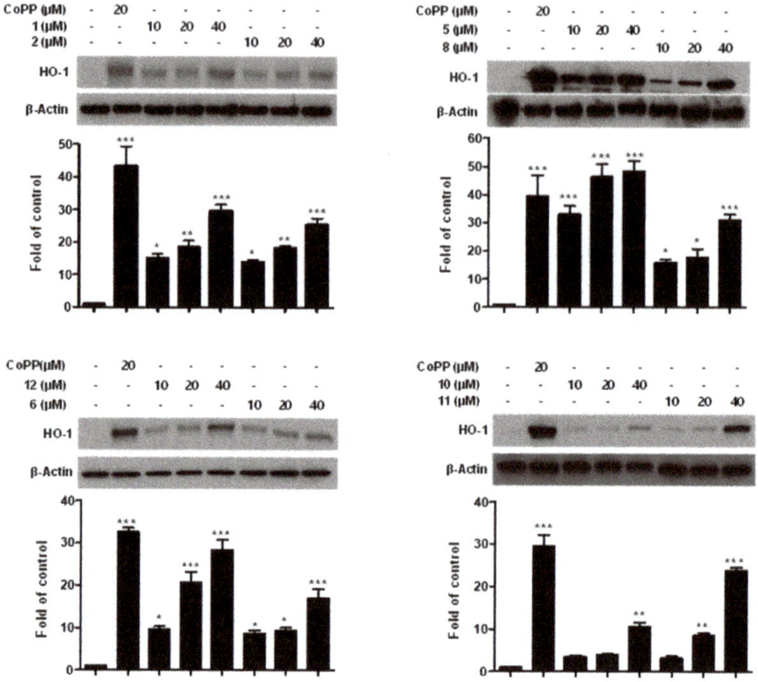

Figure 4. Effects of compounds **1**, **2**, **5**, **6**, **8**, and **10–12** on HO-1 expression in mouse hippocampal HT22 cells. Mouse hippocampal HT22 cells were treated with compounds **1**, **2**, **5**, **6**, **8**, and **10–12** at three concentrations (10, 20, and 40 μM) and then cultured for 12 h. Expression of HO-1 was measured by Western blot analysis. Cobalt protoporphyrin (CoPP, 20 μM) was used as a positive control. Representative blots of three independent experiments are shown. * $p < 0.05$, ** $p < 0.01$, *** $p < 0.001$ vs. non-treated control.

2.6. *Effects of Compounds* **1**, **2**, **5**, **6**, **8**, *and* **10–12** *on Cell Viability through HO Signaling Pathway*

Compounds **1**, **2**, **5**, **6**, **8**, and **10–12**, which exhibited cytoprotective effects, also increased HO-1 expression (Figures 3 and 4). To investigate whether HO-1 expression regulates cell viability, we assessed the protective effects of compounds **1**, **2**, **5**, **6**, **8**, and **10–12** when tin protoporphyrin IX (SnPP) was used as a HO-1 activity inhibitor. Cells were treated with compounds **1**, **2**, **5**, **6**, **8**, and **10–12** (40 μM) in the presence or absence of SnPP (50 μM) and then exposed to glutamate (5 mM) for 12 h. When cells were pre-treated with SnPP, the protective effects of the compounds decreased (Figure 5); that is, cell viability was significantly lower in SnPP-pretreated cells than in the cells not treated with SnPP. These results indicate that compounds **1**, **2**, **5**, **6**, **8**, and **10–12** inhibited oxidative-stress-induced cell damage by increasing HO-1 protein expression.

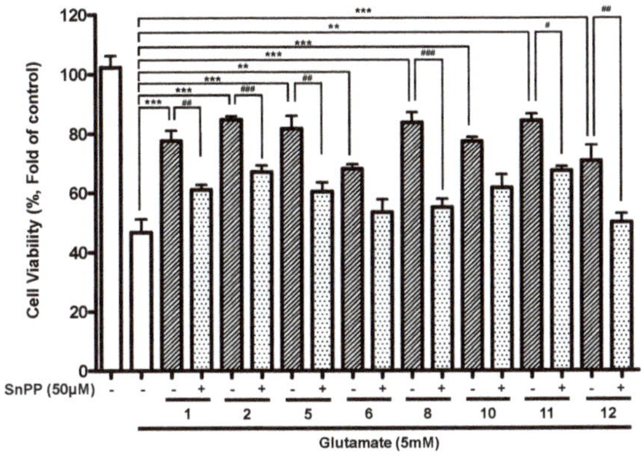

Figure 5. Effects of HO-1 expression induced by compounds **1**, **2**, **5**, **6**, **8**, and **10–12** on glutamate-induced oxidative cell damage in mouse hippocampal HT22 cells. Mouse hippocampal HT22 cells were treated with compounds **1**, **2**, **5**, **6**, **8**, and **10–12** (40 μM) in the presence or absence of tin protoporphyrin IX (SnPP, 50 μM) and then exposed to glutamate (5 mM) for 12 h. Data are presented as the mean ± standard deviation of three independent experiments. ** $p < 0.01$, *** $p < 0.001$. # $p < 0.05$, ## $p < 0.01$, ### $p < 0.001$.

3. Materials and Methods

3.1. Plant Materials

The flowers of *C. retusus* Lindl. And Paxton were gathered near Kyung Hee University, Yong-In, South Korea, in August 2014, and were identified by Prof. Dae-Keun Kim, College of Pharmacy, Woosuk University, Jeonju, South Korea. A voucher specimen (KHU-NPCL-201408) has been deposited at the Natural Products Chemistry Laboratory, Kyung Hee University.

3.2. General Experimental Procedures

The equipment and chemicals used to isolate and identify flavonoids from *C. retusus* flowers and evaluate their neuroprotective activity were obtained from the literature [32–35].

3.3. Isolation Procedure of Flavonoids (1–12) from C. retusus Flowers

Dried *C. retusus* flowers (315 g) were extracted in 80% aqueous MeOH (22.5 L × 4) at room temperature for 24 h, and then filtered and concentrated in vacuo. The concentrated MeOH extracts (145 g) were poured into H_2O (2.0 L) and successively extracted with EtOAc (2.0 L × 3) and *n*-BuOH (1.8 L × 3). Each layer was concentrated under reduced pressure to obtain EtOAc (CFE, 27 g), *n*-BuOH (CFB, 24 g), and H_2O (CFH, 94 g). Frs. CFE (27 g) was subjected to SiO_2 c.c. (Φ 11 × 12 cm) and eluted with $CHCl_3$-MeOH (CM; 40:1 → 10:1 → 5:1 → 2:1 → 1:1, 600 mL of each), with monitoring by TLC, yielding 15 frs (CFE-1 to CFE-15).

CFE-5 (3.2 g, *Ve/Vt* 0.360–0.415) was subjected to ODS c.c. (Φ 5.5 × 7 cm, MeOH-H_2O [MH] = 4:1, 1.7 L) to yield 12 Frs (CFE-5-1 to CFE-5-12). CFE-5-1 (1.0 g, *Ve/Vt* 0.000–0.110) was subjected to ODS c.c. (Φ 4.0 × 7 cm, MH = 1:1, 1.5 L) to yield 9 Frs (CFE-5-1-1 to CFE-5-1-9). CFE-5-1-3 (95.0 mg, *Ve/Vt* 0.150–0.260) was subjected to Sephadex LH-20 c.c. (Φ 1.5 × 60 cm, 80% MeOH, 560 mL) to yield 8 Frs (CFE-5-1-3-1 to CFE-5-1-3-8), along with purified compound **9** (CFE-5-1-3-4, 2.8 mg, *Ve/Vt* 0.550–0.560, TLC [SiO_2] R_f 0.37, CM = 10:1, TLC [ODS] R_f 0.58, MH = 2:1).

CFE-7 (2.4 g, *Ve/Vt* 0.430–0.480) was subjected to Sephadex LH-20 c.c. (Φ 3 × 50 cm, 80% MeOH, 1.3 L) to yield 15 Frs (CFE-7-1 to CFE-7-15), along with purified compound **8** (CFE-7-10, 77.4 mg, *Ve/Vt* 0.488-0.542, TLC [SiO_2] R_f 0.45, $CHCl_3$-MeOH-H_2O [CMH] = 10:3:1, TLC [ODS] R_f 0.60, MH = 3:2) and purified compound **1** (CFE-7-15, 14.6 mg, *Ve/Vt* 0.885-1.000, TLC [SiO_2] R_f 0.47, CMH = 10:3:1, TLC [ODS] R_f 0.74, MH = 4:1). CFE-7-12 (68.5 mg, *Ve/Vt* 0.650–0.720) was subjected to ODS c.c. (Φ 2.0 × 7 cm, MH = 1:1, 620 mL) to yield 3 Frs (CFE-7-12-1 to CFE-7-12-3), along with purified compound **5** (CFE-7-12-2, 17.4 mg, *Ve/Vt* 0.194–0.677, TLC [SiO_2] R_f 0.50, CMH = 10:3:1, TLC [ODS] R_f 0.50, MH = 4:1).

CFE-9 (2.2 g, *Ve/Vt* 0.580–0.610) was subjected to Sephadex LH-20 c.c. (Φ 3 × 50 cm, 80% MeOH, 2.2 L) to yield 14 Frs (CFE-9-1 to CFE-9-14). CFE-9-8 (36.5 mg, *Ve/Vt* 0.480-0.510) was subjected to ODS c.c. (Φ 2.0 × 5 cm, MH = 2:3, 200 mL) to yield 4 Frs (CFE-9-8-1 to CFE-9-8-4), along with purified compound **3** (CFE-9-8-2, 10.0 mg, *Ve/Vt* 0.125–0.425, TLC [SiO_2] R_f 0.50, CM = 4:1, TLC [ODS] R_f 0.65, MH = 3:1).

CFE-12 (2.1 g, *Ve/Vt* 0.710-0.790) was subjected to Sephadex LH-20 c.c. (Φ 3.0 × 50 cm, 70% MeOH, 2.3 L) to yield 14 Frs (CFE-12-1 to CFE-12-14). CFE-12-5 (200.0 mg) was subjected to SiO_2 c.c. (Φ 3.5 × 14 cm) and eluted with CMH = 10:3:1 (560 mL), with monitoring by TLC, yielding 6 Frs (CFE-12-5-1 to CFE-12-5-6), along with purified compound **10** (CFE-12-5-3, 119.4 mg, *Ve/Vt* 0.102–0.250, TLC [SiO_2] R_f 0.50, CMH = 65:35:10, TLC [ODS] R_f 0.50, MH = 2:3). CFE-12-8 (330.0 mg, *Ve/Vt* 0.370–0.410) was subjected to ODS c.c. (Φ 3.0 × 5 cm, MH = 2:3, 1.2 L) to yield 8 Frs (CFE-12-8-1 to CFE-12-8-8), along with purified compound **12** (CFE-12-8-1, 115.5 mg, *Ve/Vt* 0.000–0.058, TLC [SiO_2] R_f 0.50, CMH = 65:35:10, TLC [ODS] R_f 0.70, MH = 3:2). CFE-12-10 (240.0 mg, *Ve/Vt* 0.460-0.550) was subjected to ODS c.c. (Φ 3.0 × 5 cm, MH = 2:3, 840 mL) to yield 6 Frs (CFE-12-10-1 to CFE-12-10-6), along with purified compound **6** (CFE-12-10-4, 72.0 mg, *Ve/Vt* 0.286–0.414, TLC [SiO_2] R_f 0.55, CMH = 65:35:10, TLC [ODS] R_f 0.45, MH = 3:2).

CFE-13 (3.2 g, *Ve/Vt* 0.710-0.790) was subjected to Sephadex LH-20 c.c. (Φ 3.0 × 50 cm, 70% MeOH, 2.3 L) to yield 16 Frs (CFE-13-1 to CFE-13-16), along with purified compound **2** (CFE-13-16, 29.0 mg, *Ve/Vt* 0.846–0.912, TLC [SiO_2] R_f 0.50, CM = 5:1, TLC [ODS] R_f 0.40, MH = 3:1). CFE-13-6 (70.0 mg, *Ve/Vt* 0.270–0.320) was subjected to ODS c.c. (Φ 2.5 × 6 cm, MH = 2:3, 740 mL) to yield six Frs (CFE-13-6-1 to CFE-13-6-6), along with purified compound **7** (CFE-13-6-4, 12.0 mg, *Ve/Vt* 0.657–0.730, TLC [SiO_2] R_f 0.50, CM = 2:1, TLC [ODS] R_f 0.55, MH = 3:2). CFE-13-7 (890.0 mg, *Ve/Vt* 0.330-0.480) was subjected to ODS c.c. (Φ 5.5 × 4 cm, MH = 2:3, 2.6 L) to yield 7 Frs (CFE-13-7-1 to CFE-13-7-7), along with purified compound **11** (CFE-13-7-1, 368.0 mg, *Ve/Vt* 0.000–0.102, TLC [SiO_2] R_f 0.50, CM = 2:1, TLC [ODS] R_f 0.55, MH = 1:3) and compound **4** (CFE-13-7-6, 341.0 mg, *Ve/Vt* 0.512–0.923, TLC [SiO_2] R_f 0.50, CM = 2:1, TLC [ODS] R_f 0.65, MH = 3:2) (Scheme 1).

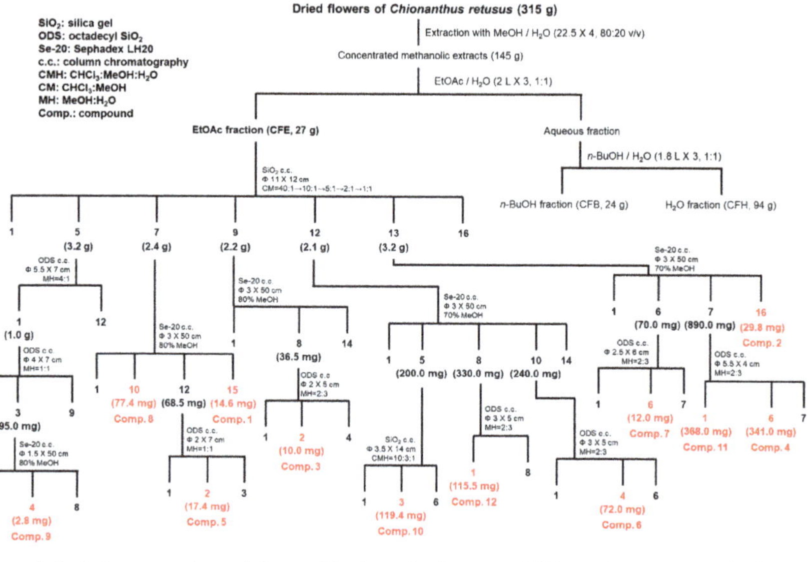

Scheme 1. Isolation procedure of flavonoids from the flowers of *Chionanthus retusus*. Words in red indicate fraction number, quantity, and compound number of isolated flavonoids.

quercetin (**1**): Yellowish powder (MeOH); m.p. 276–277 °C; ultraviolet (UV) (MeOH) λ_{max} (nm) 370, 305, 267, 255; infrared (IR) (KBr) ν_{max} 3350, 1680, 1615 cm^{-1}; positive FAB/MS *m/z* 303 $[M + H]^+$.

kaempferol (**2**): Yellowish powder (MeOH); m.p. 278–279 °C; UV (MeOH) λ_{max} (nm) 364, 320, 294, 265, 254; IR (KBr) ν_{max} 3345, 1658, 1605 cm^{-1}; positive FAB/MS *m/z* 309 $[M + Na]^+$.

astragalin (**3**): Yellowish powder (MeOH); m.p. 230–231 °C; $[\alpha]_D^{21}$ +16.0 (*c* 0.1, MeOH); UV (MeOH) λ_{max} (nm) 348, 259; IR (KBr) ν_{max} 3350, 2930, 2365, 1655, 1610 cm^{-1}; positive FAB/MS *m/z* 471 $[M + Na]^+$.

nicotiflorin (**4**): Yellowish powder (MeOH); m.p. 268–269°C; $[\alpha]_D^{21}$ −15.0 (*c* 1.0, MeOH); UV (MeOH) λ_{max} (nm) 365, 267, 254; IR (KBr) ν_{max} 3365, 2940, 2360, 1655, 1600, 1515 cm^{-1}; positive FAB/MS *m/z* 639 $[M + Na]^+$.

luteolin (**5**): Yellowish powder (MeOH); m.p. 329–330 °C; UV (MeOH) λ_{max} (nm) 349, 269, 254; IR (KBr) ν_{max} 3320, 2930, 1600, 1520 cm^{-1}; positive FAB/MS *m/z* 271 $[M + H]^+$.

luteolin 4′-O-β-D-glucopyranoside (**6**): Yellowish powder (MeOH); m.p. 178–179 °C; UV (MeOH) λ_{max} (nm) 341, 272; IR (KBr) ν_{max} 3320, 2930, 1600, 1520, 1510, 1480 cm^{-1}; positive FAB/MS *m/z* 449 $[M + H]^+$.

isorhoifolin (**7**): Yellowish needles; m.p. 269–270 °C; $[\alpha]_D^{21}$ −96.7 (*c* 1.0, MeOH); UV (MeOH) λ_{max} (nm) 331, 266; IR (KBr) ν_{max} 3365, 2360, 1635, 1600, 1515 cm^{-1}; positive FAB/MS *m/z* 579 $[M + H]^+$.

taxifolin (**8**): Yellowish powder (MeOH); m.p. 236–237 °C; $[\alpha]_D^{21}$ +23.1 (*c* 0.1, MeOH); UV (MeOH) λ_{max} (nm) 330, 280; IR (KBr) ν_{max} 3415, 1625, 1515, 1472 cm^{-1}; positive FAB/MS *m/z* 327 $[M + Na]^+$.

aromadendrin (**9**): White powder; m.p. 216–217 °C; $[\alpha]_D^{21}$ +58.5 (*c* 0.3, MeOH); UV (MeOH) λ_{max} (nm) 329, 292, 228; IR (KBr) ν_{max} 3420, 1655, 1518 cm^{-1}; positive FAB/MS *m/z* 311 $[M + Na]^+$.

aromadendrin 7-O-β-D-glucopyranoside (**10**): Yellowish powder (MeOH); m.p. 172–173 °C; $[\alpha]_D^{21}$ −18.7 (*c* 0.2, MeOH); UV (MeOH) λ_{max} (nm) 321, 285; IR (KBr) ν_{max} 3435, 1645, 1520, 1365 cm^{-1}; positive FAB/MS *m/z* 473 $[M + Na]^+$.

taxifolin 7-O-β-D-glucopyranoside (**11**): Yellowish powder (MeOH); m.p. 169–170 °C; $[\alpha]_D^{21}$ −48.2 (*c* 0.2, MeOH); UV (MeOH) λ_{max} (nm) 331, 283; IR (KBr) ν_{max} 3420, 1635, 1450, 1510, 1390 cm^{-1}; positive FAB/MS *m/z* 467 $[M + H]^+$.

eriodictyol 7-O-β-D-glucopyranoside (**12**): Yellowish powder (MeOH); m.p. 173–174 °C; $[\alpha]_D^{21}$ −35.5 (*c* 0.2, MeOH); UV (MeOH) λ_{max} (nm) 283, 233; IR (KBr) ν_{max} 3455, 1690, 1595, 1510 cm^{-1}; positive FAB/MS *m/z* 451 $[M + H]^+$.

^{1}H-NMR (400 MHz, δ_H) and ^{13}C-NMR (100 MHz, δ_C) spectroscopic data of flavonoids **1–12**, see Tables 3 and 4.

Table 3. ^{1}H-NMR data of flavonoids **1–12** (δ_H in ppm, coupling pattern, *J* in Hz).

No.	1 [(a)]	2 [(a)]	3 [(a)]	4 [(a)]	5 [(b)]	6 [(b)]	7 [(b)]	8 [(a)]	9 [(a)]	10 [(a)]	11 [(a)]	12 [(a)]
2	-	-	-	-	-	-	-	5.09, d, 12.0	5.05, d, 11.6	5.03, d, 12.0	5.09, d, 11.6	5.14, dd, 12.4, 2.8
3	-	-	-	-	6.91, s	6.90, s	6.80, s	4.71, d, 12.0	4.58, d, 11.6	4.59, d, 12.0	4.71, d, 11.6	3.00, dd, 16.4, 12.4 2.63, dd, 16.4, 2.8
6	6.17, brs	6.18, brs	6.20, brs	6.20, brs	6.75, d, 1.6	6.75, brs	6.68, brs	6.05, d, 1.2	5.87, d, 2.0	6.15, d, 1.6	6.05, d, 1.2	5.09, brs
8	6.38, brs	6.38, brs	6.36, brs	6.37, brs	6.76, d, 1.6	6.77, brs	6.78, brs	6.13, d, 1.2	5.89, d, 2.0	6.21, d, 1.6	6.09, d, 1.2	6.00, brs
2′	7.73, d, 1.8	8.10, d, 8.4	8.06, d, 8.0	8.07, d, 8.0	7.83, brs	7.85, d, 1.2	7.87, d, 7.6	7.20, d, 1.2	7.29, d, 8.4	7.38, d, 8.0	7.18, d, 1.2	6.73, brs
3′	-	6.91, d, 8.4	6.90, d, 8.0	6.90, d, 8.0	-	-	7.19, d, 7.6	-	6.80, d, 8.4	6.85, d, 8.0	-	-
5′	6.88, d, 8.4	6.91, d, 8.4	6.90, d, 8.0	6.90, d, 8.0	7.39, d, 8.0	7.36, d, 8.0	7.19, d, 7.6	7.01, d, 7.6	6.80, d, 8.4	6.85, d, 8.0	7.01, d, 7.6	6.61, d, 8.0
6′	7.63, dd, 8.4, 1.8	8.10, d, 8.4	8.06, d, 8.0	8.07, d, 8.0	7.61, brd, 8.4	7.63, dd, 8.0, 1.2	7.87, d, 7.6	7.08, dd, 7.6, 1.2	7.29, d, 8.4	7.38, d, 8.0	7.05, dd, 7.6, 1.2	6.62, brd, 8.0
glc-1	-	-	5.28, d, 7.6	5.12, d, 7.6	-	5.24, d, 8.0	5.09, d, 7.6	-	-	5.05, d, 7.6	5.20, d, 8.0	5.09, d, 7.6
glc-2	-	-	3.51, O	3.27–3.79, O	-	3.59, O	3.25–3.81, O	-	-	3.57, O	3.50, O	3.47, dd, 7.6, 7.2
glc-3	-	-	3.28, O	3.27–3.79, O	-	3.35, O	3.25–3.81, O	-	-	3.33, O	3.49, O	3.39, dd, 7.2, 7.2
glc-4	-	-	3.38, dd, 7.6, 8.0	3.27–3.79, O	-	3.45, dd, 7.6, 8.0	3.25–3.81, O	-	-	3.63, O	3.37, O	3.34, O
glc-5	-	-	3.50, O	3.27–3.79, O	-	3.48, O	3.25–3.81, O	-	-	3.55, O	3.37, O	3.33, O
glc-6	-	-	3.74, dd, 12.4, 6.0 3.60, dd, 12.4, 2.4	3.27–3.79, O 3.27–3.79, O	-	3.68, dd, 11.6, 5.4 3.55, dd, 11.6, 2.0	3.25–3.81, O 3.25–3.81, O	-	-	3.75, dd, 12.0, 4.8 3.65, dd, 12.0, 1.6	3.86, dd, 11.6, 4.4 3.64, dd, 11.6, 1.2	3.85, dd, 11.6, 5.2 3.67, dd, 11.6, 1.8
rha-1	-	-	-	4.50, brs	-	-	4.49, brs	-	-	-	-	-
rha-2	-	-	-	3.27–3.79, O	-	-	3.25–3.81, O	-	-	-	-	-
rha-3	-	-	-	3.27–3.79, O	-	-	3.25–3.81, O	-	-	-	-	-
rha-4	-	-	-	3.27–3.79, O	-	-	3.25–3.81, O	-	-	-	-	-
rha-5	-	-	-	3.27–3.79, O	-	-	3.25–3.81, O	-	-	-	-	-
rha-6	-	-	-	1.10, d, 6.0	-	-	1.10, d, 6.0	-	-	-	-	-

[(a)] CD_3OD, 400 MHz; [(b)] pyridine-d_5, 400 MHz; glc: β-D-glucopyranosyl; rha: α-L-rhamnopyranosyl; O: overlapped.

Table 4. ^{13}C-NMR data of flavonoids **1–12**.

No.	**1** [a]	**2** [a]	**3** [a]	**4** [a]	**5** [b]	**6** [b]	**7** [b]	**8** [a]	**9** [a]	**10** [a]	**11** [a]	**12** [a]
2	158.7	158.0	158.1	158.1	163.3	163.2	164.2	83.8	82.9	84.9	83.8	80.9
3	135.8	137.2	135.2	135.3	104.5	104.1	103.8	72.4	71.5	73.9	72.4	44.3
4	179.4	177.3	179.3	161.1	182.0	181.6	182.4	196.9	197.9	201.1	196.9	198.7
5	163.2	162.5	162.6	162.6	162.5	162.2	163.0	164.0	166.9	165.0	164.0	165.0
6	100.0	99.3	99.5	99.8	99.7	99.6	98.9	96.2	96.0	98.4	96.2	98.0
7	166.1	165.5	165.9	165.6	166.4	166.0	166.5	167.4	163.3	167.2	167.4	166.9
8	94.9	94.5	94.8	94.7	94.5	94.4	94.9	95.1	95.0	96.8	95.1	97.0
9	159.2	158.2	159.0	159.1	157.9	157.9	158.1	163.2	162.6	166.5	163.2	164.6
10	105.7	104.5	105.8	105.4	104.0	103.8	116.5	100.5	100.4	103.0	100.5	103.8
1′	123.4	123.7	122.5	122.4	127.0	126.9	121.9	128.4	127.6	129.3	128.4	121.6
2′	116.1	130.7	132.1	132.1	114.6	114.3	128.9	114.7	129.5	130.5	114.8	116.1
3′	150.2	116.3	116.1	115.9	150.5	150.5	104.9	145.7	114.9	116.1	145.7	147.0
4′	145.8	160.5	161.5	161.1	149.4	149.3	162.7	144.9	157.8	160.1	144.8	144.5
5′	117.8	116.3	116.1	115.9	117.3	117.5	104.9	114.8	114.9	116.1	115.0	117.2
6′	123.2	130.7	132.1	132.1	119.8	119.7	128.9	119.7	129.5	130.5	119.8	119.0
glc-1	-	-	104.0	104.5	-	104.1	104.5	-	-	101.2	103.9	101.1
glc-2	-	-	75.5	75.5	-	75.7	75.8	-	-	75.9	75.7	74.7
glc-3	-	-	78.1	77.9	-	78.3	77.8	-	-	78.2	78.1	77.9
glc-4	-	-	71.1	73.7	-	71.1	74.0	-	-	71.8	71.3	70.9
glc-5	-	-	77.8	76.9	-	77.8	76.8	-	-	78.0	78.0	78.1
glc-6	-	-	62.6	68.4	-	62.3	68.7	-	-	62.4	62.4	62.4
rha-1	-	-	-	102.1	-	-	102.1	-	-	-	-	-
rha-2	-	-	-	71.8	-	-	72.1	-	-	-	-	-
rha-3	-	-	-	72.1	-	-	72.3	-	-	-	-	-
rha-4	-	-	-	71.2	-	-	71.5	-	-	-	-	-
rha-5	-	-	-	69.5	-	-	70.1	-	-	-	-	-
rha-6	-	-	-	17.9	-	-	18.0	-	-	-	-	-

[a] CD_3OD, 100 MHz; [b] pyridine-d_5, 100 MHz; glc: β-D-glucopyranosyl; rha: α-L-rhamnopyranosyl.

3.4. Cell Culture and MTT Assay

Mouse hippocampal HT22 cells were donated by Wonkwang University, Iksan, Korea (Prof. Youn-Chul Kim). Cytoprotective activity assay was performed, as per the previously described method [35]. Cell viability was evaluated using the MTT assay reported in the literature [36].

3.5. Macrophage RAW 264.7 Culture, Viability Assay, and NO Measurement

Macrophage RAW 264.7 culture, viability assay, and NO measurement were carried out as per the previously described method [35].

3.6. Determination of Total Phenols and Flavonoids Contents in C. retusus Flower

Determination of the total phenolic and flavonoid contents of *C. retusus* flower was carried out as per the previously described method [37].

3.7. Western Blot Analysis

Pelleted HT22 cells were washed with PBS and lysed with an RIPA buffer from Sigma Chemical Co. The same amount of protein from each sample was mixed into a sample loading buffer, subjected to SDS-PAGE, and transferred to a membrane.

3.8. Statistical Analysis

Statistical analysis was performed with GraphPad Prism 5 software (ver. 3.03, San Diego, CA, USA). Data are presented as the mean ± standard deviation of 3 independent experiments. The mean differences were derived using one-way ANOVA and Tukey's multiple comparison test, and statistical significance was defined as $p < 0.05$, $p < 0.01$, and $p < 0.001$.

4. Conclusions

In conclusion, four flavonols (**1**–**4**), three flavones (**5**–**7**), four flavanonols (**8**–**11**), and one flavanone (**12**) were isolated from *C. retusus* flowers. Flavonoids **4**–**6** and **8**–**11** were isolated from the flowers of *C. retusus* for the first time in this study. Flavonoids **1**, **2**, **5**, **6**, **8**, and **10**–**12** exhibited significant anti-inflammatory and neurocytoprotective activity, and effectively increased HO-1 protein expression. The flavonoids that displayed significant neuroprotective activity were found to recover oxidative stress-induced cell damage by increasing HO-1 protein expression. The relationships between the structural characteristics of these flavonoids and their anti-inflammatory and neuroprotective activity were revealed. Further studies are needed to investigate the potential therapeutic effects of flavonoids in innovative anti-inflammatory and neuroprotective strategies.

Author Contributions: Conceptualization, Y.-G.L. and N.-I.B.; methodology, Y.-G.L., D.-S.L., and N.-I.B.; formal analysis, Y.-G.L. and N.-I.B.; investigation, Y.-G.L., H.L., D.-S.L., and N.-I.B.; resources, Y.-G.L., J.-W.J., K.-H.S., D.Y.L., H.-G.K., J.-H.K., and N.-I.B.; data curation, Y.-G.L., J.-W.J., K.-H.S., D.Y.L., H.-G.K., J.-H.K., and N.-I.B.; writing-original draft preparation, Y.-G.L.; writing-review and editing, supervision, project administration, and funding acquisition, N.-I.B.

Funding: This work was carried out with the support of the "Cooperative Research Program for Agriculture Science & Technology Development (Project No. PJ01420403)", Rural Development Administration, Republic of Korea.

Conflicts of Interest: The authors declare no conflict of interest.

Abbreviations

c.c.	column chromatography
CoPP	cobalt protoporphyrin
Fr	fraction
HO	heme oxygenase
IR	infrared
SnPP	tin protoporphyrin IX
SiO_2	silica gel
ODS	octadecyl SiO_2
PC	positive control
TLC	thin layer chromatography
UV	ultraviolet
Ve/Vt	elution volume/total volume

References

1. Han, A.R.; Park, S.A.; Ahn, B.E. Reduced stress and improved physical functional ability in elderly with mental health problems following a horticultural therapy program. *Complement. Ther. Med.* **2018**, *38*, 19–23. [CrossRef] [PubMed]
2. Nabavi, S.F.; Braidy, N.; Habtemariam, S.; Orhan, I.E.; Daglia, M.; Manayi, A.; Gortzi, O.; Nabavi, S.M. Neuroprotective effects of chrysin: From chemistry to medicine. *Neurochem. Int.* **2015**, *90*, 224–231. [CrossRef] [PubMed]
3. Levi, M.S.; Brimble, M.A. A review of neuroprotective agents. *Curr. Med. Chem.* **2004**, *11*, 2383–2397. [CrossRef] [PubMed]
4. Lee, A.Y.; Lee, M.H.; Lee, S.H.; Cho, E.J. Alpha-linolenic acid regulates amyloid precursor protein processing by mitogen-activated protein kinase pathway and neuronal apoptosis in amyloid beta-induced SH-SY5Y neuronal cells. *Appl. Biol. Chem.* **2018**, *61*, 61–71. [CrossRef]
5. Vauzour, D.; Vafeiadou, K.; Rodriguez-Mateos, A.; Rendeiro, C.; Spencer, J.P.E. The neuroprotective potential of flavonoids: A multiplicity of effects. *Genes Nutr.* **2008**, *3*, 91. [CrossRef] [PubMed]
6. Dajas, F.; Rivera-Megret, F.; Blasina, F.; Arredondo, F.; Abin-Carriquiry, J.A.; Costa, G.; Echeverry, C.; Lafon, L.; Heizen, H.; Ferreira, M.; et al. Neuroprotection by flavonoids. *Braz. J. Med. Biol. Res.* **2003**, *36*, 1613–1620. [CrossRef] [PubMed]
7. Spagnuolo, C.; Moccia, S.; Russo, G.L. Anti-inflammatory effects of flavonoids in neurodegenerative disorders. *Eur. J. Med. Chem.* **2018**, *153*, 105–115. [CrossRef] [PubMed]
8. Dok-Go, H.; Lee, K.H.; Kim, H.J.; Lee, E.H.; Lee, J.Y.; Song, Y.S.; Lee, Y.H.; Jin, C.B.; Lee, Y.S.; Cho, J.S. Neuroprotective effects of antioxidative flavonoids, quercetin, (+)-dihydroquercetin and quercetin 3-methyl ether, isolated from *Opuntia ficus-indica* var. *saboten*. *Brain Res.* **2003**, *965*, 130–136. [CrossRef]
9. Hwang, S.L.; Shih, P.H.; Yen, G.C. Neuroprotective effects of citrus flavonoids. *J. Agric. Food Chem.* **2012**, *60*, 877–885. [CrossRef]
10. Kang, H.M.; Kim, J.H.; Lee, M.Y.; Son, K.H.; Yang, D.C.; Baek, N.I.; Kwon, B.M. Relationship between flavonoid structure and inhibition of farnesyl protein transferase. *Nat. Prod. Res.* **2004**, *18*, 349–356. [CrossRef]
11. Yang, H.J.; Shin, Y.J. Antioxidant compounds and activities of edible roses (*Rosa hybrida* spp.) from different cultivars grown in Korea. *Appl. Biol. Chem.* **2017**, *60*, 129–136. [CrossRef]
12. Lee, J.M.; Rodriguez, J.P.; Quilantang, N.G.; Lee, M.H.; Cho, E.J.; Jacinto, S.D.; Lee, S.H. Determination of flavonoids from *Perilla frutescens* var. *japonica* seeds and their inhibitory effect on aldose reductase. *Appl. Biol. Chem.* **2017**, *60*, 155–162. [CrossRef]
13. Rodriguez, J.P.; Lee, J.M.; Park, J.Y.; Kang, K.S.; Hahm, D.H.; Lee, S.C.; Lee, S.H. HPLC-UV analysis of sample preparation influence on flavonoid yield from *Cirsium japonicum* var. *maackii*. *Appl. Biol. Chem.* **2017**, *60*, 519–525. [CrossRef]
14. Dötterl, S.; Vereecken, N.J. The chemical ecology and evolution of bee-flower interactions: A review and perspectives. *Can. J. Zool.* **2010**, *88*, 668–697. [CrossRef]

15. Raguso, R.A. Wake up and smell the roses: The ecology and evolution of floral scent. *Annu. Rev. Ecol. Evol. Syst.* **2008**, *39*, 549–569. [CrossRef]
16. Wright, G.A.; Schiestl, F.P. The evolution of floral scent: The influence of olfactory learning by insect pollinators on the honest signalling of floral rewards. *Funct. Ecol.* **2009**, *23*, 841–851. [CrossRef]
17. Sun, X.M.; Li, X.F.; Deng, R.X.; Liu, Y.Q.; Hou, X.W.; Xing, Y.P.; Liu, P. Extraction technology and antioxidant activity of total flavonoids from the flower of *Chionanthus retusa*. *Food Sci.* **2015**, *36*, 266–271, 278.
18. Lee, Y.N.; Jeong, C.H.; Shim, K.H. Isolation of antioxidant and antibrowning substance from *Chionanthus retusa* leaves. *J. Korean Soc. Food Sci. Nutr.* **2004**, *33*, 1419–1425.
19. Deng, R.X.; Zhang, C.F.; Liu, P.; Duan, W.L.; Yin, W.P. Separation and identification of flavonoids from Chinese Fringetree Flowers (*Chionanthus retusa* Lindl et Paxt). *Food Sci.* **2014**, *35*, 74–78.
20. Kwak, J.H.; Kang, M.W.; Roh, J.H.; Choi, S.U.; Zee, O.P. Cytotoxic phenolic compounds from *Chionanthus retusus*. *Arch. Pharm. Res.* **2009**, *32*, 1681–1687. [CrossRef]
21. Baek, Y.S.; Song, N.Y.; Nam, T.G.; Kim, D.O.; Kang, H.C.; Kwon, O.K.; Baek, N.I. Flavonoids from *Fragaria ananassa* calyx and their antioxidant capacities. *Appl. Biol. Chem.* **2015**, *58*, 787–793. [CrossRef]
22. Han, J.T.; Bang, M.H.; Chun, O.K.; Kim, D.O.; Lee, C.Y.; Baek, N.I. Flavonol glycosides from the aerial parts of *Aceriphyllum rossii* and their antioxidant activities. *Arch. Pharm. Res.* **2004**, *27*, 390–395. [CrossRef] [PubMed]
23. Braca, A.; Tommasi, N.D.; Bari, L.D.; Pizza, C.; Politi, M.; Morelli, I. Antioxidant principles from *Bauhinia tarapotensis*. *J. Nat. Prod.* **2001**, *64*, 892–895. [CrossRef] [PubMed]
24. Weber, B.; Herrmann, M.; Hartmann, B.; Joppe, H.; Schmidt, C.O.; Bertram, H.J. HPLC/MS and HPLC/NMR as hyphenated techniques for accelerated characterization of the main constituents in Chamomile (*Chamomilla recutita* [L.] Rauschert). *Eur. Food Res. Technol.* **2008**, *226*, 755–760. [CrossRef]
25. Shrestha, S.; Lee, D.Y.; Park, J.H.; Cho, J.G.; Lee, D.S.; Li, B.; Kim, Y.C.; Jeon, Y.J.; Yeon, S.W.; Baek, N.I. Flavonoids from the fruits of Nepalese sumac (*Rhus parviflora*) attenuate glutamate-induced neurotoxicity in HT22 cells. *Food Sci. Biotechnol.* **2013**, *22*, 895–902. [CrossRef]
26. Hiep, N.T.; Kwon, J.Y.; Kim, D.W.; Hong, S.G.; Guo, Y.Q.; Hwang, B.Y.; Kim, N.H.; Mar, W.C.; Lee, D. Neuroprotective constituents from the fruits of *Maclura tricuspidata*. *Tetrahedron* **2017**, *73*, 2747–2759. [CrossRef]
27. Latté, K.P.; Ferreira, D.; Venkatraman, M.S.; Kolodziej, H. *O*-Galloyl-*C*-glycosylflavones from *Pelargonium reniforme*. *Phytochemistry* **2002**, *59*, 419–424. [CrossRef]
28. Simonian, N.A.; Coyle, J.T. Oxidative stress in neurodegenerative diseases. *Annu. Rev. Pharmacol. Toxicol.* **1996**, *36*, 83–106. [CrossRef] [PubMed]
29. Lee, D.S.; Jeong, G.S. Butein provides neuroprotective and anti-neuroinflammatory effects through Nrf2/ARE-dependent haem oxygenase 1 expression by activating the PI3K/Akt pathway. *Br. J. Pharmacol.* **2016**, *173*, 2894–2909. [CrossRef]
30. Lee, M.S.; Lee, J.N.; Kwon, D.Y.; Kim, M.S. Ondamtanggamibang protects neurons from oxidative stress with induction of heme oxygenase-1. *J. Ethnopharmacol.* **2006**, *108*, 294–298. [CrossRef]
31. Choi, B.M.; Kim, H.J.; Oh, G.S.; Pae, H.O.; Oh, H.C.; Jeong, S.J.; Kwon, T.O.; Kim, Y.M.; Chung, H.T. 1,2,3,4,6-Penta-*O*-galloyl-beta-D-glucose protects rat neuronal cells (Neuro 2A) from hydrogen peroxide-mediated cell death via the induction of heme oxygenase-1. *Neurosci. Lett.* **2002**, *328*, 185–189. [CrossRef]
32. Nhan, N.T.; Song, H.S.; Oh, E.J.; Lee, Y.G.; Ko, J.H.; Kwon, J.E.; Kang, S.C.; Lee, D.Y.; Baek, N.I. Phenylpropanoids from *Lilium* Asiatic hybrid flowers and their anti-inflammatory activities. *Appl. Biol. Chem.* **2017**, *60*, 527–533.
33. Lee, Y.G.; Lee, D.G.; Gwag, J.E.; Kim, M.S.; Kim, M.J.; Kim, H.G.; Ko, J.H.; Yeo, H.J.; Kang, S.H.; Baek, N.I. A 1,1′-biuracil from *Epidermidibacterium keratini* EPI-7 shows anti-aging effects on human dermal fibroblasts. *Appl. Biol. Chem.* **2019**, *62*, 14. [CrossRef]
34. Lee, Y.G.; Rodriguez, I.; Nam, Y.H.; Gwag, J.E.; Woo, S.H.; Kim, H.G.; Ko, J.H.; Hong, B.N.; Kang, T.H.; Baek, N.I. Recovery effect of lignans and fermented extracts from *Forsythia koreana* flowers on pancreatic islets damaged by alloxan in zebrafish (*Danio rerio*). *Appl. Biol. Chem.* **2019**, *62*, 7. [CrossRef]
35. Lee, Y.G.; Seo, K.H.; Lee, D.S.; Gwag, J.E.; Kim, H.G.; Ko, J.H.; Park, S.H.; Lee, D.Y.; Baek, N.I. Phenylethanoid glycoside from *Forsythia koreana* (Oleaceae) flowers shows a neuroprotective effect. *Braz. J. Bot.* **2018**, *41*, 523–528. [CrossRef]

36. Mosmann, T. Rapid colorimetric assay for cellular growth and survival: Application to proliferation and cytotoxicity assays. *J. Immunol. Methods* **1983**, *65*, 55–63. [CrossRef]
37. Lee, Y.G.; Lee, J.H.; Lee, N.Y.; Kim, N.K.; Jung, D.W.; Wang, W.; Kim, Y.S.; Kim, H.G.; Nguyen, N.T.; Park, H.S.; et al. Evaluation for the flowers of compositae plants as whitening cosmetics functionality. *J. Appl. Biol. Chem.* **2017**, *60*, 5–11. [CrossRef]

International Journal of
Molecular Sciences

MDPI

Article

Auraptene Mitigates Parkinson's Disease-Like Behavior by Protecting Inhibition of Mitochondrial Respiration and Scavenging Reactive Oxygen Species

Yunseon Jang [1,2,3,†], Hyosun Choo [1,3,†], Min Joung Lee [1,2,3,†], Jeongsu Han [1,3], Soo Jeong Kim [1,3], Xianshu Ju [1,2,3], Jianchen Cui [1,2,3], Yu Lim Lee [1,2,3], Min Jeong Ryu [1,4], Eung Seok Oh [1,5], Song-Yi Choi [1,6], Woosuk Chung [2,7,8,9], Gi Ryang Kweon [1,2,4,*] and Jun Young Heo [1,2,3,*]

1 Department of Biochemistry, Chungnam National University School of Medicine, Daejeon 35015, Korea
2 Department of Medical Science, Chungnam National University School of Medicine, Daejeon 35015, Korea
3 Infection Control Convergence Research Center, Chungnam National University School of Medicine, Daejeon 35015, Korea
4 Research Institute for Medical Science, Chungnam National University School of Medicine, Daejeon 35015, Korea
5 Department of Neurology, Chungnam National University Hospital, Daejeon 35015, Korea
6 Department of Pathology, Chungnam National University School of Medicine, Daejeon 35015, Korea
7 Department of Anesthesiology and pain medicine, Chungnam National University Hospital, Daejeon 35015, Korea
8 Department of Anesthesiology and pain medicine, Chungnam National University, Daejeon 35015, Korea
9 Brain Research Institute, Chungnam National University School of Medicine, Daejeon 35015, Korea
* Correspondence: mitochondria@cnu.ac.kr (G.R.K.); junyoung3@gmail.com (J.Y.H.); Tel.: +82-42-580-8226 (G.R.K.); +82-42-580-8222 (J.Y.H.); Fax: +82-42-580-8121 (G.R.K.&J.Y.H.)
† Authors contribute equally to this work.

Received: 17 June 2019; Accepted: 9 July 2019; Published: 11 July 2019

Abstract: Current therapeutics for Parkinson's disease (PD) are only effective in providing relief of symptoms such as rigidity, tremors and bradykinesia, and do not exert disease-modifying effects by directly modulating mitochondrial function. Here, we investigated auraptene (AUR) as a potent therapeutic reagent that specifically protects neurotoxin-induced reduction of mitochondrial respiration and inhibits reactive oxygen species (ROS) generation. Further, we explored the mechanism and potency of AUR in protecting dopaminergic neurons. Treatment with AUR significantly increased the viability of substantia nigra (SN)-derived SN4741 embryonic dopaminergic neuronal cells and reduced rotenone-induced mitochondrial ROS production. By inducing antioxidant enzymes AUR treatment also increased oxygen consumption rate. These results indicate that AUR exerts a protective effect against rotenone-induced mitochondrial oxidative damage. We further assessed AUR effects in vivo, investigating tyrosine hydroxylase (TH) expression in the striatum and substantia nigra of MPTP-induced PD model mice and behavioral changes after injection of AUR. AUR treatment improved movement, consistent with the observed increase in the number of dopaminergic neurons in the substantia nigra. These results demonstrate that AUR targets dual pathogenic mechanisms, enhancing mitochondrial respiration and attenuating ROS production, suggesting that the preventative potential of this natural compound could lead to improvement in PD-related neurobiological changes.

Keywords: auraptene; dopamine neuron; Parkinson's disease; neuroprotection; antioxidant; mitochondria

1. Introduction

Current therapeutics for Parkinson's disease (PD) lack neuroprotective properties and are only effective in providing symptom relief [1]. To overcome the limitations of PD drugs, researchers have focused on early pathological changes in PD [1–3], with the goal of developing strategies for early interventions, prior to the onset of severe motor symptoms, such as bradykinesia, rigidity and resting tremors, in patients with preclinical or prodromal stage PD [4].

Oxidative stress on dopaminergic neurons causes neurodegeneration and induces behavioral symptoms of PD. More than 90% of intracellular reactive oxygen species (ROS) are produced by aberrant electron transfer during mitochondrial respiration [5,6]. There is some evidence to suggest that mitochondrial alterations lead to PD-like pathologies. For example, genetic mutations in the PD-related genes, *Parkin, DJ-1* or *PTEN-induced kinase 1 (PINK1)*, cause mitochondrial dysfunction in offspring of familial-type PD patients, and 1-methyl-4-phenyl-1,2,3,6-tetrahydropyridine (MPTP) and rotenone, which are known to be PD-inducing toxins, inhibit mitochondrial complex I [6]. These two neurotoxins are suitable to show the effects of auraptene (AUR) in PD models, which results from mitochondrial dysfunction because both toxins lead to PD by inducing oxidative stress. The accumulation of α-synuclein, which has neurotoxic effects prior to the onset of PD symptoms, can also cause mitochondrial alterations and ROS production [7]. Therefore, modulating mitochondrial function during the pathogenesis of PD could be an effective preventive therapeutic strategy in prodromal stage PD.

Auraptene (AUR) is a 7-geranyloxylated coumarin isolated from citrus fruit [8]. Natural compounds such as AUR might generally be expected to offer advantages of safety and minimal adverse effects [9]; notably, AUR is able to cross the blood-brain barrier [10]. We previously showed that AUR inhibits progression of renal cell carcinoma by altering mitochondrial metabolism [11]. In addition to its anticancer effects, AUR has been used in conjunction with various toxins, including N-methyl-D-aspartate, lipopolysaccharide (LPS) and scopolamine, to study the neuroprotective effects of AUR against various neurotoxic defects (e.g., cerebral ischemia and neurodegenerative diseases), focusing on movement disorders and memory impairments [12–15]. Although AUR treatment inhibits microglial activation and prevents dopaminergic neuronal loss in an LPS mouse model [14], the molecular and cellular mechanisms for the protective effects of AUR in PD models are not yet clear, and the effects of AUR on motor function in PD have not yet been investigated.

In the context of cancer, biosynthetic substrates and energy supplied by mitochondria support cancer cell proliferation and metastasis. Because AUR treatment suppresses mitochondrial function, it leads to inhibition of cancer proliferation. However, in the context of neurodegeneration, maintenance or protection of neurotoxin-induced reduction in mitochondrial respiration increases neuronal activity and survival. In order to clarify the antioxidative effect by treatment with AUR, we investigated the alteration of cell viability, antioxidant enzyme expression and ROS generation by using rotenone, MPP^+ in SN4741 cell line. We demonstrated that pretreatment with AUR improves movement deficits in association with an increase in the number of dopamine neurons in the substantia nigra (SN) of MPTP-induced PD mouse models which inhibits the mitochondrial complex I. On the basis of these findings, we suggest that AUR pretreatment acts through protection of a decrease in mitochondrial respiration by neurotoxins and down-regulation of ROS of dopaminergic neurons to produce its beneficial PD-related neurobiological changes.

2. Results

2.1. *AUR Increases Cell Viability and Protects Against Neurotoxin-Induced Inhibition of Mitochondrial Respiration*

Rotenone and 1-methyl-4-phenylpyridinium (MPP^+), the active metabolite of MPTP, are commonly used neurotoxins in PD models [16,17]. Accordingly, we examined the protective effect of AUR on neurotoxin-induced cell death in dopaminergic neuron-like SN4741 cells. Using sulforhodamine B (SRB) assays to assess the viability of SN4741 cells after rotenone or MPP^+ treatment, we found that these toxins caused cell death in a dose-dependent manner (Figure 1A,B). Notably, AUR pretreated SN4741 cells were resistant to the neurotoxicity of both rotenone and MPP^+ compared to cells without

AUR treatment (Figure 1A,B). At a concentration of 1 μM, AUR alone had no effect on cell viability, as shown in Figure S1.

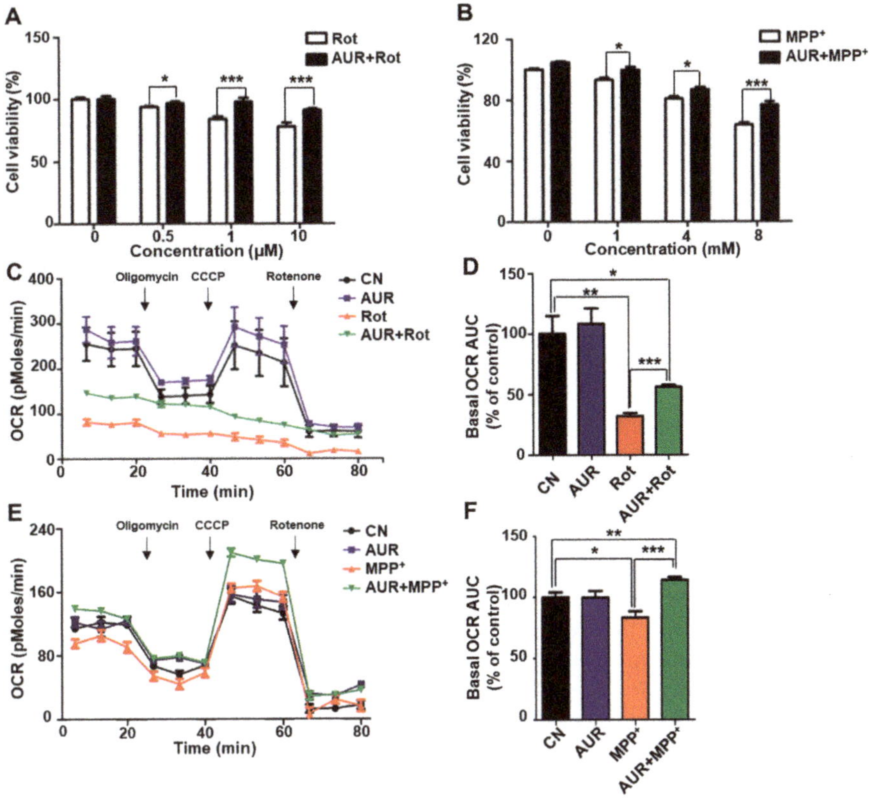

Figure 1. Auraptene (AUR) increases SN4741 cell viability and oxygen consumption rate (OCR) in the presence of neurotoxins. (**A**,**B**) SN4741 cells (5×10^3) plated in 96-well plates were incubated in media containing different concentrations (0, 0.5, 1 or 10 μM) of rotenone (Rot) for 6 h or MPP^+ (0, 1, 4, or 8 mM) for 24 h in the presence or absence of AUR (1 μM). Cell viability was measured by sulforhodamine B (SRB) assay after 6 or 24 h of drug treatment. (**C**–**F**), OCR was measured in SN4741 cells cultured with rotenone (**C**,**D**) or MPP^+ (**E**,**F**), with or without treatment with AUR. (**D**,**F**) Basal OCR area under the curve was calculated using XF24 analyzer software. Values are presented as means ± SD (bars) of triplicate samples (* $P < 0.05$, ** $P < 0.01$, *** $P < 0.001$ vs. corresponding controls). CN, control.

It has previously been reported that AUR affects mitochondrial complex I and inhibits mitochondrial respiration in RCC4 renal cell carcinoma cells [8,11]. In this context, effects of AUR on mitochondrial oxygen consumption rate (OCR), shown in Figure 1C,D, are somewhat counterintuitive. In these experiments SN4741 cells were pretreated with AUR and then incubated with 0.25 μM rotenone for 24 h, after which the effects of AUR on mitochondrial respiration were determined by measuring OCR using an XF24 analyzer. Incubation with rotenone alone for 24 h led to a 67.8% reduction in the basal OCR area under the curve compared with that of the control group. Notably, treatment with AUR prior to rotenone treatment attenuated these effects, blunting the inhibitory effect of rotenone by 24.1% (Figure 1C,D). AUR alone and short-term cotreatment with AUR and rotenone did not change basal OCR level (Figure S2). Similar results were obtained following MPP^+ treatment. The group treated with MPP^+ only exhibited a 17% decrease in basal OCR (Figure 1E), whereas the AUR pretreated group showed a basal OCR that was 14.2% higher than that of controls (Figure 1F). Extracellular acidification rate (ECAR) was also increased in the AUR pretreated, rotenone-exposed group compared with the rotenone-only group, but was unchanged in the MPP^+ group (Figure S3). Taken together, these results suggest that AUR protects against decreases in cell viability and suppression of mitochondrial respiration induced by neurotoxins in dopaminergic neuronal cells.

2.2. AUR Induces Antioxidant Enzyme Expression in a Rotenone-Treated Cell Model

Antioxidant compounds protect against cellular responses to ROS, which cause oxidative cellular damage in PD [18–23]. Given previously reported antioxidant effects of AUR on lymphocytes treated with H_2O_2 [24], we hypothesized that AUR affects antioxidant enzyme expression in dopaminergic neuronal cells. As a first step in determining the effect of AUR on antioxidant systems, we measured the levels of NRF2 (nuclear factor, erythroid 2 like 2), a transcription factor inducing antioxidant-related gene [25] in SN4741 cells. We observed that NRF2 protein levels were significantly increased in rotenone or MPP^+-treated cells pretreated with AUR compared with those in cells treated with either neurotoxin alone (Figure 2A–D). These results indicate that AUR treatment induces NRF2 protein expression in cells.

To determine whether AUR alters expression of ROS scavengers, we quantified the expression of transcripts of genes encoding antioxidant enzymes and those involved in glutathione (GSH) production and recycling using quantitative reverse transcription-polymerase chain reaction (RT-qPCR) [23,26,27]. Specifically, we analyzed transcript levels of Nrf2, Nqo1, Gpx1, Gst, Gclc, Gclm and Gr, as well as transcript levels of mitochondrial antioxidant enzymes, including Sod1 and Sod2. Nrf2, Nqo1, and Gpx1 mRNA levels were increased in AUR pretreated cells subsequently treated with rotenone or MPP^+ (Figure 2E,F). In the case of enzymes involved in GSH production and regeneration, Gclc mRNA was induced by AUR in the presence of MPP^+, but not in the presence of rotenone (Figure 2G,H). In SN4741 cells incubated in the presence of MPP^+ for 24 h, both Sod2 mRNA and protein levels were comparable to those of controls, regardless of AUR pretreatment (Figure S4). Taken together, these results suggest that AUR prevents neurotoxin-induced oxidative damage in dopaminergic neurons by enhancing antioxidant enzyme expression.

2.3. AUR Inhibits Rotenone-Induced Cytosolic ROS Production

Rotenone induces ROS production by inhibiting mitochondrial complex I [28]. Because AUR treatment significantly induced the expression of antioxidant enzyme transcripts, we investigated whether AUR prevents rotenone-induced ROS production in dopaminergic neuronal cells using the fluorescent dye DCFDA, which detects cytosolic ROS. We observed a 21.6% decrease in ROS levels in rotenone-exposed cells pretreated with 1 μM AUR compared with cells treated with rotenone only (Figure 3A,B), as assessed by flow cytometry. We then examined whether AUR treatment altered rotenone-induced mitochondrial superoxide production in SN4741 cells by adding the red fluorescent dye MitoSOX™ (which specifically targets mitochondrial superoxide) to rotenone- and AUR-treated cells, and quantified the results using flow cytometry. As shown in Figure 3C,D, mitochondrial superoxide levels in cells treated with rotenone only were comparable to those in AUR pretreated cells. These results are consistent with qPCR analyses, which showed that AUR specifically increased the transcription factor NRF2 and expression of its downstream targets, including Nqo1 and Gpx1, without affecting mitochondrial ROS scavenging enzymes, such as Sod1 and Sod2 (Figure S4). We found that AUR differentially regulates Gclc expression in the presence of rotenone or MPP^+. We pretreated AUR for 1 h before treatment of neurotoxins to induce antioxidant enzyme expression. Although both rotenone and MPP^+ targets complex I, rotenone showed higher inhibitory effect on mitochondrial respiration of SN4741 cells than MPP^+, causing more ROS generation than MPP^+. Increased ROS could offset against Gclc induction in rotenone treated cells. These results suggest that AUR induces expression of antioxidant enzymes, which act to effectively remove cellular ROS in dopaminergic neurons in the presence of neurotoxins, without altering mitochondrial ROS.

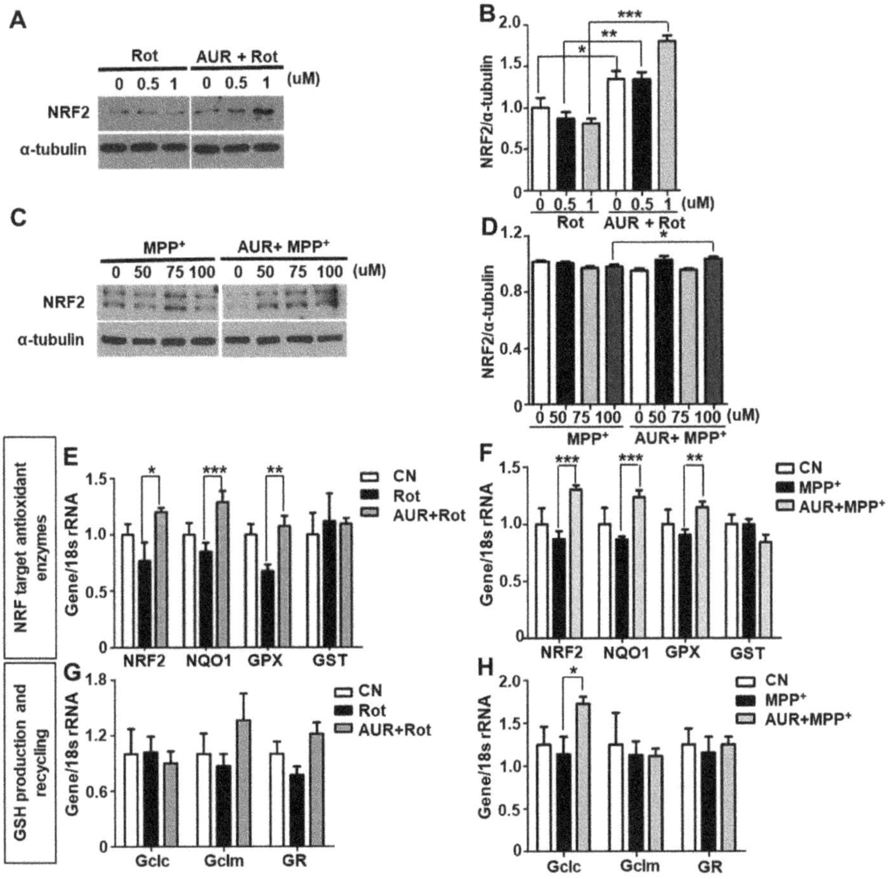

Figure 2. AUR induces expression of genes encoding antioxidant enzymes. (**A–D**) SN4741 cells were incubated in media containing different concentrations (0, 0.5 or 1 μM) of rotenone (Rot) or MPP^+ (0, 50, 75 or 100 μM), with or without pretreatment for 1 h with 10 μM AUR or DMSO. NRF2 protein expression was determined by Western blotting after 24 h (**A**) or 6 h (**C**) of drug treatment. The band intensity of NRF2 was measured using the ImageJ program (**B**,**D**). (**E–H**) Expression of mRNA for NRF2 target antioxidant enzymes (**E**,**F**) and GSH recycling-related genes (**G**,**H**) were assessed after a 24 h drug treatment using qPCR. Values are presented as means ± SD (bars) of triplicate samples (* $P < 0.05$, ** $P < 0.01$, *** $P < 0.001$ vs. corresponding controls).

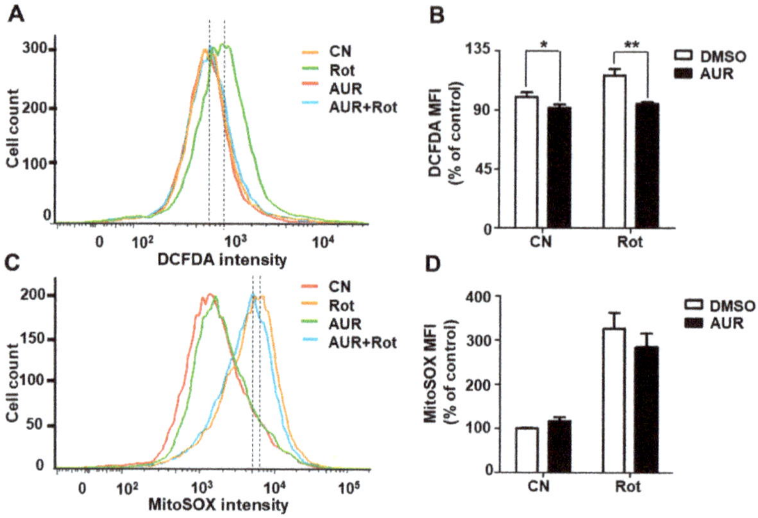

Figure 3. AUR protects against rotenone-induced ROS production. (**A–D**) SN4741 cells were incubated with rotenone (Rot) for 6 h, with or without AUR pretreatment for 1 h. Cells were stained with DCFDA or MitoSOX™, and fluorescence intensity was measured by flow cytometry. Total ROS was determined by measuring DCFDA-stained cells (**A**,**B**), and mitochondrial ROS was determined by measuring MitoSOX™-stained cells (**C**,**D**). Median fluorescence intensity (MIF) values are presented as means ± SD of three experiments (* $P < 0.05$, ** $P < 0.01$ vs. corresponding controls). CN, control.

2.4. AUR Protects Neurotoxin-Induced Loss of Tyrosine Hydroxylase Expression

Tyrosine hydroxylase (TH) expression in the SN and projections of TH neurons to the striatum is reduced in association with progression of PD [29]. It has also been shown that MPTP-induced PD animal models show a loss of TH-positive neurons [30]. Accordingly, we determined whether AUR treatment protects against the loss of TH expression in the SN and striatum of MPTP-induced PD mice. AUR (25 mg/kg) or DMSO (vehicle control) was intraperitoneally injected into B6 mice 1 day before MPTP treatment (20 mg/kg, four times a day), and was then injected for two additional days. Using a brain slice preparation, we found a significant decrease in TH immunoreactivity in both the SN and striatum of mice injected with MPTP for 7 days compared with saline-injected mice. In contrast, TH immunoreactivity was preserved in AUR-pretreated mice (Figure 4A–D). Specifically, the number of TH-positive neurons was decreased by 43.4% in MPTP-injected mice compared with saline-injected mice, and was increased by 32% in AUR-treated mice compared with DMSO injected mice (Figure 4D).

It is known that AUR significantly decreases inflammation in the SN region of LPS-injected mice [14]. Because the number of reactive astrocytes in the SN is increased in MPTP-induced PD model mice [31], we examined whether AUR alleviates astrogliosis by immunofluorescence staining for the astrocyte marker, glial fibrillary acidic protein (GFAP). Because it is clear to show the neuroinflammation with astrocyte activation in this model as we previously reported [32], we chose the GFAP as a maker of neuroinflammation by MPTP. Whereas the relative GFAP intensity in the MPTP-only group was 3.3-fold higher than that in control mice, it was only 2.8-fold higher in the AUR-treated group, indicating a decrease in the number of reactive astrocytes (Figure 4E). These results suggest that AUR protects against the MPTP-induced reduction in TH expression and astrocyte activation.

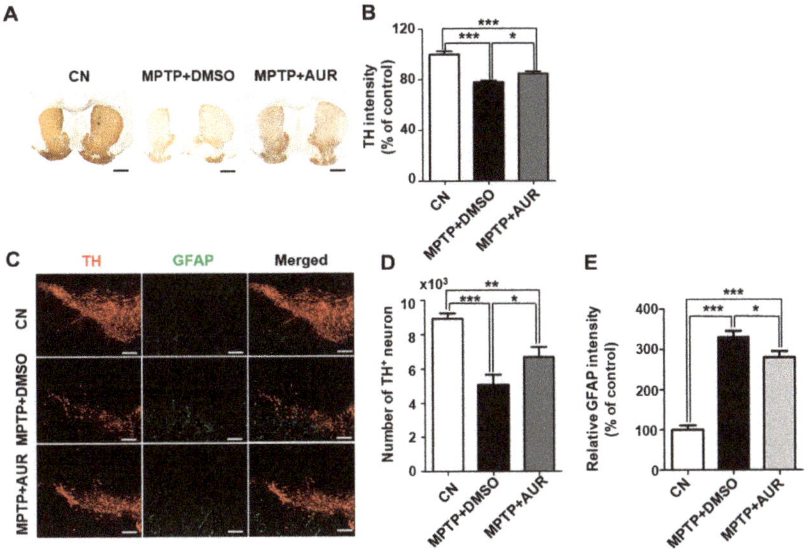

Figure 4. Pretreatment with AUR reduces MPTP-induced loss of TH expression in the SN and striatum. (**A**) Immunohistochemical detection of TH in the striatum of C57BL/6 mice injected with MPTP (20 mg/kg, i.p.) or saline, together with AUR (25 mg/kg, i.p.) or DMSO. Scale bars: 50 μm. (**B**) TH expression was decreased in MPTP-injected mice, an effect that was attenuated by AUR cotreatment. TH intensity was measured using ImageJ, and results are presented as a percentage of control values. (**C**) Immunofluorescence detection of TH in the SN region. TH-positive dopaminergic neurons (red) and astrocytes (green) were visualized by confocal microscopy. (**D**,**E**) Number of TH-positive neurons was calculated, and relative GFAP intensity was measured using ImageJ. Data are presented as means ± SD of three experiments (n = 10/group; * $P < 0.05$, ** $P < 0.01$, *** $P < 0.001$ vs. corresponding controls). CN, control. Scale bars: 500 μm.

2.5. AUR Ameliorates MPTP-Induced Motor Deficits

The nigrostriatal dopamine pathway is responsible for motor control, and TH activity is necessary for the release of dopamine, which regulates movement [33,34]. Because we found that AUR induces TH expression, we investigated the effect of AUR on movement deficits in MPTP-induced PD mice (Figure 5A). AUR-treated mice showed improved movement after MPTP injection compared with DMSO-treated mice, determined by monitoring behavior for 1 h in an open-field test (Figure 5B). Specifically, the total distance moved was decreased by 20.6% in MPTP-injected mice after 5 days compared with saline-injected mice (Figure 5C), whereas AUR-treated mice showed a 15.3% increase in movement distance compared with DMSO-treated mice (Figure 5C). Results presented in heat map form showed that AUR treatment significantly reduced residence time in the corner of the arena compared with that observed in mice treated with MPTP only (Figure S5). To further assess motor dysfunction, we performed vertical-grid tests of MPTP-injected and AUR-treated mice, as described by Kim et al. [35]. As shown in Figure 5D,E, MPTP-injected mice required 20 s longer to turn and a total of 25 s more time than control mice to complete the task. The time required to climb down was decreased by 5 s in MPTP-injected mice because of a 2-fold increase in missed steps compared with the control mice (Figure 5F,G). We found that AUR injection had no effect on the time to turn or total time, but restored the time to climb down to normal levels by decreasing missed steps observed in MPTP-only mice by 7% (Figure 5F,G). These findings suggest that AUR improves grip strength reduced by MPTP treatment.

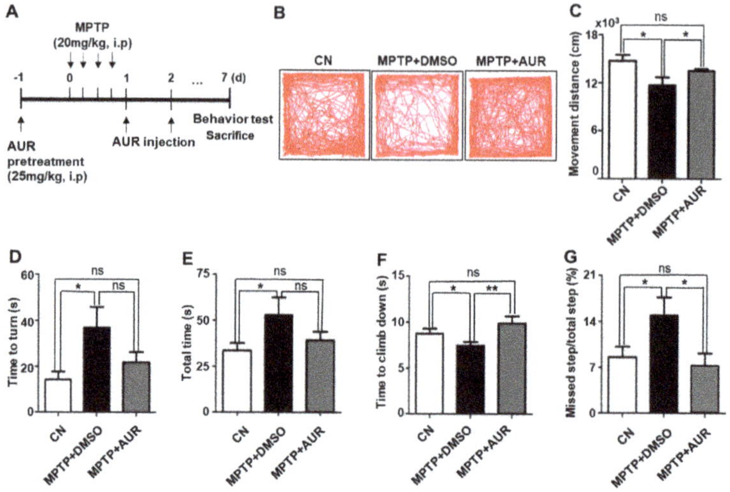

Figure 5. AUR improves MPTP-induced movement disorders. (**A**) Experimental timeline of AUR injection into the MPTP-induced mouse model of PD and behavioral tests. Mice were intraperitoneally injected with MPTP (20 mg/kg) 24 h after AUR (25 mg/kg) administration; AUR was further injected 24 h and 48 h after MPTP injection. Open-field and vertical-grid tests were performed after 7 days of MPTP injection. (**B**) Tracks visualizing mouse movements for 1 h are presented. Eight-week-old MPTP-induced PD mice showed a decrease in movement compared with control mice, whereas AUR-cotreated mice showed improveed movement (n = 5/group). (**C**) Total distance moved in 1 h was determined using EthoVision software and is presented as means ± SD. (**D–G**) Mice were placed at the bottom of the vertical grid and allowed to climb upward while movement was recorded. Time to turn (**D**), total climbing time (**E**), time to climb down (**F**), and percentage of total steps missed (**G**) were calculated. Values are presented as means ± SD (n = 5/group; * $P < 0.05$, ** $P < 0.01$ vs. corresponding controls; ns, not significant). CN, control.

Taken together, these results suggest that AUR mitigates motor dysfunction in MPTP-induced PD mice. As shown in Figure 6, we propose that AUR attenuates the effect of PD-related toxins on dopaminergic neurons through induction of NRF2 and expression of its target genes encoding antioxidant enzymes. AUR also increases mitochondrial respiration, which is suppressed in the presence of PD-related toxins (Figure 6). These protective effects of AUR on dopaminergic neurons consequently improve neurotoxin-induced motor deficits through preservation of TH expression.

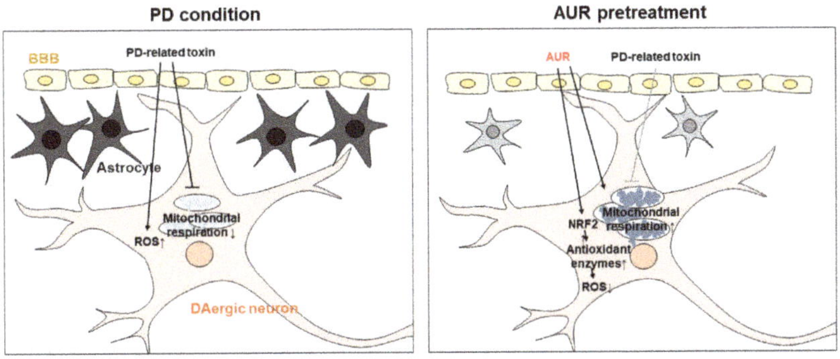

Figure 6. Schematic representation of the dopaminergic neuron-protective mechanism of AUR in a PD model. AUR alleviates neurotoxin-induced oxidative stress in dopaminergic neurons by stimulating the transcription factor NRF2 and inducing expression of downstream genes encoding antioxidant enzymes. Inhibition of mitochondrial respiration by PD-related toxins is mitigated by AUR treatment. AUR protects dopaminergic neurons against neurotoxins and ameliorates PD-like behavior.

3. Discussion

The complexity of PD and the variety of causative factors that contribute to its development create difficulties in identifying specific targets for effective treatments that might achieve complete disease remission. In the present study, we focused on modulation of mitochondrial energy metabolism and inhibition of ROS production by damaged mitochondria using the natural compound AUR. We postulate a dual preventive mechanism of AUR: (1) Induction of expression of genes encoding antioxidant enzymes, which protect against ROS, and (2) reduction of mitochondrial respiration by neurotoxins.

The lack of available treatment options for preventing or slowing the progression of PD has driven increased efforts to delay the occurrence of PD symptoms—the primary concept in current drug development strategies [36]. One disease-modifying agent, vitamin E, counteracts oxidative stress, and its intake is inversely correlated with PD occurrence [37]. In addition, the green tea polyphenol, (–)-epigallocatechine-3-gallate [38], and two Mediterranean plant-based extracts, *Padina pavonica* (EPP) and *Opuntia ficus-indica* (EOFI), ameliorate neurodegeneration in PD [39]. However, the mechanisms by which these treatments affect PD pathogenesis have not been identified. Unlike these latter studies, we focused specifically on mitochondrial respiration—considered the first target of environmental causative factors such as paraquat—and ROS overproduction by damaged mitochondria [36]. We assessed the protective effect of AUR by measuring mitochondrial oxygen consumption rate (OCR) and antioxidant enzyme expression levels in a neuronal cell line model of mitochondrial toxicity. We found that the overall changes in cellular metabolism induced by AUR are just a slight change in mitochondrial respiration. In the AUR-pretreated and MPP^+-treated groups, basal OCR was higher than that of the control. However, there was no significant difference in behavioral tests such as the open-field test and the vertical grid test between control and AUR-treated groups (Figure 5). These results suggest that AUR increases OCR of dopaminergic neurons in the presence of MPP^+ and it is consequently sufficient to improve MPTP-induced PD-like behavior to a normal level. But, additive beneficial effects on behavior or hypermobility were not found. Therefore, AUR could be used for prevention purposes by reducing adverse effects. Thus, our findings suggest that AUR, a coumarin from a source as simple and natural as citrus peel oil, could assist in preventing PD.

In general, enhancing mitochondrial respiration is expected to increase ROS generation, because the mitochondrial respiratory chain is a major source of intracellular ROS production and many enzymes that convert molecular oxygen to ROS are present in mitochondria [40]. Impairment of mitochondrial respiration plays a major role in the pathogenesis of PD, and increased ROS levels are known to be among the important causes of PD [40]. The key strength of AUR is its dual function described above, which enables AUR to protect a decrease in mitochondrial respiration caused by neurotoxins without increasing cellular ROS, although how these two effects are linked is not yet clear.

In a previous study, we reported that AUR suppresses mitochondrial respiration in the renal cell carcinoma cell line, RCC4 [11]. It has also been reported that AUR acts as a mitochondrial poison in the T-47D human breast cancer cell line [8]. However, our study suggests that AUR increases mitochondrial function in PD-like conditions. Although these two observations are seemingly at odds, they might actually be compatible, given that cancer cells possess exceptional cellular pathways compared with normal cells. Activation of NRF2 has been reported in several types of cancer cells [41]. NRF2, which is responsive to oxidative stress, is constitutively expressed in normal cells, but its protein level is low because of KEAP1-mediated ubiquitination and degradation [42]. Considering that AUR acts, at least in part, through induction of NRF2, its actions on cellular pathways could be different in cancer cells and normal cells. It is also worth noting that the AUR concentration range was significantly different between these two studies. In the cancer cell study, cellular metabolism was targeted by inhibiting translation of the HIF-1α transcription factor using an AUR concentration of 100 μM. At a high concentration, AUR reduced basal OCR to 67% of that in untreated cancer cells, which show immature mitochondrial function. In the current study, we tested AUR at a concentration of 1 μM, and found that it increased basal OCR in dopaminergic neuron-like cells in the presence of neurotoxins. Notable in this context, some antioxidants, including EGCG, have been reported to show neuroprotective activity at low concentrations, but pro-oxidant activity at high concentrations [38].

We also suggest the potential of AUR in trials of combined therapy with levodopa. Levodopa is one of the main drugs used for relief of PD symptoms, but it should be used with caution in younger patients with early PD [36,43]. If there were a drug that could prevent progression of the disease, it should be used starting as early as possible. Although drugs currently used in combination with levodopa, such as benserazide and carbidopa, reduce the peripheral effects of levodopa and increase levodopa concentrations in the brain [36], combination therapy with AUR would provide additional neuronal protective effects through a different pathway. If an early diagnosis of pre-symptomatic PD patients is possible in the near future, AUR could be beneficial to delay the loss of dopaminergic neurons and PD-behavior symptoms. Combining these drugs in a single therapeutic regimen would seek to relieve symptoms while delaying disease progression.

4. Materials and Methods

4.1. Cell Culture

SN4741 mouse embryonic substantial nigra dopaminergic neuronal cell line was cultured in RF media containing Dulbecco's modified Eagle's medium (DMEM, Welgene, Korea), 10% FBS (Hyclone, MA, USA), 1% penicillin and streptomycin (Hyclone, MA, USA), 0.6% D-glucose and 0.7% 200 mM·L-glutamine at 33 °C under 5% CO_2 and 21% O_2 condition.

4.2. Measurement of Cell Viability

In the sulforhodamine B assay, SN4741 cells (5×10^3 cells per well) were seeded in triplicate in 96-well plates and incubated overnight. Added to each well were media containing Rot (0, 0.5, 1 and 10 uM, Sigma-Aldrich, MO, USA) for 6 h or MPP^+ (0, 1, 4, 8 mM, Sigma-Aldrich, MO, USA) for 24 h in the presence or absence of AUR 1 uM (Sigma-Aldrich, MO, USA). The media were removed and cells were fixed with 10% TSA at 4 °C for 1 h. After washing, the cells were incubated with 0.4% SRB (Sigma-Aldrich, MO, USA) solution at room temperature for 20 min. The wells were washed with 1% acetic acid five times and dried in air. After resolving the proteins with 10 mM unbuffered Tris, absorbance was read at 490 nm using a Multiskan Ascent plate reader.

4.3. Flow Cytometry

For analyzing ROS generation, the fluorescent dye, MitoSOX™ red reagent (Invitrogen, CA, USA) and DCFDA (Invitrogen, CA, USA) were used following the manufacturers' instructions. SN4741 cells (2–4 × 10^5 cells in 60 mm dish) were incubated with Rot for 6 h and AUR was pretreated for 1 h. Media was discarded and washed with HBSS and incubated for 30 min in the dark with DCFDA or MitoSOX™ (5 μM final concentration). Cells were washed with PBS and trypsinized, then resuspended in PBS/EDTA. After washed with PBS, cells were collected and kept on ice in the dark for immediate detection with the flow cytometer. Fluorescence was measured on a FACScan (BD Biosciences, NJ, USA) using excitation/emission wavelengths of 485/535 nm, and 510/580 nm for DCFDA and MitoSOX™, respectively. The values were expressed as mean fluorescence of the cell population.

4.4. Measurement of Oxygen Consumption Rate (OCR)

SN4741 cells cultured with rotenone or MPP^+ ± treatment with AUR 2 uM were plated 2 × 10^4 cells at each well. Basal OCR was analyzed by XF24 analyzer (Seahorse, MA, USA). Then, 20 μg/mL of oligomycin A (an ATPase inhibitor, Sigma-Aldrich, MO, USA), 50 μM of carbonyl cyanide 3-chlorophenylhydrazone (CCCP, an uncoupler, Sigma-Aldrich, MO, USA) and 20 μM rotenone (a mitochondrial complex I inhibitor, Sigma-Aldrich, MO, USA) were sequentially added into each well and OCR was measured at 37 °C.

4.5. RNA Isolation and Real Time PCR

Total RNA was isolated using Trizol from SN4741 cells treated with Rot (0, 0.5 or 1 uM) or MPP^+ (0, 50, 75 or 100 uM) and AUR for 24 h. cDNA was synthesized from total RNA with 5× RT premix. After mixing cDNA, primers and SYBR mix, mRNA expression was analyzed using a Rotor Gene 6000 system (Corbett Life Science, Venlo, Netherlands) and normalized to 18s rRNA. Primers used in this study: NRF2, 5′-CCAGAAGCCACACTGACAGA-3′ (forward) and 5′-GGAGAGGATGCTGCTGAAAG-3′ (reverse); NQO1, 5′-TTCTCTGGCCGATTCAGAGT-3′ (forward) and 5′-GGCTGCTTGGAGCAAAATAG-3′ (reverse); GPX, 5′- GTCCACCGTGTATGCC TTCT-3′ (forward) and 5′-TCTGCAGATCGTTCATCTCG-3′ (reverse); GST, 5′-GGCATCTGAAG CCTTTTGAG-3′ (forward) and 5′-GAGCCACATAGGCAGAGAGC-3′ (reverse); Gclc, 5′-AGGC TCTCTGCACCATCACT-3′ (forward) and 5′- TGGCACATTGATGACAACCT-3′ (reverse); Gclm, 5′-TGGAGCAGCTGTATCAGTGG -3′ (forward) and 5′-AGAGCAGTTCTTTCGGGTCA-3′ (reverse); GR, 5′-CACGACCATGATTCCAGATG-3′ (forward) and 5′-CAGCATAGACGCCTTTGACA-3′ (reverse); 18s rRNA, 5′-CGACCAAAGGAACCATAACT-3′ (forward) and 5′-CTGGTTGATCC TGCCAGTAG-3′ (reverse).

4.6. Animal Experiments

Temperature was maintained to 22 °C and light condition was adjusted to a 12 h light-dark cycle. Animal experiments were approved by the Institutional Animal Care and Use Committee of Chungnam National University. The ethical approval number is CNU-00912 and approval date is March-1-2017. To establish the MPTP-induced PD mouse model, C57BL/6 mice (8-week-old, male) were intraperitoneally injected with MPTP (1-methyl-4phenyl-1.2.3.6-tetrahydropyridine, Sigma-Aldrich, MO, USA, 2 mg/mL in saline, 20 mg/kg for one injection) four times with 2 h intervals in a day. Control mice were injected with saline. Before 24 h and 48 h of MPTP injection, auraptene (25 mg/kg) was injected intraperitoneally.

4.7. Immunofluorescence Staining and Immunohistochemistry

Saline and MPTP injected Mice were perfused and fixed with 4% paraformaldehyde (PFA). The whole brain was dipped in the 4% PFA and then moved to 30% sucrose solution to dehydrate for three days. The samples were frozen and sectioned, 25 μm of each slice. For the immunofluorescence staining, after 15 min of PBS washing, sections were blocked for 1.5 h with 3% donkey serum (Dako, Glostrup, Denmark) and 0.3% triton x-100 with PBS. Then, sections were incubated with anti-TH antibody (Millipore, MA, USA), anti-GFAP (1:1000, Abcam, Cambridge, UK) diluted with blocking solution overnight at 4 °C. Sections were washed with PBS and incubated with anti-mouse Alexa 594 and anti-chicken Alexa 488-conjugated anti-IgG secondary antibodies containing solution for 1 h at room temperature. For immunohistochemistry, brain slices were incubated with anti-TH antibody for overnight at 4 °C and then incubated with a secondary antibody (Dako EnVision$^+$ system-HRP, CA, USA) for 1 h. The slices were reacted with DAB$^+$ substrate buffer. After mounting with mounting medium (Dako North America Inc., CA, USA), the slides were visualized using an IX70 confocal microscope (Olympus, Tokyo, Japan).

4.8. Protein Isolation and Western Blotting

The protein of mice tissues and SN4741 cells, treated with Rot (0, 0.5 or 1 uM) or MPP$^+$ (0, 50, 75 or 100 uM) and pretreated with 10 uM Auraptene or DMSO for 1 h, were extracted using RIPA buffer (1% Nonidet P-40, 0.1% SDS, 150 mM NaCl, 50 mM Tris–HCl pH 7.5 and 0.5% deoxycholate) with 10% of phosphatase inhibitor and protease inhibitor (Roche, Basel, Switzerland). Equal amounts of proteins were loaded on SDS-PAGE gel and run by electrophoresis. After, they were transferred to polyvinylidene fluoride (PVDF) membrane, blocked by 5% skim milk for 1 h. Then, membranes were incubated with primary antibody including anti-NRF2 (Santa Cruz Biotechnology, CA, USA) and anti-α-Tubulin (Santa Cruz Biotechnology, CA, USA) antibody at 4 °C overnight. Anti-IgG horseradish peroxidase antibody (Pierce Biotechnology, MA, USA) correspond with the host of primary antibody was used as secondary antibody. Protein bands were detected by ECL system (Thermo Scientific, MA, USA).

4.9. Behavior Test

Open-field test: Mice were placed in a 40 × 40 × 40-box respectively. Movement was recorded for 1 h and analyzed with EthoVision XT 11.5 software.

Vertical grid test: The vertical grid test was performed following the previous study [35]. For performing the vertical grid test, mice were habituated to the apparatus. After habituation for 3 days, a mouse was placed inside the apparatus and was allowed to turn and climb down. The movement was recorded.

4.10. Statistical Analysis

All data are represented as mean values ± SEM (error bars). The statistical analysis of data was performed using Prizm version 5 software (Graphpad, CA, USA). Significance of differences between two groups were analyzed by one-tailed student's t-test. A P value <0.05 was considered statistically significant.

Supplementary Materials: Supplementary materials can be found at http://www.mdpi.com/1422-0067/20/14/3409/s1.

Author Contributions: Conceptualization, Y.J., H.C., M.J.L., G.R.K. and J.Y.H.; Formal analysis, M.J.L. and J.H.; Investigation, J.C. and Y.L.L.; Methodology, M.J.R.; Supervision, G.R.K. and J.Y.H.; Validation, S.J.K. and X.J.; Writing—original draft, Y.J. and H.C.; Writing—review & editing, E.S.O., S.-Y.C. and W.C.

Funding: This work was supported by the National Research Foundation of the Republic of Korea (NRF) grant funded by the Ministry of Science, ICT & Future Planning (MSIP) (2014R1A6A1029617, 2016R1A2B4010398, 2016R1D1A1B03932766, 2017R1A5A2015385, 2017R1A6A3A11029367) and by the Research Fund of Chungnam National University.

Conflicts of Interest: The authors have declared that no conflicts of interest exist.

References

1. Kalia, L.V.; Lang, A.E. Parkinson disease in 2015: Evolving basic, pathological and clinical concepts in PD. *Nat. Rev. Neurol.* **2016**, *12*, 65. [CrossRef] [PubMed]
2. Postuma, R.B.; Berg, D.; Stern, M.; Poewe, W.; Olanow, C.W.; Oertel, W.; Obeso, J.; Marek, K.; Litvan, I.; Lang, A.E. MDS clinical diagnostic criteria for Parkinson's disease. *Mov. Disord.* **2015**, *30*, 1591–1601. [CrossRef] [PubMed]
3. Postuma, R.B.; Berg, D.; Adler, C.H.; Bloem, B.R.; Chan, P.; Deuschl, G.; Gasser, T.; Goetz, C.G.; Halliday, G.; Joseph, L. The new definition and diagnostic criteria of Parkinson's disease. *Lancet Neurol.* **2016**, *15*, 546–548. [CrossRef]
4. Dal Ben, M.; Bongiovanni, R.; Tuniz, S.; Fioriti, E.; Tiribelli, C.; Moretti, R.; Gazzin, S. Earliest Mechanisms of Dopaminergic Neurons Sufferance in a Novel Slow Progressing Ex Vivo Model of Parkinson Disease in Rat Organotypic Cultures of Substantia Nigra. *Int. J. Mol. Sci.* **2019**, *20*, 2224. [CrossRef] [PubMed]
5. Blesa, J.; Trigo-Damas, I.; Quiroga-Varela, A.; Jackson-Lewis, V.R. Oxidative stress and Parkinson's disease. *Front. Neuroanat.* **2015**, *9*, 91. [CrossRef] [PubMed]
6. Perier, C.; Vila, M. Mitochondrial biology and Parkinson's disease. *Cold Spring Harb. Perspect. Med.* **2012**, *2*, a009332. [CrossRef] [PubMed]
7. Ganguly, G.; Chakrabarti, S.; Chatterjee, U.; Saso, L. Proteinopathy, oxidative stress and mitochondrial dysfunction: Cross talk in Alzheimer's disease and Parkinson's disease. *Drug Des. Dev. Ther.* **2017**, *11*, 797. [CrossRef]
8. Li, J.; Mahdi, F.; Du, L.; Jekabsons, M.B.; Zhou, Y.-D.; Nagle, D.G. Semisynthetic studies identify mitochondria poisons from botanical dietary supplements—Geranyloxycoumarins from Aegle marmelos. *Bioorg. Med. Chem.* **2013**, *21*, 1795–1803. [CrossRef]
9. Ghanbarabadi, M.; Iranshahi, M.; Amoueian, S.; Mehri, S.; Motamedshariaty, V.S.; Mohajeri, S.A. Neuroprotective and memory enhancing effects of auraptene in a rat model of vascular dementia: Experimental study and histopathological evaluation. *Neurosci. Lett.* **2016**, *623*, 13–21. [CrossRef]
10. Okuyama, S.; Morita, M.; Kaji, M.; Amakura, Y.; Yoshimura, M.; Shimamoto, K.; Ookido, Y.; Nakajima, M.; Furukawa, Y. Auraptene acts as an anti-inflammatory agent in the mouse brain. *Molecules* **2015**, *20*, 20230–20239. [CrossRef]
11. Jang, Y.; Han, J.; Kim, S.J.; Kim, J.; Lee, M.J.; Jeong, S.; Ryu, M.J.; Seo, K.-S.; Choi, S.-Y.; Shong, M. Suppression of mitochondrial respiration with auraptene inhibits the progression of renal cell carcinoma: Involvement of HIF-1α degradation. *Oncotarget* **2015**, *6*, 38127. [CrossRef] [PubMed]
12. Epifano, F.; Molinaro, G.; Genovese, S.; Ngomba, R.T.; Nicoletti, F.; Curini, M. Neuroprotective effect of prenyloxycoumarins from edible vegetables. *Neurosci. Lett.* **2008**, *443*, 57–60. [CrossRef] [PubMed]
13. Okuyama, S.; Minami, S.; Shimada, N.; Makihata, N.; Nakajima, M.; Furukawa, Y. Anti-inflammatory and neuroprotective effects of auraptene, a citrus coumarin, following cerebral global ischemia in mice. *Eur. J. Pharmacol.* **2013**, *699*, 118–123. [CrossRef] [PubMed]
14. Okuyama, S.; Semba, T.; Toyoda, N.; Epifano, F.; Genovese, S.; Fiorito, S.; Taddeo, V.A.; Sawamoto, A.; Nakajima, M.; Furukawa, Y. Auraptene and other prenyloxyphenylpropanoids suppress microglial activation and dopaminergic neuronal cell death in a lipopolysaccharide-induced model of Parkinson's disease. *Int. J. Mol. Sci.* **2016**, *17*, 1716. [CrossRef] [PubMed]
15. Tabrizian, K.; Yaghoobi, N.S.; Iranshahi, M.; Shahraki, J.; Rezaee, R.; Hashemzaei, M. Auraptene consolidates memory, reverses scopolamine-disrupted memory in passive avoidance task, and ameliorates retention deficits in mice. *Iran. J. Basic Med. Sci.* **2015**, *18*, 1014. [PubMed]
16. Bove, J.; Perier, C. Neurotoxin-based models of Parkinson's disease. *Neuroscience* **2012**, *211*, 51–76. [CrossRef] [PubMed]
17. Martinez, T.N.; Greenamyre, J.T. Toxin models of mitochondrial dysfunction in Parkinson's disease. *Antioxid. Redox Signal.* **2012**, *16*, 920–934. [CrossRef] [PubMed]
18. Manoharan, S.; Guillemin, G.J.; Abiramasundari, R.S.; Essa, M.M.; Akbar, M.; Akbar, M.D. The role of reactive oxygen species in the pathogenesis of Alzheimer's disease, Parkinson's disease, and Huntington's disease: A mini review. *Oxid. Med. Cell. Longev.* **2016**. [CrossRef] [PubMed]
19. Ransohoff, R.M.; Perry, V.H. Microglial physiology: Unique stimuli, specialized responses. *Annu. Rev. Immunol.* **2009**, *27*, 119–145. [CrossRef] [PubMed]

20. Campolo, M.; Casili, G.; Biundo, F.; Crupi, R.; Cordaro, M.; Cuzzocrea, S.; Esposito, E. The neuroprotective effect of dimethyl fumarate in an MPTP-mouse model of Parkinson's disease: Involvement of reactive oxygen species/nuclear factor-κB/nuclear transcription factor related to NF-E2. *Antioxid. Redox Signal.* **2017**, *27*, 453–471. [CrossRef]
21. Zhang, Y.; Dawson, V.L.; Dawson, T.M. Oxidative stress and genetics in the pathogenesis of Parkinson's disease. *Neurobiol. Dis.* **2000**, *7*, 240–250. [CrossRef] [PubMed]
22. Orth, M.; Schapira, A. Mitochondrial involvement in Parkinson's disease. *Neurochem. Int.* **2002**, *40*, 533–541. [CrossRef]
23. Gorrini, C.; Harris, I.S.; Mak, T.W. Modulation of oxidative stress as an anticancer strategy. *Nat. Rev. Drug Discov.* **2013**, *12*, 931. [CrossRef] [PubMed]
24. Soltani, F.; Mosaffa, F.; Iranshahi, M.; Karimi, G.; Malekaneh, M.; Haghighi, F.; Behravan, J. Auraptene from Ferula szowitsiana protects human peripheral lymphocytes against oxidative stress. *Phytother. Res. Int. J. Devoted Pharmacol. Toxicol. Eval. Nat. Prod. Deriv.* **2010**, *24*, 85–89.
25. Tonelli, C.; Chio, I.I.C.; Tuveson, D.A. Transcriptional Regulation by Nrf2. *Antioxid. Redox Signal.* **2018**, *29*, 1727–1745. [CrossRef]
26. Mates, J. Effects of antioxidant enzymes in the molecular control of reactive oxygen species toxicology. *Toxicology* **2000**, *153*, 83–104. [CrossRef]
27. Espinosa-Diez, C.; Miguel, V.; Mennerich, D.; Kietzmann, T.; Sánchez-Pérez, P.; Cadenas, S.; Lamas, S. Antioxidant responses and cellular adjustments to oxidative stress. *Redox Biol.* **2015**, *6*, 183–197. [CrossRef]
28. Li, N.; Ragheb, K.; Lawler, G.; Sturgis, J.; Rajwa, B.; Melendez, J.A.; Robinson, J.P. Mitochondrial complex I inhibitor rotenone induces apoptosis through enhancing mitochondrial reactive oxygen species production. *J. Biol. Chem.* **2003**, *278*, 8516–8525. [CrossRef]
29. Cheng, H.C.; Ulane, C.M.; Burke, R.E. Clinical progression in Parkinson disease and the neurobiology of axons. *Ann. Neurol.* **2010**, *67*, 715–725. [CrossRef]
30. Meredith, G.E.; Rademacher, D.J. MPTP mouse models of Parkinson's disease: An update. *J. Parkinsons Dis.* **2011**, *1*, 19–33.
31. Ghosh, A.; Kanthasamy, A.; Joseph, J.; Anantharam, V.; Srivastava, P.; Dranka, B.P.; Kalyanaraman, B.; Kanthasamy, A.G. Anti-inflammatory and neuroprotective effects of an orally active apocynin derivative in pre-clinical models of Parkinson's disease. *J. Neuroinflamm.* **2012**, *9*, 241. [CrossRef] [PubMed]
32. Kim, S.J.; Ryu, M.J.; Han, J.; Jang, Y.; Lee, M.J.; Ju, X.; Ryu, I.; Lee, Y.L.; Oh, E.; Chung, W.; et al. Non-cell autonomous modulation of tyrosine hydroxylase by HMGB1 released from astrocytes in an acute MPTP-induced Parkinsonian mouse model. *Lab. Investig.* **2019**. [CrossRef] [PubMed]
33. Jang, Y.; Lee, M.J.; Han, J.; Kim, S.J.; Ryu, I.; Ju, X.; Ryu, M.J.; Chung, W.; Oh, E.; Kweon, G.R.; et al. A High-fat Diet Induces a Loss of Midbrain Dopaminergic Neuronal Function That Underlies Motor Abnormalities. *Exp. Neurobiol.* **2017**, *26*, 104–112. [CrossRef] [PubMed]
34. Korchounov, A.; Meyer, M.F.; Krasnianski, M. Postsynaptic nigrostriatal dopamine receptors and their role in movement regulation. *J. Neural Transm.* **2010**, *117*, 1359–1369. [CrossRef] [PubMed]
35. Kim, S.T.; Son, H.J.; Choi, J.H.; Ji, I.J.; Hwang, O. Vertical grid test and modified horizontal grid test are sensitive methods for evaluating motor dysfunctions in the MPTP mouse model of Parkinson's disease. *Brain Res.* **2010**, *1306*, 176–183. [CrossRef] [PubMed]
36. Connolly, B.S.; Lang, A.E. Pharmacological treatment of Parkinson disease: A review. *JAMA* **2014**, *311*, 1670–1683. [CrossRef] [PubMed]
37. Schirinzi, T.; Martella, G.; Imbriani, P.; Di Lazzaro, G.; Franco, D.; Colona, V.L.; Alwardat, M.; Sinibaldi Salimei, P.; Mercuri, N.B.; Pierantozzi, M.; et al. Dietary Vitamin E as a Protective Factor for Parkinson's Disease: Clinical and Experimental Evidence. *Front. Neurol.* **2019**, *10*, 148. [CrossRef] [PubMed]
38. Weinreb, O.; Mandel, S.; Youdim, M.B. Gene and protein expression profiles of anti-and pro-apoptotic actions of dopamine, R-apomorphine, green tea polyphenol (−)-epigallocatechine-3-gallate, and melatonin. *Ann. N. Y. Acad. Sci.* **2003**, *993*, 351–361. [CrossRef] [PubMed]
39. Briffa, M.; Ghio, S.; Neuner, J.; Gauci, A.J.; Cacciottolo, R.; Marchal, C.; Caruana, M.; Cullin, C.; Vassallo, N.; Cauchi, R.J. Extracts from two ubiquitous Mediterranean plants ameliorate cellular and animal models of neurodegenerative proteinopathies. *Neurosci. Lett.* **2017**, *638*, 12–20. [CrossRef] [PubMed]
40. Kalogeris, T.; Bao, Y.; Korthuis, R.J. Mitochondrial reactive oxygen species: A double edged sword in ischemia/reperfusion vs. preconditioning. *Redox Biol.* **2014**, *2*, 702–714. [CrossRef] [PubMed]

41. Sporn, M.B.; Liby, K.T. NRF2 and cancer: The good, the bad and the importance of context. *Nat. Rev. Cancer* **2012**, *12*, 564. [CrossRef] [PubMed]
42. Lu, K.; Alcivar, A.L.; Ma, J.; Foo, T.K.; Zywea, S.; Mahdi, A.; Huo, Y.; Kensler, T.W.; Gatza, M.L.; Xia, B. NRF2 Induction Supporting Breast Cancer Cell Survival Is Enabled by Oxidative Stress-Induced DPP3-KEAP1 Interaction. *Cancer Res.* **2017**, *77*, 2881–2892. [CrossRef] [PubMed]
43. Kalia, L.V.; Lang, A.E. Parkinson's disease. *Lancet* **2015**, *386*, 896–912. [CrossRef]

International Journal of
Molecular Sciences

MDPI

Article

Centella asiatica Protects D-Galactose/AlCl$_3$ Mediated Alzheimer's Disease-Like Rats via PP2A/GSK-3β Signaling Pathway in Their Hippocampus

Samaila Musa Chiroma [1,2], Mohamad Taufik Hidayat Baharuldin [1], Che Norma Mat Taib [1], Zulkhairi Amom [3], Saravanan Jagadeesan [1,4], Mohd Ilham Adenan [5], Onesimus Mahdi [1,6] and Mohamad Aris Mohd Moklas [1,*]

1 Department of Human Anatomy, Faculty of Medicine and Health Sciences, Universiti Putra Malaysia, Serdang 43400, Selangor, Malaysia; musasamailachiroma@yahoo.com (S.M.C.); taufikb@upm.edu.my (M.T.H.B.); chenorma@upm.edu.my (C.N.M.T.); mljsaravanan@gmail.com (S.J.); omahdi2010@gmail.com (O.M.)
2 Department of Human Anatomy, Faculty of Basic Medical Sciences, University of Maiduguri, Maiduguri 600230, Borno State, Nigeria
3 Faculty of Health Sciences, Universiti Teknologi Mara (UiTM) Kampus Puncak Alam, Bandar Puncak Alam 42300, Selangor, Malaysia; zulkha2992@puncakalam.uitm.edu.my
4 Department of Human Anatomy, Universiti Tunku Abdul Rahman (UTAR), Bandar Sungai Long Cheras 43000, Selangor, Malaysia
5 Atta-ur-Rahman Institute for Natural Product Discovery, Universiti Teknologi Mara (UiTM) Kampus Puncak Alam, Bandar Puncak Alam 42300, Selangor, Malaysia; mohdilham@puncakalam.uitm.edu.my
6 Department of Human Anatomy, College of Medical Sciences, Gombe State University, Gombe 760211, Gombe State, Nigeria
* Correspondence: aris@upm.edu.my; Tel.: +60-193-387-042

Received: 23 February 2019; Accepted: 5 March 2019; Published: 16 April 2019

Abstract: Alzheimer's disease (AD) is a progressive neurodegenerative disorder more prevalent among the elderly population. AD is characterised clinically by a progressive decline in cognitive functions and pathologically by the presence of neurofibrillary tangles (NFTs), deposition of beta-amyloid (Aβ) plaque and synaptic dysfunction in the brain. *Centella asiatica* (CA) is a valuable herb being used widely in African, Ayurvedic, and Chinese traditional medicine to reverse cognitive impairment and to enhance cognitive functions. This study aimed to evaluate the effectiveness of CA in preventing D-galactose/aluminium chloride (D-gal/AlCl$_3$) induced AD-like pathologies and the underlying mechanisms of action were further investigated for the first time. Results showed that co-administration of CA to D-gal/AlCl$_3$ induced AD-like rat models significantly increased the levels of protein phosphatase 2 (PP2A) and decreased the levels of glycogen synthase kinase-3 beta (GSK-3β). It was further observed that, CA increased the expression of mRNA of Bcl-2, while there was minimal effect on the expression of caspase 3 mRNA. The results also showed that, CA prevented morphological aberrations in the connus ammonis 3 (CA 3) sub-region of the rat's hippocampus. The results clearly demonstrated for the first time that CA could alleviate D-gal/AlCl$_3$ induced AD-like pathologies in rats via inhibition of hyperphosphorylated tau (P-tau) bio-synthetic proteins, anti-apoptosis and maintenance of cytoarchitecture.

Keywords: Alzheimer's disease; *Centella asiatica*; hippocampus; protein poshophatase 2; glycogen synthase kinase 3; B-cell lymphoma 2

1. Introduction

Alzheimer's disease (AD), is an irreversible neurodegenerative disorder prevalent among the older age-group of the population around the globe for which there is no cure. With increasing life expectancy globally and the resulting increase in aging population, AD is becoming a global healthcare problem [1]. AD is characterised clinically by progressive decline in cognitive functions such as memory loss and learning ability, and pathologically by the presence of neurofibrillary tangles (NFTs), deposition of beta amyloid (Aβ) plaque and synaptic dysfunction in the brain [2]. D-galactose induced ageing model in animals are characterised by pathological changes which closely resemble those seen in clinically diagnosed AD patients, including cognitive impairment, cholinergic dysfunction, oxidative stress and neurodegeneration [3]. Aluminium (Al), a neurotoxic agent has been linked to pathogenesis of AD, as chronic administration of Al has shown to produce oxidative damage, cholinergic dysfunction and cognitive impairment in rat brain [4]. Recent studies have reported that co-administration of D-gal/$AlCl_3$ resulted in hyperphosphorylation of tau, oxidative stress, cholinergic dysfunction, memory impairment, apoptosis, and hippocampal neurodegeneration in brain of rats [5–9]. Hence, rats which are continuously co-administered with D-gal and $AlCl_3$ could serve as good model for investigating AD-like pathologies and for drug screening.

Although, accumulation of Aβ and hyperphosphorylation of tau proteins are involved in the progression of AD [10], there is a growing evidence showing a major role played by P-tau in pathogenesis and progression of AD through impairment of the axonal transport of neurotransmitters and subcellular organelles [11]. Hyperphosphorylation of tau protein is one of the suggested theories explaining the pathogenesis of AD in humans and experimental animal models. A balance between the activities of glycogen synthase kinase-3 beta (GSK-3β) which is the main tau kinase and protein phosphatase 2A as the main tau phosphatase has been described as key contributor in defining tau phosphorylation/deposphorylation status [12]. Several reports of post-mortem from brains of AD patients have supported this theory as they demonstrated high level of GSK-3β and reduced activity of PP2A in tangles bearing neurons [13]. Further, increased phosphorylation of PP2A at Tyr 307 has also been reported in tangle bearing neurons in the brains of AD patients [14].

Centella asiatica (CA), locally known as "pegaga" in Malaysia is one of the valuable herbal medicine widely used in the treatment of various chronic ailments and also is proved to be safe and effective [15,16]. It is used in Ayurveda and Chinese traditional medicine to reverse/treat cognitive impairment and to enhance cognitive functions. These effects of CA have been well documented by studies conducted on healthy human subjects [17] and in those with mild cognitive deficits [18]. Further, the neuroprotective and cognitive enhancing effects of CA is well documented on in vitro and in multiple rodents' models of neurodegenerative diseases as well as in the settings of cognitive impairments due to variety of neurotoxic insults [19–23]. It has been recently reported that CA improves learning and memory in rats by increasing expression of, α-amino-3-hydroxy-5-methyl-4-isoxazolepropionic acid receptor (AMPAR) GluA1 and GluA2 subunits, and NMDAR GluN2B subunits, while reducing the *N*-methyl-D-aspartate receptor (NMDAR) GluN2 A subunits in their hippocampus and entorhinal cortex [24]. The present study describes the effectiveness of CA in preventing D-gal/$AlCl_3$ mediated AD-like neurotoxicity in rats via PP2A/GSK-3β and apoptosis pathways.

2. Results

2.1. CA Increased the Activity of PP2A and Decreased the Activity of GSK-3β in Hippocampus of Rats Exposed to D-Gal and $AlCl_3$

Expressions of PP2A from the hippocampus of the rats were assessed by western blot analysis (Figure 1A). One way ANOVA showed statistically significant differences in the levels of PP2A expression in the hippocampus of the various rat groups (F (5, 12) = 12.79, p = 0.0002) (Figure 1B). Tukey's post hoc revealed decrease in PP2A activities in the hippocampus of model group of rats (0.43 ± 0.02, p = 0.0001), when compared to control group (1 ± 0). Increased PP2A activities were observed in the donepezil (0.68 ± 0.05, p = 0.004), CA 200 (0.70 ± 0.04, p = 0.02), CA 400 (0.73 ± 0.14, p = 0.01) and CA 800 (0.76 ± 0.13, p = 0.005) groups of rats, when compared to the model group (0.43 ± 0.02). The expression of GSK-3β in the hippocampus of the rats groups were also assessed, which showed statistically significant differences by one way ANOVA (F (5, 12) = 9.344, p = 0.008)

(Figure 1C). Tukeys' post hoc revealed increases in GSK-3β activities in the hippocampus of model group of rats (1.4 ± 0.07), when compared to the control group (1 ± 0). Further, decreases in GSK-3β activities were observed in the donepezil (0.62 ± 0.11, $p = 0.0001$), CA 200 (0.76 ± 0.17, $p = 0.0002$), CA 400 (0.92 ± 0.32, $p = 0.0008$) and CA 800 (0.84 ± 0.08, $p = 0.0004$) groups of rats, when compared to the model group (1.4 ± 0.07).

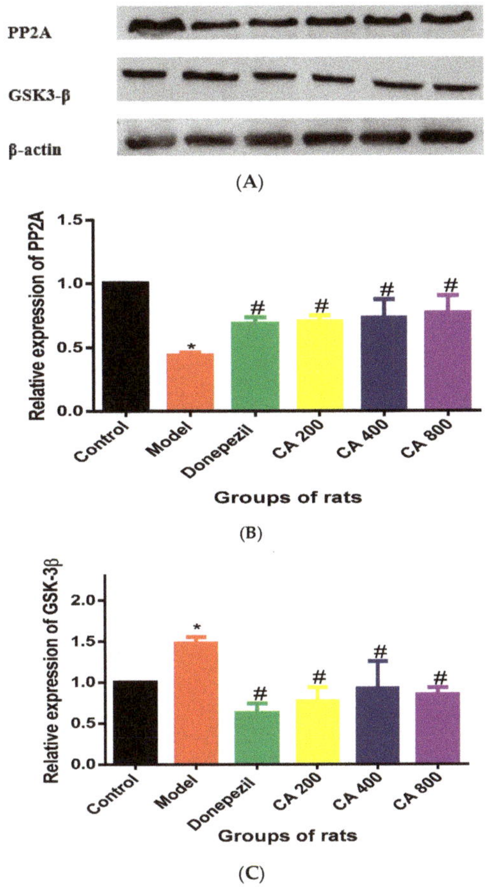

Figure 1. Expressions of PP2A and GSK3-β in rat's hippocampus. (**A**) Immunoblots of Levels of PP2A and GSK3-β in D-gal and $AlCl_3$ induced rats. (**B**) Immunoblot analysis showed dose-dependent increases in PP2A activities. (**C**) Immunoblot analysis showed decreases of GSK3-β activities. ImageJ software (NIH, Bethesda, MD, USA) was used for densitometry. Values are expressed as mean ± SD ($n = 3$), * $p < 0.05$ vs. control, # $p < 0.05$ vs. the model group of rats.

2.2. *Effects of CA on Intrinsic Mitochondria Mediated Apoptosis Related Genes of Rat Hippocampus Exposed to D-Gal and $AlCl_3$*

During the intrinsic mitochondria-mediated apoptotic pathway process, Bcl-2 is an anti-apoptotic factor. In the present study, mRNA expressions of Bcl-2 were assessed using RT-PCR. One way ANOVA showed statistical significant differences in the expressions of Bcl-2 mRNA ($F (5, 12) = 51.58$, $p = 0.0001$) in the hippocampus of the various rats groups. Tukey's post hoc revealed fold change decreases in the expression of Bcl-2 mRNA in the model group of rats (0.17 ± 0.09, $p = 0.0001$), when compared to the control group (1 ± 0). Further, increased fold change in the expressions of Bcl-2 mRNA were observed in the rat groups administered with donepezil (0.53 ± 0.001, $p = 0.004$), CA 200 (0.89 ± 0.19, $p = 0.0001$), CA 400 (0.71 ± 0.009, $p = 0.0001$), and CA 800 (0.59 ± 0.10, $p = 0.0001$), when compared to the model group of rats (0.17 ± 0.009) (Figure 2).

In the intrinsic mitochondria-mediated apoptotic pathway, Caspase-3 was one of the major proteases responsible for initiating a caspase cascade leading to apoptosis. In the current study, expressions of caspase-3 mRNA were determined using RT-PCR. One way ANOVA showed no statistically significant differences in the expressions of caspase-3 mRNA ($F (5, 12) = 0.956$, $p = 0.48$) in the hippocampus of the various rats groups (Figure 3). Although there was 2.3-fold change increase in the expression of caspase-3 mRNA in the model group, when compared to the control group, slight fold change decreases were observed in the CA administered groups of rats.

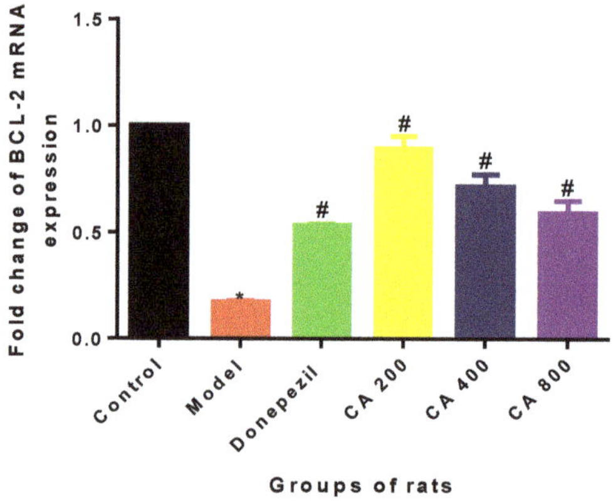

Figure 2. Effects of CA on mRNA expression of Bcl-2 in the hippocampus of rats. Donepezil and CA effectively increased Bcl-2 mRNA expressions. Values are expressed as mean ± SD ($n = 3$). * $p < 0.05$ vs. Control, # $p < 0.05$ vs. Model group of rats.

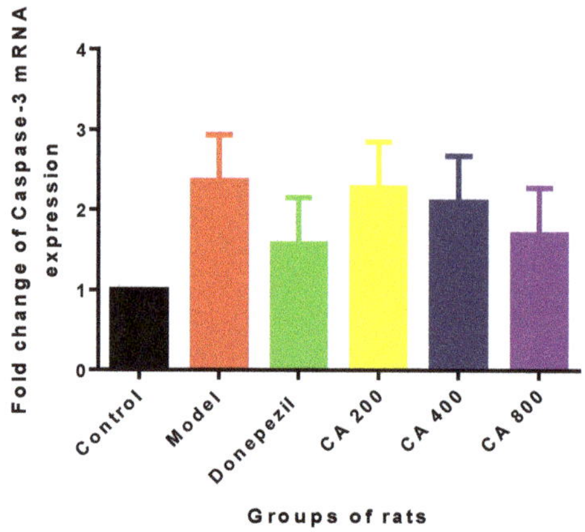

Figure 3. Effects of CA on mRNA expression of caspase-3 in the hippocampus of rats. No statistically significant differences were observed, even if there were fold change increases or decreases in the expressions of caspase-3 mRNA. Values are expressed as mean ± SEM ($n = 3$).

2.3. CA Protects against D-Gal and $AlCl_3$ Induced Pyramidal Cells Loss in CA 3 Subregions of Hippocampus of Rats

As shown in Figure 4A, the observation of CA 3 sub-region of hippocampus of rats of control group showed cells with well-defined nuclear membrane, clearly visible nucleolus and fewer abnormalities. Noticeable changes were observed in the CA 3 sub-region of hippocampus of rats of the model group, which included cells with indistinct nuclear membrane as well as no prominent nucleolus, besides being darkly stained. Further, the number of normal pyramidal cells were also reduced in the CA 3 sub-regions of the hippocampus in the model group of rats. Interestingly, these pathological changes observed in the hippocampus of model group of rats were altered in groups where D-gal and $AlCl_3$ were co-administered with donepezil 1 mg/kg·bwt or CA at doses of 200, 400 and 800 mg/kg·bwt. The extent of histopathological changes observed in the CA 3 sub-regions of the rat's hippocampus were estimated semi quantitatively. One way ANOVA was used to analyse the population of normal pyramidal neurons (F (5, 474) = 36.15, p = 0.0001) (Figure 4B). A statistically significant reductions in the number of normal pyramidal cells in CA 3 sub-regions of hippocampus were observed in the model group (13 ± 3.25, p = 0.0001) of rats when compared to control (21 ± 4.20), as revealed by Tukey's post hoc test. Whereas, the scenario was reversed in rats administered with donepezil (18.19 ± 4.33, p = 0.0001), CA 200(17.9 ± 3.90, p = 0.0001), CA 400(20.7 ± 4.65, p = 0.0001), and CA 800(19.9 ± 3.64, p = 0.0001) when compared to the model group of rats (13 ± 3.25).

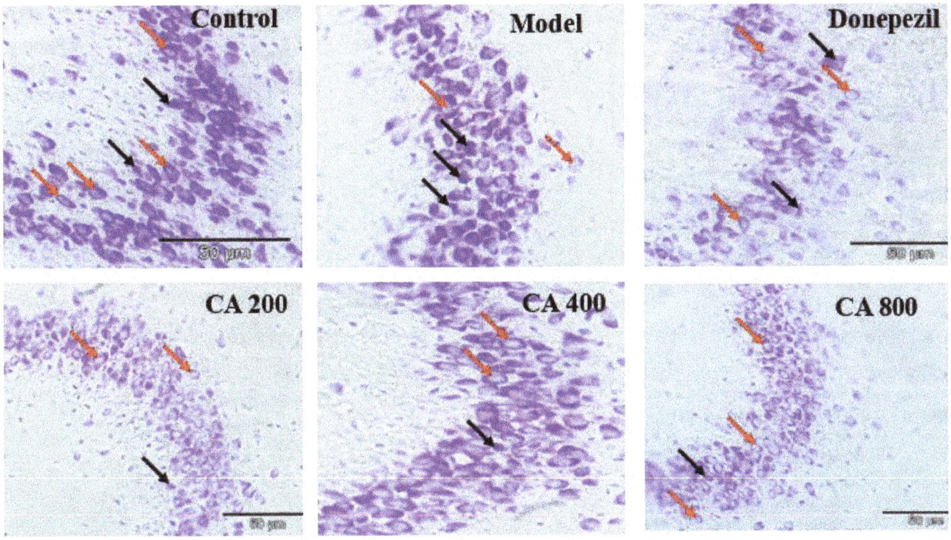

(A)

Figure 4. *Cont.*

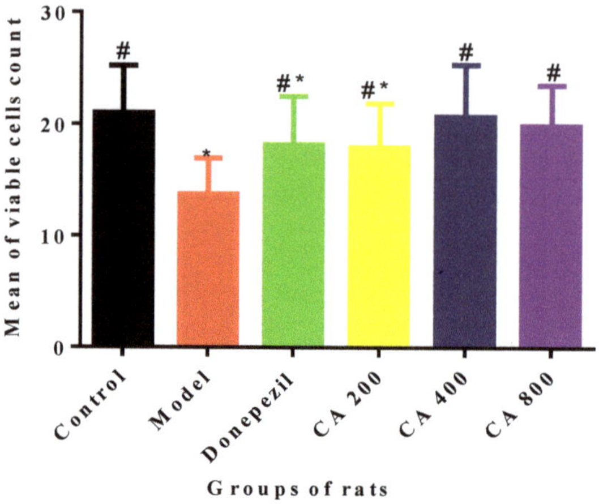

(**B**)

Figure 4. Protective effects of CA against D-gal and $AlCl_3$ induced neurodegeneration in CA3 sub region of the rat's hippocampus. (**A**) Cresyl violet stain showing the control and treatment groups. Red arrows pointing to normal pyramidal cells while black arrows pointing dead pyramidal cells. (**B**) Semi quantitative analysis of the number of normal pyramidal cells in the CA3 region of the hippocampus of all the rats groups.

3. Discussion

Our previous studies have shown that CA extract can attenuate cognitive deficits in rats induced by D-gal and $AlCl_3$ and can also prevent morphological aberrations in the CA1 region of their hippocampus. These effects were confirmed as rats co-administered with CA and D-gal/$AlCl_3$ showed a better performance in both spatial and non-spatial memory tests. Further, observation of the ultrastructure also revealed that CA protects the rat's hippocampus by preventing morphological alterations of the pyramidal cells and their intracellular organelles [25]. Results of the present study indicating that CA inhibited P-tau biosynthetic proteins in the hippocampus could be another mechanism through which CA improves learning and memory in D-gal/$AlCl_3$ mediated AD-like rats' model.

Hyperphosphorylation of tau protein is among the top reported factors in AD pathophysiology [26]. Earlier studies have reported that rodents exposed to D-gal/$AlCl_3$ exhibited AD-like features such as Aβ accumulation, hyperphosphorylation of tau protein and increased acetylcholinesterase (AChE) activities in their brains [6,8,9,27]. A balance between the activities of PP2A and GSK-3β, the main phosphatase and kinase has been reported to be the key contributing factor in describing tau dephosphorylation/phosphorylation status [12,28]. The aforementioned findings have been reinforced by reports from numerous post-mortem studies done on brains of AD patients which demonstrated that tangles bearing neurons were associated with decreased activities of PP2A due to increased phosphorylation at Tyr307 and the presence of high levels of GSK-3β [13,29]. Hence, it seems likely that PP2A and GSK-3β could be involved in enhancement of the aggregation of tau in the brains of AD patients [30]. In the present study, exposure of D-gal and $AlCl_3$ to rats has led to decreased PP2A activities and increases the levels of GSK-3β in the hippocampus of rats in the model group. Co-administration of CA to D-gal and $AlCl_3$ exposed rat's reverses these changes as there were increases in PP2A activities and decreases in GSK-3β levels in the rat's hippocampus. Hence, from these results it can be observed that the levels of P-tau in the rats' hippocampus could be altered by the actions of PP2A and GSK-3β. Although, few studies have reported that some phosphorylated residues of tau in the brains of AD patients were not sensitive to the actions of PP2A and GSK-3β [30,31]. As phosphorylation of tau could be achieved through other kinases, including cyclin-dependent kinase 5 (cdk5) and protein kinase A (PKA) [32].

There is growing evidence that neuronal apoptosis plays an important role in the pathogenesis of AD [33,34]. Among the other conditions which induce apoptosis, production of reactive oxygen species (ROS), nitric oxide (NO), glucocorticoids and over expression of Bax are known to be major contributory factors for the release of cytochrome c (Cty c) [2,35]. The Bcl-2 family of proteins, which include pro-apoptotic proteins like Bax and anti-apoptotic proteins like Bcl-2, strictly regulates the release of Cyt c [36,37]. Cyt c binds and activates the cytosolic protein Apaf-1 as well as procaspase-9, and together with adenosine triphosphate (ATP) they form "apoptosome" [38]. Balance between pro-apoptotic and anti-apoptotic proteins in the cell regulates the activation of intrinsic mitochondria-mediated apoptotic pathway [1,39]. Initiation of intrinsic mitochondria-mediated apoptotic pathway via pro-apoptotic Bcl-2 proteins is able to initiate different pathways for cell death [40]. The main upstream events leading to the initiation of these various pathways is mitochondrial outer membrane permeabilisation (MOMP). The process is activated by insertion and oligomerization of pro-apoptotic members BAK and BAX into the membrane, which lead to the subsequent release of apoptotic activating factors such as Cyt c from mitochondrial inter-membrane space to the cytosol. On the other hand, anti-apoptotic Bcl-2 proteins are integral intracellular membrane proteins notably present in the mitochondrial outer membrane (MOM), where they act by inhibiting the process of MOMP through binding with pro-apoptotic Bcl-2 proteins, thereby preventing apoptosis [41]. Surgucheva reported that decreased concentration of γ-synuclein (Syn G) in retinal ganglion cells (RGC-5) triggers mitochondrial pathway apoptosis via interaction of dephosphorylated Bad protein with pro-survival Bcl-2 family members, such as Bcl-2 and BcL-XL [42]. Activation of upstream caspases, such as caspase-9, will trigger downstream effector caspases, such as caspase-3, which can, in turn, cleave nuclear and cytoskeletal proteins to produce apoptosis [2,43]. For evaluating the extent of apoptosis in the hippocampus of rats exposed to D-gal and $AlCl_3$ and the protective effects of CA, the expressions of Bcl-2 and caspase-3 were assessed in the present study by RT PCR. Genetic expression analyses of hippocampus of various rat groups showed that expression of Bcl-2 was reduced when there was two-fold change increases in casepase-3 expression in rats exposed to D-gal and $AlCl_3$, when compared to the control group of rats. Similar findings were reported earlier in mice by Yang [1]. Co-administration of CA to rats exposed to D-gal and $AlCl_3$ ameliorated mRNA expressions of Bcl-2, while it had less effects on mRNA of caspase-3. The present study is limited in its scope to the sole use of genetic expressions of intrinsic mitochondria-mediated apoptosis proteins, and so additional research is required to confirm if the genetic expression changes actually reflect the expressions of Bcl-2 and caspase-3 proteins in the rat hippocampus.

Neurodegenerative diseases, such as AD, are morphologically featured by progressive loss of neurons in specific vulnerable regions of the central nervous system. The mechanisms of neurodegeneration is believed to be multifactorial which includes, mitochondrial dysfunction, oxidative stress, defective protein degradation and aggregation, genetic, environmental, and endogenous factors [44,45]. In the present study, exposure to D-gal and $AlCl_3$ readily led to significant morphological aberrations in the CA 3 sub-regions of the rat's hippocampus. Such changes includes increased number of pyknotic cells, alterations of the pyramidal cellular arrangement, and disruption of the nucleus. These changes could be due to enhanced GSK-3β levels and decreased PP2A levels, besides enhanced mRNA expression of caspase-3 and decreased mRNA expressions of Bcl-2 in D-gal and $AlCl_3$ induced rats. However, the neuroprotective role of CA prevents these degeneration at the maximum. Thus, co-administration of CA together with D-gal/$AlCl_3$ can alleviate the aforesaid degenerative changes (diminished pyknotic neurons, defective alignment of pyramidal cell layers, and increased density of normal neurons). Results from the present study clearly suggest that CA has cytoprotective effects and helped to maintain the normal cytoarchitecture of the CA 3 sub-region in the rat hippocampus.

Numerous approaches have been employed in the treatment of AD, such as the use of compounds that can prevent or clear Aβ generation [46], the use of antioxidants that elevates antioxidants defence system or reduces the levels of ROS to protect neurons from Aβ-induced toxicity [47] and the use of therapeutics that targets the cholinergic system [48]. Others focused on prevention of tau phosphorylation [9] while some concentrate on apoptotic pathways [49]. It can be observed that, the common trend among all these strategies for the prevention and cure for AD is ascribed to neuronal protection which could be achieved by enhancing oxidative defence system. It could be observed that most of the strategies focused on treating advanced stages of AD and symptomatic management of AD [50]. Only strategies that can prevent neuronal degeneration at early stage can prevent progression of AD. In this study the neuronal degeneration was prevented by co-administration of CA with D-gal/$AlCl_3$. Studies are also being conducted to evaluate the oxidative defence capacity and anti-cholinesterase activities of CA on the D-gal/$AlCl_3$ induced AD-like rat models as well. This study was limited by not measuring the concentration of $AlCl_3$ in the rats' brains. Deloncle [51] reported that $AlCl_3$ toxicity was mainly due to its ability to cross the blood brain barrier and its accumulation in the rat's brain. Does CA and its compounds has the potential to form coordination compounds with aluminium to remove it from the system?

In summary, results from the present study demonstrated that CA protected against D-gal and $AlCl_3$ induced toxicity and neurodegeneration in the hippocampus of rats. These effects of CA can be attributed to its ability to enhance the expression of PP2A and inhibits the levels of GSK-3β in the hippocampus, increase the expression of Bcl-2 mRNA and the maintenance of the cytoarchitecture of pyramidal neurons in the CA 3 sub-region of the rats' hippocampus (Figure 5).

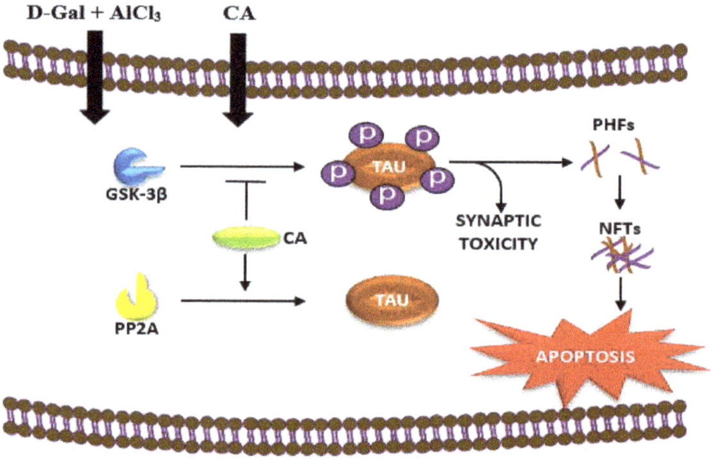

Figure 5. Proposed mechanism of protective effects of CA against D-gal and $AlCl_3$ induced neurotoxicity in rats, via the inhibition of GSK-3β and enhancing the expression of PP2A in the hippocampus of the rats. D-gal/$AlCl_3$ enhances phosphorylation of tau protein, which leads to paired helical forms (PHFs) formation and subsequently aggregates to form neurofibrillary tangles (NFTs), eventually leading to the death of the neuron. CA blocks the action of GSK-3β and enhances the activities of PP2A.

4. Materials and Methods

4.1. Ethics Statement

The study protocol was reviewed and approved by the Institutional Animal Care and Use Committee of the Universiti Putra Malaysia on 20 March 2017, with project identification code UPM/IACUC/AUP-R096/2016. A total of 36 male albino wistar rats, 2–3 months old (250–300 g) were obtained from a local vendor (Bistari International, Serdang, Malaysia). They were kept under constant temperature (25 ± 2 °C), 12-h light/dark cycle (lights on 7:00 AM–7:00 PM) and with free access to food and water. All the experimental procedures were strictly followed as recommended by the animal ethics committee guide lines, Universiti Putra Malaysia.

4.2. Chemicals and Reagents

Antibodies for western blotting (PP2A, GSK-3β and Beta actin) were purchased from Cell Signalling Technology (Danvers, MA, USA). The RNeasy mini kit was purchased from Qiagen (Hilden, Germany), the RNALater purchased from Thermo Fisher Scientific (Carlsbad, CA, USA), while the qPCRBIO cDNA synthesis kit and the qPCRBIO SyGreen Mix were purchased from PCR Biosystems Ltd. (London, UK). Aluminium chloride, D-galactose, donepezil, and cresyl violet were purchased from Sigma Aldreich (St. Louis, MO, USA), while standardised 60% aqueous ethanol extract of CA (ref. no. AuRins-MIA-1-0) [24,52] was made available by Prof. Mohd Ilham Adenan from Atta-ur-Rahman Institute for Natural Product Discovery, Universiti Technology Mara, Puncak Alam, Malaysia. All other chemicals used were of analytical grades.

4.3. Experimental Design and Treatment Protocol

After one week of acclimatisation, the rats were randomly divided in to six groups ($n = 6$) and administered with different treatments for 10 consecutive weeks (Table 1). D-gal, $AlCl_3$, donepezil and CA were all dissolved in distilled water, the experimental design together with treatments protocol were previously published [25]. At the end of the experiment, the rats were euthanised by decapitation so as to avoid contamination of brain tissues by anaesthetics and gases [53]. The rats brains were removed, rinsed in ice cold saline and kept in −80 °C for molecular studies while the remaining brains were fixed in 10% formalin for cresyl violet staining.

Table 1. $AlCl_3$, D-gal, donepezil and CA treated groups and the control.

Groups	Description	Treatment i.p	Treatment p.o
I	Control	Saline	Distilled water
II	Model	D-gal 60 mg/kg·bwt	$AlCl_3$ 200 mg/kg·bwt
III	Donepezil	D-gal 60 mg/kg·bwt	$AlCl_3$ 200 mg/kg·bw + Done 1 mg/kg·bwt
IV	CA 200	D-gal 60 mg/kg·bwt	$AlCl_3$ 200 mg/kg·bw + CA 200 mg/kg·bwt
V	CA 400	D-gal 60 mg/kg·bwt	$AlCl_3$ 200 mg/kg·bw + CA 400 mg/kg·bwt
VI	CA 800	D-gal 60 mg/kg·bwt	$AlCl_3$ 200 mg/kg·bw + CA 800 mg/kg·bwt

4.4. Protein Estimation

The total protein concentration in the hippocampal tissues were measured using bicinchoninic assay (BCA). Bovine serum albumin (BSA) (2 mg/mL) was used as a standard with a working range between 20–2000 µg/mL.

4.5. Western Blotting Analysis

The hippocampal tissues of the rats were homogenized on ice with AgileGrinder™ tissue homogenizer ACTGene, Inc. (Piscataway, NJ, USA) using radioimmunoprecipitatation assay (RIPA) buffer supplemented with phosphatase and protease inhibitors at a ratio of 1:500 and 1:1000 respectively and spun at 15,000× g for 15 min at 4 °C. For SDS-PAGE preparation, 4% of stacking gel (0.65 mL of 30% acrylamide, 3.05 mL of ddH_2O, 1.25 mL of stacking buffer, 0.05 mL of 10% SDS, 0.025 mL of 10% APS, 0.005 mL of TEMED), and 10% of resolving gel (1.65 mL of 30% acrylamide, 2.05 mL of ddH_2O, 1.25 mL of Resolving buffer, 0.05 mL of 10% SDS, 0.025 mL of 10% APS, 0.005 mL of TEMED) were used. Twenty microlitres of the 20 µg of the rat brain samples were added to 20 µL of laemmlli sample buffer supplemented with 1:19 dilution of β-mercaptoethanol and heated at 95 °C for 5 min. The samples were vortexed, centrifuged at 1000 rpm for 1 min, and loaded into the SDS-PAGE 20 µL per well. The electrophoresis procedure was initially run using 1-times running buffer (25 mM Trizma, glysine 192 mM, 0.1% SDS) at 100 V for 60 min, before the voltage was increased to 150 V for 30 min. The separated proteins were then transferred to 0.25 µM thick polyvinylidene difluoride (PVDF) membranes (Merck Millipore, Darmstadt, Germany) using 1-times transfer buffer ((10% (*v*/*v*) methanol, 25 mM Trizma, glysine 192 mM) at 20 V for 2 h. The PVDF membranes were stained with Ponceau S to observe and confirm the transfer of protein bands, before being incubated for 1 h at room temperature, with blocking buffer (5% (*w*/*v*) skimmed milk or 5% BCA in TBS-Tween 20) to prevent non-specific proteins binding. The membranes were then incubated overnight at 4 °C with primary antibodies (PP2A, dilution 1:1000, GSK-3β, dilution 1:1000 and β-actin, dilution 1:1000) diluted in blocking buffer. After the overnight incubation, membranes were washed three times with washing buffer (TBS-Tween 20) for 5 min each and probed using anti rabbit secondary antibodies (diluted in blocking buffer (1:2000)) for 1 h. After probing the membranes were then washed three times (5 min for each wash) with washing buffer and subsequently developed in a dark room by incubating it for 2 min in chemiluminescence HRP substrate (1:1 of WesternBright ECL and WesternBright peroxide). Gel documentation equipment was used to view the membranes and the image bands of the proteins of interest were obtained and subsequently analysed using ImageJ software 1.8.0 (NIH, Bethesda, MD, USA).

4.6. RNA Extraction and cDNA Synthesis

The Qiagen RNeasy mini kit was used for the isolation of RNA from rat hippocampus following the manufacturer's manual. The concentration and the purity of the total RNA samples were measured using Nanodrop spectrophotometer, while their integrity (28S/18S ribosomal RNA ratio) were checked by agarose gel electrophoresis. The total RNA (100 μg) was then reverse-transcribed into cDNA using a qPCRBIO cDNA synthesis kit, Biosystems Ltd. (London, UK) adhering strictly to the user's guide.

4.7. Reverse Transcriptase-Polymerase Chain Reaction (RT-PCR)

To detect the expression of Bcl-2 and caspase-3 in the rats' hippocampus, RT-PCR were performed. The primers for the genes of interest (GOI) and reference genes (RG) were designed with Primer 5.6 software according to the sequence in GenBank and manufactured by iDNA Technology (Table 2). Using 20 μL mixed system PCR reactions were performed, including 10 μL of 2x qPCRBIO SyGreen Blue Mix, 0.8 μL of forward primer, 0.8 μL of reverse primer, 2 μL of cDNA and 6.4 μL of RNases free water. An Eppendorf Mastercycler *ep* realplex *4S* PCR was used to perform the RT-PCR based on, heat activation at 95 °C for 2 min, followed by 40 cycles of 15 s denaturation at 95 °C, 30 s annealing at 59 °C and 30 s extension at 72 °C, while the fluorescence signals were detected at 59 °C. Using the obtained C_T values, the fold change of gene expressions were analysed using the Livak method [54]. The average C_T values of each GOI ($C_T{}^{AVG\ GOI}$) were normalised with the average C_T values of the reference genes ($C_T{}^{AVG\ RG}$) ($\Delta C_T = C_T{}^{AVG\ GOI} - C_T{}^{AVG\ RG}$). The $\Delta\Delta C_T$ ($\Delta C_T{}^{TREATMENT} - \Delta C_T{}^{CONTROL}$) were calculated and the fold change of each gene among the various rat groups were expressed as $2^{-(\Delta\Delta CT)}$.

Table 2. The nucleotide sequence of PCR primers for amplification and sequence-specific detection of cDNA (obtained from the GenBank database).

Accession No.	Gene Symbol	Primer	Sequence	Length	Tm	Amplicon Size
L14680.1	Bcl-2	Forward	5′-GGTGGACAACATCGCTCT-3′	18	57.01	143 bp
		Reverse	5′-GAGACAGCCAGGAGAAATCA-3′	20	57.94	
NM_012922.2	Caspase-3	Forward	5′-GAGCGTAAGGAAAGGAGAGG-3′	20	58.15	140 bp
		Reverse	5′-GACATCATCCACACAGACCAG-3′	21	58.96	
AY618569.1	B-Actin	Forward	5′-TGGCTCTGTGGCTTCTACTG-3′	20	58.16	192 bp
		Reverse	5′-TACCTTCCCAACTCCTCACC-3′	20	58.97	

4.8. Cresyl Violet Staining and Scoring

Cresyl violet stain was used to evaluate the protective effects of CA on cell survival in the CA3 region of hippocampus in rats. The protocol followed for the staining procedures as well as the methods for scoring was published earlier [8,55].

4.9. Statistical Analyses

The statistical significance was evaluated using one way analysis of variance (ANOVA) by Graghpad Prism version 6 (ISI, San Diego, CA, USA) software. Tukey's post hoc analyses was used for comparisons where applicable and data were presented mean ± SD, $p < 0.05$ were considered significant.

5. Conclusions

For the past few decades, anti-AD therapeutic research were focused on targeting one factor at a time, but that could not result in to any efficient drug to yet cure the disease. Since AD is a complex neurodegenerative disease with multiple causative factors, research shifted attention to targeting more than one factor at a time. Hence, it is necessary to search for natural products that can focus on multiple causative factors of AD at a time. This work reported for the first time that, CA extract showed multiple beneficial effects in D-gal/$AlCl_3$ mediated AD-like rat models. Outside this study, it can be postulated that CA could be used as a source of chemical compounds which could be further developed in to efficient anti-AD therapeutics.

Author Contributions: M.A.M.M., C.N.M.T., M.T.H.B., Z.A., M.I.A., O.M., and S.M.C. conceived and designed the experiment; S.J. and S.M.C. performed the experiments. The manuscript was drafted by S.M.C. and S.J. and approved by all authors.

Funding: The research presented in this paper was financially supported by Department of Anatomy, Universiti Putra Malaysia (grant no.GP-IPS 9535400).

Acknowledgments: The authors appreciate Nasiru Mohammed Wana and Sharif Alhassan, PhD candidates from Parasitology Laboratory, Faculty of Medicine and Health Sciences, Universiti Putra Malaysia, for their inputs during the gene expression studies.

Conflicts of Interest: The authors declare no conflict of interest.

Abbreviations

AD	Alzheimer's disease
$AlCl_3$	Aluminium chloride
AMPAR	α-Amino-3-hydroxy-5-methyl-4-isoxazolepropionic acid receptor
ANOVA	Analysis of variance
ATP	Adenosine triphosphate
BAK	Bcl-2 antagonist/killer 1
Bax	BCL2-Associated X Protein
Bcl-2	B-cell lymphoma 2
CA	Centella asiatica
CA3	Connus ammonis 3
Caspase 3	Cysteine-aspartic acid protease 3
Cdk5	Cyclin-Dependent Kinase 5
Cyt c	Cytochrome c
ddH_2O	Double distilled water
D-Gal	D-galactose
GSK-3β	Glycogen synthase kinase-3 beta
HRP	Horseradish peroxidase
mRNA	Messenger Ribonucleic Acid
MOM	Mitochondrial outer membrane
MOMP	Mitochondrial outer membrane permealisation
n	Number of rats per group
NFTs	Neutofibrillary tangles
NO	Nitric oxide
PP2A	Protein phosphatase 2
PVDF	Polyvinylidene difluoride
ROS	Reactive oxygen species

RT PCR	Real time polymerase chain reaction
SD	Standard deviation
SDS	Sodium dodecyl sulfate
TBST	Tri-buffered saline, 0.1% tween 20
SDS-PAGE	Sodium dodecyl sulfate polyacrilmide gel electrophoresis
Tyr	Tyrosine

References

1. Yang, W.; Shi, L.; Chen, L.; Zhang, B.; Ma, K.; Liu, Y.; Qian, Y. Protective effects of perindopril on D-galactose and aluminum trichloride induced neurotoxicity via the apoptosis of mitochondria-mediated intrinsic pathway in the hippocampus of mice. *Brain Res. Bull.* **2014**, *109*, 46–53. [CrossRef]
2. Xing, Z.; He, Z.; Wang, S.; Yan, Y.; Zhu, H.; Gao, Y.; Zhao, Y.; Zhang, L. Ameliorative effects and possible molecular mechanisms of action of fibrauretine from Fibraurea recisa Pierre on D-galactose/$AlCl_3$-mediated Alzheimer's disease. *RSC Adv.* **2018**, *8*, 31646–31657. [CrossRef]
3. Yang, W.N.; Han, H.; Hu, X.D.; Feng, G.F.; Qian, Y.H. The effects of perindopril on cognitive impairment induced by D-galactose and aluminum trichloride via inhibition of acetylcholinesterase activity and oxidative stress. *Pharmacol. Biochem. Behav.* **2013**, *114–115*, 31–36. [CrossRef]
4. Kumar, A.; Dogra, S.; Prakash, A. Protective effect of curcumin (*Curcuma longa*), against aluminium toxicity: Possible behavioral and biochemical alterations in rats. *Behav. Brain Res.* **2009**, *205*, 384–390. [CrossRef]
5. Zhang, Y.; Pi, Z.; Song, F.; Liu, Z. Ginsenosides attenuate D-galactose- and $AlCl_3$-inducedspatial memory impairment by restoring the dysfunction of the neurotransmitter systems in the rat model of Alzheimer's disease. *J. Ethnopharmacol.* **2016**, *194*, 188–195. [CrossRef] [PubMed]
6. Zhang, Y.; Yang, X.; Jin, G.; Yang, X.; Zhang, Y. Polysaccharides from *Pleurotus ostreatus* alleviate cognitive impairment in a rat model of Alzheimer's disease. *Int. J. Biol. Macromol.* **2016**, *92*, 935–941. [CrossRef] [PubMed]
7. Bilgic, Y.; Demir, E.A.; Bilgic, N.; Dogan, H.; Tutuk, O.; Tumer, C. Detrimental effects of chia (*Salvia hispanica* L.) seeds on learning and memory in aluminum chloride-induced experimental Alzheimer's disease. *Acta Neurobiol. Exp. Wars* **2018**, *78*, 322–331. [CrossRef]
8. Chiroma, S.M.; Mohd Moklas, M.A.; Mat Taib, C.N.; Baharuldin, M.T.H.; Amon, Z. D-galactose and aluminium chloride induced rat model with cognitive impairments. *Biomed. Pharmacother.* **2018**, *103*, 1602–1608. [CrossRef]
9. Li, H.; Kang, T.; Qi, B.; Kong, L.; Jiao, Y.; Cao, Y.; Zhang, J.; Yang, J. Neuroprotective effects of ginseng protein on PI3K/Akt signaling pathway in the hippocampus of D-galactose/$AlCl_3$ inducing rats model of Alzheimer' s disease. *J. Ethnopharmacol.* **2016**, *179*, 162–169. [CrossRef] [PubMed]
10. Pascoal, T.A.; Mathotaarachchi, S.; Shin, M.; Benedet, A.L.; Mohades, S.; Wang, S.; Beaudry, T.; Kang, M.S.; Soucy, J.; Labbe, A.; et al. Synergistic interaction between amyloid and tau predicts the progression to dementia. *Alzheimer's Dement.* **2017**, *13*, 644–653. [CrossRef]
11. Reddy, P.H. Abnormal tau, mitochondrial dysfunction, impaired axonal transport of mitochondria, and synaptic deprivation in Alzheimer's disease. *Brain Res.* **2011**, *1415*, 136–148. [CrossRef]
12. Liu, F.; Grundke-Iqbal, I.; Iqbal, K.; Gong, C. Contributions of protein phosphatases PP1, PP2A, PP2B and PP5 to the regulation of tau phosphorylation. *Eur. J. Neurosci.* **2005**, *22*, 1942–1950. [CrossRef] [PubMed]
13. Hernandez, F.; Lucas, J.J.; Avila, J. GSK3 and tau: Two convergence points in Alzheimer's disease. *J. Alzheimer's Dis.* **2013**, *33*, S141–S144. [CrossRef] [PubMed]
14. Sontag, E.; Luangpirom, A.; Hladik, C.; Mudrak, I.; Ogris, E.; Speciale, S.; White, C.L., III. Altered Expression Levels of the Protein Phosphatase 2A ABαC Enzyme Are Associated with Alzheimer Disease Pathology. *J. Neuropathol. Exp. Neurol.* **2004**, *63*, 287–301. [CrossRef]
15. Subaraja, M.; Vanisree, A.J. The novel phytocomponent Asiaticoside-D isolated from *Centella asiatica* exhibits monoamine oxidase–B inhibiting potential in the rotenone degenerated cerebral ganglions of *Lumbricus terretris*. *Phytomedicine* **2019**, 152833, in press. [CrossRef] [PubMed]
16. Rasid, N.A.M.; Nazmi, N.N.M.; Isa, M.I.N.; Sarbon, N.M. Rheological, functional and antioxidant properties of films forming solution and active gelatin films incorporated with *Centella asiatica* (L.) urban extract. *Food Packag. Shelf Life* **2018**, *18*, 115–124. [CrossRef]

17. Dev, R.D.O.; Mohamed, S.; Hambali, Z.; Samah, B.A. Comparison on cognitive effects of *Centella asiatica* in healthy middle age female and male volunteers. *Eur. J. Sci. Res.* **2009**, *31*, 553–565.
18. Tiwari, S.; Singh, S.; Patwardhan, K.; Gehlot, S.; Gambhir, I.S. Effect of *Centella asiatica* on mild cognitive impairment (MCI) and other common age-related clinical problems. *Dig. J. Nanomater. Biostruct.* **2008**, *3*, 215–220.
19. Soumyanath, A.; Zhong, Y.-P.; Henson, E.; Wadsworth, T.; Bishop, J.; Gold, B.G.; Quinn, J.F. *Centella asiatica* Extract Improves Behavioral Deficits in a Mouse Model of Alzheimer's Disease: Investigation of a Possible Mechanism of Action. *Int. J. Alzheimer's Dis.* **2012**, *2012*, 381974.
20. Gupta, R.; Flora, S.J.S. Effect of *Centella asiatica* on arsenic induced oxidative stress and metal distribution in rats. *J. Appl. Toxicol.* **2006**, *26*, 213–222. [CrossRef] [PubMed]
21. Gray, N.E.; Morré, J.; Kelley, J.; Maier, C.S.; Stevens, J.F.; Quinn, J.F.; Soumyanath, A. Caffeoylquinic acids in centella asiatica protect against amyloid-β toxicity. *J. Alzheimer's Dis.* **2014**, *40*, 359–373. [CrossRef]
22. Lokanathan, Y.; Omar, N.; Ahmad Puz, N.N.; Saim, A.; Hj Idrus, R. Recent updates in neuroprotective and neuroregenerative potential of *Centella asiatica*. *Malays. J. Med. Sci.* **2016**, *23*, 4–14.
23. Gray, N.E.; Zweig, J.A.; Caruso, M.; Zhu, J.Y.; Wright, K.M.; Quinn, J.F.; Soumyanath, A. *Centella asiatica* attenuates hippocampal mitochondrial dysfunction and improves memory and executive function in β-amyloid overexpressing mice. *Mol. Cell Neurosci.* **2018**, *93*, 1–9. [CrossRef]
24. Wong, J.H.; Muthuraju, S.; Reza, F.; Senik, M.H.; Zhang, J.; Mohd Yusuf Yeo, N.A.B.; Chuang, H.G.; Jaafar, H.; Yusof, S.R.; Mohamad, H.; et al. Differential expression of entorhinal cortex and hippocampal subfields α-amino-3-hydroxy-5-methyl-4-isoxazolepropionic acid (AMPA) and *N*-methyl-D-aspartate (NMDA) receptors enhanced learning and memory of rats following administration of *Centella asiatica*. *Biomed. Pharmacother.* **2019**, *110*, 168–180. [CrossRef] [PubMed]
25. Chiroma, S.M.; Baharuldin, M.T.H.; Taib, C.N.M.; Amom, Z.; Jagadeesan, S.; Adenan, M.I.; Moklas, M.A.M. Protective effect of *Centella asiatica* against D-galactose and aluminium chloride induced rats: Behavioral and ultrastructural approaches. *Biomed. Pharmacother.* **2019**, *109*, 853–864. [CrossRef] [PubMed]
26. Hanger, D.P.; Anderton, B.H.; Noble, W. Tau phosphorylation: The therapeutic challenge for neurodegenerative disease. *Trends Mol. Med.* **2009**, *15*, 112–119. [CrossRef]
27. Xiao, F.; Li, X.-G.; Zhang, X.-Y.; Hou, J.-D.; Lin, L.-F.; Gao, Q.; Luo, H.M. Combined administration of D-galactose and aluminium induces Alzheimerlike lesions in brain. *Neurosci. Bull.* **2011**, *27*, 143–155. [CrossRef] [PubMed]
28. Eldar-Finkelman, H. Glycogen synthase kinase 3: An emerging therapeutic target. *Trends Mol. Med.* **2002**, *8*, 126–132. [CrossRef]
29. Lovell, M.A.; Xiong, S.; Xie, C.; Davies, P.; Markesbery, W.R. Induction of hyperphosphorylated tau in primary rat cortical neuron cultures mediated by oxidative stress and glycogen synthase kinase-3. *J. Alzheimer's Dis.* **2004**, *6*, 659–671. [CrossRef]
30. Hanger, D.P.; Noble, W. Functional implications of glycogen synthase kinase-3-mediated tau phosphorylation. *Int. J. Alzheimers Dis.* **2011**, *2011*, 352805. [CrossRef] [PubMed]
31. Guerra-Araiza, C.; Amorim, M.A.R.; Camacho-Arroyo, I.; Garcia-Segura, L.M. Effects of progesterone and its reduced metabolites, dihydroprogesterone and tetrahydroprogesterone, on the expression and phosphorylation of glycogen synthase kinase-3 and the microtubule-associated protein tau in the rat cerebellum. *Dev. Neurobiol.* **2007**, *67*, 510–520. [CrossRef]
32. Zhang, Q.; Li, X.; Cui, X.; Zuo, P. D-galactose injured neurogenesis in the hippocampus of adult mice. *Neurol. Res.* **2005**, *27*, 552–556. [CrossRef]
33. Tanzi, R.E.; Moir, R.D.; Wagner, S.L. Clearance of Alzheimer's Aβ peptide: The many roads to perdition. *Neuron* **2004**, *43*, 605–608. [PubMed]
34. Pompl, P.N.; Yemul, S.; Xiang, Z.; Ho, L.; Haroutunian, V.; Purohit, D.; Mohs, R.; Pasinetti, G.M. Caspase gene expression in the brain as a function of the clinical progression of Alzheimer disease. *Arch. Neurol.* **2003**, *60*, 369–376. [CrossRef] [PubMed]
35. Kroemer, G.; Blomgren, K. Mitochondrial cell death control in familial Parkinson disease. *PLoS Biol.* **2007**, *5*, e206. [CrossRef]
36. Adams, J.M.; Cory, S. The Bcl-2 protein family: Arbiters of cell survival. *Science* **1998**, *281*, 1322–1326. [CrossRef] [PubMed]

37. Wang, Y.; Li, Y.; Yang, W.; Gao, S.; Lin, J.; Wang, T.; Zhou, K.; Hu, H. Ginsenoside Rb1 inhibit apoptosis in rat model of alzheimer's disease induced by Aβ1-40. *Am. J. Transl. Res.* **2018**, *10*, 796.
38. Love, S. Apoptosis and brain ischaemia. *Prog. Neuro-Psychopharmacol. Biol. Psychiatry* **2003**, *27*, 267–282. [CrossRef]
39. Woo, R.-S.; Lee, J.-H.; Yu, H.-N.; Song, D.-Y.; Baik, T.-K. Expression of ErbB4 in the apoptotic neurons of Alzheimer's disease brain. *Anat. Cell Biol.* **2010**, *43*, 332–339. [CrossRef]
40. Kilbride, S.M.; Prehn, J.H.M. Central roles of apoptotic proteins in mitochondrial function. *Oncogene* **2013**, *32*, 2703. [CrossRef]
41. Anilkumar, U.; Prehn, J.H.M. Anti-apoptotic BCL-2 family proteins in acute neural injury. *Front. Cell Neurosci.* **2014**, *8*, 281. [CrossRef]
42. Surgucheva, I.; Shestopalov, V.I.; Surguchov, A. Effect of γ-synuclein silencing on apoptotic pathways in retinal ganglion cells. *J. Biol. Chem.* **2008**, *283*, 36377–36385. [CrossRef]
43. D'Amelio, M.; Sheng, M.; Cecconi, F. Caspase-3 in the central nervous system: Beyond apoptosis. *Trends Neurosci.* **2012**, *35*, 700–709. [CrossRef]
44. Sharma, D.R.; Wani, W.Y.; Sunkaria, A.; Kandimalla, R.J.; Sharma, R.K.; Verma, D.; Bal, A.; Gill, K.D. Quercetin attenuates neuronal death against aluminum-induced neurodegeneration in the rat hippocampus. *Neuroscience* **2016**, *324*, 163–176. [CrossRef]
45. Ravi, S.K.; Ramesh, B.N.; Mundugaru, R.; Vincent, B. Multiple pharmacological activities of *Caesalpinia crista* against aluminium-induced neurodegeneration in rats: Relevance for Alzheimer's disease. *Environ. Toxicol. Pharmacol.* **2018**, *58*, 202–211. [CrossRef]
46. Donahue, J.E.; Flaherty, S.L.; Johanson, C.E.; Duncan, J.A.; Silverberg, G.D.; Miller, M.C.; Tavares, R.; Yang, W.; Wu, Q.; Sabo, E.; et al. RAGE, LRP-1, and amyloid-beta protein in Alzheimer's disease. *Acta Neuropathol.* **2006**, *112*, 405–415. [CrossRef]
47. Smith, J.V.; Luo, Y. Elevation of oxidative free radicals in Alzheimer's disease models can be attenuated by Ginkgo biloba extract EGb 761. *J. Alzheimer's Dis.* **2003**, *5*, 287–300. [CrossRef]
48. Wilcock, G.; Howe, I.; Coles, H.; Lilienfeld, S.; Truyen, L.; Zhu, Y.; Bullock, R.; Kershaw, P. A long-term comparison of galantamine and donepezil in the treatment of Alzheimer's disease. *Drugs Aging* **2003**, *20*, 777–789. [CrossRef]
49. Mathiyazahan, D.B.; Justin Thenmozhi, A.; Manivasagam, T. Protective effect of black tea extract against aluminium chloride-induced Alzheimer's disease in rats: A behavioural, biochemical and molecular approach. *J. Funct. Foods* **2015**, *16*, 423–435. [CrossRef]
50. Chen, C.L.; Tsai, W.H.; Chen, C.J.; Pan, T.M. *Centella asiatica* extract protects against amyloidβ1–40-induced neurotoxicity in neuronal cells by activating the antioxidative defence system. *J. Tradit. Complement. Med.* **2016**, *6*, 362–369. [CrossRef]
51. Deloncle, R.; Huguet, F.; Babin, P.; Fernandez, B.; Quellard, N.; Guillard, O. Chronic administration of aluminium L-glutamate in young mature rats: Effects on iron levels and lipid peroxidation in selected brain areas. *Toxicol. Lett.* **1999**, *104*, 65–73. [CrossRef]
52. Binti Mohd Yusuf Yeo, N.A.; Muthuraju, S.; Wong, J.H.; Mohammed, F.R.; Senik, M.H.; Zhang, J.; Yusof, S.R.; Jaafar, H.; Adenan, M.L.; Mohamad, H.; et al. Hippocampal amino-3-hydroxy-5-methyl-4-isoxazolepropionic acid GluA1 (AMPA GluA1) receptor subunit involves in learning and memory improvement following treatment with *Centella asiatica* extract in adolescent rats. *Brain Behav.* **2018**, *8*, 1–14. [CrossRef]
53. van Rijn, C.M.; Krijnen, H.; Menting-Hermeling, S.; Coenen, A.M.L. Decapitation in Rats: Latency to Unconsciousness and the 'Wave of Death'. *PLoS ONE* **2011**, *6*, e16514. [CrossRef]
54. Livak, K.J.; Schmittgen, T.D. Analysis of relative gene expression data using real-time quantitative PCR and the $2^{-\Delta\Delta CT}$ method. *Methods* **2001**, *25*, 402–408. [CrossRef]
55. Adeli, S.; Zahmatkesh, M.; Tavoosidana, G.; Karimian, M.; Hassanzadeh, G. Simvastatin enhances the hippocampal klotho in a rat model of streptozotocin-induced cognitive decline. *Prog. Neuro-Psychopharmacol. Biol. Psychiatry* **2017**, *72*, 87–94. [CrossRef]

International Journal of
Molecular Sciences

MDPI

Article

Neuroprotective Effects of Thymol, a Dietary Monoterpene Against Dopaminergic Neurodegeneration in Rotenone-Induced Rat Model of Parkinson's Disease

Hayate Javed [1,*], Sheikh Azimullah [2], MF Nagoor Meeran [2], Suraiya A Ansari [3] and Shreesh Ojha [2,*]

1 Department of Anatomy, College of Medicine and Health Sciences, United Arab Emirates University, Al Ain, P.O. Box 17666, UAE
2 Department of Pharmacology and Therapeutics, College of Medicine and Health Sciences, United Arab Emirates University, Al Ain, P.O. Box 17666, UAE; azim.sheikh@uaeu.ac.ae (S.A.); nagoormeeran1985@uaeu.ac.ae (M.F.N.M.)
3 Department of Biochemistry, College of Medicine and Health Sciences, United Arab Emirates University, Al Ain, P.O. Box 17666, UAE; sansari@uaeu.ac.ae
* Correspondence: h.javed@uaeu.ac.ae (H.J.); shreeshojha@uaeu.ac.ae (S.O.); Tel.: +971-3-7137603 (H.J.); +971-3-7137525 (S.O.); Fax: +971 3 767 2033 (H.J. & S.O.)

Received: 25 February 2019; Accepted: 18 March 2019; Published: 27 March 2019

Abstract: Parkinson's disease (PD), a multifactorial movement disorder that involves progressive degeneration of the nigrostriatal system affecting the movement ability of the patient. Oxidative stress and neuroinflammation both are shown to be involved in the etiopathogenesis of PD. The aim of this study was to evaluate the therapeutic potential of thymol, a dietary monoterpene phenol in rotenone (ROT)-induced neurodegeneration in rats that precisely mimics PD in humans. Male Wistar rats were injected ROT at a dose of 2.5 mg/kg body weight for 4 weeks, to induce PD. Thymol was co-administered for 4 weeks at a dose of 50 mg/kg body weight, 30 min prior to ROT injection. The markers of dopaminergic neurodegeneration, oxidative stress and inflammation were estimated using biochemical assays, enzyme-linked immunosorbent assay, western blotting and immunocytochemistry. ROT challenge increased the oxidative stress markers, inflammatory enzymes and cytokines as well as caused significant damage to nigrostriatal dopaminergic system of the brain. Thymol treatment in ROT challenged rats appears to significantly attenuate dopaminergic neuronal loss, oxidative stress and inflammation. The present study showed protective effects of thymol in ROT-induced neurotoxicity and neurodegeneration mediated by preservation of endogenous antioxidant defense networks and attenuation of inflammatory mediators including cytokines and enzymes.

Keywords: neurodegeneration; oxidative injury; Parkinson's disease; terpenes, rotenone; thymol

1. Introduction

Parkinson's disease (PD) is pathologically described by the continued loss of dopaminergic neuronal cells in the substantia nigra pars compacta (SNc), which results in motor impairments such as loss of motion, postural and gait instability, resting tremors, and muscle rigidity [1,2]. Accumulating evidence suggests that mitochondrial dysfunction, lipid peroxidation, brain aging, and genetic susceptibility, which often involve oxidative stress and neuroinflammatory changes, play a major part in the pathogenesis of PD [3–5]. Oxidative stress and inflammation are the two central pathways in microglial cells activation that lead to progressive neuronal degeneration and represent an important therapeutic target in PD [3–7]. The activation and release of proinflammatory cytokines, such as IL-1β, IL-6, and TNFα, along with free radical generation including reactive oxygen species (ROS) and inducible nitric oxide synthase (iNOS), has detrimental effects on the existence of dopaminergic neurons in the SNc [6,7].

To ensure cellular homeostasis, a balance between pro- and antioxidant systems is typically required. Hence, the restoration of the cellular antioxidant system using antioxidants is one of the emerging therapeutic strategies to protect susceptible dopaminergic neurons from oxidative stress and subsequent inflammation. The adverse effects of anti-inflammatory agents and the pro-oxidant action of the synthetic antioxidants are of concern in therapeutics. This concern shifted the focus of drug discovery to explore plant extracts and plant-derived phytochemicals that possess antioxidant and anti-inflammatory activities for their therapeutic and preventive benefits in PD [8,9]. Therefore, in recent years, the focus of pharmacological therapy has been on the development of novel nutraceutical-based plant-derived phytochemicals that possess high antioxidant and anti-inflammatory properties, with a lesser degree of cytotoxic effects [9].

Among numerous plant-derived dietary phytochemicals, thymol has received attention due to its favorable physicochemical, pharmacokinetic, and pharmacological properties [10]. Thymol, a dietary monoterpene is chemically known as 2-isopropyl-5-methylphenol and found predominantly in many edible or culinary plants such as *Centipeda minima, Lippia multiflora, Nigella sativa, Ocimum gratissimum, Satureja hortensis, Satureja thymbra, Thymus spp. (Thymus vulgaris, Thymus pectinatus, Thymus zygis, and Thymus ciliates), Trachyspermum ammi* and *Zataria multiflora* [10]. Thymol is catalogued as 'Generally Recognized as Safe' for use as a preservative and additive in food, beverages and cosmetic products, therefore it is considered to be safe for dietary use with minimal toxicity. Thymol exhibits potent pharmacological properties including antioxidant [11], anti-inflammatory [12], antimutagenic [13], analgesic [14], and anti-microbial [15] effects. It has been approved for use as a food additive and flavoring agent in cosmetics and food preparations. Its long-time dietary use, acceptable safety profile, and low toxicity have generated interest in evaluating its possible therapeutic use in neurodegenerative diseases. Therefore, in the current study we examined the neuroprotective efficacy and underlying mechanism of thymol in a rotenone (ROT)-induced rat model of neurodegeneration mimicking PD in humans. ROT, a plant-derived insecticide, inhibits mitochondrial complex I resulting in loss of ATP production, increase in oxidative stress, inflammation, prolonged glial cell activation, and nigrostriatal degeneration that mimics human PD [16–19]. The experimental models of ROT-induced neurodegeneration in rats, fruit fly or cell lines are popularly employed to screen and evaluate agents for their potential neuroprotective potential and therapeutic efficacy [20–24].

2. Results

2.1. Thymol Preserved TH+ Dopaminergic Neurons in SNc Regions and Dopaminergic Fibers in Striatum Regions of the Brain

In the current study, thymol (Figure 1), a monoterpene phenol was used to protect the dopaminergic neuronal death caused by ROT administration. We performed the immunohistochemical analysis of TH+ neurons in the SNc and TH-ir fibers in the striatum to observe the effects of thymol on nigrostriatal dopaminergic loss. The ROT injected animals showed significant ($p < 0.001$) degeneration of dopaminergic neurons in the SNc region when compared to rats of the control group received only vehicle (Figure 2A,C). Thymol administration significantly ($p < 0.05$) protected against ROT-induced degeneration of dopaminergic neurons. Dopaminergic neurons venture their axons to the striatum region wherein the terminal fibers are consisting of the dopamine transporter (DAT). Therefore, it was essential to examine whether the degeneration of dopaminergic neurons in the SNc region is associated with the loss of dopaminergic nerve terminals as evaluated by assessing the intensity of striatal TH-ir dopaminergic nerve terminal fibers. A significant ($p < 0.001$) loss in TH-ir fibers intensity was observed in animals challenged with ROT in comparison with animals of the control group received only vehicle. However, thymol pretreatment to ROT injected animals has produced a significant ($p < 0.01$) increase in the intensity of TH-ir nerve terminals compared to animals injected with ROT alone. This observation suggests the protective effect of thymol on dopaminergic neurons and nerve fibers (Figure 2B,D).

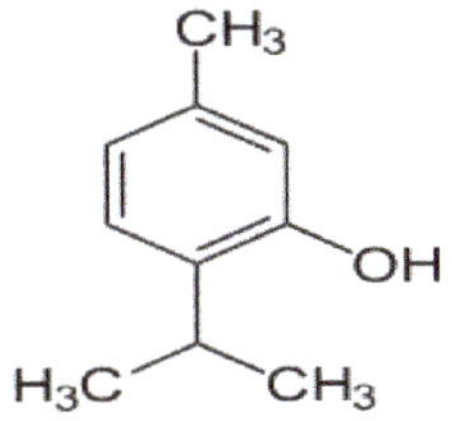

Figure 1. The chemical structure of thymol.

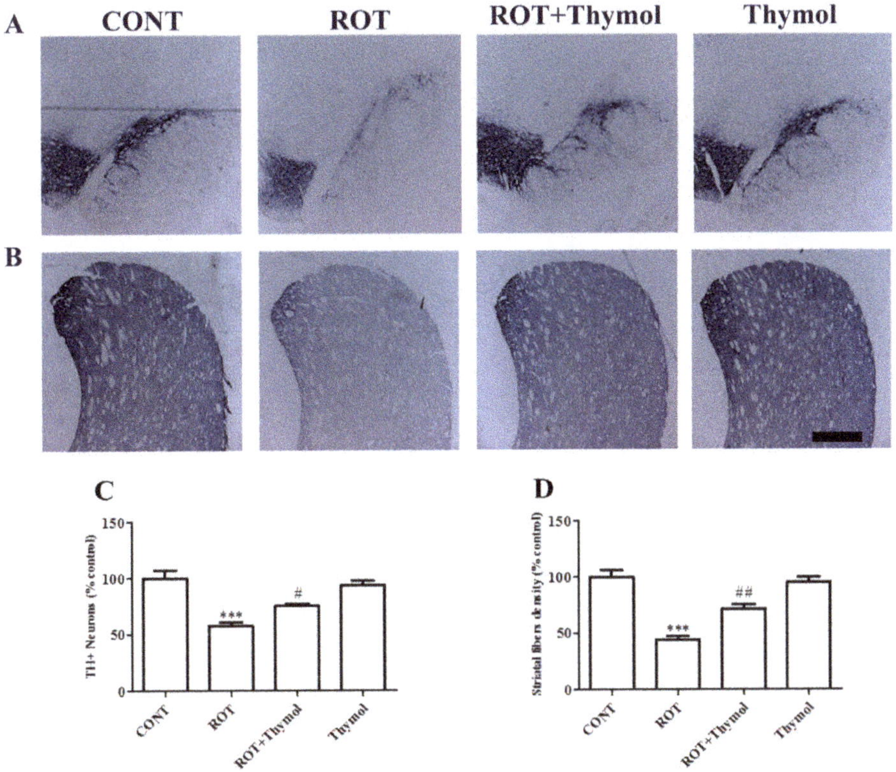

Figure 2. The illustrative photomicrograph showing expression of TH+ neurons in substantia nigra par compacta (SNc) (**A**) and TH-ir dopaminergic fibers in striatum (**B**). The scale bar is 100 µm. The expression of TH+ neurons and TH-ir fibers were reduced in the SNc region of rotenone (ROT) challenged rats as compared to vehicle injected rats in the CONT group. Thymol treatment to ROT challenged rats showed remarkable expressions of TH+ neurons and TH-ir fibers as compared to ROT injected rats. Quantification data showed significant (*** $p < 0.001$) decrease in the number of TH+ neurons and density of TH-ir fibers in ROT group rats compared to control rats. While thymol treatment to ROT injected rats showed significant (# $p < 0.05$; ## $p < 0.01$) increase in TH+ neurons and TH fibers density as compared to ROT alone injected rats (**C**,**D**).

2.2. Thymol Inhibited Lipid Peroxidation and Restored GSH and Endogenous Enzymes Activity

The markers of lipid peroxidation, such as malondialdehyde (MDA), and the endogenous tripeptide antioxidant, glutathione (GSH), endogenous antioxidant enzymes (SOD and CAT) were measured in homogenates of the mid brain tissues. ROT administration induced a significant ($p < 0.001$, Figure 3A) rise in MDA levels in comparison with rats of control group. However, thymol treatment to the ROT challenged animals produced a significant ($p < 0.01$) decline in the MDA levels. ROT challenged rats show significant ($p < 0.001$) reduction in the levels of GSH as compared to control rats (Figure 3B). In contrast, thymol treatment significantly ($p < 0.01$) increase the GSH levels in ROT-injected rats compared to animals injected with ROT alone. Moreover, ROT injection also significantly decreases ($p < 0.05$) endogenous antioxidant enzyme activity such as: SOD and CAT in the ROT injected rats compared to control rats. However, thymol treatment significantly ($p < 0.05$) enhanced activity of SOD (Figure 3C) and CAT (Figure 3D) compared to ROT-injected animals. Further, thymol alone injected animals did not show any remarkable changes in the antioxidant enzymes activity.

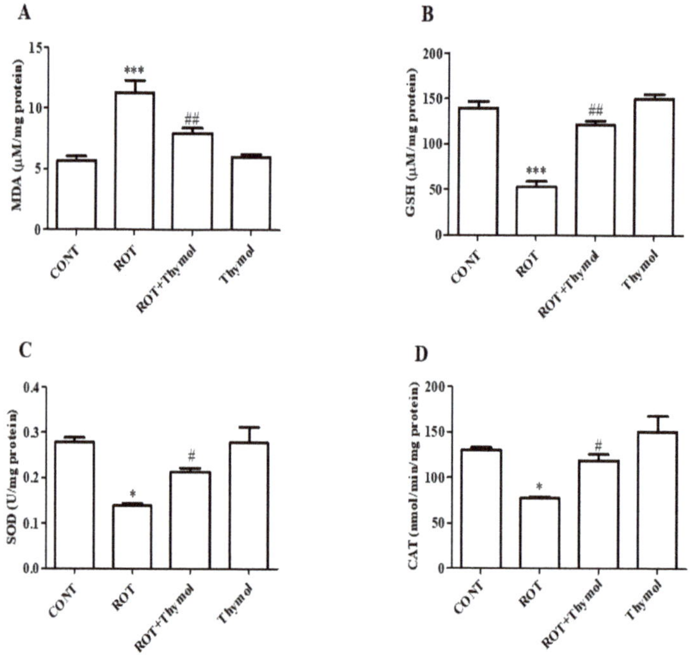

Figure 3. The levels of MDA, GSH and enzymatic activity of SOD and CAT were determined in the mid brain tissues of rats from different experimental groups. ROT treated rats showed significant (*** $p < 0.01$) increase in MDA (**A**) and decrease in GSH (**B**) levels when compared to control rats. Thymol treatment to ROT administered rats showed significantly (## $p < 0.05$) decreased level of MDA and increased (## $p < 0.01$) level of GSH. Moreover, ROT challenge also showed significant (* $p < 0.05$) decreased enzymatic activity of SOD (**C**) and CAT (**D**) when compared CONT rats. Thymol treatment to ROT challenged rats significantly (# $p < 0.05$) increased the activities of SOD and CAT when compared to ROT alone injected rats. The values are presented as mean ± SEM (n = 6–8).

2.3. Thymol Inhibited Activation of Glial Cells

Prolonged and sustained activation of the glial cells induces the release of inflammatory mediators including proinflammatory cytokines and inflammatory enzymes, which amplifies the neuroinflammatory process. We examined ROT-induced glial cells activation (astrocytes and microglia) in the striatum region. ROT injections significantly ($p < 0.001$) enhanced the expression of glial fibrillary acidic protein (GFAP) and ionized calcium binding adaptor protein (Iba-1) markers, which represent the number of activated astrocytes and microglial cells, respectively (Figure 4A–D). The increased expressions of GFAP and Iba-1 are considered the indices of inflammatory response following the activation of astrocytes and microglia. ROT administration caused a significant ($p < 0.001$) rise in the number of activated astrocytes and microglia as compared to rats received vehicle in control group. However, thymol treatment to ROT-administered rats led to a significant ($p < 0.05$) decrease in the quantity of activated astrocytes and microglial cells. Rats treatment with thymol alone did not exhibit notable activation of astrocytes and microglia when compared to the control animals, that is reasonable suggestive of its relative safety on astrocytes and microglia and aid in to the neuroprotective actions on the neurons.

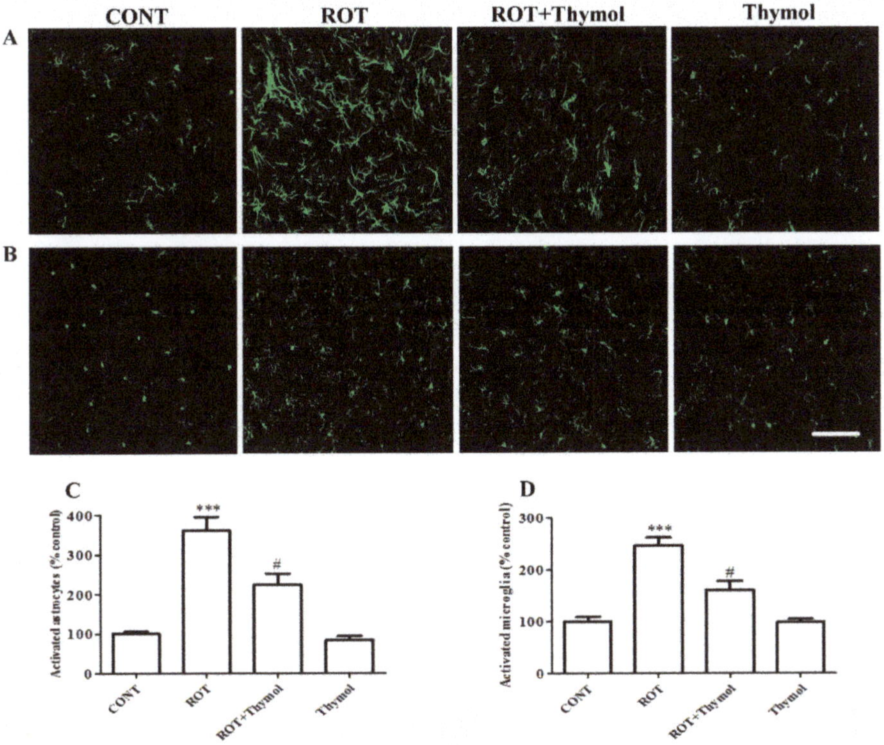

Figure 4. The immunofluorescence staining of GFAP (**A**) and Iba1 (**B**) in the striatum region of different experimental groups. Intense immunoreactivity of GFAP positive astrocytes (**A**) and Iba-1 positive microglia (**B**) were observed in the ROT challenged rats as compared to CONT rats. Thymol treatment to ROT challenged rats exhibited modest staining of GFAP and Iba-1 when compared to rats injected ROT (Scale bar 200 μm). Quantification data showed significant (*** $p < 0.001$) increased percentage number of activated astrocytes (**C**) and microglia (**D**) in ROT injected animals when compared to CONT rats. However, thymol treatment to ROT injected rats showed significantly (# $p < 0.05$) reduced percentage number of activated astrocytes and microglia as compared to rats injected with ROT alone. The values are presented as percent mean± SEM ($n = 3$).

2.4. Thymol Attenuated Activation of Proinflammatory Cytokines

The increased secretion of inflammatory mediators, including proinflammatory cytokines, plays a key role in the etiopathogenesis and progression of PD. Therefore, the level of proinflammatory cytokines such as IL-1β, IL-6, and TNF-α, were quantified in ROT-challenged rats. A significant ($p < 0.001$) increase in the levels of IL-1β, IL-6, and TNF-α, were observed in ROT challenged rats compared to vehicle treated control rats (Figure 5A–C). However, thymol treatment to ROT injected rats significantly reduced the levels of IL-1β ($p < 0.01$), IL-6 ($p < 0.05$), and TNF-α ($p < 0.01$) compared to ROT alone injected animals (Figure 5A–C). The rats received thymol only did not cause substantial change in the level of proinflammatory cytokines compared to vehicle treated control animals.

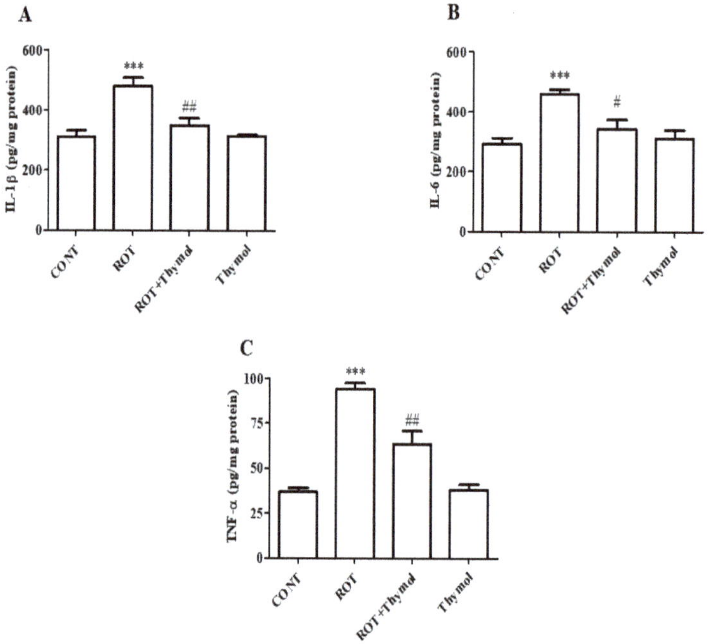

Figure 5. ELISA was used to quantify the level of proinflammatory cytokines; IL-1β, IL-6 and TNF-α in the mid brain tissues of rats from different experimental groups. The levels of IL-1β (**A**), IL-6 (**B**) and TNF-α (**C**) were significantly (*** $p < 0.001$) enhanced in ROT challenged rats when compared to CONT group rats. Thymol treatment to ROT challenged rats showed a significant (## $p < 0.01$; # $p < 0.05$) decrease in the levels of ROT-induced rise of proinflammatory cytokines. Additionally, the cytokines levels did no show significant difference in the rats of CONT and thymol alone groups. The values are presented as mean ± SEM (n = 6–8).

2.5. Thymol Attenuated Expression Levels of COX-2 and iNOS

We also examined the protein expression of inflammatory enzyme mediators such as COX-2 and iNOS by western blotting (Figure 6A–C). A significant ($p < 0.001$) rise in the expression of COX-2 and iNOS was observed in the striatal tissues of rats challenged with ROT in comparison with the vehicle injected rats in CONT group. Thymol treatment to ROT injected rats showed significantly reduced expression of COX-2 ($p < 0.05$) and iNOS ($p < 0.01$) when compared to rats challenged with ROT alone. However, the rats received thymol only was not found to produce significant alteration in the expression of COX-2 and iNOS compared to vehicle injected rats in CONT group.

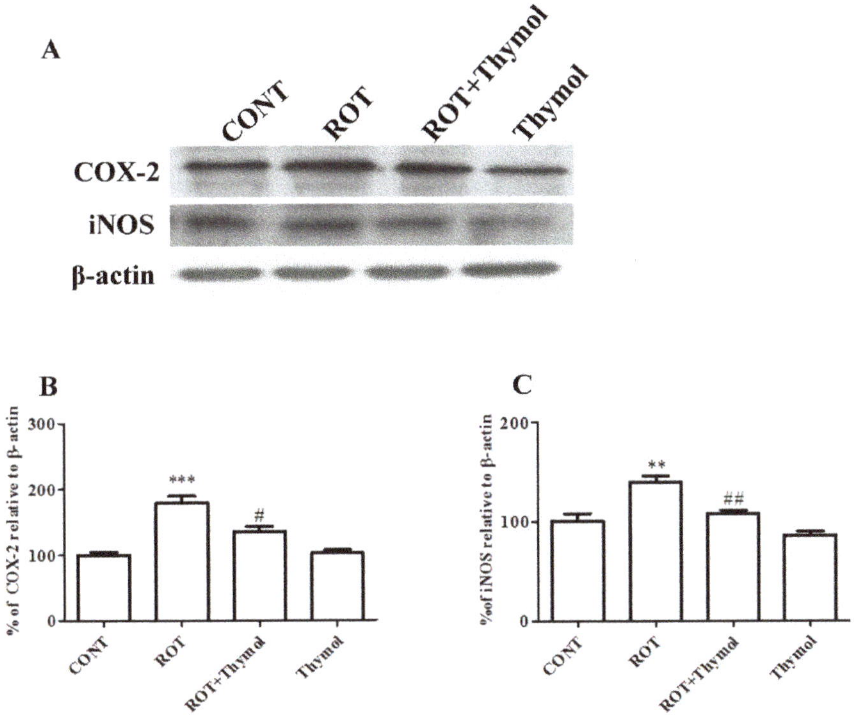

Figure 6. Striatal tissues were used to determine the expression levels of COX-2 and iNOS using western blotting (**A**). ROT challenge causes significant (*** $p < 0.001$) increase in COX-2 and iNOS levels when compared to CONT rats. Thymol treatment to ROT challenged rats exhibited significant (# $p < 0.05$; ## $p < 0.01$) decrease in the expression levels ofCOX-2 and iNOS as compared to rats received only ROT (**B**,**C**). Thymol alone treatment did not exhibit noteworthy change in the expression of COX-2 and iNOS when compared to vehicle injected rats of CONT group ($n = 4$).

3. Discussion

The results of the present study demonstrate that thymol protect against ROT-induced neurodegeneration, mediating antioxidant and anti-inflammatory actions. The ROT model of neurodegeneration in rats is seemingly used as an experimental model for the assessment of agents for preventive and therapeutic efficacy and understanding the pathogenesis of PD [20,21]. The widespread activation of the microglia was observed in both the SNc and striatum following ROT challenge [18] and this appears consistent with the biochemical changes in the inflammatory mediators found in idiopathic PD [25,26], supporting the ROT model of PD. ROT induces nigrostriatal dopaminergic toxicity to mimic most of the pathological features of human PD including dopaminergic neurons loss, oxidative and nitrosative stress, impairment of the ubiquitin proteasome system and mitochondrial function along with α-synuclein aggregation and behavioral abnormalities [16,17,19].

Experimental and epidemiological studies suggest the health promoting properties and therapeutic benefits of numerous plant extracts, as well as their bioactive constituents, popularly known as phytochemicals, against various human diseases including PD [8,9,23,24]. Many phytochemicals have been found effective in treating numerous neurodegenerative diseases including PD [9,23,24,27]. Despite numerous pharmacological studies, there is no report available for the preventive or therapeutic potential of thymol against ROT induced neurodegeneration in in rats as an experimental model of PD. Additionally, thymol was found to inhibit β-amyloid (Aβ)-induced cognitive impairments in rats [28] that suggests thymol crosses the blood brain barrier and achieve the concentrations sufficient enough to exert its therapeutic effects on neurons. Therefore, in the current study we examined the neuroprotective role of thymol against ROT-induced neurodegeneration.

In the present study, a four-week regimen of ROT injections induced a significant degeneration of TH+ dopaminergic neurons in the SNc region and dopaminergic nerve fibers in the striatum of brain. TH+ neurons in the SNc region project their nerve terminals to the striatum. Therefore, the degeneration of dopaminergic neurons in the SNc area results in the diminution of dopaminergic nerve fibers/terminal in the striatum region. The loss of dopaminergic neurons and nerve terminals is reflected as one of the main pathological indices of PD. Importantly, thymol treatment protected the ROT-injected animals from the diminution of dopaminergic neurons and nerve terminals that is clearly suggestive of the neuroprotective effects of thymol against ROT-induced neurodegeneration.

ROT being highly lipophilic in nature easily crosses the blood brain barrier independent of any transporter and diffuses into neurons, accumulates in mitochondria and inhibits complex I. Mitochondrial complex I inhibition leads to loss of ATP production and subsequent rise in the ROS levels resulting oxidative stress [18,20,21]. Over generation of free radicals including ROS causes lipid peroxidation that is considered a crucial event in the etiopathogenesis of PD and an abnormal rise in the formation of MDA, a stable lipid peroxidation product, that has been shown in experimental and human studies [29,30]. Considerably, the brain tissues are highly susceptible to oxidative damage due to higher fatty acid contents, increased ROS level, and lessened endogenous enzymatic and non-enzymatic antioxidant defense components. We observed that thymol treatment significantly inhibited lipid peroxidation evidenced by reduced MDA levels in the midbrain tissues, which was induced by ROT injections and is suggestive of thymol's lipid peroxidation inhibitory activity. The perturbation of endogenous non-enzymatic and enzymatic antioxidant defenses, such as GSH and SOD or CAT, has been well demonstrated in the brain tissues of experimental models and human PD [30]. The imbalance between the endogenous antioxidant defense system and ROS-induced oxidative stress is often linked with a simultaneous reduction in the GSH levels in the brain tissues with a concomitant fall in the activity of the intracellular antioxidant enzymes, SOD and CAT. To demonstrate the action of thymol on antioxidant defenses, we measured the activity of enzymatic antioxidants, SOD and CAT, and the level of non-enzymatic antioxidants, GSH. The administration of ROT induced a significant depletion of the levels of GSH and reduction in the activities of antioxidant enzymes SOD and CAT, whereas thymol treatment significantly restored the activity of antioxidant enzymes evidenced by improved antioxidant activity and prevented the depletion of GSH. This is suggestive of that thymol mitigates ROT-induced oxidative damage in brains attributed to its potent antioxidant and free radical scavenging properties. The reason for potent antioxidant and free radical scavenging property of thymol is ascribed to the presence of a phenolic hydroxyl group in its chemical structure that is believed to accountable for absorbing or neutralizing free radicals and augmenting endogenous antioxidants in protection against the deleterious effects of free radicals [31].

Chronic low grade sustained neuroinflammation is a contributing element of many neurodegenerative diseases including PD [4]. Neuroinflammation involves the activation of glial cells and secretion of classic inflammatory mediators such as proinflammatory cytokines and inflammatory enzymes; COX-2 and iNOS [32]. Given the crucial role of neuroinflammation in the onset and progression of PD, numerous studies so far have demonstrated the potential usefulness of anti-inflammatory drugs to decrease the development of neurodegeneration and lessen the risk factors for the individuals developing PD [33,34]. Though, the potential adverse effects of anti-inflammatory drugs limit their therapeutic use. Thymol has been shown to reduce inflammation by mitigating the onset and progression of the inflammatory processes in different experimental models of human diseases and appear safe in terms of adverse effects [35–37].

Therefore, we measured the levels of proinflammatory cytokines (IL-1β, IL-6, and TNF-α) in brain tissues of rats challenged with ROT. We observed that thymol treatment significantly reduced the release and activation of proinflammatory cytokines as evidenced by reduced levels in brain tissues of the rats challenged with ROT. Elevation in the activity and secretions of the proinflammatory cytokines, TNF-α, IL-1β, and IL-6 showed to participate in dopaminergic neurotoxicity and amplify the deleterious cascade of neurodegeneration in PD [38]. We also observed that thymol treatment significantly decreased the number of activated astrocytes and microglia in the striatum region in ROT-injected animals. The reduction in the number of glial cells following thymol treatment in ROT challenged rats is suggestive of its anti-inflammatory effects. Additionally, we also measured the expression of inflammatory enzymes mediators such as iNOS and COX-2, which rises following the induction of proinflammatory cytokines and increase in NF-κB, a transcription factor, in PD brains [39]. The COX-2 enzyme, an important physiologic and constitutive component of arachidonic acid metabolism pathway leads to the oxidation of dopamine to form dopamine-quinone conjugate that react with cysteinyl residues in proteins causes the alterations in protein structure and function [40]. These alterations further result in to neuronal cell death and suggested to be one of the probable explanations for the protective effect of COX-2 ablation [41].

Furthermore, activated glial cells, which express iNOS, are believed to enhance the levels of nitric oxide (NO) [5,42]. NO causes inhibition of the activity of several enzymes of the mitochondrial electron transport chain and leads the augmented generation of ROS. The crucial role of NO in PD pathogenesis is convincingly demonstrated in immunohistochemical studies performed on postmortem brain tissues that displays enhanced expression of iNOS in basal ganglia structures [43]. The current study findings shows that ROT injections elicited a remarkable increase in the expression of COX-2 and iNOS in the striatum, compared to control animals. However, the animals that received thymol treatment exhibited reduced expression of COX-2 and iNOS that is clearly suggestive of the potent anti-inflammatory effects of thymol.

4. Materials and Methods

4.1. Drugs and Chemicals

The antibodies used in this study included polyclonal rabbit anti-tyrosine hydroxylase (Novus Biologicals, Littleton, CO, USA), polyclonal rabbit anti-inducible nitric oxide synthase (iNOS), anti-cyclooxygenase-2 (COX-2), and anti-glial fibrillary acidic protein (GFAP) (Abcam, Cambridge, MA, USA), polyclonal rabbit anti-ionized calcium binding adaptor molecule-1 (Iba-1) (Wako Chemicals, Richmond, VA, USA), biotinylated secondary anti-rabbit antibody (Jackson Immunoresearch, West Grove, PA, USA), and Alexa fluor 488-conjugated goat anti-rabbit secondary antibodies (Life Technologies, Grand Island, NY, USA). The test compound, thymol was procured from Santa Cruz Biotechnology Inc, CA, USA. ROT, the chemical to induce PD in rats were purchased from Sigma Aldrich, St. Louis, MO, USA. The ELISA assay kits for antioxidant enzymes and glutathione (GSH) as well as other analytical grade reagents were also obtained from Sigma Aldrich, St. Louis, MO, USA.

4.2. Experimental Animals

The animal experiments were performed on five to six months old male adult albino Wistar rats weighing between 280–300 g. All the animals used in this study were provided by the animal research facility of College of Medicine and Health Sciences, United Arab Emirates University, Al Ain, United Arab Emirates. The animals were housed in polyacrylic cages under standard experimental animal housing conditions. The animals were maintained on a 12 h light/dark cycle and food and water was fed ad libitum. The animal experiments were performed following the guidelines and approval of Animal Ethics Committee of United Arab Emirates University, United Arab Emirates (ERA_2017_5500).

4.3. Experimental Design

In order to induce PD in rats, ROT was injected intraperitoneally once daily for 4 weeks with a dosage of 2.5 mg/kg body weight. The doses and schedule of ROT used for PD induction in rats in the present study was similar to that previously described and published report with slight modifications [22–24]. Briefly, a stock solution of 50× ROT was prepared in dimethyl sulfoxide and was used at a concentration of 2.5 mg/mL after dilution of stock in sunflower oil as vehicle. Thymol was prepared after dilution in sunflower oil at a concentration of 50 mg/2mL. The dose of thymol was selected based on previous studies [44,45] and was used at 50 mg/kg body weight through intraperitoneal injection 30 min prior to ROT challenge, once in a daily for a total of 4 weeks. The animals injected with same amount of oil only (vehicle) were designated as controls.

The animals were grouped in the following four categories as independent experimental groups of eight rats each. Group I: rats received vehicle injections, designated as normal control group (CONT). Group II: rats received rotenone and vehicle injections, designated as ROT group (ROT). Group III: rats received thymol 30 min prior to rotenone and vehicle injections, designated as thymol-treated group (ROT + Thymol). Group IV: rats received thymol injections alone, designated as thymol group (Thymol).

4.4. Tissue Collection

Animals of all the experimental groups were euthanized 48 h after the final administration of thymol or ROT to ensure a sufficient washout period. Prior to their sacrifice, animals received intraperitoneal injections of anesthesia pentobarbital (40 mg/kg body weight) followed by cardiac perfusion using phosphate-buffered saline (0.01 M, pH 7.4) to wash out the blood. Following perfusion, the brain was removed quickly, and the two hemispheres were separated. The midbrain and the striatum region were dissected out on ice from one of the hemisphere and the tissue was snap frozen under liquid nitrogen until further use. The other hemisphere was fixed with 4% paraformaldehyde solution for 48 h and subsequently exchanged with 10% sucrose solution three times a day for three consecutive days at 4 °C prior to cryostat sectioning.

4.5. Sample Preparation for Biochemical Studies

Tissue samples (mid brain) were prepared after lysis of the frozen midbrain tissues in KCL buffer supplemented with cocktail of protease and phosphatase inhibitor using a hand held tissue homogenizer separately for each group. The homogenate of each sample was centrifuged at 14,000 *g* for 20 min at 4 °C to get the post-mitochondrial supernatant for the quantification of endogenous enzymatic and no-enzymatic antioxidants, markers of lipid peroxidation, and levels of proinflammatory cytokines employing spectrophotometric assessment and enzyme-linked immunosorbent assay (ELISA).

4.6. Assessment of Lipid Peroxidation and Glutathione

The markers of lipid peroxidation, malondialdehyde (MDA) (North West Life science Vancouver, WA, USA) and glutathione (Sigma Aldrich, St. Louis, MO, USA), were estimated following the manufacturer's protocol provided with the kit. The data are presented as μM/mg protein.

4.7. Assessment of Antioxidant Enzymes Activity

The activities of endogenous antioxidant enzymes such as superoxide dismutase (SOD) and catalase (CAT) were estimated following the protocols prescribed in manufacturer's kits (Cayman Chemicals Company, Ann Arbor, MI, USA). The activities of SOD and CAT are expressed as U/mg protein, and nmol/min/mg protein, respectively.

4.8. Estimation of Proinflammatory Cytokines

ELISA assays were carried out in order to determine the quantity of proinflammatory cytokines such as interleukin-1β (IL-1β), interleukin-6 (IL-6), and tumor necrosis factor-alpha (TNF-α) in midbrain tissues, following the manufacturer's protocol provided with the kits (R&D Systems, Minneapolis, MN, USA). The data are presented as pg/mg protein.

4.9. Immunohistochemistry of Tyrosine Hydroxylase (TH)

The immunohistochemical staining for TH was performed as published before [23,24]. Briefly, 14-μm thick coronal brain sections were sliced out at the level of the striatum and SNc using a cryostat (Leica, Wetzlar, Germany) and TH+ neurons in the SNc and TH-ir fibers in the striatum were evaluated following a method as described previously [23,24]. The loss of TH+ neurons in the SNc area after ROT administration was determined by enumerating the TH+ neurons at three different levels (section) (−4.8, −5.04, and −5.28 mm from the bregma) of the SNc region from each rat. A total three sections of each level and three rats per group were included in the analysis and the average count for each group is represented as a percentage. Therefore, in total, nine sections per group were analyzed for TH+ neurons. The differences in the optical density of TH-ir dopaminergic fibers in the striatum was measured using Image J software (NIH, Bethesda, MD, USA) in three different fields of each section (three sections/rat $n = 3$) with equal areas (adjacent to 0.3 mm from the bregma). An average of the three sections was calculated and is presented as a percentage compared to the control group. As background, the optical density was measured from the overlying cortex and the values obtained were subtracted from the values obtained for striatum. An investigator who was masked to the experimental groups and treatment was assigned to perform the enumeration of TH+ neurons and measurement of the optical density of the TH-ir fibers.

4.10. Immunofluorescence Staining of GFAP and Iba-1

Immunofluorescence microscopy was employed on 14-μm thick striatum sections to examine GFAP positive astrocytes and Iba-1 positive microglia using previously published protocols [23,24].

4.11. Determination of Activated Astrocytes and Microglia in the Striatum

In order to analyze the number of activated astrocytes and microglia, at least three coronal sections from a similar size of striatum from each animal and total three animals per group were utilized. The enumeration of activated astrocytes and microglia was undertaken based on the immunostaining intensity for GFAP and Iba-1 respectively and exhibiting morphological characteristics of hypertrophy and extended glial processes. The quantification of activated astrocytes and microglia were performed using Image J software (NIH, Bethesda, MD, USA) on the three randomly chosen equal area of different fields in each section.

4.12. Western Blot Analysis of COX-2 and iNOS

The tissues dissected from striatum of each experimental group were homogenized in 1X RIPA buffer supplemented with cocktail inhibitor of protease and phosphatase. The crude lysate was centrifuged at 14,000 rpm for 20 min in a refrigerated micro-centrifuge. A total of 35 μg of protein from each tissue sample was electrophoresed on a 10% SDS-polyacrylamide gel following a protocol as published before [23,24]. The blots were quantitated using image J software (NIH, Bethesda, USA).

4.13. Protein Estimation

The quantity of protein in samples were measured employing the Pierce BCA protein assay following the manufacturer's instructions provided with the kit (Thermo Fisher Scientific, Rockford, IL, USA).

4.14. Statistical Analyses

The results are presented as the mean ± SEM. Statistical analysis were made using one-way analysis of variance (ANOVA) followed by Tukey's test to calculate the statistical significance of differences between various groups including immunohistochemical cell/fiber count data. The data with *p*-values < 0.05 were considered significant.

5. Conclusions

Taken altogether, the present study clearly demonstrates that thymol provides protection against ROT-induced dopaminergic neurodegeneration, and the neuroprotective effects are attributed to the antioxidant and anti-inflammatory properties of thymol. Based on the findings of this study, it can be suggested that thymol or the herbs rich in thymol could be useful in the prevention of neurodegeneration in PD. Nonetheless, the translation of beneficial effects in humans and identification of the exact molecular mechanisms require further investigation.

Author Contributions: All the authors provided significant scholarly input from ideation to conduct and manuscript writing, reviewed the contents and approved the manuscript. Conceptualized the idea: S.O., S.A.A. Conducted the animal experiments, collected and processed samples and analyzed the data: M.F.N.M., S.A., H.J. First draft of the manuscript: H.J., S.A.A., S.O. Revision of the manuscript: H.J., S.A.A., S.O.

Funding: This research was funded by the United Arab Emirates University, Al Ain, UAE (center based interdisciplinary grant # 31R127).

Conflicts of Interest: The authors declare no conflict of interest. The funders have no role in study design, analysis and report writing.

References

1. Cacabelos, R. Parkinson's Disease: From Pathogenesis to Pharmacogenomics. *Int. J. Mol. Sci.* **2017**, *18*, 551. [CrossRef]
2. Van Bulck, M.; Sierra-Magro, A.; Alarcon-Gil, J.; Perez-Castillo, A.; Morales-Garcia, J.A. Novel Approaches for the Treatment of Alzheimer's and Parkinson's Disease. *Int. J. Mol. Sci.* **2019**, *20*, 719. [CrossRef]
3. Moreira, P.I.; Zhu, X.; Wang, X.; Lee, H.G.; Nunomura, A.; Petersen, R.B.; Perry, G.; Smith, M.A. Mitochondria: A therapeutic target in neurodegeneration. *Biochem. Biophys. Acta* **2010**, *1802*, 212–220. [CrossRef] [PubMed]
4. Glass, C.K.; Saijo, K.; Winner, B.; Marchetto, M.C.; Gage, F.H. Mechanisms underlying inflammation in neurodegeneration. *Cell* **2010**, *140*, 918–934. [CrossRef] [PubMed]
5. Mander, P.; Brown, G.C. Activation of microglial NADPH oxidase is synergistic with glial iNOS expression in inducing neuronal death: A dual-key mechanism of inflammatory neurodegeneration. *J. Neuroinflamm.* **2005**, *2*, 20. [CrossRef]
6. Sanchez-Pernaute, R.; Ferree, A.; Cooper, O.; Yu, M.; Liisa Brownell, A.; Isacson, O. Selective COX-2 inhibition prevents progressive dopamine neuron degeneration in a rat model of Parkinson's disease. *J. Neuroinflamm.* **2004**, *1*, 6. [CrossRef]
7. Hirsch, E.C.; Breidert, T.; Rousselet, E.; Hunot, S.; Hartmann, A.; Michel, P.P. The role of glial reaction and inflammation in Parkinson's disease. *Ann. N. Y. Acad Sci.* **2003**, *991*, 214–228. [CrossRef]
8. Kujawska, M.; Jodynis-Liebert, J. Polyphenols in Parkinson's Disease: A Systematic Review of In Vivo Studies. *Nutrients* **2018**, *10*, 642. [CrossRef] [PubMed]
9. Shahpiri, Z.; Bahramsoltani, R.; Hosein Farzaei, M.; Farzaei, F.; Rahimi, R. Phytochemicals as future drugs for Parkinson's disease: A comprehensive review. *Rev. Neurosci.* **2016**, *27*, 651–668. [CrossRef] [PubMed]
10. Nagoor Meeran, M.F.; Javed, H.; Al Taee, H.; Azimullah, S.; Ojha, S.K. Pharmacological Properties and Molecular Mechanisms of Thymol: Prospects for Its Therapeutic Potential and Pharmaceutical Development. *Front. Pharmacol.* **2017**, *8*, 380. [CrossRef]
11. Yanishlieva, N.V.; Marinova, E.M.; Gordon, M.H.; Ranevaa, V.G. Antioxidant activity and mechanism of action of thymol and carvacrol in two lipid systems. *Food Chem.* **1999**, *64*, 59–66. [CrossRef]
12. Aeschbach, R.; Loliger, J.; Scott, B.C.; Murcia, A.; Butler, J.; Halliwell, B.; Aruoma, O.I. Antioxidant actions of thymol, carvacrol, 6-gingerol, zingerone and hydroxytyrosol. *Food Chem. Toxicol.* **1994**, *32*, 31–36. [CrossRef]

13. Zahin, M.; Ahmad, I.; Aqil, F. Antioxidant and antimutagenic activity of Carum copticum fruit extracts. *Toxicol. In Vitro* **2010**, *24*, 1243–1249. [CrossRef] [PubMed]
14. Ozen, T.; Demirtas, I.; Aksit, H. Determination of antioxidant activities of various extracts and essential oil compositions of *Thymus praecox* subsp. skorpilii var. skorpilii. *Food Chem.* **2011**, *124*, 58–64. [CrossRef]
15. Karpanen, T.J.; Worthington, T.; Hendry, E.R.; Conway, B.R.; Lambert, P.A. Antimicrobial efficacy of chlorhexidine digluconate alone and in combination with eucalyptus oil, tea tree oil and thymol against planktonic and biofilm cultures of Staphylococcus epidermidis. *J. Antimicrob. Chemother.* **2008**, *62*, 1031–1036. [CrossRef]
16. Littlejohn, D.; ManFano, E.; Clarke, M.; Bobyn, J.; Moloney, K.; Haylay, S. Inflammatory mechanisms of neurodegeneration in toxin-based models of Parkinson's disease. *Parkinsons Dis.* **2010**, *2011*, 713517.
17. Cannon, J.R.; Tapias, V.; Na, H.M.; Honick, A.S.; Drolet, R.E.; Greenamyre, J.T. A highly reproducible rotenone model of Parkinson's disease. *Neurobiol. Dis.* **2011**, *34*, 279–290. [CrossRef]
18. Sherer, T.B.; Betarbet, R.; Kim, J.H.; Greenamyre, J.T. Selective microglial activation in the rat rotenone model of Parkinson's disease. *Neurosci. Lett.* **2003**, *341*, 87–90. [CrossRef]
19. Betarbet, R.; Sherer, T.B.; MacKenzie, G.; Garcia-Osuna, M.; Panov, A.V.; Greenamyre, J.T. Chronic systemic pesticide exposure reproduces features of Parkinson's disease. *Nat. Neurosci.* **2000**, *3*, 1301–1306. [CrossRef]
20. Saravanan, K.S.; Sindhu, K.M.; Senthilkumar, K.S.; Mohanakumar, K.P. L-deprenyl protects against rotenone-induced, oxidative stress-mediated dopaminergic neurodegeneration in rats. *Neurochem. Int.* **2006**, *49*, 28–40. [CrossRef]
21. Inden, M.; Kitamura, Y.; Tamaki, A.; Yanagida, T.; Shibaike, T.; Yamamoto, A.; Takata, K.; Yasui, H.; Taira, T.; Ariga, H.; et al. Neuroprotective effect of the antiparkinsonian drug pramipexole against nigrostriatal dopaminergic degeneration in rotenone-treated mice. *Neurochem. Int.* **2009**, *55*, 760–767. [CrossRef]
22. Fujikawa, T.; Kanada, N.; Shimada, A.; Ogata, M.; Suzuki, I.; Hayashi, I.; Nakashima, K. Effect of sesamin in Acanthopanax senticosus HARMS on behavioral dysfunction in rotenone-induced parkinsonian rats. *Biol. Pharm. Bull.* **2005**, *28*, 169–172. [CrossRef]
23. Javed, H.; Azimullah, S.; Haque, M.E.; Ojha, S.K. Cannabinoid Type 2 (CB2) Receptors Activation Protects against Oxidative Stress and Neuroinflammation Associated Dopaminergic Neurodegeneration in Rotenone Model of Parkinson's Disease. *Front. Neurosci.* **2016**, *10*, 321. [CrossRef]
24. Ojha, S.; Javed, H.; Azimullah, S.; Haque, M.E. β-Caryophyllene, a phytocannabinoid attenuates oxidative stress, neuroinflammation, glial activation, and salvages dopaminergic neurons in a rat model of Parkinson disease. *Mol. Cell. Biochem.* **2016**, *418*, 59–70. [CrossRef] [PubMed]
25. Tansey, M.G.; Goldberg, M.S. Neuroinflammation in Parkinson's disease: Its role in neuronal death and implications for therapeutic intervention. *Neurobiol. Dis.* **2010**, *37*, 510–518. [CrossRef]
26. Gerhard, A.; Pavese, N.; Hotton, G.; Turkheimer, F.; Hammers, M.; Es, A.; Eggert, K.; Oertel, W.; Banati, R.B.; Brooks, D.J. In vivo imaging of microglial activation with [11C] (R)-PK11195 PET in idiopathic Parkinson's disease. *Neurobiol. Dis.* **2006**, *21*, 404–412. [CrossRef]
27. Chen, G.; Liu, J.; Jiang, L.; Ran, X.; He, D.; Li, Y.; Huang, B.; Wang, W.; Fu, S. Galangin Reduces the Loss of Dopaminergic Neurons in an LPS-Evoked Model of Parkinson's Disease in Rats. *Int. J. Mol. Sci.* **2017**, *19*, 12. [CrossRef] [PubMed]
28. Azizi, Z.; Ebrahimi, S.; Saadatfar, E.; Kamalinejad, M.; Majlessi, N. Cognitive enhancing activity of thymol and carvacrol in two rat models of dementia. *Behav. Pharmacol.* **2012**, *23*, 241–249. [CrossRef] [PubMed]
29. Jenner, P.; Olanow, C.W. Oxidative stress and the pathogenesis of Parkinson's disease. *Neurology* **1996**, *47*, S161–S170. [CrossRef] [PubMed]
30. Pohl, F.; Kong Thoo Lin, P. The Potential Use of Plant Natural Products and Plant Extracts with Antioxidant Properties for the Prevention/Treatment of Neurodegenerative Diseases: In Vitro, In Vivo and Clinical Trials. *Molecules* **2018**, *23*, 3283. [CrossRef]
31. Wojdylo, A.; Oszmianski, J.; Czemerys, R. Antioxidant activity and phenolic compounds in 32 selected herbs. *Food Chem.* **2007**, *105*, 940–949. [CrossRef]
32. Barone, F.C.; Feuerstein, G.Z. Inflammatory Mediators and Stroke: New Opportunities for Novel Therapeutics. *J. Cereb. Blood Flow Metab.* **1999**, *19*, 819–834. [CrossRef] [PubMed]
33. Sita, G.; Hrelia, P.; Tarozzi, A.; Morroni, F. Isothiocyanates Are Promising Compounds against Oxidative Stress, Neuroinflammation and Cell Death that May Benefit Neurodegeneration in Parkinson's Disease. *Int. J. Mol. Sci.* **2016**, *17*, 1454. [CrossRef]

34. Song, I.U.; Kim, Y.D.; Cho, H.J.; Chung, S.W. Is Neuroinflammation Involved in the Development of Dementia in Patients with Parkinson's disease? *Intern. Med.* **2013**, *52*, 1787–1792. [CrossRef]
35. Nagoor Meeran, M.F.; Jagadeesh, G.S.; Selvaraj, P. Thymol, a dietary monoterpene phenol abrogates mitochondrial dysfunction in β-adrenergic agonist induced myocardial infarcted rats by inhibiting oxidative stress. *Chem. Biol. Interact.* **2016**, *244*, 159–168. [CrossRef] [PubMed]
36. Nagoor Meeran, M.F.; Jagadeesh, G.S.; Selvaraj, P. Thymol attenuates inflammation in isoproterenol induced myocardial infarcted rats by inhibiting the release of lysosome enzymes and downregulating the expressions of proinflammatory cytokines. *Eur. J. Pharmacol.* **2015**, *753*, 153–161. [CrossRef] [PubMed]
37. Riella, K.R.; Marinho, R.R.; Santos, J.S.; Pereira-Filho, R.N.; Cardoso, J.C.; Albuquerque-Junior, R.L.; Thomazzi, S.M. Anti-inflammatory and cicatrizing activities of thymol, a monoterpene of the essential oil from *Lippia gracilis*, in rodents. *J. Ethnopharmacol.* **2012**, *143*, 656–663. [CrossRef]
38. Boyko, A.A.; Troyanova, N.I.; Kovalenko, E.I.; Sapozhnikov, A.M. Similarity and Differences in Inflammation-Related Characteristics of the Peripheral Immune System of Patients with Parkinson's and Alzheimer's Diseases. *Int. J. Mol. Sci.* **2017**, *18*, 2633. [CrossRef]
39. Hunot, S.; Brugg, B.; Ricard, D.; Michel, P.P.; Muriel, M.P.; Ruberg, M.; Faucheux, B.A.; Agid, Y.; Hirsch, E.C. Nuclear translocation of NF-kappaB is increased in dopaminergic neurons of patients with parkinson disease. *Proc. Natl. Acad. Sci. USA* **1997**, *94*, 7531–7536. [CrossRef]
40. Hastings, T.G. Enzymatic oxidation of dopamine: The role of prostaglandin H synthase. *J. Neurochem.* **1995**, *64*, 919–924. [CrossRef] [PubMed]
41. Teismann, P.; Tieu, K.; Choi, D.K.; Wu, D.C.; Naini, A.; Hunot, S.; Vila, M.; Jackson-Lewis, V.; Przedborski, S. Cyclooxygenase-2 is instrumental in Parkinson's disease neurodegeneration. *Proc. Natl. Acad. Sci. USA* **2003**, *100*, 5473–5478. [CrossRef]
42. Bal-Price, A.; Brown, G.C. Inflammatory neurodegeneration mediated by nitric oxide from activated glia-inhibiting neuronal respiration, causing glutamate release and excitotoxicity. *J. Neurosci.* **2001**, *21*, 6480–6491. [CrossRef]
43. Eve, D.J.; Nisbet, A.P.; Kingsbury, A.E.; Hewson, E.L.; Daniel, S.E.; Lees, A.J.; Marsden, C.D.; Foster, O.J. Basal ganglia neuronal nitric oxide synthase mRNA expression in Parkinson's disease. *Brain Res. Mol. Brain Res.* **1998**, *63*, 62–71. [CrossRef]
44. Aliabadi, A.; Izadi, M.; Rezvani, M.E.; Esmaeili-Dehaj, M. Effects of thymol on serum biochemical and antioxidant indices in kindled rats. *Int. J. Med. Lab.* **2016**, *3*, 43–49.
45. Mottawie, H.; Ibrahiem, A.; Amer, H. Role of some phytochemicals in the treatment of bronchial asthma. *Med. J. Cairo Univ.* **2011**, *79*, 75–80.

International Journal of *Molecular Sciences*

MDPI

Review

Natural Products as Modulators of the Proteostasis Machinery: Implications in Neurodegenerative Diseases

Karina Cuanalo-Contreras [1,*,†] and Ines Moreno-Gonzalez [1,2,3,*]

1 The Mitchell Center for Alzheimer's Disease and Related Brain Disorders, Department of Neurology, The University of Texas Houston Health Science Center at Houston, Houston, TX 77030, USA
2 Departamento Biologia Celular, Genetica y Fisiologia, Instituto de Investigacion Biomedica de Malaga-IBIMA, Facultad de Ciencias, Universidad de Malaga, 28031 Madrid, Spain
3 Networking Research Center on Neurodegenerative Diseases (CIBERNED), 28031 Madrid, Spain
* Correspondence: karina.contreras@northwestern.edu (K.C.-C.); inesmoreno@uma.es (I.M.-G.); Tel.: +847-491-3714 (K.C.-C.); +34-952-131-935 (I.M.-G.); Fax: +847-491-4461 (K.C.-C.); +34-952-131-937 (I.M.-G.)
† Current address: Department of Molecular Biosciences, Rice Institute for Biomedical Research, Northwestern University, Evanston, IL 60208, USA.

Received: 28 August 2019; Accepted: 15 September 2019; Published: 20 September 2019

Abstract: Proteins play crucial and diverse roles within the cell. To exert their biological function they must fold to acquire an appropriate three-dimensional conformation. Once their function is fulfilled, they need to be properly degraded to hamper any possible damage. Protein homeostasis or proteostasis comprises a complex interconnected network that regulates different steps of the protein quality control, from synthesis and folding, to degradation. Due to the primary role of proteins in cellular function, the integrity of this network is critical to assure functionality and health across lifespan. Proteostasis failure has been reported in the context of aging and neurodegeneration, such as Alzheimer's and Parkinson's disease. Therefore, targeting the proteostasis elements emerges as a promising neuroprotective therapeutic approach to prevent or ameliorate the progression of these disorders. A variety of natural products are known to be neuroprotective by protein homeostasis interaction. In this review, we will focus on the current knowledge regarding the use of natural products as modulators of different components of the proteostasis machinery within the framework of age-associated neurodegenerative diseases.

Keywords: proteostasis; neurodegeneration; chaperones; autophagy; ubiquitin-proteasome; unfolded protein response; natural compounds

---

1. Proteostasis Failure in Aging and Neurodegenerative Diseases

The proteostasis network is composed of a series of interconnected elements that assure correct protein functionality and degradation [1]. It starts when polypeptide chains are synthetized in the ribosome and fold with the help of chaperones and co-chaperones. Newly folded proteins are transported to their appropriate locations and once their life cycle finishes, they are degraded either by the ubiquitin proteasome system (UPS) or the autophagy machinery. Proteostasis network imbalance plays a key -if not causative- role in many age-related pathologies [2]. Age is the most relevant risk factor for neurodegenerative diseases including Alzheimer's disease (AD), Parkinson disease (PD), frontotemporal dementia (FTD) and several other forms of proteinopathies [3]. Although there is no consensus in the field regarding the molecular mechanisms that explain their augmented incidence in the elderly brain, a common feature of all these diseases is the accumulation of abnormal protein aggregates in the form of oligomers and inclusions, suggesting that general mechanisms controlling proteostasis may underlay the etiology of these diseases [4]. Recent hypotheses suggest that a progressive reduction in the repair capacity of the proteostasis network may generate a "pathological aging" that results in protein aggregation and higher incidence of neurodegenerative disease [5–8]. Cerebral aging involves a range of cellular and molecular alterations related to proteostasis impairment such as increased oxidative stress [9], altered autophagy machinery [10], accumulation of ubiquitinated protein aggregates [11], and impaired signaling by numerous neurotransmitters and neurotrophic factors [12]. The endoplasmic reticulum (ER) is an essential compartment of the proteostasis network, which is also disturbed by the aging process [4]. Importantly, functional studies indicate that altered proteostasis at the level of the ER is one of the major contributors to aging [4,13]. Several harmful stimuli, such oxidative stress and disturbances in the secretory pathway may lead to accumulation of unfolded or misfolded proteins at the ER lumen, thus activating the ER stress response [14].

The most prominent pathological hallmarks of AD are the extracellular accumulation of amyloid β (Aβ) peptides in the form of plaques and the intracellular accumulation of hyper-phosphorylated tau (ptau) proteins as neurofibrillary tangles (NFTs) [15], whereas in PD, α-synuclein tends to misfold and accumulate inside dopaminergic neurons, leading to Lewy bodies formation [16]. Formation of misfolded proteins as oligomers, proto-fibrils and fibrils leads to the accumulation of amyloid deposition and spreading to affected areas [17,18]. Several intrinsic and extrinsic factors that alter proteostasis cause a decreased protein quality control, contributing to the accumulation of damaged proteins. If not rescued, this condition can lead to protein misfolding disorders, such as AD and PD [19–23]. For instance, a growing amount of evidence indicate that the activity of the molecular chaperones -Hsp60, Hsp70 and Hsp90- is compromised in age-related neurodegenerative diseases [24,25]. The fact that the expression of Hsp60 and Hsp70 is decreased in AD animal models [26], suggests that impairments in the folding pathways play a key role in promoting age-related neurodegeneration. In prion diseases, reduction of the molecular chaperone GRP78/BiP expression leads to the acceleration of the pathology [27]. Alterations in the major protein degradation pathways have a major involvement as well. For instance, the reduction in the activity of the UPS through the manipulation of various UPS components (Rpt2, Rpt3, ubiquitin) causes deposition of pathological misfolded proteins and subsequent neurodegeneration in experimental models, resembling what is observed in AD and PD [28–30]. In addition, neurodegenerative diseases have in common autophagic failure [31,32]. The inhibition of the autophagy response is known to exacerbate protein toxicity and accelerate disease progression [33–35]. The genetic and pharmacological activation of the autophagy has shown to improve the clearance of AD and PD misfolded aggregated proteins [36–38]. Therefore, one can conclude that boosting up the elements of the proteostasis machinery is a promising broad-spectrum therapeutic approach, with the potential to treat or revert not only age-associated neurodegeneration, but a variety of protein misfolding disorders.

2. Chaperone System

Chaperones are highly conserved proteins that assist and mediate the achievement of the proper three-dimensional conformation of proteins. They bind and stabilize unfolded polypeptides, aiding their folding during synthesis and inter-organelle transport [39]. Chaperones play important roles during stress response, hence they are known as heat shock proteins (Hsp). Hsp are classified by their molecular mass (Hsp32, Hsp27, Hsp40, Hsp60, Hsp70, Hsp90). They possess a substrate binding domain that transiently binds to hydrophobic regions of polypeptides, shielding them from undesired intermolecular interactions that could interfere with their adequate folding [40]. The capacity of the Hsp is overloaded during chronic cellular stress, proteotoxic conditions and disease. For instance, Hsp failure has been observed in the context of neurodegenerative disorders, such as AD, PD and Huntington's disease [41]. Notably, the solely over-expression of different Hsp members has been able to rescue in vivo neuronal toxicity in different models [42–45]. With this in mind, the pharmacological activation of Hsp represents an interesting therapeutic approach to treat neurodegeneration.

To date, several natural products have been identified as Hsp modulators. Among them, the potent phytochemical curcumin, a polyphenol of the plant *Curcumin longa*, has shown the ability to induce the in vitro (rat glioma cells, rat liver cells, and mouse fibroblasts) and in vivo (heat-stressed rats) expression of Hsp27 and Hsp70 under proteotoxic conditions, through the formation of an intermediate form of Hsf1 (heat shock factor 1) [46,47]. Although the administration of curcumin in animal models of neurodegenerative diseases has proven to be beneficial [48,49] and has no major side effects in humans, some clinical trials show no evidence of efficacy in ameliorating memory impairment nor reducing levels of amyloid in blood, suggesting a low bioavailability of curcumin following oral administration [50,51]. Several other nutraceuticals have the ability to boost the chaperone system, such as the proanthocyanidins present in cranberry extract. When administered to an AD nematode model, they delayed Aβ toxicity through the activation of Hsf1, which is a master regulator of Hsp expression [52]. Another interesting phytochemical is celastrol (extracted from the thunder god vine, *Tripterygium wilfordii*). Celastrol administration to aged mature cortical cultures induced the expression of Hsp70, Hsp32 and Hsp27 [53]. To highlight the in vivo neuroprotective activity of this natural product, intraperitoneal as well as subcutaneous administration in AD mice reduced Aβ pathology [54]. No clinical trials have been performed using cranberry extract or celastrol to treat AD.

Paeoniflorin is an herbal compound isolated from the perennial flowering plant *Paeonia lactiflora* and the fern *Salvinia molesta*. This phytochemical bears the ability to induce Hsp expression through activation of Hsf1 and promotes thermotolerance in mammalian cell culture as well [55]. Another major constituent of the same herbal medicines is Glycyrrhizin, which can be found in the liquorice root. Several properties have been attributed to Glycyrrhizin, such as antiviral, anti-inflammatory, and anti-allergic. In fact, it has been tested in over 20 different clinical trials related with liver diseases with positive outcomes, but none of them evaluated its effect in neurodegenerative diseases. In the case of the heat shock response, Glycyrrhizin is not able to promote the expression of Hsp itself, however it enhances their induction, making it an interesting compound that could potentially be used in combination with activators of the heat shock response [55]. Some natural occurring antibiotics have Hsp induction properties too. Geldanamycin is a 1,4-benzoquinone ansamycin natural antibiotic compound isolated from the bacterial species *Streptomyces hygroscopicus*. When administered to mammalian cells expressing huntingtin exon 1 protein, it induces the expression of Hsp40, Hsp70 and Hsp90. The consequent activation of the heat shock response causes a marked inhibition on huntingtin aggregation [56]. In patients with primary brain tumor or brain metastases, geldanamycin induces Hsp70 with minimal toxicity [57]. Therefore, this compound bears the potential to treat disease-associated protein aggregation. Another antibiotic compound isolated from *Streptomyces* is herbimycin-A. Herbimicyn-A has the ability to induce the expression of Hsp72 thereby protecting cell cultures from heat stress [58]. Radicicol is a natural macrocyclic compound biosynthesized and isolated from the nematophagous fungi *Pochonia chlamydosporia*. This compound protects primary cell cultures against stressful conditions, by inducing the heat shock response in a HSF-1 related manner, following a similar mechanism than Herbimicyn and Geldanamycin [59]. In addition to this natural antibiotics, several other compounds have shown the ability to boost the chaperone system. While there are not reports yet on their action in the context of neurodegeneration or even clinical trials, they represent promising candidates to restore proteostasis balance and may have potential to delay the onset or treat diseases such AD or PD. One example is withaferin, a lactone derived from the plant *Vassobia breviflora*. Withaferin enhanced the heat shock response through Hsp70, Hsp32 and Hsp27 upregulation in a cancer model [60] and it is reported to ameliorate symptoms in schizophrenia patients with minimal side effects [61]. Shikonin is another potential candidate to treat proteinopaties. Its ability to induce Hsp70 in a human lymphoma cell model was discovered through a screening of chemical inducers derived from medicinal plants. Shikonin is present in the roots of *Lithospermum erythrorhizon* and it bears antibacterial, anti-inflammatory and anticancer activities as well [62]. Edible gastropods seem to be an interesting source of compounds with potential to modulate proteostasis response. As an example, the derivative 6-bromoindirubin-3-oxime, an indirubin present in mollusks, increased proteasome subunits and Hsp70 expression, with a consequent increase in healthspan and lifespan in *Drosophila* [63]. Few clinical trials have tested the efficacy and safety of indirubins, such as indigo naturalis extract. Although it can be considered a safe therapy [64], these studies have been tested in psoriasis patients and therefore, its bioavailability remains unknown.

3. Autophagy

Autophagy is a highly conserved homeostatic clearance mechanism. Is in charge of the degradation of damaged proteins, cytosolic components and organelles. It involves the lysosomal system and contributes to the regulation of metabolism, healthspan and longevity. Cellular autophagy activity is present at basal levels, however is particularly stimulated under stress conditions, as a protective mechanism to assure survival and homeostasis [65,66]. Autophagy impairment has been reported in several pathologies, from neurodegeneration to cancer [67,68]. Autophagy targets the degradation of misfolded aggregated proteins considered hallmarks of different proteinopathies [67,69]. However, during disease, the autophagy machinery fails, with deleterious cellular consequences [31]. Several studies have pinpointed a downregulation of important components of the autophagy pathway during AD and PD, such as Beclin 1 [70], as well as alteration in vesicle trafficking and inhibition of autophagic vesicles [71]. Notably, the genetic and pharmacological induction of autophagy has the ability to reduce the accumulation of misfolded proteins and has been associated to amelioration of these disorders [36,37,72]. In this regard, polyphenolic compounds are known potent activators of the autophagy response. As an illustration, the red wine polyphenol quercetin prevents Aβ associated aggregation and its obnoxious consequences through modulation of autophagy, both in nematodes [73] and murine models of AD [74] and PD [75]. Currently, there is a clinical trial to determine the brain penetration of quercetin to potentially treat AD patients using a senolytic therapy. Kaempferol is another potent polyphenol found in different dietary sources such as grapes and tomatoes. In vitro kaempferol treatment increases LC3-II, an autophagosome-bound microtubule-associated protein, and preserved the stratial glutamatergic response in a rat model of PD, positioning this natural product as an important enhancer of autophagy with promising therapeutic applications [76]. Interestingly, caffeine elevates LC3-II levels as well and has proven protective actions against AD and PD [77,78]. In fact, some studies suggest that drinking coffee may be associated with a decreased risk to develop AD and PD [79–81], however, no evidence has been obtained from randomized controlled trials about the beneficial effect of caffeine in neurodegenerative diseases to our knowledge. Resveratrol is another compound of interest present in grapes and berries. The fact that it can cross the blood-brain-barrier makes it an interesting candidate to treat neurodegeneration [82]. Among the many reported activities of resveratrol, it activates autophagy by up-regulating Sirtuin 1, a potent inductor of autophagy [83,84]. Moreover, in a clinical trial performed in AD patients, resveratrol modulates Aβ deposition and reduces inflammatory markers with no side effects [85,86]. In addition to this dietary sources, there is a growing amount of evidence demonstrating the beneficial effects of Mediterranean diet on age-associated neurodegeneration [87]. Olive oil is a significant component of this dietary regimen. Olive oil is enriched with the polyphenol oleuropein aglycone. The administration of oleuropein aglycone improved cognition and reduced amyloid deposition in a transgenic AD mouse model, mainly through activation of the autophagy [88]. A multitude of studies have study the effect of olive oil in combination with Mediterranean diet in an effort to evaluate its effect in patients with cognitive decline and dementia [89], including AD and PD, but none of them analyzed the capability of oleuropein aglycone to cross the blood-brain barrier (BBB), tolerance, biodistribution or its effect in treating neurodegenerative disorders. Another dietary molecule, present in high quantities on mushrooms and aged cheese, is spermidine. This compound induces autophagy and delays aging, the main risk factor for AD and PD, in humans and mice [90,91]. Glycoconjugate metabolites isolated from traditional medicine remedies are an interesting group of phytochemical compounds with properties to activate autophagy. For example, the ginseng derived steroid glycoside Rg2 is a potent inducer of in vitro and in vivo autophagy in an AMPK-ULK1 dependent [92]. In the same line, a derivative chemical compound from the root ginseng, 1-(3,4-dimethoxyphenethyl)-3-(3-dehydroxyl-20(s)-protopanaxadiol-3β-yl)-urea (DDPU), improved cognition and promoted neuroprotection in the APP/PS1 mouse model of AD. No clinical trials have been reported. DDPU targets different branches of the proteostasis network, as it has activity on both the ER stress and autophagy [93]. Berberine is a natural alkaloid isolated from *Rhizoma coptidis*, a traditional Chinese herbal medicine, with high distribution when administered orally, including the CNS, in pre-clinical studies [94]. When berberine was orally administered to a triple-transgenic AD mouse model, it promoted Aβ clearance through autophagy by increasing the levels of LC3-II. Phenotipically, berberine treatment significantly improved spatial learning and memory retention in the treated animals [95]. Corynoxine B joins the list of natural alkaloid molecules with autophagy-inducer properties in cellular and mouse AD models. Corynoxine B is an oxindole alkaloid present in the medicinal plant *Uncaria rhynchophylla*, a widely used Chinese traditional remedy. This compound was tested in cells expressing the APP_{Swe} mutation and intraperitoneally administered once a day to Tg2576 mice at 8 months of age. Corynoxine B treatment reduced Aβ levels by increasing LC3-II, lysosomal activation and changes in APP [96]. Surprisingly, the source of compounds with potential anti-neurodegenerative capacity is not limited to the ground. The study of marine organisms has helped to identify several compounds with the ability to modulate proteostasis. Among them, chromomycin A2, psammaplin A, and ilimaquinone induced the expression of autophagy, in the context of cancer [97]. It would be extremely interesting to test their effect on neurodegeneration, both in vitro and *in vivo*, as it will expand the sources of therapeutic molecules. As stated, autophagy is a major player in the cellular response to stress and turnover of damaged proteins. In view of its potential, targeting autophagy through the use of natural products is an emerging and promising field that requires further exploration.

4. Ubiquitin Proteasome System

The ubiquitin proteasome system (UPS) is the main responsible for degrading intracellular damaged proteins. Briefly, a subset of enzymes is involved in ubiquitin-tag the proteins that need to be degraded, this tag is then recognized by the proteasome -a multi-subunit barrel complex- for its proteolytic degradation [98]. Several natural compounds have been widely explored for their ability to decrease the activity of the UPS, especially in the context of cancer research [99]. However, relatively few have been studied for their capacity to activate the UPS. The mechanisms of action among them vary, for example, the natural compounds olein, linoleic acid, linolenic acid, ceramides, and oleuropein increase proteasome activity by exerting conformational changes that promote the entry of the substrate into the proteolytic chamber [100]. A derivative of linoleic acid has been reported to cross the BBB, tolerable, and safe, but specific studies to determine its potential to treat dementia are still needed [101]. On the other hand, dietary intake of linolenic acid seems to have no effect in other brain disorders such as stroke [102]. It remains to be determined whether this is due to a low brain penetrance of the compound in the CNS or lack of therapeutic potential in this specific disorder. Other natural molecules activate the proteasome by enhancing its catalytic activities, such as the lipid fraction of the algae *Phaeodactylum tricornutum* and the triterpene betulinic acid [103–105]. Two clinical trials are currently ongoing to determine the safety, tolerability and effectiveness of betulinic acid. The compounds present in the Chinese traditional herb *Corydalis bungeana* boost in vivo proteasomal activity by upregulation of the regulatory subunits [106]. In the same line, the polyphenol resveratrol enhances proteasome activity through increase on the expression of proteasome subunits and proteolysis in the brain of AD transgenic mice, protecting them against memory loss and enhancing cognition [107]. Quercetin is another polyphenolic compound that exhibits in vivo enhancing proteasome activity [108] and reduces Aβ-induced toxicity in a dose-dependent manner when administered to a *Caenorhabditis elegans* AD model [73]. Since impaired UPS activity is one of the main features present in all protein misfolding disorders, it will be interesting to explore the natural chemical space in the lookout for more activators.

5. Unfolded Protein Response

Three branches of a conserved signaling pathway collectively termed as the unfolded protein response (UPR) are triggered in response to the ER stress: ATF6 (activating transcription factor 6), PERK (PKR-like kinase), and IRE1 (inositol-requiring enzyme 1) [14]. UPR activation results in global protein synthesis reduction [109] and upregulation of genes involved in protein folding [14], which facilitates proper protein folding, therefore arresting protein aggregation. In the brain of AD, PD and FTD patients, levels of UPR markers are elevated [110,111]. This could represent an emergency response triggered by the ability of misfolded proteins to induce neuronal ER stress and activate the UPR [112]. However, when the ER response is chronically activated, proteostasis cannot be restored with devastating consequences for the brain, leading to synaptic impairment and neurodegeneration. In this line of thoughts, recent studies indicate that reduction of ER stress with chemical chaperones alleviate synapse and memory loss in experimental models of AD [111,112]. Levels of eIF2α phosphorylation are elevated in AD brains. PERK regulation decreases eIF2α phosphorylation levels and ameliorates memory impairment in AD and prion-infected mice [113,114]. On the other hand, activation of PERK increases tau phosphorylation [115], as well as ptau activates UPR [116]. IRE1 leads to the expression of XBP1 (X-box binding protein 1) that upregulates the expression of chaperones, increasing the size of the ER and promoting the degradation of misfolded proteins through the proteasome system [14,117]. It has been recently described that IRE1 signaling promotes AD progression whereas its deletion ameliorates learning and memory impairment as well as reduces amyloid deposition [118]. Furthermore, tau and Aβ accumulation has been also found associated with UPR activation by inhibition of ATF6 and ER-associated degradation (ERAD), likely through soluble oligomeric forms [116,119,120].

Few natural occurring compounds have been explored in regards to their ability to modulate the UPR in the context of neurodegeneration. Among them, Bajijiasu, a dimeric fructose isolated from the Chinese medicinal herb *Morinda officinalis*, has shown to exert protection against Aβ induced neurotoxicity by attenuation of ER stress in the hippocampus and cortex of APP/PS1 mice [121]. No clinical trials have been reported evaluating this compound. Kaempferol is phytoestrogen and one of the main components of *Ginkgo biloba* extract with the ability to inhibit ER stress and protect cells against apoptosis by upregulation of CHOP mRNA levels in vitro [122]. Clinical trial using *G. biloba* extracts indicate its symptomatic beneficial effects in patients with MCI, AD, and related dementia [123–125]. Honokiol is a promising biphenolic lignan isolated from the Magnolia tree that can cross the blood brain barrier and therefore represents an interesting candidate to treat neurodegeneration due to its high bioavailability. This lignan modulated ER-stress in the brain of mice, and reduced the levels of proinflammatory cytokines as well [126]. Its tolerance, safeness, biodistribution, and effectiveness has not been tested yet to treat brain disorders. More research is needed to evaluate the effect on neurodegeneration of other known modulators of ER-stress, as the current literature is quite limited. A special focus should be made on compounds with the ability to cross the blood brain barrier that can effectively target the cells that are compromised in this diseases.

6. Conclusions

Aging is the main risk factor for a variety of neurodegenerative disorders, such as AD and PD. Recent studies indicate that there is a dramatic age-associated collapse of proteostasis responses, leaving the cells vulnerable to physiological and environmental stressors, and more susceptible to disease. In the case of diseases associated with protein misfolding, the proteostasis machinery takes initial care of the aberrant protein aggregates. However, as the clearance ability gets compromised, the accumulated aggregates cause cellular toxicity, tissue dysfunction, and disease. Therefore, boosting up the proteostasis machinery by the use of natural compounds emerges as a potent pharmacological tool with promising effects to treat and protect against neurodegenerative disorders. In this study we compile a list of natural modulators of the proteostasis network (Figure 1). Not surprisingly, majority of them are of plant-origin. However it is remarkable to note that we report some compounds of marine-animal-origin as well. It is indeed necessary to explore more alternative sources of natural compounds. In addition, further studies are required to understand the precise mechanism of action of the natural proteostasis activators, their off-target effects and their in vivo bioavailability. We foresee that the development of innovative, natural and safe therapeutic strategies to tackle the accumulation of misfolded protein aggregates through the modulation of the proteostasis machinery, will have exceptional effects to prevent and treat disorders related to age-dependent protein aggregation.

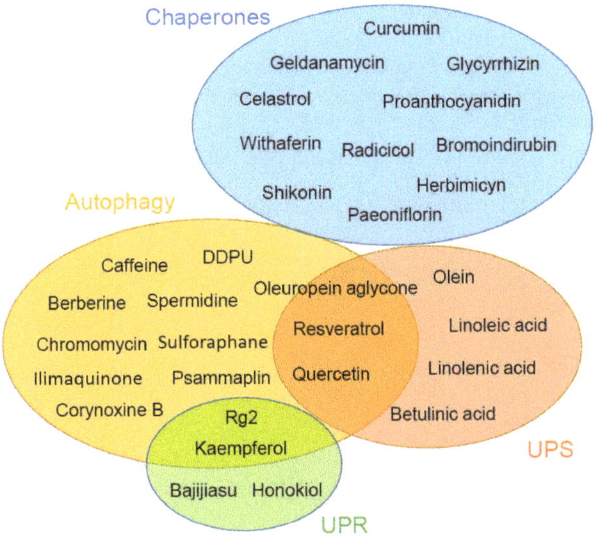

Figure 1. Schematic representation of natural compounds that positively regulate different elements of the proteostasis machinery. There is an extensive heterogeneity of chemical classes that compose the proteostasis-enhancing compounds, however we observed an enrichment in polyphenolic molecules. It is noted that oleuropein aglycone, resveratrol, and quercetin target the autophagy and the UPS, suggesting that they could be used as strong activators to restore the proteostasis network during aging and disease, whereas chaperones' modifiers seem to exclusively interfere with this pathway.

Author Contributions: Conceptualization, K.C.-C. and I.M.-G.; writing—original draft preparation, K.C.-C. writing—review and editing, I.M.-G.; funding acquisition, I.M.-G."

Funding: This research was funded by the Texas Alzheimer's Council on Disease and Related Disorders 2018-51-93-JI to I.M.-G., the Alzheimer's Association New Investigator Research Grant NIRG-394284 to I.M.-G. and the Alzheimer's Association Postdoctoral Fellowship AARFD-19-564718 to K.C.-C.

Conflicts of Interest: The authors declare no conflict of interest.

References

1. Sala, A.J.; Bott, L.C.; Morimoto, R.I. Shaping proteostasis at the cellular, tissue, and organismal level. *J. Cell Biol.* **2017**, *216*, 1231–1241. [CrossRef] [PubMed]
2. Klaips, C.L.; Jayaraj, G.G.; Hartl, F.U. Pathways of cellular proteostasis in aging and disease. *J. Cell Biol.* **2018**, *217*, 51–63. [CrossRef] [PubMed]
3. Douglas, P.M.; Dillin, A. Protein homeostasis and aging in neurodegeneration. *J. Cell Biol.* **2010**, *190*, 719–729. [CrossRef] [PubMed]
4. Kaushik, S.; Cuervo, A.M. Proteostasis and aging. *Nat. Med.* **2015**, *21*, 1406–1415. [CrossRef] [PubMed]
5. Martínez, G.; Duran-Aniotz, C.; Cabral-Miranda, F.; Vivar, J.P.; Hetz, C. Endoplasmic reticulum proteostasis impairment in aging. *Aging Cell* **2017**, *16*, 615–623. [CrossRef] [PubMed]
6. López-Otín, C.; Blasco, M.A.; Partridge, L.; Serrano, M.; Kroemer, G. The hallmarks of aging. *Cell* **2013**, *153*. [CrossRef] [PubMed]

7. Labbadia, J.; Morimoto, R.I. The Biology of Proteostasis in Aging and Disease. *Annu. Rev. Biochem.* **2015**, *84*, 435–464. [CrossRef]
8. Morimoto, R.I.; Cuervo, A.M. Proteostasis and the aging proteome in health and disease. *J. Gerontol Ser. A Biol. Sci. Med. Sci.* **2014**, *69*, S33–S38. [CrossRef]
9. Scheff, S.W.; Ansari, M.A.; Mufson, E.J. Oxidative stress and hippocampal synaptic protein levels in elderly cognitively intact individuals with Alzheimer's disease pathology. *Neurobiol. Aging* **2016**, *42*, 1–12. [CrossRef]
10. Webb, A.E.; Brunet, A. FOXO transcription factors: Key regulators of cellular quality control. *Trends Biochem. Sci.* **2014**, *39*, 159–169. [CrossRef]
11. Gavilán, M.P.; Pintado, C.; Gavilán, E.; Jiménez, S.; Ríos, R.M.; Vitorica, J.; Castaño, A.; Ruano, D. Dysfunction of the unfolded protein response increases neurodegeneration in aged rat hippocampus following proteasome inhibition. *Aging Cell* **2009**, *8*, 654–665. [CrossRef] [PubMed]
12. Leal, S.L.; Yassa, M.A. Neurocognitive Aging and the Hippocampus across Species. *Trends Neurosci.* **2015**, *38*, 800–812. [CrossRef] [PubMed]
13. Labbadia, J.; Morimoto, R.I. Proteostasis and longevity: When does aging really begin? *F1000Prime Rep.* **2014**, *6*. [CrossRef]
14. Hetz, C. The unfolded protein response: Controlling cell fate decisions under ER stress and beyond. *Nat. Rev. Mol. Cell Biol.* **2012**, *13*, 89–102. [CrossRef] [PubMed]
15. Serrano-Pozo, A.; Frosch, M.P.; Masliah, E.; Hyman, B.T. Neuropathological alterations in Alzheimer disease. *Cold Spring Harb. Perspect. Med.* **2011**, *1*, a006189. [CrossRef] [PubMed]
16. Schlossmacher, M.G.; Frosch, M.P.; Gai, W.P.; Medina, M.; Sharma, N.; Forno, L.; Ochiishi, T.; Shimura, H.; Sharon, R.; Hattori, N.; et al. Parkin localizes to the Lewy bodies of Parkinson disease and dementia with Lewy bodies. *Am. J. Pathol.* **2002**, *160*, 1655–1667. [CrossRef]
17. Moreno-Gonzalez, I.; Soto, C. Misfolded protein aggregates: Mechanisms, structures and potential for disease transmission. *Semin. Cell Dev. Biol.* **2011**, *22*, 482–487. [CrossRef]
18. Duran-Aniotz, C.; Moreno-Gonzalez, I.; Morales, R. Agregados amiloides: Rol en desórdenes de conformación proteica. *Rev. Med. Chil.* **2013**, *141*, 495–505. [CrossRef]
19. Cheng, J.; North, B.J.; Zhang, T.; Dai, X.; Tao, K.; Guo, J.; Wei, W. The emerging roles of protein homeostasis-governing pathways in Alzheimer's disease. *Aging Cell* **2018**, *17*, e12801. [CrossRef]
20. Scheper, W.; Nijholt, D.A.T.; Hoozemans, J.J.M. The unfolded protein response and proteostasis in Alzheimer disease: Preferential activation of autophagy by endoplasmic reticulum stress. *Autophagy* **2011**, *7*, 910–911. [CrossRef]
21. Lehtonen, Š.; Sonninen, T.-M.; Wojciechowski, S.; Goldsteins, G.; Koistinaho, J. Dysfunction of Cellular Proteostasis in Parkinson's Disease. *Front. Neurosci.* **2019**, *13*, 457. [CrossRef] [PubMed]
22. Ben-Zvi, A.; Miller, E.A.; Morimoto, R.I. Collapse of proteostasis represents an early molecular event in Caenorhabditis elegans aging. *Proc. Natl. Acad. Sci. USA* **2009**, *106*, 14914–14919. [CrossRef] [PubMed]
23. Morawe, T.; Hiebel, C.; Kern, A.; Behl, C. Protein homeostasis, aging and Alzheimer's disease. *Mol. Neurobiol.* **2012**, *46*, 41–54. [CrossRef]
24. Campanella, C.; Pace, A.; Bavisotto, C.C.; Marzullo, P.; Gammazza, A.M.; Buscemi, S.; Piccionello, A.P. Heat Shock Proteins in Alzheimer's Disease: Role and Targeting. *Int. J. Mol. Sci.* **2018**, *19*, 2603. [CrossRef] [PubMed]
25. Lackie, R.E.; Maciejewski, A.; Ostapchenko, V.G.; Marques-Lopes, J.; Choy, W.Y.; Duennwald, M.L.; Prado, V.F.; Prado, M.A.M. The Hsp70/Hsp90 chaperone machinery in neurodegenerative diseases. *Front. Neurosci.* **2017**, *11*, 1–23. [CrossRef]
26. Jiang, Y.-Q.; Wang, X.-L.; Cao, X.-H.; Ye, Z.-Y.; Li, L.; Cai, W.-Q. Increased heat shock transcription factor 1 in the cerebellum reverses the deficiency of Purkinje cells in Alzheimer's disease. *Brain Res.* **2013**, *1519*, 105–111. [CrossRef]
27. Park, K.W.; Eun Kim, G.; Morales, R.; Moda, F.; Moreno-Gonzalez, I.; Concha-Marambio, L.; Lee, A.S.; Hetz, C.; Soto, C. The Endoplasmic Reticulum Chaperone GRP78/BiP Modulates Prion Propagation in vitro and in vivo. *Sci. Rep.* **2017**, *7*, 44723. [CrossRef]
28. Bedford, L.; Hay, D.; Devoy, A.; Paine, S.; Powe, D.G.; Seth, R.; Gray, T.; Topham, I.; Fone, K.; Rezvani, N.; et al. Depletion of 26S proteasomes in mouse brain neurons causes neurodegeneration and Lewy-like inclusions resembling human pale bodies. *J. Neurosci.* **2008**, *28*, 8189–8198. [CrossRef]

29. Tashiro, Y.; Urushitani, M.; Inoue, H.; Koike, M.; Uchiyama, Y.; Komatsu, M.; Tanaka, K.; Yamazaki, M.; Abe, M.; Misawa, H.; et al. Motor neuron-specific disruption of proteasomes, but not autophagy, replicates amyotrophic lateral sclerosis. *J. Biol. Chem.* **2012**, *287*, 42984–42994. [CrossRef]
30. Fischer, D.F.; van Dijk, R.; van Tijn, P.; Hobo, B.; Verhage, M.C.; van der Schors, R.C.; Li, K.W.; van Minnen, J.; Hol, E.M.; van Leeuwen, F.W. Long-term proteasome dysfunction in the mouse brain by expression of aberrant ubiquitin. *Neurobiol. Aging* **2009**, *30*, 847–863. [CrossRef]
31. Wong, E.; Cuervo, A.M. Autophagy gone awry in neurodegenerative diseases. *Nat. Neurosci.* **2010**, *13*, 805–811. [CrossRef] [PubMed]
32. Sanchez-Varo, R.; Trujillo-Estrada, L.; Sanchez-Mejias, E.; Torres, M.; Baglietto-Vargas, D.; Moreno-Gonzalez, I.; De Castro, V.; Jimenez, S.; Ruano, D.; Vizuete, M.; et al. Abnormal accumulation of autophagic vesicles correlates with axonal and synaptic pathology in young Alzheimer's mice hippocampus. *Acta Neuropathol.* **2012**, *123*, 53–70. [CrossRef] [PubMed]
33. Pickford, F.; Masliah, E.; Britschgi, M.; Lucin, K.; Narasimhan, R.; Jaeger, P.A.; Small, S.; Spencer, B.; Rockenstein, E.; Levine, B.; et al. The autophagy-related protein beclin 1 shows reduced expression in early Alzheimer disease and regulates amyloid beta accumulation in mice. *J. Clin. Invest.* **2008**, *118*, 2190–2199. [PubMed]
34. Komatsu, M.; Waguri, S.; Chiba, T.; Murata, S.; Iwata, J.; Tanida, I.; Ueno, T.; Koike, M.; Uchiyama, Y.; Kominami, E.; et al. Loss of autophagy in the central nervous system causes neurodegeneration in mice. *Nature* **2006**, *441*, 880–884. [CrossRef]
35. Hara, T.; Nakamura, K.; Matsui, M.; Yamamoto, A.; Nakahara, Y.; Suzuki-Migishima, R.; Yokoyama, M.; Mishima, K.; Saito, I.; Okano, H.; et al. Suppression of basal autophagy in neural cells causes neurodegenerative disease in mice. *Nature* **2006**, *441*, 885–889. [CrossRef]
36. Berger, Z.; Ravikumar, B.; Menzies, F.M.; Oroz, L.G.; Underwood, B.R.; Pangalos, M.N.; Schmitt, I.; Wullner, U.; Evert, B.O.; O'Kane, C.J.; et al. Rapamycin alleviates toxicity of different aggregate-prone proteins. *Hum. Mol. Genet.* **2006**, *15*, 433–442. [CrossRef] [PubMed]
37. Ravikumar, B.; Vacher, C.; Berger, Z.; Davies, J.E.; Luo, S.; Oroz, L.G.; Scaravilli, F.; Easton, D.F.; Duden, R.; O'Kane, C.J.; et al. Inhibition of mTOR induces autophagy and reduces toxicity of polyglutamine expansions in fly and mouse models of Huntington disease. *Nat. Genet.* **2004**, *36*, 585–595. [CrossRef] [PubMed]
38. Sarkar, S.; Davies, J.E.; Huang, Z.; Tunnacliffe, A.; Rubinsztein, D.C. Trehalose, a novel mTOR-independent autophagy enhancer, accelerates the clearance of mutant huntingtin and alpha-synuclein. *J. Biol. Chem.* **2007**, *282*, 5641–5652. [CrossRef]
39. Kim, Y.E.; Hipp, M.S.; Bracher, A.; Hayer-Hartl, M.; Hartl, F.U. Molecular Chaperone Functions in Protein Folding and Proteostasis. *Annu. Rev. Biochem.* **2013**, *82*, 323–355. [CrossRef]
40. Saibil, H. Chaperone machines for protein folding, unfolding and disaggregation. *Nat. Rev. Mol. Cell Biol.* **2013**, *14*, 630–642. [CrossRef]
41. Winklhofer, K.F.; Tatzelt, J.; Haass, C. The two faces of protein misfolding: Gain- and loss-of-function in neurodegenerative diseases. *EMBO J.* **2008**, *27*, 336–349. [CrossRef] [PubMed]
42. Fonte, V.; Kipp, D.R.; Yerg, J.; Merin, D.; Forrestal, M.; Wagner, E.; Roberts, C.M.; Link, C.D. Suppression of in vivo β-amyloid peptide toxicity by overexpression of the HSP-16.2 small chaperone protein. *J. Biol. Chem.* **2008**, *283*, 784–791. [CrossRef] [PubMed]
43. Akbar, M.T.; Lundberg, A.M.C.; Liu, K.; Vidyadaran, S.; Wells, K.E.; Dolatshad, H.; Wynn, S.; Wells, D.J.; Latchman, D.S.; de Belleroche, J. The neuroprotective effects of heat shock protein 27 overexpression in transgenic animals against kainate-induced seizures and hippocampal cell death. *J. Biol. Chem.* **2003**, *278*, 19956–19965. [CrossRef] [PubMed]
44. Klucken, J.; Shin, Y.; Masliah, E.; Hyman, B.T.; McLean, P.J. Hsp70 Reduces alpha-Synuclein Aggregation and Toxicity. *J. Biol. Chem.* **2004**, *279*, 25497–25502. [CrossRef] [PubMed]
45. Magrané, J.; Smith, R.C.; Walsh, K.; Querfurth, H.W. Heat shock protein 70 participates in the neuroprotective response to intracellularly expressed beta-amyloid in neurons. *J. Neurosci.* **2004**, *24*, 1700–1706. [CrossRef] [PubMed]
46. Kato, K.; Ito, H.; Kamei, K.; Iwamoto, I. Stimulation of the stress-induced expression of stress proteins by curcumin in cultured cells and in rat tissues in vivo. *Cell Stress Chaperones* **1998**, *3*, 152–160. [CrossRef]
47. Maiti, P.; Manna, J.; Veleri, S.; Frautschy, S. Molecular Chaperone Dysfunction in Neurodegenerative Diseases and Effects of Curcumin. *Biomed Res. Int.* **2014**, *2014*, 1–14. [CrossRef] [PubMed]

48. Begum, A.N.; Jones, M.R.; Lim, G.P.; Morihara, T.; Kim, P.; Heath, D.D.; Rock, C.L.; Pruitt, M.A.; Yang, F.; Hudspeth, B.; et al. Curcumin structure-function, bioavailability, and efficacy in models of neuroinflammation and Alzheimer's disease. *J. Pharmacol. Exp. Ther.* **2008**, *326*, 196–208. [CrossRef] [PubMed]
49. Maiti, P.; Dunbar, G.L. Use of Curcumin, a Natural Polyphenol for Targeting Molecular Pathways in Treating Age-Related Neurodegenerative Diseases. *Int. J. Mol. Sci.* **2018**, *19*, 1637. [CrossRef] [PubMed]
50. Ringman, J.M.; Frautschy, S.A.; Teng, E.; Begum, A.N.; Bardens, J.; Beigi, M.; Gylys, K.H.; Badmaev, V.; Heath, D.D.; Apostolova, L.G.; et al. Oral curcumin for Alzheimer's disease: Tolerability and efficacy in a 24-week randomized, double blind, placebo-controlled study. *Alzheimer's Res. Ther.* **2012**, *4*, 43. [CrossRef] [PubMed]
51. Baum, L.; Lam, C.W.; Cheung, S.K.; Kwok, T.; Lui, V.; Tsoh, J.; Lam, L.; Leung, V.; Hui, E.; Ng, C.; et al. Six-month randomized, placebo-controlled, double-blind, pilot clinical trial of curcumin in patients with Alzheimer's disease. *J. Clin. Psychopharmacol.* **2008**, *28*, 110–113. [CrossRef]
52. Guo, H.; Cao, M.; Zou, S.; Ye, B.; Dong, Y. Cranberry Extract Standardized for Proanthocyanidins Alleviates β-Amyloid Peptide Toxicity by Improving Proteostasis Through HSF-1 in Caenorhabditis elegans Model of Alzheimer's Disease. *J. Gerontol. A Biol. Sci. Med. Sci.* **2016**, *71*, 1564–1573. [CrossRef]
53. Chow, A.M.; Tang, D.W.F.; Hanif, A.; Brown, I.R. Induction of heat shock proteins in cerebral cortical cultures by celastrol. *Cell Stress Chaperones* **2013**, *18*, 155–160. [CrossRef]
54. Paris, D.; Ganey, N.J.; Laporte, V.; Patel, N.S.; Beaulieu-Abdelahad, D.; Bachmeier, C.; March, A.; Ait-Ghezala, G.; Mullan, M.J. Reduction of beta-amyloid pathology by celastrol in a transgenic mouse model of Alzheimer's disease. *J. Neuroinflamm.* **2010**, *7*, 17. [CrossRef]
55. Yan, D.; Saito, K.; Ohmi, Y.; Fujie, N.; Ohtsuka, K. Paeoniflorin, a novel heat shock protein-inducing compound. *Cell Stress Chaperones* **2004**, *9*, 378–389. [CrossRef]
56. Sittler, A. Geldanamycin activates a heat shock response and inhibits huntingtin aggregation in a cell culture model of Huntington's disease. *Hum. Mol. Genet.* **2001**, *10*, 1307–1315. [CrossRef]
57. Banerji, U.; O'Donnell, A.; Scurr, M.; Pacey, S.; Stapleton, S.; Asad, Y.; Simmons, L.; Maloney, A.; Raynaud, F.; Campbell, M.; et al. Phase I pharmacokinetic and pharmacodynamic study of 17-allylamino, 17-demethoxygeldanamycin in patients with advanced malignancies. *J. Clin. Oncol.* **2005**, *23*, 4152–4161. [CrossRef]
58. Murakami, Y.; Uehara, Y.; Yamamoto, C.; Fukazawa, H.; Mizuno, S. Induction of hsp 72/73 by herbimycin A, an inhibitor of transformation by tyrosine kinase oncogenes. *Exp. Cell Res.* **1991**, *195*, 338–344. [CrossRef]
59. Griffin, T.M.; Valdez, T.V.; Mestril, R. Radicicol activates heat shock protein expression and cardioprotection in neonatal rat cardiomyocytes. *Am. J. Physiol. Heart Circ. Physiol.* **2004**, *287*, H1081–H1088. [CrossRef]
60. Grogan, P.T.; Sleder, K.D.; Samadi, A.K.; Zhang, H.; Timmermann, B.N.; Cohen, M.S. Cytotoxicity of withaferin A in glioblastomas involves induction of an oxidative stress-mediated heat shock response while altering Akt/mTOR and MAPK signaling pathways. *Invest. New Drugs* **2013**, *31*, 545–557. [CrossRef]
61. Chengappa, K.N.R.; Brar, J.S.; Gannon, J.M.; Schlicht, P.J. Adjunctive use of a standardized extract of withania somnifera (ashwagandha) to treat symptom exacerbation in schizophrenia: A randomized, double-blind, placebo-controlled study. *J. Clin. Psychiatry* **2018**, *79*. [CrossRef]
62. Ahmed, K.; Furusawa, Y.; Tabuchi, Y.; Emam, H.F.; Piao, J.-L.; Hassan, M.A.; Yamamoto, T.; Kondo, T.; Kadowaki, M. Chemical inducers of heat shock proteins derived from medicinal plants and cytoprotective genes response. *Int. J. Hyperth.* **2012**, *28*, 1–8. [CrossRef]
63. Tsakiri, E.N.; Gaboriaud-Kolar, N.; Iliaki, K.K.; Tchoumtchoua, J.; Papanagnou, E.-D.; Chatzigeorgiou, S.; Tallas, K.D.; Mikros, E.; Halabalaki, M.; Skaltsounis, A.-L.; et al. The Indirubin Derivative 6-Bromoindirubin-3′-Oxime Activates Proteostatic Modules, Reprograms Cellular Bioenergetic Pathways, and Exerts Antiaging Effects. *Antioxid. Redox Signal.* **2017**, *27*, 1027–1047. [CrossRef]
64. Lin, Y.K.; Chang, Y.C.; Hui, R.C.Y.; See, L.C.; Chang, C.J.; Yang, C.H.; Huang, Y.H. A Chinese herb, indigo naturalis, extracted in oil (Lindioil) used topically to treat psoriatic nails: A randomized clinical trial. *JAMA Dermatol.* **2015**, *151*, 672–674. [CrossRef]
65. Cuanalo-Contreras, K.; Mukherjee, A.; Soto, C. Role of protein misfolding and proteostasis deficiency in protein misfolding diseases and aging. *Int. J. Cell Biol.* **2013**, *2013*, 638083. [CrossRef]

66. Giampieri, F.; Afrin, S.; Forbes-Hernandez, T.Y.; Gasparrini, M.; Cianciosi, D.; Reboredo-Rodriguez, P.; Varela-Lopez, A.; Quiles, J.L.; Battino, M. Autophagy in Human Health and Disease: Novel Therapeutic Opportunities. *Antioxid. Redox Signal.* **2019**, *30*, 577–634. [CrossRef]
67. Menzies, F.M.; Fleming, A.; Caricasole, A.; Bento, C.F.; Andrews, S.P.; Ashkenazi, A.; Füllgrabe, J.; Jackson, A.; Jimenez Sanchez, M.; Karabiyik, C.; et al. Autophagy and Neurodegeneration: Pathogenic Mechanisms and Therapeutic Opportunities. *Neuron* **2017**, *93*, 1015–1034. [CrossRef]
68. Mathew, R.; Karantza-Wadsworth, V.; White, E. Role of autophagy in cancer. *Nat. Rev. Cancer* **2007**, *7*, 961–967. [CrossRef]
69. Ciechanover, A.; Kwon, Y.T. Degradation of misfolded proteins in neurodegenerative diseases: Therapeutic targets and strategies. *Exp. Mol. Med.* **2015**, *47*, e147. [CrossRef]
70. Nascimento-Ferreira, I.; Santos-Ferreira, T.; Sousa-Ferreira, L.; Auregan, G.; Onofre, I.; Alves, S.; Dufour, N.; Colomer Gould, V.F.; Koeppen, A.; Déglon, N.; et al. Overexpression of the autophagic beclin-1 protein clears mutant ataxin-3 and alleviates Machado–Joseph disease. *Brain* **2011**, *134*, 1400–1415. [CrossRef]
71. Volpicelli-Daley, L.A.; Gamble, K.L.; Schultheiss, C.E.; Riddle, D.M.; West, A.B.; Lee, V.M.-Y. Formation of α-synuclein Lewy neurite–like aggregates in axons impedes the transport of distinct endosomes. *Mol. Biol. Cell* **2014**, *25*, 4010–4023. [CrossRef] [PubMed]
72. Ravikumar, B. Aggregate-prone proteins with polyglutamine and polyalanine expansions are degraded by autophagy. *Hum. Mol. Genet.* **2002**, *11*, 1107–1117. [CrossRef] [PubMed]
73. Regitz, C.; Dußling, L.M.; Wenzel, U. Amyloid-beta ($A\beta_{1\text{-}42}$)-induced paralysis in Caenorhabditis elegans is inhibited by the polyphenol quercetin through activation of protein degradation pathways. *Mol. Nutr. Food Res.* **2014**, *58*, 1931–1940. [CrossRef] [PubMed]
74. Hayakawa, M.; Itoh, M.; Ohta, K.; Li, S.; Ueda, M.; Wang, M.; Nishida, E.; Islam, S.; Suzuki, C.; Ohzawa, K.; et al. Quercetin reduces eIF2α phosphorylation by GADD34 induction. *Neurobiol. Aging* **2015**, *36*, 2509–2518. [CrossRef] [PubMed]
75. El-Horany, H.E.; El-Latif, R.N.A.; ElBatsh, M.M.; Emam, M.N. Ameliorative Effect of Quercetin on Neurochemical and Behavioral Deficits in Rotenone Rat Model of Parkinson's Disease: Modulating Autophagy (Quercetin on Experimental Parkinson's Disease). *J. Biochem. Mol. Toxicol.* **2016**, *30*, 360–369. [CrossRef]
76. Filomeni, G.; Graziani, I.; de Zio, D.; Dini, L.; Centonze, D.; Rotilio, G.; Ciriolo, M.R. Neuroprotection of kaempferol by autophagy in models of rotenone-mediated acute toxicity: Possible implications for Parkinson's disease. *Neurobiol. Aging* **2012**, *33*, 767–785. [CrossRef]
77. Sinha, R.A.; Farah, B.L.; Singh, B.K.; Siddique, M.M.; Li, Y.; Wu, Y.; Ilkayeva, O.R.; Gooding, J.; Ching, J.; Zhou, J.; et al. Caffeine stimulates hepatic lipid metabolism by the autophagy-lysosomal pathway in mice. *Hepatology* **2014**, *59*, 1366–1380. [CrossRef]
78. Yenisetti, S.; Muralidhara, M. Beneficial Role of Coffee and Caffeine in Neurodegenerative Diseases: A Minireview. *AIMS Public Heal.* **2016**, *3*, 407–422.
79. Eskelinen, M.H.; Kivipelto, M. Caffeine as a protective factor in dementia and Alzheimer's disease. *J. Alzheimer's Dis.* **2010**, *20*, S167–S174. [CrossRef]
80. Santos, C.; Costa, J.; Santos, J.; Vaz-Carneiro, A.; Lunet, N. Caffeine intake and dementia: Systematic review and meta-analysis. *J. Alzheimer's Dis.* **2010**, *20*, S187–S204. [CrossRef]
81. Palacios, N.; Gao, X.; McCullough, M.L.; Schwarzschild, M.A.; Shah, R.; Gapstur, S.; Ascherio, A. Caffeine and risk of Parkinson's disease in a large cohort of men and women. *Mov. Disord.* **2012**, *27*, 1276–1282. [CrossRef] [PubMed]
82. Lange, K.W.; Li, S. Resveratrol, pterostilbene, and dementia. *BioFactors* **2018**, *44*, 83–90. [CrossRef] [PubMed]
83. Lee, I.H.; Cao, L.; Mostoslavsky, R.; Lombard, D.B.; Liu, J.; Bruns, N.E.; Tsokos, M.; Alt, F.W.; Finkel, T. A role for the NAD-dependent deacetylase Sirt1 in the regulation of autophagy. *Proc. Natl. Acad. Sci. USA* **2008**, *105*, 3374–3379. [CrossRef] [PubMed]
84. Borra, M.T.; Smith, B.C.; Denu, J.M. Mechanism of Human SIRT1 Activation by Resveratrol. *J. Biol. Chem.* **2005**, *280*, 17187–17195. [CrossRef] [PubMed]
85. Moussa, C.; Hebron, M.; Huang, X.; Ahn, J.; Rissman, R.A.; Aisen, P.S.; Turner, R.S. Resveratrol regulates neuro-inflammation and induces adaptive immunity in Alzheimer's disease. *J. Neuroinflamm.* **2017**, *14*, 1. [CrossRef]

86. Zhu, C.W.; Grossman, H.; Neugroschl, J.; Parker, S.; Burden, A.; Luo, X.; Sano, M. A randomized, double-blind, placebo-controlled trial of resveratrol with glucose and malate (RGM) to slow the progression of Alzheimer's disease: A pilot study. *Alzheimer's Dement. Transl. Res. Clin. Interv.* **2018**, *4*, 609–616. [CrossRef] [PubMed]
87. McEvoy, C.T.; Guyer, H.; Langa, K.M.; Yaffe, K. Neuroprotective Diets Are Associated with Better Cognitive Function: The Health and Retirement Study. *J. Am. Geriatr. Soc.* **2017**, *65*, 1857–1862. [CrossRef] [PubMed]
88. Grossi, C.; Rigacci, S.; Ambrosini, S.; ed Dami, T.; Luccarini, I.; Traini, C.; Failli, P.; Berti, A.; Casamenti, F.; Stefani, M. The Polyphenol Oleuropein Aglycone Protects TgCRND8 Mice against Aβ Plaque Pathology. *PLoS ONE* **2013**, *8*, e71702. [CrossRef]
89. Omar, S.H. Mediterranean and MIND Diets Containing Olive Biophenols Reduces the Prevalence of Alzheimer's Disease. *Int. J. Mol. Sci.* **2019**, *20*, 2797. [CrossRef]
90. Madeo, F.; Carmona-Gutierrez, D.; Kepp, O.; Kroemer, G. Spermidine delays aging in humans. *Aging (Albany. NY)* **2018**, *10*, 2209–2211. [CrossRef]
91. Eisenberg, T.; Abdellatif, M.; Schroeder, S.; Primessnig, U.; Stekovic, S.; Pendl, T.; Harger, A.; Schipke, J.; Zimmermann, A.; Schmidt, A.; et al. Cardioprotection and lifespan extension by the natural polyamine spermidine. *Nat. Med.* **2016**, *22*, 1428–1438. [CrossRef] [PubMed]
92. Fan, Y.; Wang, N.; Rocchi, A.; Zhang, W.; Vassar, R.; Zhou, Y.; He, C. Identification of natural products with neuronal and metabolic benefits through autophagy induction. *Autophagy* **2017**, *13*, 41–56. [CrossRef] [PubMed]
93. Guo, X.; Lv, J.; Lu, J.; Fan, L.; Huang, X.; Hu, L.; Wang, J.; Shen, X. Protopanaxadiol derivative DDPU improves behavior and cognitive deficit in AD mice involving regulation of both ER stress and autophagy. *Neuropharmacology* **2018**, *130*, 77–91. [CrossRef] [PubMed]
94. Imenshahidi, M.; Hosseinzadeh, H. Berberine and barberry (Berberis vulgaris): A clinical review. *Phyther. Res.* **2019**, *33*, 504–523. [CrossRef] [PubMed]
95. Huang, M.; Jiang, X.; Liang, Y.; Liu, Q.; Chen, S.; Guo, Y. Berberine improves cognitive impairment by promoting autophagic clearance and inhibiting production of β-amyloid in APP/tau/PS1 mouse model of Alzheimer's disease. *Exp. Gerontol.* **2017**, *91*, 25–33. [CrossRef]
96. Durairajan, S.; Huang, Y.; Chen, L.; Song, J.; Liu, L.; Li, M. Corynoxine isomers decrease levels of amyloid-β peptide and amyloid-β precursor protein by promoting autophagy and lysosome biogenesis. *Mol. Neurodegener.* **2013**, *8*, P16. [CrossRef]
97. Ratovitski, E.A. Tumor Protein (TP)-p53 Members as Regulators of Autophagy in Tumor Cells upon Marine Drug Exposure. *Mar. Drugs* **2016**, *14*, 154. [CrossRef]
98. Papaevgeniou, N.; Chondrogianni, N. The ubiquitin proteasome system in Caenorhabditis elegans and its regulation. *Redox Biol.* **2014**, *2*, 333–347. [CrossRef]
99. Yang, H.; Landis-Piwowar, K.R.; Chen, D.; Milacic, V.; Dou, Q.P. Natural compounds with proteasome inhibitory activity for cancer prevention and treatment. *Curr. Protein Pept. Sci.* **2008**, *9*, 227–239. [CrossRef]
100. Chondrogianni, N.; Gonos, E.S. Proteasome activation as a novel antiaging strategy. *IUBMB Life* **2008**, *60*, 651–655. [CrossRef]
101. Zesiewicz, T.; Heerinckx, F.; de Jager, R.; Omidvar, O.; Kilpatrick, M.; Shaw, J.; Shchepinov, M.S. Randomized, clinical trial of RT001: Early signals of efficacy in Friedreich's ataxia. *Mov. Disord.* **2018**, *33*, 1000–1005. [CrossRef] [PubMed]
102. Bork, C.S.; Venø, S.K.; Lundbye-Christensen, S.; Jakobsen, M.U.; Tjønneland, A.; Schmidt, E.B.; Overvad, K. Dietary intake of α-linolenic acid is not appreciably associated with risk of ischemic stroke among middle-aged Danish men and women. *J. Nutr.* **2018**, *148*, 952–958. [CrossRef] [PubMed]
103. Grune, T.; Reinheckel, T.; Davies, K.J. Degradation of oxidized proteins in mammalian cells. *FASEB J.* **1997**, *11*, 526–534. [CrossRef] [PubMed]
104. Chondrogianni, N.; Petropoulos, I.; Grimm, S.; Georgila, K.; Catalgol, B.; Friguet, B.; Grune, T.; Gonos, E.S. Protein damage, repair and proteolysis. *Mol. Aspects Med.* **2014**, *35*, 1–71. [CrossRef]
105. Huang, L.; Ho, P.; Chen, C.-H. Activation and inhibition of the proteasome by betulinic acid and its derivatives. *FEBS Lett.* **2007**, *581*, 4955–4959. [CrossRef] [PubMed]
106. Fu, R.-H.; Wang, Y.-C.; Chen, C.-S.; Tsai, R.-T.; Liu, S.-P.; Chang, W.-L.; Lin, H.-L.; Lu, C.-H.; Wei, J.-R.; Wang, Z.-W.; et al. Acetylcorynoline attenuates dopaminergic neuron degeneration and α-synuclein aggregation in animal models of Parkinson's disease. *Neuropharmacology* **2014**, *82*, 108–120. [CrossRef]

107. Corpas, R.; Griñán-Ferré, C.; Rodríguez-Farré, E.; Pallàs, M.; Sanfeliu, C. Resveratrol Induces Brain Resilience Against Alzheimer Neurodegeneration Through Proteostasis Enhancement. *Mol. Neurobiol.* **2019**, *56*, 1502–1516. [CrossRef]
108. Fitzenberger, E.; Deusing, D.J.; Marx, C.; Boll, M.; Lüersen, K.; Wenzel, U. The polyphenol quercetin protects the *mev-1* mutant of *Caenorhabditis elegans* from glucose-induced reduction of survival under heat-stress depending on SIR-2.1, DAF-12, and proteasomal activity. *Mol. Nutr. Food Res.* **2014**, *58*, 984–994. [CrossRef]
109. Yamamoto, K.; Sato, T.; Matsui, T.; Sato, M.; Okada, T.; Yoshida, H.; Harada, A.; Mori, K. Transcriptional Induction of Mammalian ER Quality Control Proteins Is Mediated by Single or Combined Action of ATF6α and XBP1. *Dev. Cell* **2007**, *13*, 365–376. [CrossRef]
110. Roussel, B.D.; Kruppa, A.J.; Miranda, E.; Crowther, D.C.; Lomas, D.A.; Marciniak, S.J. Endoplasmic reticulum dysfunction in neurological disease. *Lancet Neurol.* **2013**, *12*, 105–118. [CrossRef]
111. Scheper, W.; Hoozemans, J.J.M. The unfolded protein response in neurodegenerative diseases: A neuropathological perspective. *Acta Neuropathol.* **2015**, *130*, 315–331. [CrossRef] [PubMed]
112. Lourenco, M.V.; Clarke, J.R.; Frozza, R.L.; Bomfim, T.R.; Forny-Germano, L.; Batista, A.F.; Sathler, L.B.; Brito-Moreira, J.; Amaral, O.B.; Silva, C.A.; et al. TNF-α mediates PKR-dependent memory impairment and brain IRS-1 inhibition induced by Alzheimer's β-amyloid oligomers in mice and monkeys. *Cell Metab.* **2013**, *18*, 831–843. [CrossRef] [PubMed]
113. Ma, T.; Trinh, M.A.; Wexler, A.J.; Bourbon, C.; Gatti, E.; Pierre, P.; Cavener, D.R.; Klann, E. Suppression of eIF2α kinases alleviates Alzheimer's disease-related plasticity and memory deficits. *Nat. Neurosci.* **2013**, *16*, 1299–1305. [CrossRef] [PubMed]
114. Moreno, J.A.; Halliday, M.; Molloy, C.; Radford, H.; Verity, N.; Axten, J.M.; Ortori, C.A.; Willis, A.E.; Fischer, P.M.; Barrett, D.A.; et al. Oral treatment targeting the unfolded protein response prevents neurodegeneration and clinical disease in prion-infected mice. *Sci. Transl. Med.* **2013**, *5*, 206ra138. [CrossRef] [PubMed]
115. Ho, Y.S.; Yang, X.; Lau, J.C.; Hung, C.H.; Wuwongse, S.; Zhang, Q.; Wang, J.; Baum, L.; So, K.F.; Chang, R.C. Endoplasmic reticulum stress induces tau pathology and forms a vicious cycle: Implication in Alzheimer's disease pathogenesis. *J. Alzheimers. Dis.* **2012**, *28*, 839–854. [CrossRef]
116. Abisambra, J.F.; Jinwal, U.K.; Blair, L.J.; O'Leary, J.C.; Li, Q.; Brady, S.; Wang, L.; Guidi, C.E.; Zhang, B.; Nordhues, B.A.; et al. Tau Accumulation Activates the Unfolded Protein Response by Impairing Endoplasmic Reticulum-Associated Degradation. *J. Neurosci.* **2013**, *33*, 9498–9507. [CrossRef]
117. Lee, A.-H.; Iwakoshi, N.N.; Glimcher, L.H. XBP-1 Regulates a Subset of Endoplasmic Reticulum Resident Chaperone Genes in the Unfolded Protein Response. *Mol. Cell. Biol.* **2003**, *23*, 7448–7459. [CrossRef]
118. Duran-Aniotz, C.; Cornejo, V.H.; Espinoza, S.; Ardiles, Á.O.; Medinas, D.B.; Salazar, C.; Foley, A.; Gajardo, I.; Thielen, P.; Iwawaki, T.; et al. IRE1 signaling exacerbates Alzheimer's disease pathogenesis. *Acta Neuropathol.* **2017**, *134*, 489–506. [CrossRef]
119. Fonseca, A.C.; Oliveira, C.R.; Pereira, C.F.; Cardoso, S.M. Loss of proteostasis induced by amyloid beta peptide in brain endothelial cells. *Biochim. Biophys. Acta* **2014**, *1843*, 1150–1161. [CrossRef]
120. Soejima, N.; Ohyagi, Y.; Nakamura, N.; Himeno, E.; Iinuma, K.M.; Sakae, N.; Yamasaki, R.; Tabira, T.; Murakami, K.; Irie, K.; et al. Intracellular accumulation of toxic turn amyloid-β is associated with endoplasmic reticulum stress in Alzheimer's disease. *Curr. Alzheimer Res.* **2013**, *10*, 11–20.
121. Xu, T.T.; Zhang, Y.; He, J.Y.; Luo, D.; Luo, Y.; Wang, Y.J.; Liu, W.; Wu, J.; Zhao, W.; Fang, J.; et al. Bajijiasu Ameliorates β-Amyloid-Triggered Endoplasmic Reticulum Stress and Related Pathologies in an Alzheimer's Disease Model. *Cell. Physiol. Biochem.* **2018**, *46*, 107–117. [CrossRef] [PubMed]
122. Abdullah, A.; Ravanan, P. Kaempferol mitigates Endoplasmic Reticulum Stress Induced Cell Death by targeting caspase. *Sci. Rep.* **2018**, *8*, 2189. [CrossRef] [PubMed]
123. Van Dongen, M.C.J.M.; van Rossum, E.; Kessels, A.G.H.; Sielhorst, H.J.G.; Knipschild, P.G. The efficacy of ginkgo for elderly people with dementia and age-associated memory impairment: New results of a randomized clinical trial. *J. Am. Geriatr. Soc.* **2000**, *48*, 1183–1194. [CrossRef] [PubMed]
124. Savaskan, E.; Mueller, H.; Hoerr, R.; von Gunten, A.; Gauthier, S. Treatment effects of Ginkgo biloba extract EGb 761® on the spectrum of behavioral and psychological symptoms of dementia: Meta-analysis of randomized controlled trials. *Int. Psychogeriatr.* **2018**, *30*, 285–293. [CrossRef] [PubMed]

125. Snitz, B.E.; O'Meara, E.S.; Carlson, M.C.; Arnold, A.M.; Ives, D.G.; Rapp, S.R.; Saxton, J.; Lopez, O.L.; Dunn, L.O.; Sink, K.M.; et al. Ginkgo biloba for preventing cognitive decline in older adults a randomized trial. *J. Am. Med. Assoc.* **2009**, *302*, 2663–2670. [CrossRef] [PubMed]
126. Jangra, A.; Dwivedi, S.; Sriram, C.S.; Gurjar, S.S.; Kwatra, M.; Sulakhiya, K.; Baruah, C.C.; Lahkar, M. Honokiol abrogates chronic restraint stress-induced cognitive impairment and depressive-like behaviour by blocking endoplasmic reticulum stress in the hippocampus of mice. *Eur. J. Pharmacol.* **2016**, *770*, 25–32. [CrossRef] [PubMed]

International Journal of
Molecular Sciences

MDPI

Review

Amelioration of Mitochondrial Quality Control and Proteostasis by Natural Compounds in Parkinson's Disease Models

Bongki Cho [1,†], Taeyun Kim [2,3,†], Yu-Jin Huh [2,3,†], Jaemin Lee [2,*] and Yun-Il Lee [1,3,*]

1 Division of Biotechnology, Daegu Gyeongbuk Institute of Science and Technology, Daegu 42988, Korea; cbk34@dgist.ac.kr
2 Department of New Biology, Daegu Gyeongbuk Institute of Science and Technology, Daegu 42988, Korea; xoxsdfaa7876@dgist.ac.kr (T.K.); yujin.huh@dgist.ac.kr (Y.-J.H.)
3 Well Aging Research Center, Daegu Gyeongbuk Institute of Science and Technology, Daegu 42988, Korea
* Correspondence: jaeminlee@dgist.ac.kr (J.L.); ylee56@dgist.ac.kr (Y.-I.L.)
† These authors contributed equally to this work.

Received: 4 October 2019; Accepted: 17 October 2019; Published: 21 October 2019

Abstract: Parkinson's disease (PD) is a well-known age-related neurodegenerative disorder associated with longer lifespans and rapidly aging populations. The pathophysiological mechanism is a complex progress involving cellular damage such as mitochondrial dysfunction and protein homeostasis. Age-mediated degenerative neurological disorders can reduce the quality of life and also impose economic burdens. Currently, the common treatment is replacement with levodopa to address low dopamine levels; however, this does not halt the progression of PD and is associated with adverse effects, including dyskinesis. In addition, elderly patients can react negatively to treatment with synthetic neuroprotection agents. Recently, natural compounds such as phytochemicals with fewer side effects have been reported as candidate treatments of age-related neurodegenerative diseases. This review focuses on mitochondrial dysfunction, oxidative stress, hormesis, proteostasis, the ubiquitin-proteasome system, and autophagy (mitophagy) to explain the neuroprotective effects of using natural products as a therapeutic strategy. We also summarize the efforts to use natural extracts to develop novel pharmacological candidates for treatment of age-related PD.

Keywords: Parkinson's disease (PD); mitochondrial dysfunction; dynamics; hormesis; proteostasis; ubiquitin-proteasome system (UPS); autophagy; mitophagy; natural compounds

1. Introduction

Parkinson's disease (PD) is the second most common neurodegenerative disease. Approximately 1% of the elderly population above 60 years of age suffers from PD, and the prevalence of the disease increases to 4% in the highest age group [1]. Because the incidence of PD depends strongly on age, the number of PD patients is estimated to dramatically increase as lifespans also increase. The economic burden of PD was estimated to be $14.4 billion in the United States in 2010 [2]. However, it increased up to $51.9 billion in 2017 [3], and is expected to increase more dramatically in the future. The most effective therapeutic option is the administration of l-3,4-dihydroxyphenylalanine (L-DOPA), which can cross the blood-brain barrier and be metabolized to dopamine [4]. However, all currently available drugs, including L-DOPA, only modulate dopamine levels in PD patients' brains and are of limited effectiveness in the initial stages of the disease, which can last for 1–5 years [5]. Novel strategies are therefore needed to prevent and manage PD in the later stages.

PD is histologically characterized by the progressive loss of dopaminergic (DA) neurons in the substantia nigra pars compacta (SNpc), which innervates basal ganglia and regulates motor control through the release of dopamine. The loss of DA neurons occurs before the onset of motor symptoms [6]. At the end stage of PD, neuronal degeneration become widespread, resulting in various symptoms. Another notable characteristic of PD is Lewy pathology (LP), particularly within the brain stem and olfactory system during early-stage PD. As the disease progresses, LP spreads to the limbic and neocortical regions of the brain. LP is usually observed in PD patients' brains using histopathological methods [7]. However, LP is also observed in non-PD human brains, making LP a poor predictor of PD [8].

1.1. Major Pathological Mechanisms of Neurodegeneration in PD

The mechanism of PD pathogenesis has been studied extensively, although questions remain [9,10]. In brief, impairment of quality control in mitochondria and proteins by oxidative stress, and α-synuclein accumulation, is the primary mechanism associated with degeneration of DA neurons in PD with neuroinflammation [10]. Because this pathological mode is a common characteristic in other neurodegenerative diseases, including Alzheimer's disease and amyotrophic lateral sclerosis, we will discuss PD-specific pathological mechanisms of mitochondrial quality control and proteostasis.

1.2. Impairment of Mitochondrial Quality Control

Several genes have been identified to be related with early-onset PD, and their physiological roles have been extensively studied. Parkin and PINK1 are major components for autophagy-mediated degradation of mitochondria (mitophagy), and their genetic mutations are closely related with accumulation of dysfunctional mitochondria in early-onset PD [11,12]. In addition, DJ-1 is critical for the antioxidant process against oxidative stress, which is induced by Ca^{2+} oscillation in autonomously pacemaking DA neurons [13,14], and its autosomal recessive mutation is also related with early-onset PD [15]. These observations suggest a pathological role of mitochondrial dysfunction in early-onset and potentially sporadic PD. Especially, decreased activity of mitochondrial respiratory chain complex I has been observed in post-mortem SNpc of sporadic PD patients [16]. Neurotoxins, such as 6-hydroxydopamine (6-OHDA), 1-methyl-4-phenyl-1,2,3,6-tetrahydropyridine/ 1-methyl-4-phenyl-pyridinium (MPTP/MPP^{+}) and rotenone have been frequently used for experimental PD model. They inhibit the activity of mitochondrial respiratory chain complex I, and aberrantly induce mitochondrial dysfunction by oxidative stress, thereby mimicking selective loss of DA neurons in SNpc [17,18]. These indicate that impairment of mitochondrial function is linked with PD pathology.

Impairment on mitochondrial turnover also appears in PD [19]. Mitochondrial turnover is mediated by two pathways; 1) morphological balance between fusion and fission, and 2) qualitative and quantitative balance between biogenesis and mitophagy. Mitochondrial fragmentation has been well known as a common phenomenon in early stage of neuropathology including PD [20]. And reversely, mdivi-1, a synthetic blockade of mitochondrial fission as an inhibitor of Dynamin-related protein 1 (DRP1) [21], efficiently rescues DA neurons in a genetically- and chemically-induced PD model [22,23], emphasizing a critical contribution of mitochondrial dynamics in PD pathology. In addition, level of genes controlled by proliferator-activated receptor gamma coactivator 1-alpha (PGC1α), which is a master transcription factor for mitochondrial biogenesis, are downregulated in the brains of PD patients [24]. Reversely, activation of PGC-1α signaling efficiently reduces α-synuclein toxicity [25]. Furthermore, overexpression of Parkin prevented degeneration of DA neuron in PD model through activating mitophagy [26]. Those studies suggest that the activation of mitochondrial quality control can be a strategy to prevent and manage sporadic PD.

1.3. Impairment of Proteostasis

The second pathological mechanism of PD is abnormal accumulation of misfolded proteins by impairment of proteostasis. α-synuclein has been reported to be a major component in Lewy bodies in PD patients, and its mutation is involved in early-onset PD [10], raising the possibility that α-synuclein aggregates may play a critical role in PD pathogenesis. Although the physiological role of α-synuclein remains to be understood, the detrimental outcome of α-synuclein oligomers and aggregates has been widely studied. In pathological conditions, α-synuclein can oligomerize and form insoluble fibrils [27]. The α-synuclein oligomer induces aberrant generation of reactive oxygen species by inhibiting mitochondrial respiratory complex I, and leads to mitochondrial dysfunction [28]. Enhancement of proteostasis of α-synuclein by preventing aggregation and/or clearing aggregates can therefore be an effective strategy to cope with PD. A study in transgenic mice expressing human α-synuclein demonstrated that both the ubiquitin-proteasome system (UPS) and autophagy-lysosome pathway are responsible for the degradation of α-synuclein in neurons [29]. Rapamycin, an inhibitor of the mammalian target of rapamycin, consistently promotes degradation of wild-type and mutant α-synuclein [30] and rescues loss of DA neurons and parkinsonism in a 6-OHDA-induced PD mouse model [31]. These observations suggest that the activation of proteostasis mechanisms can be an effective strategy to manage PD through α-synuclein clearance.

Parkin plays a critical role in mitophagy [26] and gene transcription [32] as a PD-related multifunctional E3 ligase. Parkin targets, ubiquitinates, and degrades other proteins as well as the substrates involved in mitophagy. For instance, the genetic inactivation of Parkin leads to the accumulation of ZNF746 (PARIS), a substrate of Parkin, and this process represses PGC-1α signaling, leading to the degeneration of DA neurons [33]. PARIS accumulates excessively and consistently in familiar and sporadic PD patients' brain, indicating a pathophysiological role in PD. Parkin also ubiquitinates and degrades the aminoacyl-tRNA synthetase complex interacting multifunctional protein-2, which activates poly(ADP-ribose) polymerase-1 and promotes PAR polymerization, resulting in neuronal death via "parthanatos" [34,35]. These studies suggest a crucial role for E3 ligase activity of Parkin in the PD-related degeneration of DA neurons. Activation of UPS by Parkin or other E3 ligase may therefore also offer a crucial neuroprotective effect against PD.

2. Compounds from Natural Products Alleviating Mitochondrial Dysfunction in PD

2.1. Recovery of Redox Homeostasis

We list 84 lead compounds isolated from natural products that have neuroprotective effect in vitro and/or in vivo experimental PD models according to their chemical class with effect summary (Table 1). Among them, the reaction of some natural compounds in mitochondrial quality control is summarized in Figure 1. Oxidative stress has been proposed as a main initial factor in mitochondrial dysfunction, which appears as an early pathological event in neurodegenerative diseases, including PD [36]. Mitochondria are the main endogenous source of various free radicals, including reactive oxygen species/reactive nitrogen species (ROS/RNS) via oxidative phosphorylation and are removed by redox enzymes including catalase, superoxide dismutase, and heme oxigenase-1 with intracellular antioxidants such as glutathione (GSH) [37]. However, the failure of redox homeostasis induces excessive levels of ROS/RNS, leading to mitochondrial dysfunction [36]. Neurotoxins in experimental PD models, such as 6-OHDA, MPP^+/MPTP, rotenone, and paraquat, impair redox homeostasis by reducing the amount of antioxidants and activity of redox enzymes [38]. Traditionally, many compounds from natural products that recover redox homeostasis have been suggested for mitochondrial quality control in PD. Pre- or cotreatment of the compounds efficiently reduces levels of ROS/RNS against PD-related neurotoxins. Although the compounds, which are classified as polyphenols, terpenes, saponins, alkaloids, and other classes, exhibit anti-oxidizing activity in vitro, they may work as cellular activators and/or messengers by increasing the amount of GSH and by enhancing the activity of redox enzymes. Some mechanistic studies have revealed that nuclear factor erythroid 2-related factor 2 (NRF2) plays a central role in activating the redox system for neuroprotection against PD. Upon oxidative stress, NRF2 is stabilized by escaping from the UPS, which is mediated by Kelch-like ECH-associated protein 1 (KEAP1) and Cullin-3 (CUL3) [39]. Therefore, it accumulates in the nucleus and binds to promoters of multiple redox enzyme genes as a transcriptional activator, leading to the expression of redox enzymes as a defensive response. This process is enhanced by the following compounds: baicalein [40], luteolin [41], naringenin [42], puerarin [43] and genistein [44], auraptene [45], resveratrol [46], 11-dehydrosinulariolide [47], tanshinone I/IIA [48,49], astaxanthin [50], notoginsenoside Rg2/Rd/Re [51,52], ligustrazine [53], fucoidan [54], gastrodin [55], 3,4-dihydroxyphenyl-lactic acid [56], and salidroside [57]. However, some compounds induce expression of DJ-1, which promotes the recovery of the redox system via SOD1 and NRF2 signaling [58]. Among them are naringenin [59], sesamol [59], 11-dehydrosinulariolide [47], salidroside [57], rutin [60], and isoquercitrin [60]. Previous studies have demonstrated that various polyphenols and terpenes can evoke NRF2 signaling in other cellular contexts and environments [54]. This implies that other listed compounds can also activate NRF2 signaling, and their mechanistic study in PD models should be pursued. Taken together, we suggest that recovery of redox homeostasis is a basic property of natural compounds in PD treatment.

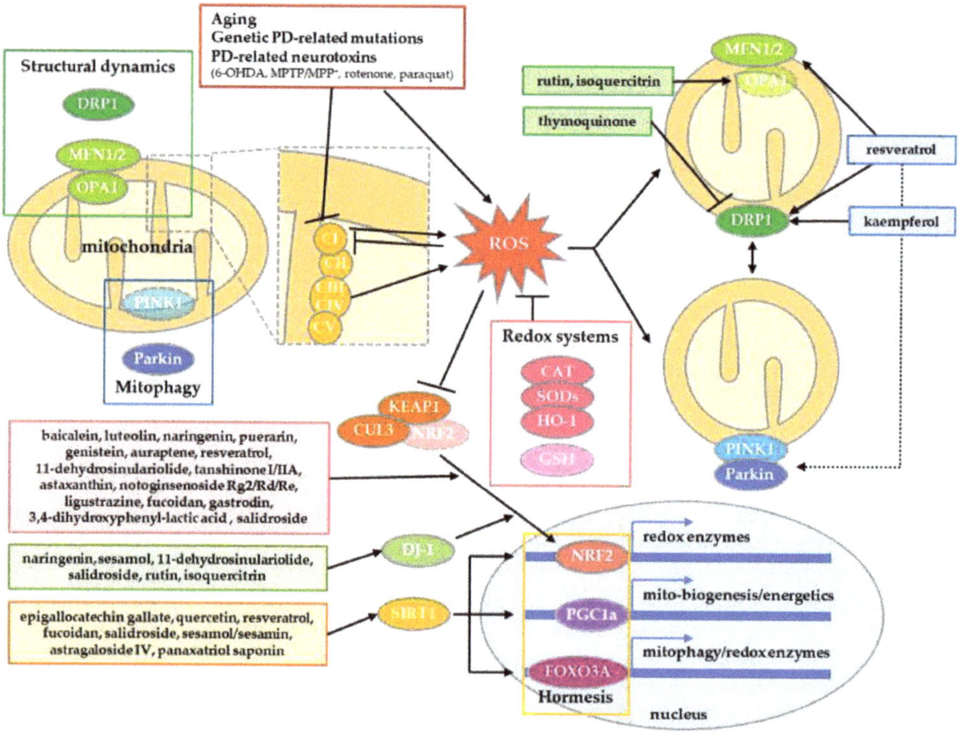

Figure 1. Neuroprotective compounds via mitochondrial quality control in PD. Mitochondrial quality is controlled by redox systems, structural dynamics, and mitophagy. In addition, it can be enhanced by hormetic adaptive stress responses. Some natural compounds revert and/or enhance redox system by NRF2 signaling, and improve mitochondrial quality by controlling structural dynamics and mitophagy. In addition, some compounds evoke adaptive stress responses mediated by SIRT1, which induce gene expression involved in redox enzymes, mitochondrial biogenesis/energetics and mitophagy. Therefore, these compounds protect DA neurons in PD.

Table 1. Lead compounds from natural products having neuroprotective effect in experimental PD model.

Class	Compounds	MitoQC	ProteoQC	Refs.	Class	Compounds	MitoQC	ProteoQC	Refs.
polyphenol/flavonoid	Epigallocatechin gallate	○	○	[61,62]	terpene/diterpene	11-Dehydrosinulariolide	△	△	[47]
	Apigenin	○	○	[63,64]		Tanshinone I	○	○	[48,65]
	Baicalein	○	○	[40,66]		Tanshinone IIA	○	○	[49,65]
	Luteolin	○	○	[41,67]		Triptolide		○	[68]
	Naringenin	○	○	[42,59]	terpene/triterpene	Celastrol	○	○	[69,70]
	Puerarin	○	○	[43,71]		Ursolic acid	○		[72]
	Quercetin	○	○	[73,74]		Asiaticoside A	○		[75]
	Rutin	○	△	[60,76]	terpene/sesquiterpene	Nerolidol	○		[77]
	Isoquercitrin	○	△	[60]	saponin	Astragaloside IV	○	○	[78,79]
	Kaempferol	○	△	[80,81]		Gypenosides	○		[82]
	Isoliquiritigenin	○		[83]		Notoginsenoside Rg1	○		[51]
	Genistein	○		[44]		Panaxatriol saponin	○		[84]
	Biochanin A	○		[85]		Onjisaponin B	△	○	[86]
	Hesperidin	○		[87]		Ginsenoside Rb1		○	[88]
	Morin	○		[89]		Ginsenoside Rd	○		[52]
	Myricetin	○		[90,91]		Ginsenoside Re	○		[52]
	Dihydromyricetin	○	○	[92,93]		Ginsenoside Rg1	○	○	[94,95]
	Troxerutin	○		[96]	alkaloid	Ligustrazine	○	○	[53]
	Liquiritigenin	○	○	[97]		Isorhynchophylline	○	○	[98,99]
polyphenol/coumarin	Auraptene	○		[45]		Conophylline	△	○	[100]
	Fraxetin	○		[101]		Amurensin G		○	[102]
	Esculin	○		[103]		6-Hydroxy-N-acetyl-β-oxotryptamine	○		[104]
	Esculetin	○		[105]	diketo-piperazine	Mactanamide	○		[104]

Table 1. *Cont.*

Class	Compounds	MitoQC	ProteoQC	Refs.	Class	Compounds	MitoQC	ProteoQC	Refs.
polyphenol/cinnamate	Chlorogenic acid	○	○	[106,107]	polyketide	8-Methoxy-3,5-dimethylisochroman-6-ol	○		[104]
	Curcumin	○	○	[90,108]		3-O-Methylorsellinic acid	○		[104]
	Rosmarinic acid	○	○	[109,110]	dibenzofuran	Candidusin A	○		[104]
polyphenol/stilbene	Resveratrol	○	○	[46,111–113]		4″-Dehydroxycandidusin A	○		[104]
	Piceatannol	○		[105]	mannose	Mannosylglycerate		○	[114]
	2,3,5,4′-tetrahydr-oxystilbene-2-O-β-D-glucoside	○	○	[115,116]	deoxy-adenosine	Cordycepin	○		[117]
	Salvianolic acid A	○	△	[118,119]	polysaccharide	Sulfated hetero-polysaccharides	○		[120]
	Salvianolic acid B	○	○	[93,121]		Sulfated galactofucan polysaccharides	○		[120]
	Polydatin	○		[122]		Fucoidan	○	△	[54,123,124]
polyphenol/xanthone	Mangiferin	○		[125]	quinone	Thymoquinone	○	○	[126,127]
polyphenol/lignan	Sesamol	○	○	[59,128]		2-methoxy-6-acetyl-7-methyljuglone	○		[129]
	Sesamin	○		[128]	anisole	β-asarone	○	○	[130]
	Magnolol	○		[131,132]	benzofurans	3-n-butylphthalide	○	○	[133]
terpene/carotenoid	Crocetin	○	○	[134,135]	glucoside	Gastrodin	○		[55]
	Crocin	○	○	[135,136]	bibenzyl	Chrysotoxine	○		[137]
	Astaxanthin	○	○	[50,138]	indolizine	Corynoxine B	△	○	[139]
terpene/monoterpene	Paeoniflorin	○	○	[140,141]	iridoid	Oleuropein	○		[76]
	Catalpol	○		[142]	lactate	3,4-dihydroxyphenyl-lactic acid	○		[56]
	Isoborneol	○		[143]	phenol-glycoside	Salidroside	△	○	[57,144,145]

We list the lead compounds in natural products having a neuroprotective effect in PD, and summarize their effects according to mitochondrial quality control (MitoQC) and protein quality control (ProteoQC), with references. Open circles or triangles indicate the existence of direct or indirect evidence in the literature, respectively.

2.2. Enhancement of Mitochondrial Turnover by Structural Dynamics

Recent papers have revealed the importance of structural quality control of mitochondria in neurodegeneration, including PD [20]. In the intra-/extracellular environment, mitochondria undergo dynamic morphological changes via controlled fusion and fission, which are mediated by fusion proteins, mitofusin1/2 and optic atrophy 1 (OPA1), and the fission protein DRP1 [146]. This process contributes to mitochondrial quality and bioenergetics by the sharing and division of metabolites and nucleoids in mitochondria (Figure 1). However, PD-related neurotoxins and genetic mutations can induce excessive fragmentation of mitochondria by enhancing fission or inhibiting fusion, resulting in excessive mitophagy and eventual mitochondria-mediated neuronal death [19]. As a result of this discovery, compounds that inhibit mitochondrial fragmentation in PD models have been proposed. Thymoquinone reverts rotenone-induced upregulation of DRP1 protein in substantia nigra and striatum in PD model rats [126]. Rutin and isoquercitrin recover the expression of OPA1 in 6-OHDA-treated PC12 cells [60]. Moreover, other compounds promote mitochondrial turnover by enhancing the overall activity of fusion/fission or mitophagy. Resveratrol upregulates the expression of both MFN1/2 and DRP1, resulting in the upscaling of mitochondrial quality by enhanced fusion/fission of mitochondria in PD models [111,112]. Kaempferol induces mitochondrial fragmentation, which contributes to efficient mitophagy, thereby protecting neurons from accumulation of abnormal mitochondria [80]. Rosmarinic acid protects membrane integrity in mitochondria against permeabilization by α-synuclein aggregates [109].

2.3. Natural Compounds Evoking Mitochondrial Hormesis

Hormesis-evoking therapeutic trials in PD have been conducted because the pathology of sporadic PD is closely linked with mitochondrial aging [147]. Hormesis is an adaptive response against severe challenges by enhancing functionality and tolerance upon preconditioned mild intracellular or extra-environmental stress [148]. Especially, mitochondrial hormesis can be evoked in response to mild mitochondrial stressors, including energetic depletion, calcium, and ROS by adaptive endoplasmic reticulum (ER)/integrated stress response and mitochondrial unfolded protein response [149]. This process promotes biogenesis, energetics, antioxidant response, protein quality control, and mitophagy of mitochondria, thereby extending lifespans with reduced metabolism via cytokine-mediated systemic regulation. Treatment with epigallocatechin gallate [61], quercetin [73], resveratrol [113] or fucoidan [123], sesamol/sesamin [128], astragaloside IV [78], panaxatriol saponin [84], or salidroside [144] in PD models activates sirtuin 1 (SIRT1) signaling, which promotes PGC1α signaling and Forkhead box O3 signaling, which are involved in the biogenesis/bioenergetics and mitophagy/redox of mitochondria, respectively [149]. In addition, rutin and oleuropein upregulate IRE1α and ATF-4 without activating CHOP, PERK, BIP, and PDI in low hormetic doses, thereby improving cell survival [76]. However, relatively high doses of panaxatriol saponin, rutin, and oleuropein inhibit cell growth and proliferation, indicating some toxic effect. Therefore, these hormesis-evoking compounds may require more intensive study on the dose-response [76,84]. SIRT1 signaling also activates the NRF2-mediated activation of the redox system via PGC1α signaling [149]. Therefore, NRF2-activating compounds may have a potential hormetic effect, but this possibility requires further study.

3. Natural Compounds Ameliorating Proteostasis Impairment in PD

The best-described pathological feature of PD is compromised proteostasis, which can be induced by oxidative or nitrosative stress resulting from misfolded protein accumulation and other exogenous neurotoxins [150,151]. In this section, we focus on two major mechanisms involved in proteostasis impairment with PD onset: UPS and autophagy. Autosomal recessive mutations of Parkin represent a large proportion of familial PD [152,153], and disruption of Parkin-mediated proteolysis leads to excessive protein misfolding, which culminates in PD [154]. On the other hand, α-synuclein forms fibril aggregates via PD-associated progressive posttranslational modifications, and it is usually degraded by autophagy-lysosome machinery. However, pathologically excessive α-synuclein aggregates impair the autophagy-lysosome machinery, leading to the vicious establishment of PD [155]. Researchers have therefore focused on ameliorating the collapsed protein quality for PD by controlling translation, chaperone-assisted folding and the degradation of protein. The regulation on proteostasis machinery by natural compounds is summarized in Figure 2.

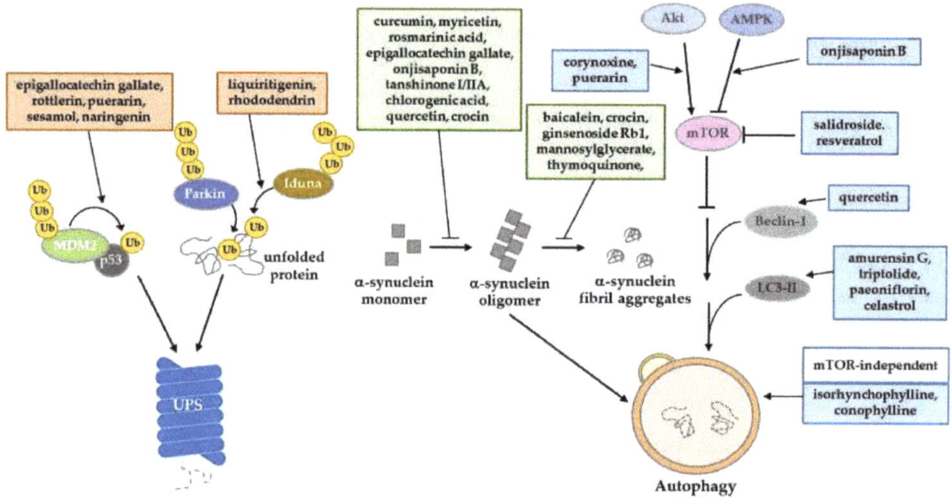

Figure 2. Summary of natural product regulation in proteostasis machinery. Natural products have a potential role to play in the amelioration of PD-induced proteostasis impairment. They regulate UPS through E3 ligase activity, increasing the autophagy-lysosome pathway, and inhibiting the posttranslational modifications of α-synuclein.

3.1. Regulation through the Ubiquitin-Proteasome System

One of the protein degradation pathways is UPS. Proteins are polyubiquitinated by E3 ligase and finally cleared by the proteasome. Some studies have tried to restore the impaired activity of UPS in PD models by using natural compounds. Salidroside decreases the level of phosphorylated α-synuclein (pSer129) by recovering proteasome activity in UPS-impaired PD models by 6-OHDA [145]. Because the E3 ligase, which catalyzes the polyubiquitination reaction, provides a key regulatory function in UPS, the regulation of its activity has been studied as a therapeutic strategy for PD. Some studies reported on the UPS-mediated regulation of p53, which is a key mediator of neuronal death in neurodegenerative diseases [156]. In PD patient brains, p53 is accumulated, and is involved in the degeneration of DA neurons [157]. Generally, MDM2, an E3 ligase, degrades p53, and could be activated by p53 in a negative feedback loop [158,159]. Upon cellular stress, including DNA damage, p53 becomes stable through its phosphorylation, mainly at the Ser-15 and -37 residues [160,161]. Due to its modification, the phosphorylated p53 destabilizes MDM2 and finally disorganizes the UPS function, leading to the aberrant protein accumulation. Some polyphenols, including flavonoids and lignans, have been reported to exhibit protective effect on impaired UPS regulating p53. Epigallocatechin gallate, rottlerin [62], puerarin [71], sesamol, and naringenin [59] inhibit the aberrant accumulation of p53 by recovering MDM2-mediated UPS, thereby suppressing p53-dependent cell death in PD models [62,71]. On the other hand, Parkin has an E3 ligase function, and its regulation has been investigated [162]. However, regulating Parkin activity through natural products is still under investigation. Another E3 ligase, IDUNA (RNF146), has PAR-dependent E3 ligase activity [163]. It protects against programmed cell death (called parthanatos) through proteasomal degradation. Recent studies have discovered that the natural products liquiritigenin and rhododendrin provide a neuroprotective effect in 6-OHDA PD models by inducing IDUNA activity. Both products bind to estrogen receptor-β stimulating transcription of IDUNA [97,164].

3.2. Regulation through the Autophagy-Lysosomal Pathway

Another major protein degradation pathway is autophagy. It is a kind of pro-survival pathway, which clears misfolded or damaged proteins that cannot be degraded by unfolded protein response. Several toxin-induced PD models have been used to simulate the epidemiology of PD. Through exogenous toxins, ER stress evoked from increased ROS generation and decreased ATP synthesis can directly impair mitochondrial respiratory complex I [165]. Many studies have reported natural products that can treat these impaired mitochondrial environments by increasing autophagy flux and targeting specific mechanisms. A well-known natural product, quercetin, is an autophagy enhancer that plays a protective role in response to ER stress in rotenone-induced PD rat models. Quercetin treatment ameliorates DNA fragmentation and decreases beclin-1 levels [74]. Triptolide [68], Amurensin G [102] and celastrol [69] induces autophagy by activating LC3-II upregulation and clears α-synuclein in vitro and in vivo PD models. Some studies have reported that natural products can elevate autophagic activity through the modulation of AKT/AMPK/mTOR signaling. An oxindole alkaloid, corynoxine, has been described as an autophagy inducer. Chen et al. (2014) suggested that corynoxine-induced autophagy can clear α-synuclein through the Akt/mTOR pathway in neuronal cells and a *Drosophila* model [139]. Furthermore, Chen et al. (2017) introduced a model of corynoxine-induced neuronal autophagy. They established a network-based algorithm of in silico kinome activity profiling, and predict phosphoproteomic data. They then suggested that corynoxine-induced autophagy could clear α-synuclein regulated by MAP2K2 and PLK1 kinase activity [166]. Onjisaponin B derived from Radix Polygalae was reported to have regulatory function of autophagy, enhancing autophagy flux by the AMPK/mTOR signaling pathway and finally removing α-synuclein A53T mutant proteins [86].

3.3. Inhibition of Protein Aggregation Formation

The most frequently described protein in the pathology of PD is α-synuclein. Aggregates of α-synuclein can be toxic in cellular environments and can lead to PD [167]. Once α-synuclein forms a fibril structure, it cannot be easily degraded through the protein degradation pathway. Inhibition of the formation of α-synuclein aggregates is therefore a promising therapeutic strategy. Some studies have reported novel natural products that control α-synuclein oligomerization. In particular, the polyphenol family has demonstrated an ability to directly or indirectly inhibit α-synuclein oligomerization. Curcumin is a well-known antioxidant that can increase the solubility of the α-synuclein form of monomers in catecholaminergic cell lines and in vivo models, thereby inhibiting oligomerization [168–171]. Pretreatment of rosmarinic acid inhibits reduction in the mitochondrial membrane potential and α-synuclein aggregation through its iron-chelating activity in an MPTP-induced PD model [110]. In addition, myricetin can inhibit α-synuclein oligomerization by directly binding to the α-synuclein N-terminal region in vitro [90]. Tanshinone I and tanshinone IIA decreased the formation of α-synuclein oligomers [65]. Ginsenoside Rb1 dissociates α-synuclein fibrillation through directly binding to α-synuclein oligomers [88]. Tea polyphenols have been shown to protect DA neurons against PD in mice models. Additionally, their therapeutic effects have been reproduced in an MPTP-induced monkey PD model that prevents α-synuclein oligomerization [172].

4. Conclusions and Future Prospects

In this review, we discussed the neuroprotective effects of lead compounds from natural products on mitochondrial quality control and proteostasis in experimental PD models. Unlike synthetic drugs that target only single molecules, some polyphenols, terpenes, and saponins have multiple and overlapped targets in other neurodegenerative diseases, including Alzheimer's disease as well as PD [173–175]. Natural compounds may serve as preventive supplements for age-related neurodegenerative diseases, and can be applied in combinatorial treatments to improve the quality of life of patients. Natural compounds have been widely tested in α-synuclein- or neurotoxin-induced PD models. However, studies testing natural compounds for therapeutic purposes may have a limitation in terms of the differences of experimental design such as the quality of the extracts and the forms of dosage [176]. This could significantly affect the efficacy and toxicity of the natural compounds tested in each setting. Thus, it is necessary to organize the design of tests of natural compounds in PD models. The main limitation is the unclear therapeutic mechanism of natural compounds. These lead compounds can be adopted to design synthetic derivatives, but intensive study is required for further drug development.

Although the bioavailability of the compounds from natural products is limited, they can be easily obtained from herbs, fruits, and marine organisms, and their intake is relatively safe, particularly via foods. Some extracts allow for the continuous absorption of multiple compounds at low doses over a lifetime, potentially evoking hormesis signaling, which may extend lifespans. Thus, further study is necessary.

Author Contributions: All authors contributed to researching the literature and writing the article.

Funding: This work was supported by DGIST R&D and the Basic Science Research Program through the National Research Foundation of Korea (NRF), funded by the Ministry of Science and ICT (MSIT) Y.-I.L. (18-LC-01, 19-BT-01) and J.L. (18-LC-01, 2017R1A2B4006200). J.L. also received funding from the Korean Institute of Oriental Medicine (KIOM), provided by the MSIT (KSN1812160).

Conflicts of Interest: The authors declare no conflict of interest.

References

1. Tysnes, O.B.; Storstein, A. Epidemiology of Parkinson's disease. *J. Neural Transm. (Vienna)* **2017**, *124*, 901–905. [CrossRef] [PubMed]
2. Kowal, S.L.; Dall, T.M.; Chakrabarti, R.; Storm, M.V.; Jain, A. The current and projected economic burden of Parkinson's disease in the United States. *Mov. Disord.* **2013**, *28*, 311–318. [CrossRef] [PubMed]
3. Economic Burden and Future Impact of Parkinson's Disease. Available online: https://www.michaeljfox.org/publication/parkinsons-disease-economic-burden-patients-families-and-federal-government-52-billion (accessed on 1 August 2019).
4. Marsot, A.; Guilhaumou, R.; Azulay, J.P.; Blin, O. Levodopa in Parkinson's Disease: A Review of Population Pharmacokinetics/Pharmacodynamics Analysis. *J. Pharm. Pharm. Sci.* **2017**, *20*, 226–238. [CrossRef] [PubMed]
5. Huynh, T. The Parkinson's disease market. *Nat. Rev. Drug Discov.* **2011**, *10*, 571–572. [CrossRef]
6. Fearnley, J.M.; Lees, A.J. Ageing and Parkinson's disease: Substantia nigra regional selectivity. *Brain* **1991**, *114 Pt 5*, 2283–2301. [CrossRef]
7. Braak, H.; Del Tredici, K.; Rub, U.; de Vos, R.A.; Jansen Steur, E.N.; Braak, E. Staging of brain pathology related to sporadic Parkinson's disease. *Neurobiol. Aging* **2003**, *24*, 197–211. [CrossRef]
8. Dijkstra, A.A.; Voorn, P.; Berendse, H.W.; Groenewegen, H.J.; Netherlands Brain, B.; Rozemuller, A.J.; van de Berg, W.D. Stage-dependent nigral neuronal loss in incidental Lewy body and Parkinson's disease. *Mov. Disord.* **2014**, *29*, 1244–1251. [CrossRef]
9. Surmeier, D.J.; Obeso, J.A.; Halliday, G.M. Selective neuronal vulnerability in Parkinson disease. *Nat. Rev. Neurosci.* **2017**, *18*, 101–113. [CrossRef]
10. Poewe, W.; Seppi, K.; Tanner, C.M.; Halliday, G.M.; Brundin, P.; Volkmann, J.; Schrag, A.E.; Lang, A.E. Parkinson disease. *Nat. Rev. Dis. Primers* **2017**, *3*, 17013. [CrossRef]
11. Pickrell, A.M.; Youle, R.J. The roles of PINK1, parkin, and mitochondrial fidelity in Parkinson's disease. *Neuron* **2015**, *85*, 257–273. [CrossRef]
12. Truban, D.; Hou, X.; Caulfield, T.R.; Fiesel, F.C.; Springer, W. PINK1, Parkin, and Mitochondrial Quality Control: What can we Learn about Parkinson's Disease Pathobiology? *J. Parkinsons Dis.* **2017**, *7*, 13–29. [CrossRef] [PubMed]
13. Guzman, J.N.; Sanchez-Padilla, J.; Wokosin, D.; Kondapalli, J.; Ilijic, E.; Schumacker, P.T.; Surmeier, D.J. Oxidant stress evoked by pacemaking in dopaminergic neurons is attenuated by DJ-1. *Nature* **2010**, *468*, 696–700. [CrossRef] [PubMed]

14. Di Nottia, M.; Masciullo, M.; Verrigni, D.; Petrillo, S.; Modoni, A.; Rizzo, V.; Di Giuda, D.; Rizza, T.; Niceta, M.; Torraco, A.; et al. DJ-1 modulates mitochondrial response to oxidative stress: Clues from a novel diagnosis of PARK7. *Clin. Genet.* **2017**, *92*, 18–25. [CrossRef] [PubMed]
15. Bonifati, V.; Rizzu, P.; van Baren, M.J.; Schaap, O.; Breedveld, G.J.; Krieger, E.; Dekker, M.C.; Squitieri, F.; Ibanez, P.; Joosse, M.; et al. Mutations in the DJ-1 gene associated with autosomal recessive early-onset parkinsonism. *Science* **2003**, *299*, 256–259. [CrossRef] [PubMed]
16. Janetzky, B.; Hauck, S.; Youdim, M.B.; Riederer, P.; Jellinger, K.; Pantucek, F.; Zochling, R.; Boissl, K.W.; Reichmann, H. Unaltered aconitase activity, but decreased complex I activity in substantia nigra pars compacta of patients with Parkinson's disease. *Neurosci. Lett.* **1994**, *169*, 126–128. [CrossRef]
17. Langston, J.W.; Ballard, P.; Tetrud, J.W.; Irwin, I. Chronic Parkinsonism in humans due to a product of meperidine-analog synthesis. *Science* **1983**, *219*, 979–980. [CrossRef] [PubMed]
18. Goldman, S.M. Environmental toxins and Parkinson's disease. *Annu. Rev. Pharmacol. Toxicol.* **2014**, *54*, 141–164. [CrossRef]
19. Park, J.S.; Davis, R.L.; Sue, C.M. Mitochondrial Dysfunction in Parkinson's Disease: New Mechanistic Insights and Therapeutic Perspectives. *Curr. Neurol. Neurosci. Rep.* **2018**, *18*, 21. [CrossRef]
20. Knott, A.B.; Perkins, G.; Schwarzenbacher, R.; Bossy-Wetzel, E. Mitochondrial fragmentation in neurodegeneration. *Nat. Rev. Neurosci.* **2008**, *9*, 505–518. [CrossRef]
21. Cassidy-Stone, A.; Chipuk, J.E.; Ingerman, E.; Song, C.; Yoo, C.; Kuwana, T.; Kurth, M.J.; Shaw, J.T.; Hinshaw, J.E.; Green, D.R.; et al. Chemical inhibition of the mitochondrial division dynamin reveals its role in Bax/Bak-dependent mitochondrial outer membrane permeabilization. *Dev. Cell* **2008**, *14*, 193–204. [CrossRef]
22. Rappold, P.M.; Cui, M.; Grima, J.C.; Fan, R.Z.; de Mesy-Bentley, K.L.; Chen, L.; Zhuang, X.; Bowers, W.J.; Tieu, K. Drp1 inhibition attenuates neurotoxicity and dopamine release deficits in vivo. *Nat. Commun.* **2014**, *5*, 5244. [CrossRef] [PubMed]
23. Filichia, E.; Hoffer, B.; Qi, X.; Luo, Y. Inhibition of Drp1 mitochondrial translocation provides neural protection in dopaminergic system in a Parkinson's disease model induced by MPTP. *Sci. Rep.* **2016**, *6*, 32656. [CrossRef] [PubMed]
24. Zheng, B.; Liao, Z.; Locascio, J.J.; Lesniak, K.A.; Roderick, S.S.; Watt, M.L.; Eklund, A.C.; Zhang-James, Y.; Kim, P.D.; Hauser, M.A.; et al. PGC-1alpha, a potential therapeutic target for early intervention in Parkinson's disease. *Sci. Transl. Med.* **2010**, *2*, 52ra73. [CrossRef] [PubMed]
25. Eschbach, J.; von Einem, B.; Muller, K.; Bayer, H.; Scheffold, A.; Morrison, B.E.; Rudolph, K.L.; Thal, D.R.; Witting, A.; Weydt, P.; et al. Mutual exacerbation of peroxisome proliferator-activated receptor gamma coactivator 1alpha deregulation and alpha-synuclein oligomerization. *Ann. Neurol.* **2015**, *77*, 15–32. [CrossRef] [PubMed]
26. Bian, M.; Liu, J.; Hong, X.; Yu, M.; Huang, Y.; Sheng, Z.; Fei, J.; Huang, F. Overexpression of parkin ameliorates dopaminergic neurodegeneration induced by 1- methyl-4-phenyl-1,2,3,6-tetrahydropyridine in mice. *PLoS ONE* **2012**, *7*, e39953. [CrossRef]
27. Melki, R. Role of Different Alpha-Synuclein Strains in Synucleinopathies, Similarities with other Neurodegenerative Diseases. *J. Parkinsons Dis.* **2015**, *5*, 217–227. [CrossRef]
28. Ludtmann, M.H.R.; Angelova, P.R.; Horrocks, M.H.; Choi, M.L.; Rodrigues, M.; Baev, A.Y.; Berezhnov, A.V.; Yao, Z.; Little, D.; Banushi, B.; et al. alpha-synuclein oligomers interact with ATP synthase and open the permeability transition pore in Parkinson's disease. *Nat. Commun.* **2018**, *9*, 2293. [CrossRef]
29. Ebrahimi-Fakhari, D.; Cantuti-Castelvetri, I.; Fan, Z.; Rockenstein, E.; Masliah, E.; Hyman, B.T.; McLean, P.J.; Unni, V.K. Distinct roles in vivo for the ubiquitin-proteasome system and the autophagy-lysosomal pathway in the degradation of alpha-synuclein. *J. Neurosci.* **2011**, *31*, 14508–14520. [CrossRef]
30. Webb, J.L.; Ravikumar, B.; Atkins, J.; Skepper, J.N.; Rubinsztein, D.C. Alpha-Synuclein is degraded by both autophagy and the proteasome. *J. Biol. Chem.* **2003**, *278*, 25009–25013. [CrossRef]
31. Liu, K.; Shi, N.; Sun, Y.; Zhang, T.; Sun, X. Therapeutic effects of rapamycin on MPTP-induced Parkinsonism in mice. *Neurochem. Res.* **2013**, *38*, 201–207. [CrossRef]
32. Da Costa, C.A.; Sunyach, C.; Giaime, E.; West, A.; Corti, O.; Brice, A.; Safe, S.; Abou-Sleiman, P.M.; Wood, N.W.; Takahashi, H.; et al. Transcriptional repression of p53 by parkin and impairment by mutations associated with autosomal recessive juvenile Parkinson's disease. *Nat. Cell Biol.* **2009**, *11*, 1370–1375. [CrossRef] [PubMed]

33. Shin, J.H.; Ko, H.S.; Kang, H.; Lee, Y.; Lee, Y.I.; Pletinkova, O.; Troconso, J.C.; Dawson, V.L.; Dawson, T.M. PARIS (ZNF746) repression of PGC-1alpha contributes to neurodegeneration in Parkinson's disease. *Cell* **2011**, *144*, 689–702. [CrossRef] [PubMed]
34. Lee, Y.; Karuppagounder, S.S.; Shin, J.H.; Lee, Y.I.; Ko, H.S.; Swing, D.; Jiang, H.; Kang, S.U.; Lee, B.D.; Kang, H.C.; et al. Parthanatos mediates AIMP2-activated age-dependent dopaminergic neuronal loss. *Nat. Neurosci.* **2013**, *16*, 1392–1400. [CrossRef] [PubMed]
35. Kam, T.I.; Mao, X.; Park, H.; Chou, S.C.; Karuppagounder, S.S.; Umanah, G.E.; Yun, S.P.; Brahmachari, S.; Panicker, N.; Chen, R.; et al. Poly(ADP-ribose) drives pathologic alpha-synuclein neurodegeneration in Parkinson's disease. *Science* **2018**, *362*, eaat8407. [CrossRef] [PubMed]
36. Lin, M.T.; Beal, M.F. Mitochondrial dysfunction and oxidative stress in neurodegenerative diseases. *Nature* **2006**, *443*, 787–795. [CrossRef]
37. Collins, Y.; Chouchani, E.T.; James, A.M.; Menger, K.E.; Cocheme, H.M.; Murphy, M.P. Mitochondrial redox signalling at a glance. *J. Cell Sci.* **2012**, *125*, 801–806. [CrossRef]
38. Puspita, L.; Chung, S.Y.; Shim, J.W. Oxidative stress and cellular pathologies in Parkinson's disease. *Mol. Brain* **2017**, *10*, 53. [CrossRef]
39. Gazaryan, I.G.; Thomas, B. The status of Nrf2-based therapeutics: Current perspectives and future prospects. *Neural Regen. Res.* **2016**, *11*, 1708–1711. [CrossRef]
40. Zhang, Z.; Cui, W.; Li, G.; Yuan, S.; Xu, D.; Hoi, M.P.; Lin, Z.; Dou, J.; Han, Y.; Lee, S.M. Baicalein protects against 6-OHDA-induced neurotoxicity through activation of Keap1/Nrf2/HO-1 and involving PKCalpha and PI3K/AKT signaling pathways. *J. Agric. Food Chem.* **2012**, *60*, 8171–8182. [CrossRef]
41. Wruck, C.J.; Claussen, M.; Fuhrmann, G.; Romer, L.; Schulz, A.; Pufe, T.; Waetzig, V.; Peipp, M.; Herdegen, T.; Gotz, M.E. Luteolin protects rat PC12 and C6 cells against MPP+ induced toxicity via an ERK dependent Keap1-Nrf2-ARE pathway. *J. Neural Transm. Suppl.* **2007**, *72*, 57–67.
42. Lou, H.; Jing, X.; Wei, X.; Shi, H.; Ren, D.; Zhang, X. Naringenin protects against 6-OHDA-induced neurotoxicity via activation of the Nrf2/ARE signaling pathway. *Neuropharmacology* **2014**, *79*, 380–388. [CrossRef] [PubMed]
43. Li, R.; Liang, T.; Xu, L.; Zheng, N.; Zhang, K.; Duan, X. Puerarin attenuates neuronal degeneration in the substantia nigra of 6-OHDA-lesioned rats through regulating BDNF expression and activating the Nrf2/ARE signaling pathway. *Brain Res.* **2013**, *1523*, 1–9. [CrossRef] [PubMed]
44. Wu, H.C.; Hu, Q.L.; Zhang, S.J.; Wang, Y.M.; Jin, Z.K.; Lv, L.F.; Zhang, S.; Liu, Z.L.; Wu, H.L.; Cheng, O.M. Neuroprotective effects of genistein on SH-SY5Y cells overexpressing A53T mutant alpha-synuclein. *Neural Regen. Res.* **2018**, *13*, 1375–1383. [CrossRef] [PubMed]
45. Jang, Y.; Choo, H.; Lee, M.J.; Han, J.; Kim, S.J.; Ju, X.; Cui, J.; Lee, Y.L.; Ryu, M.J.; Oh, E.S.; et al. Auraptene Mitigates Parkinson's Disease-Like Behavior by Protecting Inhibition of Mitochondrial Respiration and Scavenging Reactive Oxygen Species. *Int. J. Mol. Sci.* **2019**, *20*, 3409. [CrossRef] [PubMed]
46. Gaballah, H.H.; Zakaria, S.S.; Elbatsh, M.M.; Tahoon, N.M. Modulatory effects of resveratrol on endoplasmic reticulum stress-associated apoptosis and oxido-inflammatory markers in a rat model of rotenone-induced Parkinson's disease. *Chem. Biol. Interact.* **2016**, *251*, 10–16. [CrossRef]
47. Feng, C.W.; Hung, H.C.; Huang, S.Y.; Chen, C.H.; Chen, Y.R.; Chen, C.Y.; Yang, S.N.; Wang, H.D.; Sung, P.J.; Sheu, J.H.; et al. Neuroprotective Effect of the Marine-Derived Compound 11-Dehydrosinulariolide through DJ-1-Related Pathway in In Vitro and In Vivo Models of Parkinson's Disease. *Mar. Drugs* **2016**, *14*, 187. [CrossRef]
48. Jing, X.; Wei, X.; Ren, M.; Wang, L.; Zhang, X.; Lou, H. Neuroprotective Effects of Tanshinone I Against 6-OHDA-Induced Oxidative Stress in Cellular and Mouse Model of Parkinson's Disease Through Upregulating Nrf2. *Neurochem. Res.* **2016**, *41*, 779–786. [CrossRef]
49. Zhang, X.S.; Ha, S.; Wang, X.L.; Shi, Y.L.; Duan, S.S.; Li, Z.A. Tanshinone IIA protects dopaminergic neurons against 6-hydroxydopamine-induced neurotoxicity through miR-153/NF-E2-related factor 2/antioxidant response element signaling pathway. *Neuroscience* **2015**, *303*, 489–502. [CrossRef]
50. Ye, Q.; Huang, B.; Zhang, X.; Zhu, Y.; Chen, X. Astaxanthin protects against MPP(+)-induced oxidative stress in PC12 cells via the HO-1/NOX2 axis. *BMC Neurosci.* **2012**, *13*, 156. [CrossRef]
51. Meng, X.B.; Sun, G.B.; Wang, M.; Sun, J.; Qin, M.; Sun, X.B. P90RSK and Nrf2 Activation via MEK1/2-ERK1/2 Pathways Mediated by Notoginsenoside R2 to Prevent 6-Hydroxydopamine-Induced Apoptotic Death in SH-SY5Y Cells. *Evid. Based Complement. Alternat. Med.* **2013**, *2013*, 971712. [CrossRef]

52. Gonzalez-Burgos, E.; Fernandez-Moriano, C.; Lozano, R.; Iglesias, I.; Gomez-Serranillos, M.P. Ginsenosides Rd and Re co-treatments improve rotenone-induced oxidative stress and mitochondrial impairment in SH-SY5Y neuroblastoma cells. *Food Chem. Toxicol.* **2017**, *109*, 38–47. [CrossRef] [PubMed]
53. Michel, H.E.; Tadros, M.G.; Esmat, A.; Khalifa, A.E.; Abdel-Tawab, A.M. Tetramethylpyrazine Ameliorates Rotenone-Induced Parkinson's Disease in Rats: Involvement of Its Anti-Inflammatory and Anti-Apoptotic Actions. *Mol. Neurobiol.* **2017**, *54*, 4866–4878. [CrossRef] [PubMed]
54. Zhang, L.; Hao, J.; Zheng, Y.; Su, R.; Liao, Y.; Gong, X.; Liu, L.; Wang, X. Fucoidan Protects Dopaminergic Neurons by Enhancing the Mitochondrial Function in a Rotenone-induced Rat Model of Parkinson's Disease. *Aging Dis.* **2018**, *9*, 590–604. [CrossRef] [PubMed]
55. Jiang, G.; Hu, Y.; Liu, L.; Cai, J.; Peng, C.; Li, Q. Gastrodin protects against MPP(+)-induced oxidative stress by up regulates heme oxygenase-1 expression through p38 MAPK/Nrf2 pathway in human dopaminergic cells. *Neurochem. Int.* **2014**, *75*, 79–88. [CrossRef]
56. Chong, C.M.; Zhou, Z.Y.; Razmovski-Naumovski, V.; Cui, G.Z.; Zhang, L.Q.; Sa, F.; Hoi, P.M.; Chan, K.; Lee, S.M. Danshensu protects against 6-hydroxydopamine-induced damage of PC12 cells in vitro and dopaminergic neurons in zebrafish. *Neurosci. Lett.* **2013**, *543*, 121–125. [CrossRef]
57. Li, R.; Wang, S.; Li, T.; Wu, L.; Fang, Y.; Feng, Y.; Zhang, L.; Chen, J.; Wang, X. Salidroside Protects Dopaminergic Neurons by Preserving Complex I Activity via DJ-1/Nrf2-Mediated Antioxidant Pathway. *Parkinsons Dis.* **2019**, *2019*, 6073496. [CrossRef]
58. Biosa, A.; Sandrelli, F.; Beltramini, M.; Greggio, E.; Bubacco, L.; Bisaglia, M. Recent findings on the physiological function of DJ-1: Beyond Parkinson's disease. *Neurobiol. Dis.* **2017**, *108*, 65–72. [CrossRef]
59. Sonia Angeline, M.; Sarkar, A.; Anand, K.; Ambasta, R.K.; Kumar, P. Sesamol and naringenin reverse the effect of rotenone-induced PD rat model. *Neuroscience* **2013**, *254*, 379–394. [CrossRef]
60. Magalingam, K.B.; Radhakrishnan, A.; Ramdas, P.; Haleagrahara, N. Quercetin glycosides induced neuroprotection by changes in the gene expression in a cellular model of Parkinson's disease. *J. Mol. Neurosci.* **2015**, *55*, 609–617. [CrossRef]
61. Ye, Q.; Ye, L.; Xu, X.; Huang, B.; Zhang, X.; Zhu, Y.; Chen, X. Epigallocatechin-3-gallate suppresses 1-methyl-4-phenyl-pyridine-induced oxidative stress in PC12 cells via the SIRT1/PGC-1alpha signaling pathway. *BMC Complement. Altern. Med.* **2012**, *12*, 82. [CrossRef]
62. Lee, S.J.; Kim, D.C.; Choi, B.H.; Ha, H.; Kim, K.T. Regulation of p53 by activated protein kinase C-delta during nitric oxide-induced dopaminergic cell death. *J. Biol. Chem.* **2006**, *281*, 2215–2224. [CrossRef] [PubMed]
63. Patil, S.P.; Jain, P.D.; Sancheti, J.S.; Ghumatkar, P.J.; Tambe, R.; Sathaye, S. Neuroprotective and neurotrophic effects of Apigenin and Luteolin in MPTP induced parkinsonism in mice. *Neuropharmacology* **2014**, *86*, 192–202. [CrossRef] [PubMed]
64. Anusha, C.; Sumathi, T.; Joseph, L.D. Protective role of apigenin on rotenone induced rat model of Parkinson's disease: Suppression of neuroinflammation and oxidative stress mediated apoptosis. *Chem. Biol. Interact.* **2017**, *269*, 67–79. [CrossRef] [PubMed]
65. Ji, K.; Zhao, Y.; Yu, T.; Wang, Z.; Gong, H.; Yang, X.; Liu, Y.; Huang, K. Inhibition effects of tanshinone on the aggregation of alpha-synuclein. *Food Funct.* **2016**, *7*, 409–416. [CrossRef]
66. Zhu, M.; Rajamani, S.; Kaylor, J.; Han, S.; Zhou, F.; Fink, A.L. The flavonoid baicalein inhibits fibrillation of alpha-synuclein and disaggregates existing fibrils. *J. Biol. Chem.* **2004**, *279*, 26846–26857. [CrossRef]
67. Wu, Y.; Jiang, X.; Yang, K.; Xia, Y.; Cheng, S.; Tang, Q.; Bai, L.; Qiu, J.; Chen, C. Inhibition of alpha-Synuclein contributes to the ameliorative effects of dietary flavonoids luteolin on arsenite-induced apoptotic cell death in the dopaminergic PC12 cells. *Toxicol. Mech. Methods* **2017**, *27*, 598–608. [CrossRef]
68. Hu, G.; Gong, X.; Wang, L.; Liu, M.; Liu, Y.; Fu, X.; Wang, W.; Zhang, T.; Wang, X. Triptolide Promotes the Clearance of alpha-Synuclein by Enhancing Autophagy in Neuronal Cells. *Mol. Neurobiol.* **2017**, *54*, 2361–2372. [CrossRef]
69. Deng, Y.N.; Shi, J.; Liu, J.; Qu, Q.M. Celastrol protects human neuroblastoma SH-SY5Y cells from rotenone-induced injury through induction of autophagy. *Neurochem. Int.* **2013**, *63*, 1–9. [CrossRef]
70. Zhang, C.; Wang, R.; Liu, Z.; Bunker, E.; Lee, S.; Giuntini, M.; Chapnick, D.; Liu, X. The plant triterpenoid celastrol blocks PINK1-dependent mitophagy by disrupting PINK1's association with the mitochondrial protein TOM20. *J. Biol. Chem.* **2019**, *294*, 7472–7487. [CrossRef]

71. Cheng, Y.F.; Zhu, G.Q.; Wang, M.; Cheng, H.; Zhou, A.; Wang, N.; Fang, N.; Wang, X.C.; Xiao, X.Q.; Chen, Z.W.; et al. Involvement of ubiquitin proteasome system in protective mechanisms of Puerarin to MPP(+)-elicited apoptosis. *Neurosci. Res.* **2009**, *63*, 52–58. [CrossRef]
72. Rai, S.N.; Yadav, S.K.; Singh, D.; Singh, S.P. Ursolic acid attenuates oxidative stress in nigrostriatal tissue and improves neurobehavioral activity in MPTP-induced Parkinsonian mouse model. *J. Chem. Neuroanat.* **2016**, *71*, 41–49. [CrossRef] [PubMed]
73. Ay, M.; Luo, J.; Langley, M.; Jin, H.; Anantharam, V.; Kanthasamy, A.; Kanthasamy, A.G. Molecular mechanisms underlying protective effects of quercetin against mitochondrial dysfunction and progressive dopaminergic neurodegeneration in cell culture and MitoPark transgenic mouse models of Parkinson's Disease. *J. Neurochem.* **2017**, *141*, 766–782. [CrossRef] [PubMed]
74. El-Horany, H.E.; El-Latif, R.N.; ElBatsh, M.M.; Emam, M.N. Ameliorative Effect of Quercetin on Neurochemical and Behavioral Deficits in Rotenone Rat Model of Parkinson's Disease: Modulating Autophagy (Quercetin on Experimental Parkinson's Disease). *J. Biochem. Mol. Toxicol.* **2016**, *30*, 360–369. [CrossRef]
75. Xu, C.L.; Wang, Q.Z.; Sun, L.M.; Li, X.M.; Deng, J.M.; Li, L.F.; Zhang, J.; Xu, R.; Ma, S.P. Asiaticoside: Attenuation of neurotoxicity induced by MPTP in a rat model of Parkinsonism via maintaining redox balance and up-regulating the ratio of Bcl-2/Bax. *Pharmacol. Biochem. Behav.* **2012**, *100*, 413–418. [CrossRef] [PubMed]
76. Elmazoglu, Z.; Ergin, V.; Sahin, E.; Kayhan, H.; Karasu, C. Oleuropein and rutin protect against 6-OHDA-induced neurotoxicity in PC12 cells through modulation of mitochondrial function and unfolded protein response. *Interdiscip. Toxicol.* **2017**, *10*, 129–141. [CrossRef]
77. Javed, H.; Azimullah, S.; Abul Khair, S.B.; Ojha, S.; Haque, M.E. Neuroprotective effect of nerolidol against neuroinflammation and oxidative stress induced by rotenone. *BMC Neurosci.* **2016**, *17*, 58. [CrossRef]
78. Liu, Y.; Chong, L.; Li, X.; Tang, P.; Liu, P.; Hou, C.; Zhang, X.; Li, R. Astragaloside IV rescues MPP(+)-induced mitochondrial dysfunction through upregulation of methionine sulfoxide reductase A. *Exp. Ther Med.* **2017**, *14*, 2650–2656. [CrossRef]
79. Liu, X.; Zhang, J.; Wang, S.; Qiu, J.; Yu, C. Astragaloside IV attenuates the H2O2-induced apoptosis of neuronal cells by inhibiting alpha-synuclein expression via the p38 MAPK pathway. *Int. J. Mol. Med.* **2017**, *40*, 1772–1780. [CrossRef]
80. Filomeni, G.; Graziani, I.; De Zio, D.; Dini, L.; Centonze, D.; Rotilio, G.; Ciriolo, M.R. Neuroprotection of kaempferol by autophagy in models of rotenone-mediated acute toxicity: Possible implications for Parkinson's disease. *Neurobiol. Aging* **2012**, *33*, 767–785. [CrossRef]
81. Li, S.; Pu, X.P. Neuroprotective effect of kaempferol against a 1-methyl-4-phenyl-1,2,3,6-tetrahydropyridine-induced mouse model of Parkinson's disease. *Biol. Pharm. Bull.* **2011**, *34*, 1291–1296. [CrossRef]
82. Wang, P.; Niu, L.; Gao, L.; Li, W.X.; Jia, D.; Wang, X.L.; Gao, G.D. Neuroprotective effect of gypenosides against oxidative injury in the substantia nigra of a mouse model of Parkinson's disease. *J. Int. Med. Res.* **2010**, *38*, 1084–1092. [CrossRef] [PubMed]
83. Hwang, C.K.; Chun, H.S. Isoliquiritigenin isolated from licorice Glycyrrhiza uralensis prevents 6-hydroxydopamine-induced apoptosis in dopaminergic neurons. *Biosci. Biotechnol. Biochem.* **2012**, *76*, 536–543. [CrossRef]
84. Zhang, C.; Li, C.; Chen, S.; Li, Z.; Ma, L.; Jia, X.; Wang, K.; Bao, J.; Liang, Y.; Chen, M.; et al. Hormetic effect of panaxatriol saponins confers neuroprotection in PC12 cells and zebrafish through PI3K/AKT/mTOR and AMPK/SIRT1/FOXO3 pathways. *Sci. Rep.* **2017**, *7*, 41082. [CrossRef]
85. Yu, L.; Wang, X.; Chen, H.; Yan, Z.; Wang, M.; Li, Y. Neurochemical and Behavior Deficits in Rats with Iron and Rotenone Co-treatment: Role of Redox Imbalance and Neuroprotection by Biochanin A. *Front. Neurosci.* **2017**, *11*, 657. [CrossRef] [PubMed]
86. Wu, A.G.; Wong, V.K.; Xu, S.W.; Chan, W.K.; Ng, C.I.; Liu, L.; Law, B.Y. Onjisaponin B derived from Radix Polygalae enhances autophagy and accelerates the degradation of mutant alpha-synuclein and huntingtin in PC-12 cells. *Int. J. Mol. Sci.* **2013**, *14*, 22618–22641. [CrossRef]
87. Antunes, M.S.; Goes, A.T.; Boeira, S.P.; Prigol, M.; Jesse, C.R. Protective effect of hesperidin in a model of Parkinson's disease induced by 6-hydroxydopamine in aged mice. *Nutrition* **2014**, *30*, 1415–1422. [CrossRef] [PubMed]

88. Ardah, M.T.; Paleologou, K.E.; Lv, G.; Menon, S.A.; Abul Khair, S.B.; Lu, J.H.; Safieh-Garabedian, B.; Al-Hayani, A.A.; Eliezer, D.; Li, M.; et al. Ginsenoside Rb1 inhibits fibrillation and toxicity of alpha-synuclein and disaggregates preformed fibrils. *Neurobiol. Dis.* **2015**, *74*, 89–101. [CrossRef]
89. Zhang, Z.T.; Cao, X.B.; Xiong, N.; Wang, H.C.; Huang, J.S.; Sun, S.G.; Wang, T. Morin exerts neuroprotective actions in Parkinson disease models in vitro and in vivo. *Acta Pharmacol. Sin.* **2010**, *31*, 900–906. [CrossRef]
90. Takahashi, R.; Ono, K.; Takamura, Y.; Mizuguchi, M.; Ikeda, T.; Nishijo, H.; Yamada, M. Phenolic compounds prevent the oligomerization of alpha-synuclein and reduce synaptic toxicity. *J. Neurochem.* **2015**, *134*, 943–955. [CrossRef]
91. Zhang, K.; Ma, Z.; Wang, J.; Xie, A.; Xie, J. Myricetin attenuated MPP(+)-induced cytotoxicity by anti-oxidation and inhibition of MKK4 and JNK activation in MES23.5 cells. *Neuropharmacology* **2011**, *61*, 329–335. [CrossRef]
92. Ren, Z.X.; Zhao, Y.F.; Cao, T.; Zhen, X.C. Dihydromyricetin protects neurons in an MPTP-induced model of Parkinson's disease by suppressing glycogen synthase kinase-3 beta activity. *Acta Pharmacol. Sin.* **2016**, *37*, 1315–1324. [CrossRef] [PubMed]
93. Wu, J.Z.; Ardah, M.; Haikal, C.; Svanbergsson, A.; Diepenbroek, M.; Vaikath, N.N.; Li, W.; Wang, Z.Y.; Outeiro, T.F.; El-Agnaf, O.M.; et al. Dihydromyricetin and Salvianolic acid B inhibit alpha-synuclein aggregation and enhance chaperone-mediated autophagy. *Transl. Neurodegener.* **2019**, *8*, 18. [CrossRef] [PubMed]
94. Heng, Y.; Zhang, Q.S.; Mu, Z.; Hu, J.F.; Yuan, Y.H.; Chen, N.H. Ginsenoside Rg1 attenuates motor impairment and neuroinflammation in the MPTP-probenecid-induced parkinsonism mouse model by targeting alpha-synuclein abnormalities in the substantia nigra. *Toxicol. Lett.* **2016**, *243*, 7–21. [CrossRef] [PubMed]
95. Chen, X.C.; Zhou, Y.C.; Chen, Y.; Zhu, Y.G.; Fang, F.; Chen, L.M. Ginsenoside Rg1 reduces MPTP-induced substantia nigra neuron loss by suppressing oxidative stress. *Acta Pharmacol. Sin.* **2005**, *26*, 56–62. [CrossRef] [PubMed]
96. Baluchnejadmojarad, T.; Jamali-Raeufy, N.; Zabihnejad, S.; Rabiee, N.; Roghani, M. Troxerutin exerts neuroprotection in 6-hydroxydopamine lesion rat model of Parkinson's disease: Possible involvement of PI3K/ERbeta signaling. *Eur. J. Pharmacol.* **2017**, *801*, 72–78. [CrossRef]
97. Kim, H.; Ham, S.; Lee, J.Y.; Jo, A.; Lee, G.H.; Lee, Y.S.; Cho, M.; Shin, H.M.; Kim, D.; Pletnikova, O.; et al. Estrogen receptor activation contributes to RNF146 expression and neuroprotection in Parkinson's disease models. *Oncotarget* **2017**, *8*, 106721–106739. [CrossRef]
98. Li, X.M.; Zhang, X.J.; Dong, M.X. Isorhynchophylline Attenuates MPP(+)-Induced Apoptosis Through Endoplasmic Reticulum Stress- and Mitochondria-Dependent Pathways in PC12 Cells: Involvement of Antioxidant Activity. *Neuromol. Med.* **2017**, *19*, 480–492. [CrossRef]
99. Lu, J.H.; Tan, J.Q.; Durairajan, S.S.; Liu, L.F.; Zhang, Z.H.; Ma, L.; Shen, H.M.; Chan, H.Y.; Li, M. Isorhynchophylline, a natural alkaloid, promotes the degradation of alpha-synuclein in neuronal cells via inducing autophagy. *Autophagy* **2012**, *8*, 98–108. [CrossRef]
100. Sasazawa, Y.; Sato, N.; Umezawa, K.; Simizu, S. Conophylline protects cells in cellular models of neurodegenerative diseases by inducing mammalian target of rapamycin (mTOR)-independent autophagy. *J. Biol. Chem.* **2015**, *290*, 6168–6178. [CrossRef]
101. Molina-Jimenez, M.F.; Sanchez-Reus, M.I.; Andres, D.; Cascales, M.; Benedi, J. Neuroprotective effect of fraxetin and myricetin against rotenone-induced apoptosis in neuroblastoma cells. *Brain Res.* **2004**, *1009*, 9–16. [CrossRef]
102. Ryu, H.W.; Oh, W.K.; Jang, I.S.; Park, J. Amurensin G induces autophagy and attenuates cellular toxicities in a rotenone model of Parkinson's disease. *Biochem. Biophys. Res. Commun.* **2013**, *433*, 121–126. [CrossRef] [PubMed]
103. Zhao, D.L.; Zou, L.B.; Lin, S.; Shi, J.G.; Zhu, H.B. Anti-apoptotic effect of esculin on dopamine-induced cytotoxicity in the human neuroblastoma SH-SY5Y cell line. *Neuropharmacology* **2007**, *53*, 724–732. [CrossRef] [PubMed]
104. Yurchenko, E.A.; Menchinskaya, E.S.; Pislyagin, E.A.; Trinh, P.T.H.; Ivanets, E.V.; Smetanina, O.F.; Yurchenko, A.N. Neuroprotective Activity of Some Marine Fungal Metabolites in the 6-Hydroxydopamin- and Paraquat-Induced Parkinson's Disease Models. *Mar. Drugs* **2018**, *16*, 457. [CrossRef] [PubMed]
105. Angeles, D.C.; Ho, P.; Dymock, B.W.; Lim, K.L.; Zhou, Z.D.; Tan, E.K. Antioxidants inhibit neuronal toxicity in Parkinson's disease-linked LRRK2. *Ann. Clin. Transl. Neurol.* **2016**, *3*, 288–294. [CrossRef] [PubMed]

106. Shan, S.; Tian, L.; Fang, R. Chlorogenic Acid Exerts Beneficial Effects in 6-Hydroxydopamine-Induced Neurotoxicity by Inhibition of Endoplasmic Reticulum Stress. *Med. Sci. Monit.* **2019**, *25*, 453–459. [CrossRef]
107. Teraoka, M.; Nakaso, K.; Kusumoto, C.; Katano, S.; Tajima, N.; Yamashita, A.; Zushi, T.; Ito, S.; Matsura, T. Cytoprotective effect of chlorogenic acid against alpha-synuclein-related toxicity in catecholaminergic PC12 cells. *J. Clin. Biochem. Nutr.* **2012**, *51*, 122–127. [CrossRef]
108. Chen, J.; Tang, X.Q.; Zhi, J.L.; Cui, Y.; Yu, H.M.; Tang, E.H.; Sun, S.N.; Feng, J.Q.; Chen, P.X. Curcumin protects PC12 cells against 1-methyl-4-phenylpyridinium ion-induced apoptosis by bcl-2-mitochondria-ROS-iNOS pathway. *Apoptosis* **2006**, *11*, 943–953. [CrossRef]
109. Camilleri, A.; Zarb, C.; Caruana, M.; Ostermeier, U.; Ghio, S.; Hogen, T.; Schmidt, F.; Giese, A.; Vassallo, N. Mitochondrial membrane permeabilisation by amyloid aggregates and protection by polyphenols. *Biochim. Biophys. Acta* **2013**, *1828*, 2532–2543. [CrossRef]
110. Qu, L.; Xu, H.; Jia, W.; Jiang, H.; Xie, J. Rosmarinic acid protects against MPTP-induced toxicity and inhibits iron-induced alpha-synuclein aggregation. *Neuropharmacology* **2019**, *144*, 291–300. [CrossRef]
111. Lin, K.L.; Lin, K.J.; Wang, P.W.; Chuang, J.H.; Lin, H.Y.; Chen, S.D.; Chuang, Y.C.; Huang, S.T.; Tiao, M.M.; Chen, J.B.; et al. Resveratrol provides neuroprotective effects through modulation of mitochondrial dynamics and ERK1/2 regulated autophagy. *Free Radic. Res.* **2018**, *52*, 1371–1386. [CrossRef]
112. Peng, K.; Tao, Y.; Zhang, J.; Wang, J.; Ye, F.; Dan, G.; Zhao, Y.; Cai, Y.; Zhao, J.; Wu, Q.; et al. Resveratrol Regulates Mitochondrial Biogenesis and Fission/Fusion to Attenuate Rotenone-Induced Neurotoxicity. *Oxid. Med. Cell Longev.* **2016**, *2016*, 6705621. [CrossRef] [PubMed]
113. Mudo, G.; Makela, J.; Di Liberto, V.; Tselykh, T.V.; Olivieri, M.; Piepponen, P.; Eriksson, O.; Malkia, A.; Bonomo, A.; Kairisalo, M.; et al. Transgenic expression and activation of PGC-1alpha protect dopaminergic neurons in the MPTP mouse model of Parkinson's disease. *Cell Mol. Life Sci.* **2012**, *69*, 1153–1165. [CrossRef] [PubMed]
114. Faria, C.; Jorge, C.D.; Borges, N.; Tenreiro, S.; Outeiro, T.F.; Santos, H. Inhibition of formation of alpha-synuclein inclusions by mannosylglycerate in a yeast model of Parkinson's disease. *Biochim. Biophys. Acta* **2013**, *1830*, 4065–4072. [CrossRef] [PubMed]
115. Zhang, L.; Huang, L.; Li, X.; Liu, C.; Sun, X.; Wu, L.; Li, T.; Yang, H.; Chen, J. Potential molecular mechanisms mediating the protective effects of tetrahydroxystilbene glucoside on MPP(+)-induced PC12 cell apoptosis. *Mol. Cell Biochem.* **2017**, *436*, 203–213. [CrossRef] [PubMed]
116. Zhang, R.; Sun, F.; Zhang, L.; Sun, X.; Li, L. Tetrahydroxystilbene glucoside inhibits alpha-synuclein aggregation and apoptosis in A53T alpha-synuclein-transfected cells exposed to MPP^+. *Can. J. Physiol. Pharmacol.* **2017**, *95*, 750–758. [CrossRef] [PubMed]
117. Olatunji, O.J.; Feng, Y.; Olatunji, O.O.; Tang, J.; Ouyang, Z.; Su, Z. Cordycepin protects PC12 cells against 6-hydroxydopamine induced neurotoxicity via its antioxidant properties. *Biomed. Pharmacother.* **2016**, *81*, 7–14. [CrossRef] [PubMed]
118. Zhang, H.A.; Gao, M.; Zhang, L.; Zhao, Y.; Shi, L.L.; Chen, B.N.; Wang, Y.H.; Wang, S.B.; Du, G.H. Salvianolic acid A protects human SH-SY5Y neuroblastoma cells against H(2)O(2)-induced injury by increasing stress tolerance ability. *Biochem. Biophys. Res. Commun.* **2012**, *421*, 479–483. [CrossRef]
119. Wang, X.J.; Xu, J.X. Salvianic acid A protects human neuroblastoma SH-SY5Y cells against MPP+-induced cytotoxicity. *Neurosci. Res.* **2005**, *51*, 129–138. [CrossRef]
120. Wang, J.; Liu, H.; Jin, W.; Zhang, H.; Zhang, Q. Structure-activity relationship of sulfated hetero/galactofucan polysaccharides on dopaminergic neuron. *Int. J. Biol. Macromol.* **2016**, *82*, 878–883. [CrossRef]
121. Tian, L.L.; Wang, X.J.; Sun, Y.N.; Li, C.R.; Xing, Y.L.; Zhao, H.B.; Duan, M.; Zhou, Z.; Wang, S.Q. Salvianolic acid B, an antioxidant from Salvia miltiorrhiza, prevents 6-hydroxydopamine induced apoptosis in SH-SY5Y cells. *Int. J. Biochem. Cell Biol.* **2008**, *40*, 409–422. [CrossRef]
122. Chen, Y.; Zhang, D.Q.; Liao, Z.; Wang, B.; Gong, S.; Wang, C.; Zhang, M.Z.; Wang, G.H.; Cai, H.; Liao, F.F.; et al. Anti-oxidant polydatin (piceid) protects against substantia nigral motor degeneration in multiple rodent models of Parkinson's disease. *Mol. Neurodegener.* **2015**, *10*, 4. [CrossRef] [PubMed]
123. Han, Y.S.; Lee, J.H.; Lee, S.H. Fucoidan Suppresses Mitochondrial Dysfunction and Cell Death against 1-Methyl-4-Phenylpyridinum-Induced Neuronal Cytotoxicity via Regulation of PGC-1alpha Expression. *Mar. Drugs* **2019**, *17*, 518. [CrossRef] [PubMed]

124. Liang, Z.; Liu, Z.; Sun, X.; Tao, M.; Xiao, X.; Yu, G.; Wang, X. The Effect of Fucoidan on Cellular Oxidative Stress and the CatD-Bax Signaling Axis in MN9D Cells Damaged by 1-Methyl-4-Phenypyridinium. *Front. Aging Neurosci.* **2018**, *10*, 429. [CrossRef] [PubMed]
125. Feng, G.; Zhang, Z.; Bao, Q.; Zhang, Z.; Zhou, L.; Jiang, J.; Li, S. Protective effect of chinonin in MPTP-induced C57BL/6 mouse model of Parkinson's disease. *Biol. Pharm. Bull.* **2014**, *37*, 1301–1307. [CrossRef] [PubMed]
126. Ebrahimi, S.S.; Oryan, S.; Izadpanah, E.; Hassanzadeh, K. Thymoquinone exerts neuroprotective effect in animal model of Parkinson's disease. *Toxicol. Lett.* **2017**, *276*, 108–114. [CrossRef] [PubMed]
127. Ardah, M.T.; Merghani, M.M.; Haque, M.E. Thymoquinone prevents neurodegeneration against MPTP in vivo and modulates alpha-synuclein aggregation in vitro. *Neurochem. Int.* **2019**, *128*, 115–126. [CrossRef]
128. Ruankham, W.; Suwanjang, W.; Wongchitrat, P.; Prachayasittikul, V.; Prachayasittikul, S.; Phopin, K. Sesamin and sesamol attenuate H2O2 -induced oxidative stress on human neuronal cells via the SIRT1-SIRT3-FOXO3a signaling pathway. *Nutr. Neurosci.* **2019**, 1–12. [CrossRef]
129. Li, Y.B.; Lin, Z.Q.; Zhang, Z.J.; Wang, M.W.; Zhang, H.; Zhang, Q.W.; Lee, S.M.; Wang, Y.T.; Hoi, P.M. Protective, antioxidative and antiapoptotic effects of 2-methoxy-6-acetyl-7-methyljuglone from Polygonum cuspidatum in PC12 cells. *Planta Med.* **2011**, *77*, 354–361. [CrossRef]
130. Ning, B.; Zhang, Q.; Wang, N.; Deng, M.; Fang, Y. beta-Asarone Regulates ER Stress and Autophagy Via Inhibition of the PERK/CHOP/Bcl-2/Beclin-1 Pathway in 6-OHDA-Induced Parkinsonian Rats. *Neurochem. Res.* **2019**, *44*, 1159–1166. [CrossRef]
131. Muroyama, A.; Fujita, A.; Lv, C.; Kobayashi, S.; Fukuyama, Y.; Mitsumoto, Y. Magnolol Protects against MPTP/MPP(+)-Induced Toxicity via Inhibition of Oxidative Stress in In Vivo and In Vitro Models of Parkinson's Disease. *Parkinsons Dis.* **2012**, *2012*, 985157. [CrossRef]
132. Li, Y.; Wu, Z.; Gao, X.; Zhu, Q.; Jin, Y.; Wu, A.; Huang, A.C. Anchanling reduces pathology in a lactacystin-induced Parkinson's disease model. *Neural Regen. Res.* **2012**, *7*, 165–170. [CrossRef] [PubMed]
133. Huang, J.Z.; Chen, Y.Z.; Su, M.; Zheng, H.F.; Yang, Y.P.; Chen, J.; Liu, C.F. dl-3-n-Butylphthalide prevents oxidative damage and reduces mitochondrial dysfunction in an MPP(+)-induced cellular model of Parkinson's disease. *Neurosci. Lett.* **2010**, *475*, 89–94. [CrossRef] [PubMed]
134. Ahmad, A.S.; Ansari, M.A.; Ahmad, M.; Saleem, S.; Yousuf, S.; Hoda, M.N.; Islam, F. Neuroprotection by crocetin in a hemi-parkinsonian rat model. *Pharmacol. Biochem. Behav.* **2005**, *81*, 805–813. [CrossRef] [PubMed]
135. Inoue, E.; Shimizu, Y.; Masui, R.; Hayakawa, T.; Tsubonoya, T.; Hori, S.; Sudoh, K. Effects of saffron and its constituents, crocin-1, crocin-2, and crocetin on alpha-synuclein fibrils. *J. Nat. Med.* **2018**, *72*, 274–279. [CrossRef] [PubMed]
136. Rao, S.V.; Hemalatha, P.; Yetish, S.; Muralidhara, M.; Rajini, P.S. Prophylactic neuroprotective propensity of Crocin, a carotenoid against rotenone induced neurotoxicity in mice: Behavioural and biochemical evidence. *Metab. Brain Dis.* **2019**, *34*, 1341–1353. [CrossRef]
137. Song, J.X.; Shaw, P.C.; Sze, C.W.; Tong, Y.; Yao, X.S.; Ng, T.B.; Zhang, Y.B. Chrysotoxine, a novel bibenzyl compound, inhibits 6-hydroxydopamine induced apoptosis in SH-SY5Y cells via mitochondria protection and NF-kappaB modulation. *Neurochem. Int.* **2010**, *57*, 676–689. [CrossRef]
138. Lee, D.H.; Kim, C.S.; Lee, Y.J. Astaxanthin protects against MPTP/MPP+-induced mitochondrial dysfunction and ROS production in vivo and in vitro. *Food Chem. Toxicol.* **2011**, *49*, 271–280. [CrossRef]
139. Chen, L.L.; Song, J.X.; Lu, J.H.; Yuan, Z.W.; Liu, L.F.; Durairajan, S.S.; Li, M. Corynoxine, a natural autophagy enhancer, promotes the clearance of alpha-synuclein via Akt/mTOR pathway. *J. Neuroimmune Pharmacol.* **2014**, *9*, 380–387. [CrossRef]
140. Dong, H.; Li, R.; Yu, C.; Xu, T.; Zhang, X.; Dong, M. Paeoniflorin inhibition of 6-hydroxydopamine-induced apoptosis in PC12 cells via suppressing reactive oxygen species-mediated PKCdelta/NF-kappaB pathway. *Neuroscience* **2015**, *285*, 70–80. [CrossRef]
141. Sun, X.; Cao, Y.B.; Hu, L.F.; Yang, Y.P.; Li, J.; Wang, F.; Liu, C.F. ASICs mediate the modulatory effect by paeoniflorin on alpha-synuclein autophagic degradation. *Brain Res.* **2011**, *1396*, 77–87. [CrossRef]
142. Mao, Y.R.; Jiang, L.; Duan, Y.L.; An, L.J.; Jiang, B. Efficacy of catalpol as protectant against oxidative stress and mitochondrial dysfunction on rotenone-induced toxicity in mice brain. *Environ. Toxicol. Pharmacol.* **2007**, *23*, 314–318. [CrossRef] [PubMed]

143. Tian, L.L.; Zhou, Z.; Zhang, Q.; Sun, Y.N.; Li, C.R.; Cheng, C.H.; Zhong, Z.Y.; Wang, S.Q. Protective effect of (+/-) isoborneol against 6-OHDA-induced apoptosis in SH-SY5Y cells. *Cell Physiol. Biochem.* **2007**, *20*, 1019–1032. [CrossRef] [PubMed]
144. Wang, C.Y.; Sun, Z.N.; Wang, M.X.; Zhang, C. SIRT1 mediates salidroside-elicited protective effects against MPP(+) -induced apoptosis and oxidative stress in SH-SY5Y cells: Involvement in suppressing MAPK pathways. *Cell Biol. Int.* **2018**, *42*, 84–94. [CrossRef] [PubMed]
145. Li, T.; Feng, Y.; Yang, R.; Wu, L.; Li, R.; Huang, L.; Yang, Q.; Chen, J. Salidroside Promotes the Pathological alpha-Synuclein Clearance Through Ubiquitin-Proteasome System in SH-SY5Y Cells. *Front. Pharmacol.* **2018**, *9*, 377. [CrossRef] [PubMed]
146. Sanchis-Gomar, F.; Derbre, F. Mitochondrial fission and fusion in human diseases. *N. Engl. J. Med.* **2014**, *370*, 1073–1074. [CrossRef] [PubMed]
147. Calabrese, V.; Santoro, A.; Trovato Salinaro, A.; Modafferi, S.; Scuto, M.; Albouchi, F.; Monti, D.; Giordano, J.; Zappia, M.; Franceschi, C.; et al. Hormetic approaches to the treatment of Parkinson's disease: Perspectives and possibilities. *J. Neurosci. Res.* **2018**, *96*, 1641–1662. [CrossRef]
148. Calabrese, E.J.; Mattson, M.P. How does hormesis impact biology, toxicology, and medicine? *NPJ Aging Mech. Dis.* **2017**, *3*, 13. [CrossRef]
149. Quiros, P.M.; Mottis, A.; Auwerx, J. Mitonuclear communication in homeostasis and stress. *Nat. Rev. Mol. Cell Biol.* **2016**, *17*, 213–226. [CrossRef]
150. Tsang, A.H.; Chung, K.K. Oxidative and nitrosative stress in Parkinson's disease. *Biochim. Biophys. Acta* **2009**, *1792*, 643–650. [CrossRef]
151. Gu, Z.; Nakamura, T.; Lipton, S.A. Redox reactions induced by nitrosative stress mediate protein misfolding and mitochondrial dysfunction in neurodegenerative diseases. *Mol. Neurobiol.* **2010**, *41*, 55–72. [CrossRef]
152. Klein, C.; Schlossmacher, M.G. Parkinson disease, 10 years after its genetic revolution: Multiple clues to a complex disorder. *Neurology* **2007**, *69*, 2093–2104. [CrossRef] [PubMed]
153. Klein, C.; Westenberger, A. Genetics of Parkinson's disease. *Cold Spring Harb. Perspect. Med.* **2012**, *2*, a008888. [CrossRef] [PubMed]
154. Dawson, T.M.; Dawson, V.L. The role of parkin in familial and sporadic Parkinson's disease. *Mov. Disord.* **2010**, *25* (Suppl. 1), S32–S39. [CrossRef] [PubMed]
155. Winslow, A.R.; Chen, C.W.; Corrochano, S.; Acevedo-Arozena, A.; Gordon, D.E.; Peden, A.A.; Lichtenberg, M.; Menzies, F.M.; Ravikumar, B.; Imarisio, S.; et al. alpha-Synuclein impairs macroautophagy: Implications for Parkinson's disease. *J. Cell Biol.* **2010**, *190*, 1023–1037. [CrossRef]
156. Wang, D.B.; Kinoshita, C.; Kinoshita, Y.; Morrison, R.S. p53 and mitochondrial function in neurons. *Biochim. Biophys. Acta* **2014**, *1842*, 1186–1197. [CrossRef]
157. Mogi, M.; Kondo, T.; Mizuno, Y.; Nagatsu, T. p53 protein, interferon-gamma, and NF-kappaB levels are elevated in the parkinsonian brain. *Neurosci. Lett.* **2007**, *414*, 94–97. [CrossRef]
158. Wu, X.; Bayle, J.H.; Olson, D.; Levine, A.J. The p53-mdm-2 autoregulatory feedback loop. *Genes Dev.* **1993**, *7*, 1126–1132. [CrossRef]
159. Pant, V.; Lozano, G. Limiting the power of p53 through the ubiquitin proteasome pathway. *Genes Dev.* **2014**, *28*, 1739–1751. [CrossRef]
160. Haupt, Y.; Maya, R.; Kazaz, A.; Oren, M. Mdm2 promotes the rapid degradation of p53. *Nature* **1997**, *387*, 296–299. [CrossRef]
161. Sakaguchi, K.; Herrera, J.E.; Saito, S.; Miki, T.; Bustin, M.; Vassilev, A.; Anderson, C.W.; Appella, E. DNA damage activates p53 through a phosphorylation-acetylation cascade. *Genes Dev.* **1998**, *12*, 2831–2841. [CrossRef]
162. Ham, S.; Lee, Y.I.; Jo, M.; Kim, H.; Kang, H.; Jo, A.; Lee, G.H.; Mo, Y.J.; Park, S.C.; Lee, Y.S.; et al. Hydrocortisone-induced parkin prevents dopaminergic cell death via CREB pathway in Parkinson's disease model. *Sci. Rep.* **2017**, *7*, 525. [CrossRef]
163. Kang, H.C.; Lee, Y.I.; Shin, J.H.; Andrabi, S.A.; Chi, Z.; Gagne, J.P.; Lee, Y.; Ko, H.S.; Lee, B.D.; Poirier, G.G.; et al. Iduna is a poly(ADP-ribose) (PAR)-dependent E3 ubiquitin ligase that regulates DNA damage. *Proc. Natl. Acad. Sci. USA* **2011**, *108*, 14103–14108. [CrossRef] [PubMed]
164. Kim, H.; Park, J.; Leem, H.; Cho, M.; Yoon, J.H.; Maeng, H.J.; Lee, Y. Rhododendrin-Induced RNF146 Expression via Estrogen Receptor beta Activation is Cytoprotective Against 6-OHDA-Induced Oxidative Stress. *Int. J. Mol. Sci.* **2019**, *20*, 1772. [CrossRef] [PubMed]

165. Zeng, X.S.; Geng, W.S.; Jia, J.J. Neurotoxin-Induced Animal Models of Parkinson Disease: Pathogenic Mechanism and Assessment. *ASN Neuro* **2018**, *10*, 1759091418777438. [CrossRef] [PubMed]
166. Chen, L.L.; Wang, Y.B.; Song, J.X.; Deng, W.K.; Lu, J.H.; Ma, L.L.; Yang, C.B.; Li, M.; Xue, Y. Phosphoproteome-based kinase activity profiling reveals the critical role of MAP2K2 and PLK1 in neuronal autophagy. *Autophagy* **2017**, *13*, 1969–1980. [CrossRef] [PubMed]
167. Bridi, J.C.; Hirth, F. Mechanisms of alpha-Synuclein Induced Synaptopathy in Parkinson's Disease. *Front. Neurosci.* **2018**, *12*, 80. [CrossRef]
168. Zbarsky, V.; Datla, K.P.; Parkar, S.; Rai, D.K.; Aruoma, O.I.; Dexter, D.T. Neuroprotective properties of the natural phenolic antioxidants curcumin and naringenin but not quercetin and fisetin in a 6-OHDA model of Parkinson's disease. *Free Radic. Res.* **2005**, *39*, 1119–1125. [CrossRef]
169. Ono, K.; Yamada, M. Antioxidant compounds have potent anti-fibrillogenic and fibril-destabilizing effects for alpha-synuclein fibrils in vitro. *J. Neurochem.* **2006**, *97*, 105–115. [CrossRef]
170. Pandey, N.; Strider, J.; Nolan, W.C.; Yan, S.X.; Galvin, J.E. Curcumin inhibits aggregation of alpha-synuclein. *Acta Neuropathol.* **2008**, *115*, 479–489. [CrossRef]
171. Sharma, N.; Nehru, B. Curcumin affords neuroprotection and inhibits alpha-synuclein aggregation in lipopolysaccharide-induced Parkinson's disease model. *Inflammopharmacology* **2018**, *26*, 349–360. [CrossRef]
172. Chen, M.; Wang, T.; Yue, F.; Li, X.; Wang, P.; Li, Y.; Chan, P.; Yu, S. Tea polyphenols alleviate motor impairments, dopaminergic neuronal injury, and cerebral alpha-synuclein aggregation in MPTP-intoxicated parkinsonian monkeys. *Neuroscience* **2015**, *286*, 383–392. [CrossRef] [PubMed]
173. Pandareesh, M.D.; Mythri, R.B.; Srinivas Bharath, M.M. Bioavailability of dietary polyphenols: Factors contributing to their clinical application in CNS diseases. *Neurochem. Int.* **2015**, *89*, 198–208. [CrossRef] [PubMed]
174. Huang, X.; Li, N.; Pu, Y.; Zhang, T.; Wang, B. Neuroprotective Effects of Ginseng Phytochemicals: Recent Perspectives. *Molecules* **2019**, *24*, 2939. [CrossRef] [PubMed]
175. Cho, K.S.; Lim, Y.R.; Lee, K.; Lee, J.; Lee, J.H.; Lee, I.S. Terpenes from Forests and Human Health. *Toxicol. Res.* **2017**, *33*, 97–106. [CrossRef] [PubMed]
176. Cicero, A.F.G.; Fogacci, F.; Banach, M. Botanicals and phytochemicals active on cognitive decline: The clinical evidence. *Pharmacol. Res.* **2018**, *130*, 204–212. [CrossRef] [PubMed]

International Journal of
Molecular Sciences

MDPI

Review

Effectiveness of Vitamin D Supplementation in the Management of Multiple Sclerosis: A Systematic Review

Monika Berezowska, Shelly Coe * and Helen Dawes

Centre for Movement, Occupational and Rehabilitation Sciences, Oxford Brookes University, Oxford OX3 0BP, UK; mberezowska84@gmail.com (M.B.); hdawes@brookes.ac.uk (H.D.)

* Correspondence: scoe@brookes.ac.uk; Tel.: +44-1865-483-839

Received: 5 January 2019; Accepted: 5 March 2019; Published: 14 March 2019

Abstract: Objective: to examine the extent of effect vitamin D in Multiple Sclerosis (MS) on pathology and symptoms. Methods: A literature search was performed in November 2018 (CRD42018103615). Eligibility criteria: randomised control trials in English from 2012 to 2018; a clinical diagnosis of MS; interventions containing vitamin D supplementation (vitamin D3 or calcitriol) in disease activity compared to a control/placebo; improvement in: serum 25(OH)D, relapse rates, disability status by Expanded Disability Status Scale (EDSS) scores, cytokine profile, quality of life, mobility, T2 lesion load and new T2 or T1 Gd enhancing lesions, safety and adverse effects. Risk of bias was evaluated. Results: Ten studies were selected. The study size ranged from 40 to 94 people. All studies evaluated the use of vitamin D supplementation (ranging from 10 to 98,000 IU), comparing to a placebo or low dose vitamin D. The duration of the intervention ranged from 12 to 96 weeks. One trial found a significant effect on EDSS score, three demonstrated a significant change in serum cytokines level, one found benefits to current enhancing lesions and three studies evaluating the safety and tolerability of vitamin D reported no serious adverse events. Disease measures improved to a greater extent overall in those with lower baseline serum 25(OH)D levels. Conclusions: As shown in 3 out of 10 studies, improvement in disease measures may be more apparent in those with lower baseline vitamin D levels.

Keywords: Vitamin D; Multiple Sclerosis; symptom

1. Introduction

There is increasing evidence suggesting that specific environmental factors, such as exposure to infectious agents, smoking, poor diet and inadequate levels of vitamin D can influence the disease course of multiple sclerosis (MS) [1]. Adequate vitamin D status is documented as associated with reduced prevalence, activity and progression of disease in MS, and therefore high intake of vitamin D may be a useful addition to standard treatment [2]. Numerous observational studies investigating variations in sunlight exposure, latitude and diet have supported the correlation between a high serum concentration of vitamin D and reduced severity of the disease course in established MS [3,4].

Epidemiologic and experimental studies investigating the effectiveness of vitamin D supplementation in MS have shown that low serum vitamin D levels may exacerbate MS symptoms and therefore are associated with higher relapse rates, new lesions, and greater degree of disability [5–9]. Although there has been much research performed into the role of vitamin D in MS risk and progression, due to heterogeneity of study designs, there have been conflicting results. For example, baseline serum 25(OH)D levels often differ between studies. Reviews on the topic have thus far been inconclusive and are mainly focused on the role of vitamin D and risk of developing MS, rather than the outcomes after diagnosis [7]. The only two other systematic reviews to date on vitamin D for the clinical efficiency of MS did not use the full range of terms for vitamin D nor was bias assessed [10] and didn't assess cytokine outcomes nor looked at the effects of baseline Vitamin D levels on outcomes [11]. The aim of this review is to assess the evidence from existing randomised controlled trials for the clinical effectiveness of vitamin D supplementation compared to placebo supplementation in the disease and symptom management of people with MS as measured by: improvement in: serum 25(OH)D, relapse rates, disability status by Expanded Disability Status Scale (EDSS) scores, cytokine profile, quality of life, mobility, T2 lesion load and new T2 or T1 Gd enhancing lesions, safety, and adverse effects.

2. Methods

The systematic review was registered in PROSPERO (CRD42018103615). A literature search was performed in November 2018. Table 1 shows the search terms and number of hits for each database. Reference lists were hand searched for additional papers. Twenty percent of abstracts and papers were checked by a second reviewer.

Table 1. Key search databases and search terms.

Database Searched	Search Terms Used	Number of Results	Date of Search
PubMed	• "Multiple Sclerosis" or "MS" • AND • "vitamin D supplementation" OR "vitamin D" OR "cholecalciferol" OR "ergocalciferol" OR "calcitrol"	215	01/11/2017
Web of Science	• As above	197	04/11/2017
CINAHL	• As above	19	12/11/2017
Science Direct	• As above	354	12/11/2017
Total		785	

Studies were included if they met each of the following criteria: A clinical diagnosis of MS; Direct relevance of vitamin D supplementation on the management of MS compared to a low dose vitamin D or a placebo supplement; Primary outcome measurements in one or more of: serum 25(OH)D, relapse rates, disability status by EDSS scores, cytokine profile, quality of life, mobility, T2 lesion load and new T2 or T1 Gd enhancing lesions, safety and adverse effects; Randomised control trial (RCT) with a control and intervention group; Published from 2012 and in English; The published data available in full text; Only human randomised controlled clinical trials.

The Preferred Reporting Items for Systematic Reviews and Meta-Analyses (PRISMA) guidelines were followed and the flow diagram is presented in Figure 1. Bias was assessed using the RoB 2.0 tool at a study level. Data were extracted by one reviewer, and a selection of excluded abstracts and all full papers, and included papers were confirmed by a second reviewer.

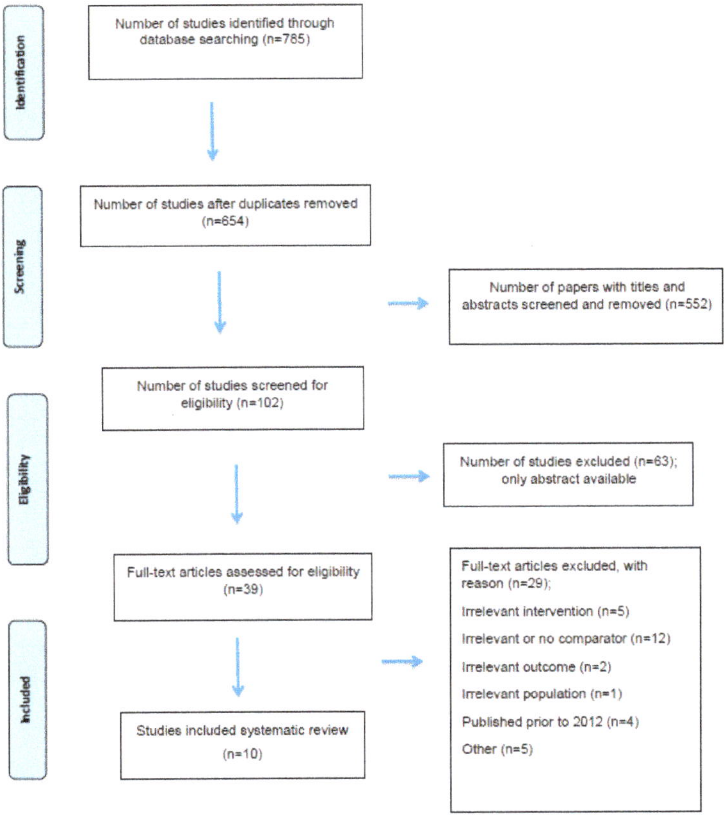

Figure 1. Preferred Reporting Items for Systematic Reviews and Meta-Analyses (PRISMA) flow diagram reporting the number of records identified, included and exclude through the different phases of a systematic review.

3. Results

Out of 785 studies, ten RCTs were identified as eligible for this review after the application of the inclusion and exclusion criteria. The information from each selected study was extracted, and detailed characteristics are shown in Table 2.

3.1. Bias

All studies were considered to have a low risk of bias and therefore systematic error was unlikely and there was no threat to validity.

3.2. General Characteristics

The studies reviewed in this report were all double-blind RCTs that focused on the role of vitamin D supplementation in the management of people with MS. Country of origin is shown in Table 2. Inclusion and exclusion criteria, in addition to other demographic information is shown in Table 3.

3.3. Participants

The studies size ranged from 40 to 94 people with MS. In these ten studies, there was a total of 627 adult participants (463 female and 164 male). Food intake of vitamin D and smoking status were not reported in any of the RCTs.

3.4. Study Objectives

Seven studies looked at the effect of vitamin D on immunological and inflammatory measures [12–18]. Outcomes related to functional ability were assessed in two studies [19,20] and relapse rate was assessed in four studies [12,19–21]. Disability and/or disease progression was assessed in five studies [12,18–21] and safety and tolerability of vitamin D supplementation was sought in four [12,13,20,21]. The studies by [16] and [18], by [15] and [19] and by [14] and [20] were based on the same trial however measured different outcomes and as such were treated in this review as separate studies.

3.5. Interventions

The intervention dose varied across studies. All studies evaluated the use of vitamin D supplements of various doses (ranging from 10 to 98,000 IU), frequency (usually delivered weekly) and formulation (vitamin D3 and calcitriol). Most studies ($n = 9$) reported concomitant immunomodulatory therapy, often interferon-β as well as different requirements relating to vitamin D and calcium supplementation that were used at baseline. The duration of the vitamin D interventions varied between studies, ranging from 12 to 96 weeks. The included studies ($n = 8$) compared vitamin D supplementation (321 participants) to placebo (264 participants) or versus low dose of vitamin D ($n = 2$; 42 participants). A variety of clinical and biochemical outcome measures were assessed at the baseline and the end of the study for intervention and control groups.

3.6. Serum 25(OH)D Levels

Nine of ten studies in this review measured the serum 25(OH)D concentration for both intervention and control group (low dose vitamin D) as an outcome parameter (Table 4). Across studies, mean improvements in cytokine profile or EDSS were seen for those with low baseline plasma Vitamin D levels ($n = 3$). Key findings and significance are shown in Table 5.

3.7. Immunologic Markers

Seven of the ten studies in the review used change in serum cytokines level as an outcome parameter with mixed results found and large heterogeneity in markers assessed across studies. Best support was found for Ashtari et al. [16] and Sotirchos et al. [13] in which significant benefits were seen in the high dose Vitamin D groups on IL-10, and on IL-17+CD4+T and CD4+T cells these were also the studies where baseline vitamin D levels were lower than normal.

Golan et al. [12] reported a significant increase in serum IL-17 concentration in people allocated to the low-dose vitamin D from a mean of 4.01 to 9.14 pg/mL at 48 weeks ($p = 0.037$) and a heterogeneous IL-17 response was observed in the high-dose vitamin D group. Therefore, there was a decrease and thus a beneficial change in 40% of participants and increase and negative change in 45% of participants after 3 months while 15% had IL-17 levels below the detection threshold at both time points. Aivo et al. [14] detected a significant increase in LAP (TGF-β) levels in the vitamin D arm after 48 weeks from a mean of 47 to 55 pg/mL ($p = 0.02$), while in those receiving placebo, this level increased but these changes were not significant ($p = 0.173$). Moreover, no significant difference in other cytokines concentration was reported in either group. Ashtari et al. [16] found that serum IL-10 concentration changed significantly in people receiving vitamin D for 12 weeks ($p = 0.015$) from a median of 12.58 to 13.76 pg/mL. Rosjo et al. [15] indicated no significant differences from baseline values for any of the inflammation markers between those receiving vitamin D or placebo after 96 weeks of treatment. Additionally, people with MS on immunomodulatory treatment (mostly consisting of IFN-b) were observed to have higher mean baseline levels of inflammation markers (IL-1Ra and CXCL16) compared to those not on therapy. However, there was no clear correlation between vitamin D supplementation and immunomodulatory treatment and its influence on the inflammation markers. Toghianifar et al. [18] showed that the proportion of cells including: nTreg, iTreg, Breg, IL4+ Th cells,

IL5, and LAP (TGFβ) was not affected by a high-dose vitamin D supplementation. No difference in IL-17 levels between those who received vitamin D, and those who received placebo were observed at 12 weeks. Muris et al. [17] found no beneficial effects of a high-dose vitamin D supplementation on the circulating regulatory immune cell compartment (the fraction of Treg as the proportion of CD4+ T cells, nTregs, IL10+ Th cells) in those with MS. Sotirchos et al. [13] detected a significant change in the proportion of proinflammatory IL-17+CD4+T cells in the high-dose group ($p = 0.016$) from a mean of 9.32% to 5.62%, while no difference was observed in the low-dose group ($p = 0.53$). Moreover, a significant difference in IL-17+CD4+T cells in the high-dose group versus low-dose group was reported ($p = 0.039$). Greater reduction in the proportion of IFN-γ+CD4+ T cells and IFN-γ+IL-17+CD4+ T cells was noted in the high-dose group versus the low-dose group but did not reach statistical significance (p = 0,12; $p = 0.14$). Also, a decreased proportion of effector memory CD4+T cell was noted after high-dose vitamin D supplementation from a mean of 40.56% to 30.69% ($p = 0.021$). The proportion of central memory and naive CD4+T cells increased significantly ($p = 0.019$; $p = 0.043$) in the high-dose group from a mean of 50.07% to 60.96% and from 38.94% to 42.2%, respectively.

3.8. Functional Measures

Only one study assessed functional measures and although there were trends for improvements in the Vitamin D groups, there were no statistically significant changes between the intervention and placebo groups. Soilu-Hänninen et al. [20] demonstrated that vitamin D supplementation resulted in fewer new T2 lesions (a mean of 0.5 compared to a mean of 1.1 in the placebo group). However, the difference between vitamin D and placebo groups was not statistically significantly different ($p = 0.286$). Participants assigned to vitamin D demonstrated lower total number of T1 Gd enhancing lesions (0.6 to 0.1) while in the placebo group no change was reported and a higher decrease in T1 enhancing lesion volume in the vitamin D group (from 57 mm^3 to 3.1 mm^3) compared with the placebo group (from 62 mm^3 to 29 mm^3) but again the difference between the treatment groups was not statistically significant ($p = 0.004$, $p = 0.320$, respectively). There were no statistically significant differences between the treatment groups in timed 10 foot tandem walk (TTW10; $p = 0.076$) (change from a mean of 11.7 to 9.7 in the vitamin D group and from 9.6 to 11.2 in the placebo group) and T25FW ($p = 0.932$) at the end of the study (change from a mean of 6.0 to 5.3 in the vitamin D group and from 4.7 to 5.1 in the placebo group).

3.9. Relapse Rate

Four of ten studies in this review investigated the effect of supplementation with vitamin D on relapse rates, with no significant differences between the vitamin D and control groups. Kampman et al. [19] demonstrated that vitamin D supplementation resulted in an increase in annualised relapse rate (ARR, calculated as the total number of relapses experienced divided by the sum of participants and duration of follow-up) from 0.11 to 0.14, whereas in placebo group a decrease from 0.15 to 0.8 was reported. The difference between vitamin D and placebo group after 96 weeks was not significant ($p = 0.25$). Shaygannejad et al. [21] documented that the relapse rate decreased significantly after 48 weeks from a mean of 1.04 to 0.32 in people who received vitamin D ($p < 0.001$) and from 1.04 to 0.40 in those who received placebo ($p < 0.001$). The study by Golan et al. [12] found an increase in ARR in patients with MS following the treatment with high-dose per day from 0.28 to 0.51 and decrease in the low-dose from 0.38 to 0.34 at week 48, but this difference was not statistically significant ($p = 0.32$). The study by Soilu-Hänninen et al. [20] found a decrease in ARR in both treatment arms: in people who received vitamin D from a mean of 0.49 to 0.26 and from 0.51 to 0.28 in those who received placebo, yet with no significant difference between groups.

Table 2. Characteristics of selected studies.

Reference	Participant Demographics	Study Design, Duration and Country of Origin	Intervention	Outcome Measures
[19]	68 participants (48f, 20m) with MS; Age mean (range) in vitamin D group 40 (21–50) and placebo 41 (26–50); BMI in vitamin D group 28 and placebo group 26	Double-blind placebo-controlled RCT; 96 weeks; Norway	35 participants received supplementation with 20,000 IU vitamin D3 (cholecalciferol)per week; comparator 33 participants received placebo	Serum levels of 25(OH)D; ARR; EDSS; MSFC components; grip strength; FSS
[21]	50 participants (424f, 6m) with RRMS; Age mean (SD) in vitamin D 38.6 (8.4) and placebo 37.9 (7.9); No BMI	Double-blind placebo-controlled RCT; 48 weeks; Iran	25 participants received 0.25 μg/d of calcitriol for 2 weeks and then 0.5 μg/d; comparator 25 participants received placebo	EDSS; relapse rate
[12]	45 participants (32f, 13m) with RRMS; Age mean in high-dose group 43.1 (21.7–63.7) and in low-dose group 43.6 (26.7–63.9); No BMI	Double-blind placebo-controlled RCT; 48 weeks; Israel	High-dose group, 24 participants received 75,000IU vitamin D3 solution every 3 weeks in addition to 800 IU vitamin D3 per day (total 4370 IU); comparator low dose group, 21 participants received placebo every 3 weeks in addition to 800 IU/d of vitamin D3	Serum levels of 25(OH)D; FLS; serum calcium, PTH, cytokine levels (IL-17, IL-10, and IFN-γ); EDSS, relapses, adverse events; QoL
[14]	59 participants (37f, 22m) with RRMS; Age mean (range) in vitamin D 38 (22–53) and in placebo 35 (24–53); BMI 24 kg/m^2	Double-blind RCT; 48 weeks; Finland	30 participants received 20,000 IU of vitamin D3 (cholecalciferol) per week; comparator 29 participants received placebo	Serum levels of 25(OH)D; inflammatory cytokine: Serum concentrations of LAP (TGF-β); IFN-γ, IL-17A, IL-2, IL-10, IL-9, IL-22, IL-6, IL-13, IL-4, IL-5, IL-1β and TNF-α
[16]	89 participants (75f, 14m) with RRMS; Age mean (SD) in vitamin D group 31.50 (7.60) and placebo 34.60 (10.12); No BMI	Double-blind placebo-controlled RCT; 12 weeks; Iran	High-dose vitamin D group, 44 participants received 50,000 IU of vitamin D3 every 5 days; comparator 45 participants received placebo	Serum levels of 25(OH)D; serum calcium; serum interleukin 10 (IL-10) levels
[15]	68 participants (48f, 20m) with RRMS; Age mean (range) in vitamin D group 40 (21–50) and placebo 41 (28–50); BMI vitamin D group 25.9 and placebo 26.5	Double-blind placebo-controlled RCT; 96 weeks; Norway	36 participants received 20,000 IU vitamin D3 per week; comparator 32 participants received placebo	Serum 25(OH)D; 11 serum markers of inflammation, bone mineral density, clinical disease activity, disease progression: ALCAMd, CCL21e, CXCL16f, IL-1Rag, MMP-9h, OPGi, OPNj, PTX3k, sFRP3l, sTNF-R1m, TGF-b1n

Table 2. *Cont.*

Reference	Participant Demographics	Study Design, Duration and Country of Origin	Intervention	Outcome Measures
[18]	89 participants (75f, 14m) with RRMS; Age mean (SD) in vitamin D group 31.50 (7.60) and placebo 34.60 (10.12); No BMI	Double-blind placebo-controlled RCT; 12 weeks; Iran	44 participants received oral vitamin D3 50,000 IU every 5 days; comparator 45 participants received placebo	Serum levels of 25(OH)D, serum calcium, IL-17
[17]	53 participants (35f, 18m) with RRMS; Age mean (SD) in vitamin D group 37.7 (7.2) and placebo 37.2 (9.6); BMI $\geq$ 25 kg/m^2	Double-blind placebo-controlled RCT; 48 weeks; Netherlands	30 participants received high-dose vitamin D3 supplementation 7000 IU/d for 4 weeks, followed by 14,000 IU/d; comparator 23 participants received placebo	Serum 25(OH)D; serum interleukin 10 (IL-10) levels; cytokine expression of IL4, IFNγ, IL17, IL22, GMCSF and TNFα by CD3+ CD8− T lymphocytes
[13]	40 participants (28f, 12m) with RRMS; Age mean (SD) in high-dose group 41.3 (8.1) and placebo 38.8 (8.8); No BMI	Double-blind RCT; 24 weeks; United States	High-dose group, 19 participants received 10,000 IU/d of cholecalciferol; comparator low-dose group, 21 participants received 400 IU/d of cholecalciferol	Serum 25(OH)D levels; adverse events, relapses, IFN-γ+ IL-17+ CD4+ T cells
[20]	66 participants (41f, 15m) with RRMS; Age median (range) in vitamin D group 39 (22–53) and placebo 35 (24–53); BMI median (range) in vitamin D group 24 (18–40) and placebo 24 (19–38)	Double-blind placebo controlled RCT; 48 weeks; Finland	34 participants received oral vitamin D3 (cholecalciferol) 20,000 IU once a week; comparator group 32 participants received placebo	Serum levels of 25(OH)D; PTH level, T2 BOD; total number of Gd enhancing T1 lesions; new/enlarging T2 lesions; Gd enhancing lesion volume; MRI activity; ARR, EDSS, T25FW and TTW10

ARR, annualised relapse rate; EDSS, Expanded Disability Status Scale; MSFC components, multiple sclerosis functional composite including (25ft timed walk; 9-hole peg test (9-HPT), paced auditory serial addition test (PASAT)); FSS, fatigue severity scale; FLS, flu-like symptoms; QoL, quality of life; T2 BOD, T2 burden of disease; T25FW, timed 25 foot walk; TTW10, timed 10 foot tandem walk; BMI, body mass index; y, years; f, female; m, male; RCT, randomised controlled trial; SD, standard deviation; MS, multiple sclerosis. RRMS, relapsing-remitting multiple sclerosis; 25(OH)D, 25-hydroxy vitamin D.

Table 3. Inclusion and exclusion criteria of reviewed studies.

Study	Age	MS Diagnosis	EDSS Score	Serum 25(OH)D Level	Other Inclusion Criteria	Exclusion Criteria
[19]	18–50 years	MS	≤4.5	n/a	n/a	Inability to walk 500 m or more; conditions or medication affecting bone health; pregnancy, lactating during the past 6 months; menopause; unwillingness to use contraception
[21]	15–60 years	RRMS	≤6	>40 ng/mL	RRMS for 1–12 years, no relapse for at least one month; continue current medications	SPMS and PPMS; other conditions; use of vitamin D supplements; pregnancy
[12]	≥18 years	RRMS	<7	<75 nmol/L or (<30 ng/mL)	IFN-β therapy or those who continue to suffer from FLS beyond 4 months of treatment with IFN-β	Abnormalities of vitamin D related hormonal system; use of medications that influence vitamin D metabolism; conditions of increased susceptibility to hypercalcemia; pregnancy

Table 3. *Cont.*

Study	Age	MS Diagnosis	EDSS Score	Serum 25(OH)D Level	Other Inclusion Criteria	Exclusion Criteria
[14]	18–55 years	RRMS	<5	<85 nmol/L or (<34 ng/mL)	IFN-β therapy for at least 1month and no neutralizing antibodies; contraception; at least one relapse during the year prior the study and/or MRI activity defined as presence of Gd-enhancing lesions on brain MRI	Serum calcium > 2.6 mmol/L; other conditions; pregnancy; use of other immunomodulatory therapy than INFB-1β; allergy to cholecalciferol or peanuts; alcohol or drug abuse
[16]	18–55 years	RRMS	<4	n/a	No relapse 30 days before inclusion; negative β-HCG test for women; calcium < 11 mg/dL	Pregnancy; lactation; other disease; receiving > 4000 IU of vitamin D, corticosteroids treatment in the previous 30 days; aspartate or alanine transaminase > 3xnormal values, ALP > 2.5xnormal values
[15]	18–50 years	RRMS	<4.5	n/a	n/a	Disease or medication affecting bone health; menopause; pregnancy; lactation; nephrolithiasis
[18]	18–55 years	RRMS	<4	<85 ng/mL	No relapse 30 days prior to study day; negative β-HCG test for women; calcium < 11 mg/dL; no relapse during the study	Pregnancy; lactation; other diseases; receiving >4000 IU of vitamin D, corticosteroids therapy in the previous 30 days; AST > 3xnormal values, ALP > 2.5xnormal values
[17]	18–50 years	RRMS	≤4	n/a	No relapse within 30 days prior to study day; first clinical event occurring within 5 years prior to screening; have had at least one relapse, or one or more Gd-enhancing or new T2 MRI lesions within the 12 months; receiving IFNβ-1a > 90 days and <12 months	Pregnancy or lactation; other diseases; use of corticosteroids or adrenocorticotrophic hormone within 30 days prior to SD1 abnormalities of vitamin D-related hormonal system; use of medications that influence vitamin D metabolism; taking N400 IU (N10 μg) of vitamin D supplement daily
[13]	18–55 years	RRMS	n/a	20–50 ng/mL	No relapse within 30 days; serum creatinine >1.5 mg/dL	Daily intake of vitamin D > 1000 IU or change of immunomodulatory therapy within the past 3 months, systemic glucocorticoid therapy; pregnancy, other condition
[20]	18–55 years	RRMS	≤5.0	<85 nmol/L	IFNB-1b use for at least 1month; no neutralising antibodies to IFNβ, as measured by the indirect myxovirus A (MxA) test, using appropriate contraceptive methods.	Pregnancy; serum calcium >2.6 mmol/L; primary hyperparathyroidism; alcohol or drug abuse; use of immunomodulatory therapy other than IFNB-1b; known allergy to cholecalciferol or peanuts; therapy with digitalis, calcitonin, vitamin D3 analogues or vitamin D; any condition predisposing to hypercalcaemia; significant hypertension (blood pressure < 180/110 mm Hg); hyperthyroidism or hypothyroidism in the year before the study began; a history of kidney stones in the previous 5 years; cardiac insufficiency or significant cardiac dysrhythmia; unstable ischaemic heart disease; depression; and inability to perform serial MRI scans.

MS, multiple sclerosis; RRMS, relapsing-remitting multiple sclerosis; EDSS, expanded disability status scale; FLS, flue like symptoms; HCG, Human chorionic gonadotropin; ALP, Alkaline phosphatase; AST, Aspartate transaminase; SPMS, secondary progressive multiple sclerosis; PPMS, primary progressive multiple sclerosis, SD1; study day 1; IFNβ, Interferon-β.

Table 4. Changes in serum 25(OH)D levels after intervention supplementation.

References	Within Group Differences	Between Group Differences
[19]	Intervention: 20,000 IU of vitamin D significantly increased serum 25(OH)D levels from a mean of 55.56 to 123.17 nmol/L. Control: there was only a minor increase from 57.33 to 61.80 nmol/L.	Significant difference in serum levels of 25(OH)D after 96 weeks between the intervention and control groups ($p < 0.001$).
[21]	n/a	n/a
[12]	Intervention: serum 25(OH)D levels significantly increased in a high-dose (4370 IU/d) groups from a mean of 48.2 to 122.6 nmol/L Control: low-dose (800 IU/d) from 48 to 68 nmol/L.	Significantly higher serum 25(OH-D) levels were reported in high dose group compared to low-dose arm after 48 weeks ($p < 0.001$).
[14]	Intervention: serum 25(OH)D levels increased significantly from a mean of 54 to 109 nmol/L. Placebo: decreased from a mean of 55 to 51 nmol/L after 48 weeks.	n/a
[16]	Intervention: Serum 25(OH)D levels rose from a median of 28.27 to 84.67 nmol/L. Placebo: fell from 39.6 to 28.66 nmol/L.	A significant difference after 12 weeks between groups ($p < 0.001$).
[15]	Intervention: serum levels of 25(OH)D significantly increased from 56 to 123nmol/L. Placebo: levels slightly increased from 57 to 63 nmol/L.	A significant difference in serum levels after 96 weeks between groups ($p < 0.001$).
[18]	Intervention: serum 25(OH)D levels significantly increased from a median of 28.27 to 84.67 ng/mL. Placebo: a decrease from 39.6 to 28.66 ng/mL.	These differences were significant between groups after 12 weeks ($p < 0.001$).
[17]	Intervention: serum 25(OH)D concentration increased significantly in the vitamin D group from 60 to 231 nmol/L. Placebo: changed to a lesser degree (54 to 60 nmol/L).	The was a significant difference after 48 weeks between the groups ($p < 0.001$).
[13]	High dose: Mean change of 34.9 ng/mL. Low dose: mean change of 6.9 ng/mL	A high dose of vitamin D resulted in significantly higher serum 25(OH-D) levels versus low-dose after 24 weeks ($p < 0.00001$).
[20]	Intervention: serum 25(OH)D levels increased from a mean of 54 to 110 nmol/L. Placebo: decreased from a mean of 56 to 50 nmol/L after 48 weeks	A significant difference between groups ($p < 0.001$).

Table 5. Key findings of reviewed studies.

Reference	Key findings	Significance	Conclusion
[19]	1. Serum 25(OH)D level significantly increased in intervention group vs control	1. $p < 0.001$	Supplementation did not result in beneficial effects on the measured MS-related outcomes; no significant difference between groups in ARR, EDSS, MSFC components, grip strength or fatigue
	2. ARR increased in intervention group vs control	2. $p = 0.25$	
	3. EDSS decreased in intervention group vs control	3. $p = 0.97$	
	4. MSFC components: 25ft timed walk decreased in intervention group vs control; 9-HPT increased in intervention group vs control; PASAT increased in intervention vs control;	4. $p = 0.87$; $p = 0.35$; $p = 0.21$	
	5. Grip strength decreased in intervention group vs control;	5. $p = 0.76$	
	6. Fatigue increased in intervention group vs control	6. $p = 0.9$	
[21]	1. Relapse rate significantly decreased in intervention and control groups; no significant difference in relapse rate between the groups;	1. $p < 0.001$; $p < 0.001$; $p > 0.05$;	No significant differences in the EDSS score or relapse rate between the vitamin D and control groups at the end of the study period; vitamin D supplementation at the doses used seems safe
	2. EDSS unchanged in intervention group and increased in control	2. N/A; $p < 0.01$	
[12]	1. Serum 25(OH)D levels increased in HDVD group vs LDVD group;	1. $p < 0.001$	Vitamin D supplementation was associated with dose-dependent changes in IL-17 serum levels, while not affecting IFN−β related FLS; vitamin D supplementation at the doses used seems safe
	2. PTH decreased in HDVD group but no significant change with LDVD;	2. $p = 0.04$; $p = 0.17$	
	3. No change in FLS	3. N/A	
	4. IL-17 levels increased in HDVD and LDVD groups;	4. $p = 0.75$; $p = 0.04$	
	5. No significant differences in relapse rate, EDSS, QoL, serum IL-10 and IFNγ;	5. $p > 0.05$	
	6. Serum calcium levels remained stable and within normal range in both dosage groups	6. $p = 0.2$; $p = 0.4$	
[14]	1. Serum levels of 25(OH)D increased in intervention group and edcreased in control;	1. N/A	Serum LAP (TGF-β) levels increased significantly in people receiving vitamin D; Therefore vitamin D might be useful in improving MRI outcomes; The levels of the other cytokines did not change significantly in either group
	2. Serum levels of LAP (TGF-β) increased in intervention and control group;	2. $p = 0.0249$; $p = 0.173$	
	3. The levels of serum IFN-gamma; IL-17A and in IL-9 increased in intervention group	3. $p = 0.0519$; $p = 0.0666$; $p = 0.0679$	

Table 5. *Cont.*

Reference	Key findings	Significance	Conclusion
[16]	1. Serum 25(OH)D levels increased in intervention group;	1. $p < 0.001$	25(OH)D levels increased significantly in those treated with vitamin D; IL-10 level increased significantly in the intervention group and its anti-inflammatory effect may play a role in improving outcomes in MS
	2. IL-10 levels increased in intervention group;	2. $p = 0.015$	
	3. No significant differences in serum calcium between groups at baseline or after 3 months	3. $p = 0.980$; $p = 0.302$	
[15]	1. Serum 25(OH)D level increased in intervention versus control;	1. $p < 0.001$	25(OH)D levels increased significantly in vitamin D group versus control; No significant differences for any inflammation markers between groups
	2. The inflammation marker averages did not differ significantly between groups	2. $p > 0.05$	
[18]	1. Serum 25(OH)D level increased in intervention versus control group;	1. $p < 0.001$	25(OH)D levels increased significantly in people in the intervention group; Significant difference in EDSS between groups; No difference in IL-17 levels between vitamin D and control group
	2. EDSS scores differ between groups;	2. $p = 0.033$	
	3. Serum levels of IL-17 changed in intervention group;	3. $p = 0.002$	
	4. No significant differences in serum calcium between groups at baseline and after 12 weeks	4. $p = 0.980$; $p = 0.302$	
[17]	1. Serum 25(OH)D level increased in intervention versus control group;	1. $p < 0.001$	25(OH)D levels increased significantly in the vitamin D group; Supplementation of vitamin D did not result in a relative increase in the total amount of lymphocytes
	2. The total amount of lymphocytes are similar between baseline and week 48;	2. $p > 0.05$	
	3. The proportion of cells in the immune regulatory cell compartment (nTreg, iTreg and Breg) did not change in either group;	3. $p > 0.05$	
	4. IL4+ Th cells decreased in the control but not the intervention group;	4. $p = 0.04$; $p = 0.92$	
	5. T cell cytokine secretion increased (IL5, LAP (TGF-β)) in the control but not the intervention group	5. $p = 0.02$; $p < 0.001$ $p = 0.06$; $p < 0.01$	
[13]	1. Serum 25(OH)D level increased in HDVD group vs LDVD;	1. $p < 0.00001$	25(OH)D levels increased significantly in the vitamin D group; Vitamin D supplementation exhibited immunomodulatory effects including reduction of interleukin-17 and decreased the proportion of effector memory CD4+ T cells with concomitant increase in central memory CD4+ T cells and naive CD4+ T cells; 10,400 IU daily is safe and tolerable
	2. The proportion of interleukin-17+CD4+ T cells, CD161+CD4+ T cells, and effector memory CD4+ T cells, the proportion of central memory CD4+ T cells and naive CD4+ T cells increased in HDVD group	2. $p = 0.016$; $p = 0.03$; $p = 0.021$; $p = 0.018$; $p = 0.04$	

Table 5. *Cont.*

Reference	Key findings	Significance	Conclusion
[20]	1. Serum 25(OH)D level significantly increased in intervention group vs control;	1. $p < 0.001$	Vitamin D3 add on treatment to IFNB reduces MRI T1 enhancing lesions. Vitamin D supplementation at the doses used seems safe.
	2. T2 BOD reduced in intervention group vs control;	2. $p = 0.105$	
	3. Total number of Gd enhancing T1 lesions significantly decreased in the intervention group vs control;	3. $p = 0.004$	
	4. Fewer new/enlarging T2 lesions in the intervention group vs control;	4. $p = 0.286$	
	5. Gd enhancing lesion volume decreased in intervention group vs control;	5. $p = 0.320$	
	6. MRI activity lower in intervention group vs control;	6. $p = 0.322$	
	7. ARR decreased in intervention group vs control;	7. N/A	
	8. EDSS decreased in intervention group vs control;	8. $p = 0.071$	
	9. TTW10 decreased in intervention group vs control;	9. $p = 0.076$	
	10. T25FW decreased in intervention group vs control	10. $p = 0.932$	

VD, vitamin D group; HDVD, high-dose vitamin D group; LDVD, low-dose vitamin D group; ARR, annualised relapse rate; EDSS, Expanded Disability Status Scale- scores range from 0 to 10; MSFC, MS functional composite including (25ft timed walk; 9-hole peg test (9-HPT), paced auditory serial addition test (PASAT)); FSS, fatigue severity scale- scores range from 1 (no fatigue) to 7; QoL, quality of life; FLS, flu-like symptoms; LAP, latency activated peptide, IFNβ, Interferon-β;.T2 BOD T25FW TTW10. T2 BOD, T2 burden of disease (BOD) on MRI scans; T25FW, timed 25 foot walk, TTW10, timed 10 foot tandem walk.

3.10. Disability

Five of ten studies reported EDSS score as outcome parameter with only one showing a benefit after supplementation with vitamin D. Kampman et al. [19] noted that EDSS score did not differ significantly between the vitamin D and placebo group after 96 weeks (p = 0.97). Shaygannejad et al. [21] found that EDSS score increased significantly ($p < 0.01$) in a placebo group from a mean of 1.7 to 1.94, whereas it did not change in people receiving vitamin D and therefore there was no significant difference in scores at the end of the trial between intervention and control groups ($p > 0.05$). Also, Golan et al. [12] demonstrated that high-dose vitamin D supplementation was not associated with reduced disability score with no significant change in EDSS score between two groups ($p = 0.26$). In contrast, Toghianifar et al. [18] showed a significant difference in EDSS scores between people allocated to vitamin D group (supplemented with 50,000 IU every five days) and placebo group after 12 weeks ($p = 0.033$) in favour of the vitamin D group, and the baseline vitamin D levels in the participants from this study was below the minimum recommendation. Soilu-Hänninen et al. [20] found no significant change in EDSS score between two groups ($p = 0.071$).

3.11. Safety and Tolerability

Four of the ten studies in this review determined the effect of high-dose vitamin D supplementation among people with MS in terms of safety and tolerability and none of the studies reported significant differences between control/placebo and vitamin D groups nor were any of the adverse events serious in either group. Shaygannejad et al. [21] showed that vitamin D treatment use up to 0.5 µg/day of calcitriol appeared to be safe and well tolerated by those with MS. The adverse events noted were mild in severity. The most frequently reported included constipation ($n = 6$ and $n = 4$), dyspepsia ($n = 6$ and $n = 2$), fatigue ($n = 4$ and $n = 5$), and headache ($n = 2$ and $n = 1$) in vitamin D and placebo groups, respectively. There were no significant differences in frequency of events between people who received vitamin D and those who received placebo. Golan et al. [12] indicated that a dose of 4370 IU/day over a 48-week period was safe in people with MS. There were no instances of hypercalcemia and no reports on new adverse events that could be vitamin D supplement related. Sotirchos et al. [13] found that a dose of 10,400 IU of cholecalciferol per day for 24 weeks was safe and tolerable in people with MS, with no serious adverse events. Soilu-Hänninen et al. [20] found no significant differences between the treatment arms in any of the other clinical chemistry parameters studied. No dose adjustments were necessary. Lack of MxA response (MxA < 50 mg/L) was detected in three people in both treatment arms at 12 months. Diarrhoea was a side effect in ($n = 5$ and $n = 2$) and fever was noted ($n = 2$ and $n = 5$) in the vitamin D group and placebo group, respectively. All other adverse events occurred in a similar number of participants in both groups. There was one serious adverse event in the vitamin D group (erysipelas in the interferon injection site treated with intravenous antibiotics in hospital) and two in the placebo group (elective hip surgery and elbow fracture).

4. Discussion

This review found some evidence for benefits of vitamin D supplementation, specifically for those with serum levels at the lower normal range in people with RRMS. Therefore, baseline serum vitamin D levels may be a predictor of improvements in disease pathology from vitamin D supplementation, cytokine profile and disability status, but possibly also relapse rate, quality of life, mobility, T2 lesions load and new T2 and T1 Gd enhancing lesions. Five out of ten studies showed improvement in: ARR(x2), EDSS(x2), IFN-gamma, IL-17A, IL-9, IL 10, 17+CD4+ T cells, CD161+CD4+ T cells, and effector memory CD4+ T cells, the proportion of central memory CD4+ T cells and naive CD4+ T, TTW10, T25FW, and MRI brain lesion markers, and these were shown in the intervention group compared with the control/placebo group. Another similar review to date differed in that cytokine outcomes were not assessed and the effects of baseline Vitamin D levels on outcome measures was not explored [11]. McLaughlin et al. [11] found that in higher dose vitamin D arms, there were actually adverse changes in ARR and EDSS and therefore although supplementation may have beneficial effects, there may be specific doses that should be considered. Jagannath et al. [22] looked at outcome measures including fatigue and HRQOL yet found conflicting results in part due to the heterogeneity of the study designs and different doses used. Zheng et al. [23] only looked at changes in ARR and EDSS score, with no beneficial effect of vitamin D as an add-on therapy on either outcome. Whilst further research is needed, this review highlights that all studies on the topic should include baseline vitamin D as part of the assessment. There was also low risk of adverse effects and low risk of bias for all studies and therefore the validity can be considered high. This review not only includes a more extensive search strategy and evaluates bias and although some of the included studies between reviews are similar, the current review is more up to date and encompasses a wider range of symptoms and pathology in MS.

The present consensus on the use of vitamin D supplementation in the management of MS is based on the hypothesis that the serum 25(OH)D is associated with prevalence and severity of the disease course in established MS. Therefore, its measurements are undertaken as part of the clinical management of MS in order to detect vitamin D insufficiency, correct it with supplementation at recommended doses and achieve the beneficial immunological effects [4]. All but one study assessed levels of serum 25(OH) and all reported a significant increase in 25(OH)D levels following vitamin D supplementation. However, the increase in 25(OH)D levels did not appear to affect all MS-related outcomes in the reviewed studies. If participants had 25(OH-D) levels at the lower end of normal at baseline, a high dose vitamin D supplement intervention may contribute to bettering of physiological mechanisms and resulting symptoms, yet if baseline levels are at the higher end of normal (i.e., 50 nmol/L) then further benefits may not be experienced. In the study by Ashtari et al. [16] and Sotirchos et al. [13] participants had levels towards the lower end of normal thereby possibly resulting in the resulting significant benefit in IL-10 and a variety of mechanistic improvements, respectively. Toghianifar et al. [18] found a resulting improvement in EDSS score which wasn't seen in other studies in this review, and again the participants in this study had baseline 25(OH-D) levels at the lower end of normal. All other studies had participants with higher baseline levels and also contained more varied results, with fewer significant changes between groups.

When looking at the immunological outcomes, the reviewed studies reported mixed effects of vitamin D supplementation. Vitamin D plays an important role in immune system function by reducing the production of proinflammatory cytokines and inducing the production of anti-inflammatory cytokines [24]. Only two selected studies detected a significant increase in levels of anti-inflammatory cytokines in the vitamin D group and therefore findings of studies evaluating the effect of the vitamin D supplementation on the reduction of proinflammatory cytokines are conflicting. The heterogeneity of intervention effects on immunologic activity reported in reviewed trials may be explained by considering possible confounding parameters including dosage and duration of administering vitamin D supplementation and supports previous findings demonstrating that a more pronounced immunologic impact of vitamin D supplementation was reported in vitamin D doses up to 40,000 IU per day [24]. Moreover, the fact that almost all participants in above trials were treated by immunomodulatory treatment, which mostly comprised interferon-beta (IFN-β) therapy (Table 6), may have altered the cytokine responses to vitamin D and/ or made it more difficult to determine the isolated effect of vitamin D supplementation and therefore beneficial effects of an increase in 25(OH)D on the outcome markers examined may be undetectable due to the strong immunomodulatory effect of IFN-β [25]. It has been suggested that type of therapy a person receives may influence the observed impact of vitamin D supplementation [26]. Notwithstanding, some studies demonstrated a synergistic immunomodulatory effect of IFN-β and vitamin D that induce favourable alterations in the inflammatory profile in people with MS [12,13]. Also, when considering the study conducted by Golan et al. [12] and Sotirchos et al. [13] including low-dose of vitamin D as a comparator may reduce the ability to notice minor differences compared to the use of a placebo. Although results of studies evaluating changes in immunological profiles in people with MS are not consistent, they suggest that supplementation of vitamin D promotes the immune regulatory cytokines and reduces proinflammatory immune parameters. Only two studies assessed changes in functional measures, and although the relationship between vitamin D and improved outcomes in participants with MS was found by Soilu-Hänninen et al. [20] T1 enhancing lesions and trends in MRI burden of disease (BOD) and EDSS, there is currently not enough clinical data to suggest the effectiveness of the treatment.

The correlation between 25(OH)D and reduced relapse rates have been found in several prospective cohort studies. The study by Laursen et al. [27] reported that the increase in serum 25(OH)D level was associated with decreases in ARR in those with RRMS. Those results were in line with a previously conducted cohort study by Simpson et al. [28] investigating a role of 25(OH)D levels in modulating MS clinical course in 145 participants with RRMS that suggests a benefit of serum 25(OH)D level on relapse rates at levels approximately 100 nmol/L. However, three reviewed studies evaluating vitamin D supplementation in management of MS have demonstrated no effect of 25(OH)D on relapse rate. Although mean serum 25(OH)D level more than doubled in the high-dose intervention groups in the study by Kampman et al. [19], Soilu-Hänninen et al. [20] and Golan et al. [12], they found no significant difference in ARR between groups at the end of the study period (96 and 48 weeks, respectively). Also, Shaygannejad et al. [21] failed to detect significant difference in relapse rate between the intervention and control groups at 48 weeks although the relapse rate decreased significantly in the vitamin D group. One possible explanation for the discrepancies between findings of above trials and previous studies may be related to eligibility criteria for included participants, vitamin D dosage and form, and duration of the intervention. Other explanations for the results in these RCTs may be related to the low ARR at baseline which could contribute to the absence of significant effects. In addition, the study conducted by Kampman et al. [19] enabled participants to continue the use of vitamin D supplements they used prior the study, which contributed to comparatively high 25(OH)D concentration in the placebo group and a difference between groups could not be detected.

High levels of 25(OH)D (>50 nmol/L) have also been shown to be associated with reduced disability measured by EDSS in MS [29]. Based on the evidence contained in this review, the effect of vitamin D supplementation on reducing disability remains unclear. Kampman et al. [19], Soilu-Hänninen et al. [20], Shaygannejad et al. [21] and Golan et al. [12] reported no significant change in EDSS score between the intervention and control groups. Conversely, a trial conducted by Toghianifar et al. [18] demonstrated a significant positive difference in EDSS scores between participants allocated to vitamin D vs placebo groups. Although the inclusion criteria were limited to participants with EDSS < 4 that indicate absence of observations in the higher EDSS range, a dose of 50,000 IU vitamin D every five days after 12 weeks was associated with less neurological disability.

Additionally, four studies looked at the safety and tolerability of high dosing regimens of vitamin D supplementation through the duration of the intervention. Through the studies observed it could be clearly recognised that vitamin D treatments were relatively safe, well-tolerated, and no concerning adverse events such as hypercalcemia and hypercalciuria triggered by high doses of vitamin D were reported. This is consistent with findings from previous studies that demonstrated safety of high-dose vitamin D below the daily limit of 10,000 IU in MS [30]. All other adverse events occurred in a similar number of participants in both groups for all studies. There was one serious adverse event in the vitamin D group (erysipelas in the interferon injection site treated with intravenous antibiotics in hospital) and two in the placebo group (elective hip surgery and elbow fracture). What can be concluded from this systematic review is that it seems participants in all studies adhered to the vitamin D interventions due to a resulting increase in serum levels in all studies ($n = 9$), and therefore the safety and tolerability of supplementation at high doses can be considered a reliable outcome.

Table 6. Dose of vitamin D and concomitant immunomodulatory therapy used in selected studies.

Study	High-Dose of Vitamin D	Low-Dose of Vitamin D	Placebo	Concomitant Immunomodulatory Therapy and Vitamin D/Calcium Supplements
[19]	20,000 IU of vitamin D3 per week	✗	✔	500 mg/d calcium; no restrictions on vitamin D supplements
[21]	0.25 μg/d of calcitriol for 2 weeks and then 0.5 μg/d	✗	✔	IFNβ (86.0% of participants), statins (10.0%), or immunosuppressive drugs (4.0%)
[12]	4370 IU/d of vitamin D3	800 IU/d of vitamin D3	✗	IFNβ
[14]	20,000 IU of vitamin D3 per week	✗	✔	IFNβ
[16]	50,000 IU of vitamin D3 every 5days	✗	✔	IFNβ; interferon-β; participants were not allowed to take any other vitamin D supplements;
[15]	20,000 IU vitamin D3 per week	✗	✔	calcium supplementation (500 mg/d); no restrictions on regular vitamin D supplementation or immunomodulatory treatment (i.e., IFN-b, glatiramer acetate, or natalizumab)
[18]	50,000 IU of vitamin D3 every 5days	✗	✔	IFNβ
[17]	7000 IU/d of vitamin D3 for 4 weeks, followed by 14,000 IU/d of vitamin D3;	✗	✔	IFNβ-1a
[13]	10,400 IU/d of vitamin D3	400 IU/d of vitamin D3	✗	89% of participants received immunomodulatory therapy; multivitamin containing 400 IU of D3 and 1000 mg/d of calcium
[20]	20,000 IU of vitamin D3 per week	✗	✔	IFNβ-1b

IFNβ, Interferon-β.

5. Limitations

As the reviewed studies took place in different geographic locations, sun exposure was different amongst groups and makes the comparison less reliable. Of note, all studies recruited participants with RRMS in order to ensure the homogeneity of the treatment groups in terms of the disease course and mechanisms. However, it has been demonstrated that immunomodulatory strategies employed for RRMS are not considered effective when applied in PPMS, suggesting cause for caution when generalising results to the greater MS population. Disease duration before the commencement of treatment varied between 4 months to 27 years and the time at which vitamin D intervention is implemented may affect the effectiveness of the treatment. Some studies assessed clinical endpoints such as relapse rates, disability scores, and physical changes, while some assessed only biomarker outcomes. As a result, heterogeneity of outcomes may have affected end-line comparisons and made doing a meta-analysis unfeasible.

6. Conclusions

Vitamin D supplementation may be a promising treatment and represents a reliable background for further exploration of potential benefit for MS regarding clinical improvements. A high dose vitamin D supplement intervention may contribute to bettering of physiological mechanisms if baseline plasma levels are at the lower end of normal. Further research addressing the matters discussed above is required before a causal association between vitamin D supplementation and disease activity in people with MS can be established.

Author Contributions: Methodology, M.B. and S.C.; validation, S.C. and H.D. formal analysis, M.B.; investigation, M.B.; resources, M.B.; writing—original draft preparation, M.B.; writing—review and editing, S.C. and H.D.; supervision, S.C.

Funding: H.D. is funded by the Elizabeth Casson Trust and the Oxford Health Biomedical Research Centre.

Conflicts of Interest: There was no conflict of interest.

References

1. Mandia, D.; Ferraro, O.E.; Nosari, G.; Montomoli, C.; Zardini, E.; Bergamaschi, R. Environmental factors and multiple sclerosis severity: A descriptive study. *Int. J. Environ. Res. Public Health* **2014**, *11*, 6417–6432. [CrossRef] [PubMed]
2. McDowell, T.Y.; Amr, S.; Culpepper, W.J.; Langenberg, P.; Royal, W.; Bever, C.; Bradham, D.D. Sun exposure, vitamin D intake and progression to disability among veterans with progressive multiple sclerosis. *Neuroepidemiology* **2011**, *37*, 52–57. [CrossRef] [PubMed]
3. Pierrot-Deseilligny, C.; Rivaud-Pechoux, S.; Clerson, P.; de Paz, R.; Souberbielle, J.C. Relationship between 25-OH-D serum level and relapse rate in multiple sclerosis patients before and after vitamin D supplementation. *Adv. Neurol. Disord.* **2012**, *5*, 187–198. [CrossRef] [PubMed]
4. Bagur, M.J.; Murcia, M.A.; Jimenez-Monreal, A.M.; Tur, J.A.; Bibiloni, M.M.; Alonso, G.L.; Martinez-Tome, M. Influence of diet in multiple sclerosis: A systematic review. *Adv. Nutr.* **2017**, *8*, 463–472. [CrossRef] [PubMed]
5. Runia, T.F.; Hop, W.C.; de Rijke, Y.B.; Buljevac, D.; Hintzen, R.Q. Lower serum vitamin D levels are associated with a higher relapse risk in multiple sclerosis. *Neurology* **2012**, *79*, 261–266. [CrossRef]
6. Wacker, M.; Holick, M.F. Sunlight and vitamin D: A global perspective for health. *Dermatoendocrinol* **2013**, *5*, 51–108. [CrossRef]
7. Duan, S.; Lv, Z.; Fan, X.; Wang, L.; Han, F.; Wang, H.; Bi, S. Vitamin D status and the risk of multiple sclerosis: A systematic review and meta-analysis. *Neurosci. Lett.* **2014**, *570*, 108–113. [CrossRef]
8. Harandi, A.A.; Harandi, A.A.; Pakdaman, H.; Sahraian, M.A. Vitamin D and multiple sclerosis. *Iran. J. Neurol.* **2014**, *13*, 1–6.
9. Fitzgerald, K.C.; Munger, K.L.; Kochert, K.; Arnason, B.G.; Comi, G.; Cook, S.; Goodin, D.S.; Filippi, M.; Hartung, H.P.; Jeffery, D.R.; et al. Association of vitamin D levels with multiple sclerosis activity and progression in patients receiving interferon beta-1b. *JAMA Neurol.* **2015**, *72*, 1458–1465. [CrossRef]
10. Pozuelo-Moyano, B.; Benito-Leon, J. Diet and multiple sclerosis. *Rev. Neurol.* **2014**, *58*, 455–464.

11. McLaughlin, L.; Clarke, L.; Khalilidehkordi, E.; Butzkueven, H.; Taylor, B.; Broadley, S.A. Vitamin D for the treatment of multiple sclerosis: A meta-analysis. *J. Neurol.* **2018**, *265*, 2893–2905. [CrossRef]
12. Golan, D.; Halhal, B.; Glass-Marmor, L.; Staun-Ram, E.; Rozenberg, O.; Lavi, I.; Dishon, S.; Barak, M.; Ish-Shalom, S.; Miller, A. Vitamin D supplementation for patients with multiple sclerosis treated with interferon-beta: A randomized controlled trial assessing the effect on flu-like symptoms and immunomodulatory properties. *BMC Neurol.* **2013**, *13*, 60. [CrossRef]
13. Sotirchos, E.S.; Bhargava, P.; Eckstein, C.; Van Haren, K.; Baynes, M.; Ntranos, A.; Gocke, A.; Steinman, L.; Mowry, E.M.; Calabresi, P.A. Safety and immunologic effects of high- vs low-dose cholecalciferol in multiple sclerosis. *Neurology* **2016**, *86*, 382–390. [CrossRef]
14. Aivo, J.; Hanninen, A.; Ilonen, J.; Soilu-Hanninen, M. Vitamin D3 administration to ms patients leads to increased serum levels of latency activated peptide (lap) of TGF-β. *J. Neuroimmunol.* **2015**, *280*, 12–15. [CrossRef]
15. Rosjo, E.; Steffensen, L.H.; Jorgensen, L.; Lindstrom, J.C.; Saltyte Benth, J.; Michelsen, A.E.; Aukrust, P.; Ueland, T.; Kampman, M.T.; Torkildsen, O.; et al. Vitamin D supplementation and systemic inflammation in relapsing-remitting multiple sclerosis. *J. Neurol.* **2015**, *262*, 2713–2721. [CrossRef]
16. Ashtari, F.; Toghianifar, N.; Zarkesh-Esfahani, S.H.; Mansourian, M. Short-term effect of high-dose vitamin D on the level of interleukin 10 in patients with multiple sclerosis: A randomized, double-blind, placebo-controlled clinical trial. *Neuroimmunomodulation* **2015**, *22*, 400–404. [CrossRef]
17. Muris, A.H.; Smolders, J.; Rolf, L.; Thewissen, M.; Hupperts, R.; Damoiseaux, J.; SOLARIUM Study Group. Immune regulatory effects of high dose vitamin D3 supplementation in a randomized controlled trial in relapsing remitting multiple sclerosis patients receiving ifnbeta; the solarium study. *J. Neuroimmunol.* **2016**, *300*, 47–56. [CrossRef]
18. Toghianifar, N.; Ashtari, F.; Zarkesh-Esfahani, S.H.; Mansourian, M. Effect of high dose vitamin D intake on interleukin-17 levels in multiple sclerosis: A randomized, double-blind, placebo-controlled clinical trial. *J. Neuroimmunol.* **2015**, *285*, 125–128. [CrossRef]
19. Kampman, M.T.; Steffensen, L.H.; Mellgren, S.I.; Jorgensen, L. Effect of vitamin D3 supplementation on relapses, disease progression, and measures of function in persons with multiple sclerosis: Exploratory outcomes from a double-blind randomised controlled trial. *Mult. Scler.* **2012**, *18*, 1144–1151. [CrossRef]
20. Soilu-Hanninen, M.; Aivo, J.; Lindstrom, B.M.; Elovaara, I.; Sumelahti, M.L.; Farkkila, M.; Tienari, P.; Atula, S.; Sarasoja, T.; Herrala, L.; et al. A randomised, double blind, placebo controlled trial with vitamin D3 as an add on treatment to interferon beta-1b in patients with multiple sclerosis. *J. Neurol. Neurosurg. Psychiatry* **2012**, *83*, 565–571. [CrossRef]
21. Shaygannejad, V.; Janghorbani, M.; Ashtari, F.; Dehghan, H. Effects of adjunct low-dose vitamin D on relapsing-remitting multiple sclerosis progression: Preliminary findings of a randomized placebo-controlled trial. *Mult. Scler. Int.* **2012**, *2012*, 452541. [CrossRef] [PubMed]
22. Jagannath, V.A.; Filippini, G.; Di Pietrantonj, C.; Asokan, G.V.; Robak, E.W.; Whamond, L.; Robinson, S.A. Vitamin D for the management of multiple sclerosis. *Cochrane Database Syst. Rev.* **2018**, *9*, CD008422. [CrossRef] [PubMed]
23. Zheng, C.; He, L.; Liu, L.; Zhu, J.; Jin, T. The efficacy of vitamin D in multiple sclerosis: A meta-analysis. *Mult. Scler. Relat. Disord.* **2018**, *23*, 56–61. [CrossRef] [PubMed]
24. Burton, J.M.; Kimball, S.; Vieth, R.; Bar-Or, A.; Dosch, H.M.; Cheung, R.; Gagne, D.; D'Souza, C.; Ursell, M.; O'Connor, P. A phase i/ii dose-escalation trial of vitamin D3 and calcium in multiple sclerosis. *Neurology* **2010**, *74*, 1852–1859. [CrossRef] [PubMed]
25. Loken-Amsrud, K.I.; Holmoy, T.; Bakke, S.J.; Beiske, A.G.; Bjerve, K.S.; Bjornara, B.T.; Hovdal, H.; Lilleas, F.; Midgard, R.; Pedersen, T.; et al. Vitamin D and disease activity in multiple sclerosis before and during interferon-beta treatment. *Neurology* **2012**, *79*, 267–273. [CrossRef] [PubMed]
26. Mosayebi, G.; Ghazavi, A.; Ghasami, K.; Jand, Y.; Kokhaei, P. Therapeutic effect of vitamin D3 in multiple sclerosis patients. *Immunol. Investig.* **2011**, *40*, 627–639. [CrossRef] [PubMed]
27. Laursen, J.H.; Sondergaard, H.B.; Sorensen, P.S.; Sellebjerg, F.; Oturai, A.B. Vitamin D supplementation reduces relapse rate in relapsing-remitting multiple sclerosis patients treated with natalizumab. *Mult. Scler. Relat. Disord.* **2016**, *10*, 169–173. [CrossRef]

28. Simpson, S., Jr.; Taylor, B.; Blizzard, L.; Ponsonby, A.L.; Pittas, F.; Tremlett, H.; Dwyer, T.; Gies, P.; van der Mei, I. Higher 25-hydroxyvitamin D is associated with lower relapse risk in multiple sclerosis. *Ann. Neurol.* **2010**, *68*, 193–203.
29. Thouvenot, E.; Orsini, M.; Daures, J.P.; Camu, W. Vitamin D is associated with degree of disability in patients with fully ambulatory relapsing-remitting multiple sclerosis. *Eur. J. Neurol.* **2015**, *22*, 564–569. [CrossRef]
30. Smolders, J.; Peelen, E.; Thewissen, M.; Cohen Tervaert, J.W.; Menheere, P.; Hupperts, R.; Damoiseaux, J. Safety and t cell modulating effects of high dose vitamin D3 supplementation in multiple sclerosis. *PLoS ONE* **2010**, *5*, e15235. [CrossRef]

International Journal of
Molecular Sciences

MDPI

Review

Natural Products in Neurodegenerative Diseases: A Great Promise but an Ethical Challenge

Marco Di Paolo, Luigi Papi, Federica Gori and Emanuela Turillazzi *

Section of Legal Medicine, Department of Surgical Pathology, Medical, Molecular and Critical Area, University of Pisa, 56124 Pisa, Italy; marco.dipaolo@unipi.it (M.D.P.); luigi.papi@unipi.it (L.P.); gori.federica@virgilio.it (F.G.)

* Correspondence: emanuela_turillazzi@inwind.it; Tel.: +39-050-2218520; Fax: +39-050-2218513

Received: 31 August 2019; Accepted: 16 October 2019; Published: 18 October 2019

Abstract: Neurodegenerative diseases (NDs) represent one of the most important public health problems and concerns, as they are a growing cause of mortality and morbidity worldwide, particularly in the elderly. Despite remarkable breakthroughs in our understanding of NDs, there has been little success in developing effective therapies. The use of natural products may offer great potential opportunities in the prevention and therapy of NDs; however, many clinical concerns have arisen regarding their use, mainly focusing on the lack of scientific support or evidence for their efficacy and patient safety. These clinical uncertainties raise critical questions from a bioethical and legal point of view, as considerations relating to patient decisional autonomy, patient safety, and beneficial or non-beneficial care may need to be addressed. This paper does not intend to advocate for or against the use of natural products, but to analyze the ethical framework of their use, with particular attention paid to the principles of biomedical ethics. In conclusion, the notable message that emerges is that natural products may represent a great promise for the treatment of many NDs, even if many unknown issues regarding the efficacy and safety of many natural products still remain.

Keywords: neurodegenerative diseases; natural products; ethics; patients' autonomy; beneficence; nonmaleficence; medical liability

---

1. Introduction

Neurodegenerative diseases (NDs) include a number of chronic progressive disorders of the central nervous system that are caused by the degradation and subsequent loss of neurons. NDs represent one of the most important public health problems and concerns, as they are a growing cause of mortality and morbidity worldwide, particularly in the elderly. The aging of the population has contributed to the increase of NDs [1,2], and age-related diseases such as NDs are becoming extremely important, due to their irreversibility, lack of effective treatment, and accompanied social and economic burdens [3].

Traditionally, classifications of NDs included Parkinson's disease, which is well characterized by a loss of dopaminergic nigrostriatal neurons; Huntington's disease, in which the loss of spiny, medium-sized striatal neurons occurs; and Alzheimer's disease (AD), due to diffuse cerebral atrophy. Other disorders such as primary dystonia or essential tremor were also referred to as NDs [4]. NDs recognize a broad, often overlapping, spectrum of symptoms, varying from memory and cognitive deficits to the impairment of a person's ability to move, speak, and breathe; they also share some clinical characteristics such as a relentless progression over years, sometimes even decades [5].

Beyond the known differences in the pathogenic mechanisms of individual diseases, neurodegeneration, understood as the chain of events leading to gradual loss of neurons' functional properties until cell death, represents the key point of this group of diseases [4], and attracts research efforts in trying to understand precise pathogenic mechanisms and achieve valid therapies.

In fact, despite remarkable breakthroughs in our understanding of NDs, there has been little success in developing effective therapies [6]. The therapies currently available seem to be inadequate for NDs, as they only act to alleviate symptoms but cannot stop the progress of the disease. The use of natural products (NPs) is growing, probably due to several factors [7,8] (Figure 1).

Factors responsible for increased use of NPs

- dissatisfaction with the results from traditional drugs
- claims on the efficacy of plant medicines
- increasing propensity of consumers for natural therapies and alternative medicines
- diffuse misconception that herbal products are safer than traditional products
- high cost of many traditional drugs
- improvements in the quality, efficacy, and safety of herbal medicines with the development of science and technology
- increased patient decision-making autonomy regarding therapeutic options and alternatives
- a movement toward self-medication

Figure 1. Factors responsible for increased use of NPs.

They may offer great potential opportunities in the prevention and therapy of NDs [9], and scientists are increasingly exploring options with herbal drugs and natural products [10]. As for many traditional drugs, many clinical concerns have arisen regarding the use of NPs, mainly focusing on the lack of scientific support or evidence for their efficacy and patient safety. These clinical uncertainties raise critical questions from a bioethical and legal point of view, as considerations relating to patient decisional autonomy, patient safety, and beneficial or non-beneficial care may need to be addressed, meaning that many intriguing points may arise regarding the use of NPs [11,12].

This paper does not intend to advocate for or against the use of natural products but to analyze the ethical framework of their use, with particular attention paid to the principles of biomedical ethics as described by Beauchamp and Childress [13], then addressing the strict intertwining of bioethics, safety, and responsibility related to the use of natural products.

2. Natural Products in Neurodegeneration

Despite specific clinical and etiopathogenic differences, NDs show some common features such as abnormal protein deposition, abnormal cellular transport, mitochondrial deficits, inflammation, intracellular $Ca2+$overload, uncontrolled generation of ROS, and excitotoxicity, thus suggesting the existence of converging pathways of neurodegeneration and reinforcing the importance of these pathways as common targets for intervention strategies [14]. Furthermore, reactive astroglia and/or microglia have been implicated in the pathogenesis of all major neurodegenerative disorders [15].

Over the years, target-based therapies such as neurotransmitter modulators, direct receptor agonists/antagonists, second messenger modulators, stem cell-based therapies, hormone replacement therapy, and neurotrophic factors, as well as regulators of mRNA synthesis and their translation into disease-causing mutant proteins, have been introduced and implemented [14].

However, there are currently no therapeutic strategies capable of either preventing or reducing the progression of NDs; many of the approved drug regimens for NDs help to treat the symptoms but do not cure the disease itself. As many of the traditional symptomatic therapies may lose their effectiveness over time, produce disruptive symptoms of their own, and show severe side effects [16], there is an urgent need to develop more effective and safer therapies that can be employed over a long duration of NDs.

Several natural agents have been proposed to complete and/or assist the traditional pharmacological agents in the treatment of neurodegenerative disorders, and the general idea of this is provided, among others, by Srivastava and Yadav [17]. Their use in NDs is widely reported in the literature [18–21], as these products show several different neuroprotective activities. Mitochondrial dysfunction, apoptosis, excitotoxicity, inflammation, oxidative stress, and protein misfolding are among the main neuroprotective targets of natural products [21–25].

Animal-based products such as omega-3 fatty acids inhibit cellular toxicity and show anti-inflammatory effects in AD [26]. Plant derived products like Lunasin, Polyphenols, Alkaloids, and Tannins are potential therapeutic candidates for AD [25]. Resveratrol and flavonoids seem to be dietary components with specific neuroprotective action and positive effects on human cognitive decline [27,28].

It is beyond the scope of this paper to analyze the pharmacological and pharmacodynamic profiles of the various natural products proposed for the treatment of NDs. However, some general concepts are necessary because, in addition to having important clinical repercussions, they reflect heavily on the bioethical—and even possibly legal—implications of the administration of natural products in NDs.

It can be said that natural products may be very promising due to their anti-neurodegenerative action, with the potential to treat a large number of patients worldwide.

A general belief that NPs are safe exists: Many patients take NPs, presumably based on the assumption that they are effective, safe, and less toxic than traditional drugs [29]. Very often, patients assess NPs as safer than biomedicine on the basis of being "natural". A factor promoting the consumption of herbal products may be the preconception that botanicals are natural and "natural is good" and consequently safe.

However, as drugs, they either may have adverse effects or may be not effective [30]. Because natural products may derive from diverse biological sources, their conversion into therapies is not trivial. Challenges may include concerns regarding their stability and neuroavailability [22], difficulty in adequately identifying and quantifying the active principle, and, finally, the difficulty of organizing large clinical trials to test these complex products.

Product characterization may be a key problem for NPs, as for many of them, the specific bioactive components have not been identified or are not fully characterized [21]. Herbal products contain complex active components or phytochemicals such as flavonoids, alkaloids, and isoprenoids. Therefore, it is frequently difficult to determine which component of the herb has the most biological activity [31]. A further complication is the fact that several natural compounds have limited stability and are easily degradable and may be metabolized to inactive products [22,32,33]. Concerns regarding compound solubility, restricted passage through the blood–brain barrier, and availability exist [22,33,34]. Furthermore, the evidence for the potential protective effects of selected herbs is generally based on experiments demonstrating a biological activity in a relevant in vitro bioassay or experiments using animal models [35]. However, in order to further widen their acceptance and use, clinical trials should be encouraged [36–38], since for many products, a translation to clinical trials may offer challenges that need to be addressed [39].

Firstly, in clinical trials, single compounds are more frequently investigated, while the investigation of plant extracts containing a variety of secondary metabolites is more common in studies prior to clinical studies. The combination of the various active principles in extracts can lead to additional or synergistic effects, giving better antioxidant/disease-modifying activity [39]. Moreover, natural antioxidants (i.e., from natural products or plant extracts) could also share this multi-target drug profile, and the combination of single compounds or extracts also needs to be further investigated.

Concerns regarding the purity and potency of herbal products exist. Product quality is influenced by many factors, including which portion of the plant is used (i.e., the root, stem, leaves, flowers), the time of harvest (i.e., young versus old plants), the handling of the product, and the proper identification of the plant. Furthermore, labelling may be inaccurate [39]. Of course, the concerns regarding the quality of herbal products cannot be generalized, and it is beyond question that high-quality products also exist in the legal market and that many manufacturers are beginning to significantly improve their quality standards. Nevertheless, the quality of herbal products still varies from one product to another, and many companies still sell low-quality products [40]. The focus on the quality of the products on the market is closely linked to the regulatory framework for such preparations [41], since a product marketed without any medical, pharmaceutical, or other regulation is bound to be very different from one which is regulated, for example, in the context of a licensing or registration scheme.

Conclusively, natural products have recently gained greater attention as alternative or integrative therapeutic agents against AD and other NDs [42]. However, critics have raised concerns that the popularity of natural products is growing without scientific support or evidence of their efficacy and safety [11].

These uncertainties justify both paying a great deal of attention to these products, and the careful monitoring of their use by clinicians. At the same time, the use of NPs often requires clinicians to make decisions under conditions of uncertainty, thus involving questions about which ethical principles clinical decision-making should rely on, what kind of information should be provided to patients, and what obligations arise on the part of physicians (Figure 2).

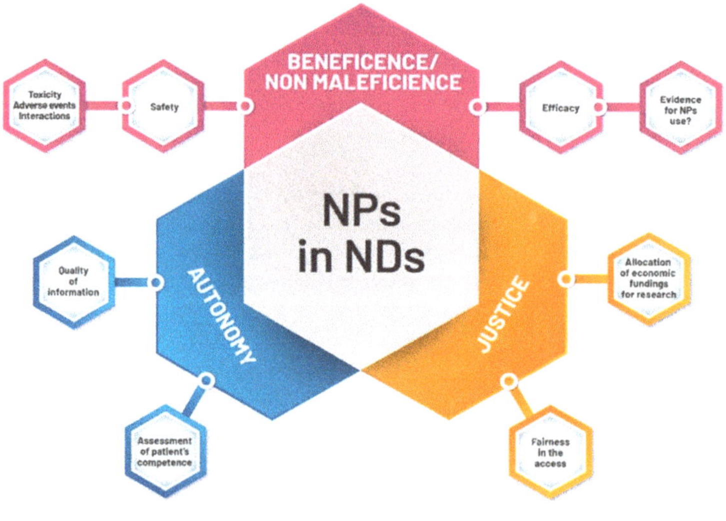

Figure 2. Main ethical topics of Natural Products (NPs) use in NDs.

3. Natural Products: On the Cutting Edge of Ethics?

3.1. Patient Autonomy

Autonomy is a fundamental bioethical principle requiring that a person has the capacity and opportunity to act autonomously; that is, to freely and voluntarily make choices. In a health-care setting, when a patient exercises her/his autonomy, she/he decides which of the options for dealing with her/his health-care problems will be best, given her/his values, concerns, and goals. A patient who makes autonomous choices is able to opt for what she/he considers will be best, all things considered [43].

It is increasingly clear that several gaps may exist regarding the use of natural products that may lead to a failure to provide adequate information regarding natural products and leave patients in the dark. The goal of the informed consent process is to provide sufficient information to a patient so that she/he can make the voluntary decision whether or not to take a natural product as an alternative or in combination with other types of drug treatment. Obtaining consent involves informing the subject about the potential risks and/or benefits of the proposed product and the alternative treatments available, if any. A description of any foreseeable risks or discomforts to the subject, an estimate of their likelihood, a description of any benefits that may reasonably be expected from the natural product, and, finally, the disclosure of any appropriate alternative procedures or courses of treatment that might be advantageous to the patient, should be the milestones of the informative process.

Unfortunately, several disturbing points about the use of natural products can compromise patient autonomy.

First of all, there is a lack or scarcity of basic information regarding the safety and effectiveness of many natural products used in NDs. In their analysis of the risks of complementary and alternative medicine, White et al. [44] clearly highlighted that natural products, such as plant extracts, may be pharmacologically complex, and so have multiple physiological effects which may represent a beneficial synergy or harmful interaction, depending on the specific context. Natural products may have pharmacological effects, just as with synthetic pharmaceuticals [45]; however, they are generally perceived as safe due to being natural, and patients themselves are less likely to report harmful incidents that may be associated with natural products than with conventional drugs [46]. Furthermore, in many cases of natural products, the associated risks are really unknown, and there is a limitation of currently available information concerning their safety and efficacy [47].

The informative process regarding the use of natural products, both as primary treatment and as adjunct treatment, may be complex and multifaceted as the evidence regarding natural products safety and efficacy is still not consolidated, even if it is growing, and many disturbing points (insufficient detailed knowledge of natural products pharmacology and drug interactions) may represent great challenges in obtaining real, informed consent regarding the use of natural products themselves, since they are not embedded in well-understood scientific paradigms. Taking these concerns to extremes, some authors even went as far as to state that it is unethical for medical professionals to offer or endorse "alternative medicine" treatments for which there is no known causal mechanism [48].

As the ethical concept of treating patients as people with respect and with the understanding of the individual's right to self-determination (autonomy) is fundamental [49,50], removing any ambiguity from the informative process is, at the moment, the only ethically correct answer. When uncertainties regarding efficacy and safety exist, as for many natural products, honest information should include telling patients about the degree of uncertainty associated with the efficacy and safety of the treatment, as well as the availability and risk–benefit ratio of other treatment options [51], so that patients could become comfortable with uncertainties and accept or refuse treatment with natural products. As such, the use of natural products in NDs could even result in a net gain in patient autonomy, as it provides patients with different therapeutic options with respect to traditional drug therapies. In other words, despite the existing gaps of knowledge on the efficacy and safety of NPs, patients affected by NDs should be able to freely choose to undergo treatments with NPs, with the hope of improving their quality of life or lengthening survival.

The final thought is that also in treatments through natural products does the culture of respect of patient autonomy, preference, and choice provide the underpinning needed to establish an effective physician–patient relationship, in which there may also be space for a conscious adherence to therapies which, although not yet rigorously validated, may nevertheless represent for the patient a beneficial alternative or supplement to traditional treatments—unfortunately, of poor effectiveness.

This paves the way for further reflection concerning the accurate assessment of patients' cognitive ability to give informed consent. A proper informative process requires that the patients "understand" the information given by the physician and appreciate the consequences of their decision. In patients affected by NDs, it may be difficult to evaluate whether the patient is able to understand the information presented and consent to or refuse the proposed treatment [52]; in other words, the assessment of the decision-making capacity, defined as the ability to understand and reason through the decision-making process, may be challenging in ND patients. This may represent a problem of great magnitude if we only consider that the numbers of older adults with cognitive impairment due to NDs is estimated to be high and rapidly increasing [53]. Case by case, the physician has the legal and ethical duty to explore the real existence of incompetence and if—and to what extent—it may affect the capacity for decision making for consent to treatment. Legal standards for decision-making capacity may be different across national jurisdictions; however, they generally include the capacity to understand the relevant information regarding a proposed treatment, its consequences, and alternatives [54,55]. A personalized, patient-targeted approach, grounded on the individual patient's characteristics and clinical situation, should guide physicians while assessing the patient's level of decisional capacity [53–55]. If the physician believes that a patient is incompetent to make a treatment decision, advance directives or legal proxies must be considered according to the existing national regulatory framework.

3.2. *Beneficence/Nonmaleficence/Justice*

The ethical concepts of beneficence, nonmaleficence, and justice warrant a shared discussion concerning the use of natural products, as they appear to be strictly interconnected.

Beneficence and the twin concept of nonmaleficence demand that patients should not be harmed by a treatment, both entreating the physician to avoid the causation of harm, and to provide benefits as well as balance benefits, burdens, and risks. The justification for the use of natural products seems to be that providing ND patients with these products provides them with benefits and could improve their quality of life, and even prolong their lives.

The central ethical challenge in the use of NPs is thus to determine when evidence has reached a sufficient level of certainty to warrant clinical introduction. In considering the available evidence, relevant factors include the scope of estimated benefit, the existence of alternative treatments, the nature and scope of potential harms, and the overall quality of evidence. When only limited evidence supports the use of natural products, the appropriate support of clinician (and patient) choice remains an ethical concern. Rarely does a single response exist. The clinically (and, of consequence, ethically) justified use of any natural product must be individualized to the patient's circumstances, including the stage of the disease, the severity of symptoms affecting the quality of life, and the existence/absence of valid therapeutic alternatives. Given the evidence gaps that still exist for many NPs, clinicians should consider what uncertainties they (and patients) are most comfortable with.

In the use of NPs, the physician's ethical obligation of beneficence/nonmaleficence (an obligation to maximize benefit and minimize harm) does not differ from standard clinical practice, where every drug is always potentially risky. When the risk of harm is disproportional to the potential benefit, providing the patient with this product may be questionable in the light of the ethical principles of beneficence/nonmaleficence.

Justice demands that all patients be treated fairly since "justice [is] fair, equitable, and appropriate treatment in light of what is due or owed to persons. Injustice involves a wrongful act or omission that denies people resources or protections to which they have a right" [13].

As mentioned above, this may be an option for ND patients who set their hopes on NPs as an alternative or supplement to the traditional synthetic drugs; however, this principle is only valid if and when patients do indeed get access to NPs.

Data from the literature recall the fact that the use of natural products within complementary and alternative medicine has been shown to be a popular choice of therapy among patients [29,56–59]. It has been reported that 80% of the world's population uses natural products, rising to 95% of the population in developing countries [56]. The use of herbal medicine was commonly reported across the European Union according to the CAMbrella consortium [57]. Patients with chronic diseases that are mostly resistant to conventional therapies tend to choose alternative and integrative therapies [58].

However, differences in NP utilization emerge for reasons that have to be discussed, as they seem to be unacceptable from an ethical point of view, thus representing red flags regarding the justice principle. Differences in utilization could be either for acceptable reasons (e.g., personal preferences and choices) or unacceptable reasons, such as costs and opportunities.

As Nissen et al. outlined [56], herbal and natural products may be the only available treatment for low-income people in developing countries; on the other hand, in high-income countries, natural products are often provided outside public healthcare services or insurance coverage, thus being mostly used by educated citizens of working age and with an above-average income. It has been reported that in industrialized societies, the use of complementary medicine has been found to be associated with higher income and higher education, and people who have lower incomes and educational levels tend not to use complementary medicine [60]. As with many other services of alternative and complementary medicine, in most countries worldwide, NPs seem to be mostly provided outside the public health system, and patients have to pay themselves for these products [56,58].

Immediate questions of ethics arise: If NPs in NDs represent a clinically prospective option, why do people who have free traditional health-care services or insurance coverage for traditional treatments have to pay out-of-pocket for NPs? Additionally, in those countries where inadequate and expensive conventional medical services exist, conventional medical care may not be accessible to the poorest minorities. In these situations, NPs are not alternative nor integrative [60]: They may represent the only therapeutic option, thus facilitating a "separate but unequal care system" [61]. In other words, the above dimensions of the use of NPs point out issues of inequity in gaining access to NPs, as whether patients affected by NDs who set their hopes on NPs are able to access them may be determined by economic and financial factors outside the patients' control, making this access unfair.

A further argument that may be subject of ethical discussion is that public funding should not be allocated to research of implausible treatments, representing a potential waste of resources both human and economic, and an additional expense over and above other healthcare costs. One of the key themes is that, given the scarcity of resources that can be allocated to research, every effort should go to those areas where reasonable, good evidence also exists [60].

However, this may represent a double-edged sword, ethically speaking: The further development of NPs is achievable only on a broad base of quality research. For example, the National Centre for Complementary and Alternative Medicine experience in the United States has shown that when funds are available and priorities are set, research on alternative and complementary medicine will grow exponentially [60,61]. The status of the research on complementary and alternative medicine in the European Union is not encouraging: Complementary and Alternative Medicine research in Europe is not well-funded by the countries or research organizations, and is in large part charitably supported [57].

This seems to feed a vicious circle, since the conduct of high-quality research on complementary and alternative medicine requires a commitment by the research community, as well as sustained financial support from governments and industry [62]. Implementing research on the use of NPs, as well as other alternative and complementary medicine, is pivotal to the achievement of "high-quality" evidence in support of the use of NPs.

4. Medical Liability Scenarios

As the use of alternative therapies and natural products grows, there is likely to be heightened concern about the liability implications of delivering these therapies. It has been reported that complementary and integrative medicine (in which the use of NPs is included) have not, until now, been the subject of serious malpractice litigation [63]; however, herbal medicines and other supplements are one of the more controversial fields of medicine [64].

Generally speaking, medical malpractice occurs if a physician fails to meet the standard of care established and negligently injures the patient, requiring three elements: (1) The patient suffered damage; (2) the negative event proceeded directly by the action or inaction of the healthcare provider; and (3) the physician was negligent, which essentially entails showing that she/he took less care than that which is customarily practiced by the average member of the profession in good standing, given the circumstances of the doctor and the patient [65].

These elements are long-established fundamental principles; however, the manner in which these principles will be applied to the use of NPs can raise important questions.

First of all, what is the standard of care for these treatments? Has the physician met the applicable standard of care while treating the patient with NPs? This may represent a difficult conceptual passage, as the issue of standard of care (i.e., the degree of care, diligence, and skill which is provided by a reasonable physician under like or similar circumstances) is not clearly defined for complementary and alternative treatments [64,66]. In some activities included in alternative or complementary practices, it may be difficult to establish the standard appropriate in the discipline, as a well-established scientific base and standardized protocols are generally lacking.

However, we can speculate that a physician who prescribes NPs would be held to a common standard in medical practice when assessing what she/he knew or should have known regarding the safety and the efficacy of the prescribed remedies [64,66]. Prescriptions of natural products, although not negligent per se, could represent a deviation from the standard of care, if it is demonstrated that this prescription was not something a reasonable physician would have done in similar circumstances and with the same patient.

The conceptual framework of potential alleged malpractice claims regarding the use of natural products focuses on some key points: (i) Whether the evidence in the scientific literature supports, does not support, or is inconclusive regarding the use of a natural product along the dimensions of safety and efficacy; and (ii) an accurate evaluation of the risk–benefit balance, including the stage and severity of the illness, the curability of the illness with conventional therapy, its potential toxicity, and adverse effects [63,64,67,68].

Although a standard of care for the use of NPs has not yet been clearly defined, as many therapies based on natural products have not yet been the subject of rigorous, controlled studies, the care provided through these products could be judged by either the clinician's duty to carefully consider findings from conventional medicine, the clinical status of that patient, the existing evidence on the natural products proposed, the reported adverse events, and, finally, the duty of carefully monitoring the clinical evolution of the patient.

As a concluding statement, it would seem appropriate to reiterate the words of Cohen, who suggested that "If evidence supports safety, but evidence regarding efficacy is inconclusive—accept but monitor; if the medical evidence supports efficacy, but evidence regarding safety is inconclusive—accept but monitor; and the medical evidence indicates either serious risk or inefficacy—avoid and discourage" [69].

5. Concluding Remarks

In conclusion, the notable message that emerges is that NPs may represent a great promise for the treatment of many NDs, where traditional therapies via synthetic drugs only act to alleviate symptoms, but cannot stop the progress of the disease, and thus are substantially inadequate.

It is, however, problematic for this message that there are still many unknown issues regarding the efficacy and safety of many natural products. There is much yet to be investigated, characterized, and learned. This message strongly underscores urgent needs. Clinicians must routinely inquire about all product use—conventional, complementary, and alternative—to promote patient safety and ethical care.

The evaluation of patient safety and of the efficacy of natural products should represent the guiding principle of physicians' conduct.

Finally, researchers should actively start to expand the knowledge base regarding natural product safety and efficacy, emphasizing that fundamental research to advance the understanding of the basic biological mechanisms of action of these products is pivotal.

Author Contributions: M.D.P., L.P., and F.G. were involved in the concept of the study; E.T. and M.D.P. were involved in the drafting and revising of the manuscript.

Conflicts of Interest: The authors declare no conflict of interest.

References

1. Heemels, M.T. Neurodegenerative diseases. *Nature* **2016**, *539*, 179. [CrossRef] [PubMed]
2. GBD. 2015 Neurological Disorders Collaborator Group. Global, regional, and national burden of neurological disorders during 1990-2015: A systematic analysis for the global burden of disease study 2015. *Lancet Neurol.* **2017**, *16*, 877–897. [CrossRef]
3. Hung, C.-W.; Chen, Y.-C.; Hsieh, W.-L.; Chiou, S.-H.; Kao, C.-L. Ageing and neurodegenerative diseases. *Ageing Res. Rev.* **2010**, *9*, S36–S46. [CrossRef] [PubMed]
4. Burgunder, J.M. Neurodegeneration. *IUBMB Life* **2003**, *55*, 291. [CrossRef] [PubMed]
5. Erkkinen, M.G.; Kim, M.O.; Geschwind, M.D. Clinical Neurology and Epidemiology of the Major Neurodegenerative Diseases. *Cold Spring Harb. Perspect. Biol.* **2018**, *10*, a033118. [CrossRef] [PubMed]
6. Katsnelson, A.; De Strooper, B.; Zoghbi, H.Y. Neurodegeneration: From cellular concepts to clinical applications. *Sci. Transl. Med.* **2016**, *8*, 364ps18. [CrossRef]
7. Ekor, M. The growing use of herbal medicines: Issues relating to adverse reactions and challenges in monitoring safety. *Front. Pharmacol.* **2013**, *4*, 177. [CrossRef]
8. Bandaranayake, W.M. Quality control, screening, toxicity, and regulation of herbal drugs. In *Modern Phytomeicine*; Turning Medicinal Plants into Drugs; Ahmad, I., Aquil, F., Owains, M., Eds.; Wiley-VCH GmbH & Co. KGaA: Weinheim, Germany, 2008; pp. 25–27.
9. Dubey, S.K.; Singhvi, G.; Krishna, K.V.; Agnihotri, T.; Saha, R.N.; Gupta, G.; Agnihitri, T. Herbal Medicines in Neurodegenerative Disorders: An Evolutionary Approach through Novel Drug Delivery System. *J. Environ. Pathol. Toxicol. Oncol.* **2018**, *37*, 199–208. [CrossRef]
10. Parvez, M.K. Natural or Plant Products for the Treatment of Neurological Disorders: Current Knowledge. *Curr. Drug Metab.* **2018**, *19*, 424–428. [CrossRef]
11. Ernst, E.; Cohen, M.H.; Stone, J. Ethical problems arising in evidence based complementary and alternative medicine. *J. Med. Ethics* **2004**, *30*, 156–159. [CrossRef]
12. Kim, H.J.; Jeon, B.; Chung, S.J. Professional ethics in complementary and alternative medicines in management of Parkinson's disease. *J. Park. Dis.* **2016**, *6*, 675–683. [CrossRef] [PubMed]
13. Beauchamp, T.; Childress, J. *Principles of Biomedical Ethics*, 5th ed.; Oxford University Press: New York, NY, USA, 2001.
14. Dadhania, V.P.; Trivedi, P.P.; Vikram, A.; Tripathi, D.N. Nutraceuticals against Neurodegeneration: A Mechanistic Insight. *Curr. Neuropharmacol.* **2016**, *14*, 627–640. [CrossRef] [PubMed]
15. Durrenberger, P.F.; Fernando, F.S.; Kashefi, S.N.; Bonnert, T.P.; Seilhean, D.; Nait-Oumesmar, B.; Schmitt, A.; Gebicke-Haerter, P.J.; Falkai, P.; Grünblatt, E.; et al. Common mechanisms in neurodegeneration and neuroinflammation: A BrainNet Europe gene expression microarray study. *J. Neural Transm. (Vienna)* **2015**, *122*, 1055–1068. [CrossRef] [PubMed]
16. Rasool, M.; Malik, A.; Qureshi, M.S.; Manan, A.; Pushparaj, P.N.; Asif, M.; Qazi, M.H.; Qazi, A.M.; Kamal, M.A.; Gan, S.H.; et al. Recent Updates in the Treatment of Neurodegenerative Disorders Using Natural Compounds. *Evid. Based Complement. Altern. Med.* **2014**, *2014*, 97973. [CrossRef]
17. Srivastava, P.; Yadav, R.S. Efficacy of Natural Compounds in Neurodegenerative Disorders. In *Glutamate and ATP at the Interface of Metabolism and Signaling in the Brain*; Springer Science and Business Media LLC: Berlin, Germany, 2016; Volume 12, pp. 107–123.
18. Lökk, J.; Nilsson, M. Frequency, type and factors associated with the use of complementary and alternative medicine in patients with Parkinson's disease at a neurological outpatient clinic. *Parkinsonism Relat. Disord.* **2010**, *16*, 540–544. [CrossRef]
19. Shan, C.S.; Zhang, H.F.; Xu, Q.Q.; Shi, Y.H.; Wang, Y.; Li, Y.; Lin, Y.; Zheng, G.Q. Herbal Medicine Formulas for Parkinson's Disease: A Systematic Review and Meta-Analysis of Randomized Double-Blind Placebo-Controlled Clinical Trials. *Front. Aging Neurosci.* **2018**, *10*, 349. [CrossRef]
20. Panda, S.S.; Jhanji, N. Natural Products as Potential anti-Alzheimer Agents. *Curr. Med. Chem.* **2018**. [CrossRef]

21. Leonoudakis, D.; Rane, A.; Angeli, S.; Lithgow, G.J.; Andersen, J.K.; Chinta, S.J. Anti-Inflammatory and Neuroprotective Role of Natural Product Securinine in Activated Glial Cells: Implications for Parkinson's Disease. *Mediat. Inflamm.* **2017**, *2017*, 8302636. [CrossRef]
22. Bagli, E.; Goussia, A.; Moschos, M.M.; Agnantis, N.; Kitsos, G. Natural Compounds and Neuroprotection: Mechanisms of Action and Novel Delivery Systems. *In Vivo* **2016**, *30*, 535–547.
23. Starkov, A.A.; Beal, F.M. Portal to Alzheimer's disease. *Nat. Med.* **2008**, *14*, 1020–1021. [CrossRef]
24. Venkatesan, R.; Ji, E.; Kim, S.Y. Phytochemicals That Regulate Neurodegenerative Disease by Targeting Neurotrophins: A Comprehensive Review. *Biomed Res. Int.* **2015**, *2015*, 1–22. [CrossRef] [PubMed]
25. Deshpande, P.; Gogia, N.; Singh, A. Exploring the efficacy of natural products in alleviating Alzheimer's disease. *Neural Regen. Res.* **2019**, *14*, 1321–1329. [PubMed]
26. Wollen, K.A. Alzheimer's disease: The pros and cons of pharmaceutical, nutritional, botanical, and stimulatory therapies, with a discussion of treatment strategies from the perspective of patients and practitioners. *Altern. Med. Rev.* **2010**, *15*, 223–244. [PubMed]
27. Cicero, A.F.; Ruscica, M.; Banach, M. Resveratrol and cognitive decline: A clinician perspective. *Arch. Med. Sci.* **2019**, *15*, 936–943. [CrossRef] [PubMed]
28. Castelli, V.; Grassi, D.; Bocale, R.; d'Angelo, M.; Antonosante, A.; Cimini, A.; Ferri, C.; Desideri, G. Diet and Brain Health: Which Role for Polyphenols? *Curr. Pharm. Des.* **2018**, *24*, 227–238. [CrossRef] [PubMed]
29. Barry, A.R. Patients' perceptions and use of natural health products. *Can. Pharm. J.* **2018**, *151*, 254–262. [CrossRef]
30. Karimi, A.; Majlesi, M.; Rafieian-Kopaei, M. Herbal versus synthetic drugs; beliefs and facts. *J. Nephropharmacol.* **2015**, *4*, 27–30.
31. Khazdair, M.R.; Anaeigoudari, A.; Hashemzehi, M.; Mohebbati, R. Neuroprotective potency of some spice herbs, a literature review. *J. Tradit. Complement. Med.* **2018**, *9*, 98–105. [CrossRef]
32. Coimbra, M.; Isacchi, B.; Van Bloois, L.; Toraño, J.S.; Ket, A.; Wu, X.; Broere, F.; Metselaar, J.M.; Rijcken, C.J.; Storm, G.; et al. Improving solubility and chemical stability of natural compounds for medicinal use by incorporation into liposomes. *Int. J. Pharm.* **2011**, *416*, 433–442. [CrossRef]
33. Shoji, Y.; Nakashima, H. Nutraceutics and delivery systems. *J. Drug Target.* **2004**, *12*, 385–391. [CrossRef]
34. Van Der Schyf, C.J.; Geldenhuys, W.J.; Youdim, M.B.H. Multifunctional drugs with different CNS targets for neuropsychiatric disorders. *J. Neurochem.* **2006**, *99*, 1033–1048. [CrossRef]
35. Pohl, F.; Kong Thoo Lin, P. The Potential Use of Plant Natural Products and Plant Extracts with Antioxidant Properties for the Prevention/Treatment of Neurodegenerative Diseases: In Vitro, In Vivo and Clinical Trials. *Molecules* **2018**, *23*, 3283. [CrossRef]
36. Parveen, A.; Parveen, B.; Parveen, R.; Ahmad, S. Challenges and guidelines for clinical trial of herbal drugs. *J. Pharm. Bioallied Sci.* **2015**, *7*, 329–333.
37. Cicero, A.F.G.; Fogacci, F.; Banach, M. Botanicals and phytochemicals active on cognitive decline: The clinical evidence. *Pharmacol. Res.* **2018**, *130*, 204–212. [CrossRef]
38. Cicero, A.F.; Bove, M.; Colletti, A.; Rizzo, M.; Fogacci, F.; Giovannini, M.; Borghi, C. Short-Term Impact of a Combined Nutraceutical on Cognitive Function, Perceived Stress and Depression in Young Elderly with Cognitive Impairment: A Pilot, Double-Blind, Randomized Clinical Trial. *Alzheimers Dis.* **2017**, *4*, 12–15.
39. Kemper, K.J.; Vohra, S.; Walls, R. Task Force on Complementary and Alternative Medicine; Provisional Section on Complementary, Holistic, and Integrative Medicine. American Academy of Pediatrics. The use of complementary and alternative medicine in pediatrics. *Pediatrics* **2008**, *122*, 1374–1386. [CrossRef]
40. Abdel-Tawab, M. Do We Need Plant Food Supplements? A Critical Examination of Quality, Safety, Efficacy, and Necessity for a New Regulatory Framework. *Planta Med.* **2018**, *84*, 372–393. [CrossRef]
41. Heinrich, M. Quality and safety of herbal medical products: Regulation and the need for quality assurance along the value chains. *Br. J. Clin. Pharmacol.* **2015**, *80*, 62–66. [CrossRef]
42. Manoharan, S.; Essa, M.M.; Vinoth, A.; Kowsalya, R.; Manimaran, A.; Selvasundaram, R. Alzheimer's Disease and Medicinal Plants: An Overview. *Adv. Neurobiol.* **2016**, *12*, 95–105.
43. Young, R. Informed consent and patient autonomy. Chapter 42. In *A Companion to Bioethics*; Khuse, H., Singer, P., Eds.; Blackwell Publishing Ltd: Hoboken, NJ, USA, 2009; p. 442.
44. White, A.; Boon, H.; Alraek, T.; Lewith, G.; Liu, J.P.; Norheim, A.J.; Steinsbekk, A.; Yamashita, H.; Fønnebø, V. Reducing the risk of complementary and alternative medicine (CAM): Challenges and priorities. *J. Integr. Med.* **2014**, *6*, 404–408. [CrossRef]

45. Chatfield, K.; Salehi, B.; Sharifi-Rad, J.; Afshar, L. Applying an Ethical Framework to Herbal Medicine. *Evid. Based Complement. Altern. Med.* **2018**, *2018*, 1903629. [CrossRef]
46. Barnes, J.; Mills, S.Y.; Abbot, N.C.; Willoughby, M.; Ernst, E. Different standards for reporting ADRs to herbal remedies and conventional OTC medicines: Face-to-face interviews with 515 users of herbal remedies. *Br. J. Clin. Pharmacol.* **1998**, *45*, 496–500. [CrossRef]
47. Ernst, E. Risks of herbal medicinal products. *Pharmacoepidemiol. Drug Saf.* **2004**, *13*, 767–771. [CrossRef]
48. Shahvisi, A. No Understanding, No Consent: The Case Against Alternative Medicine. *Bioethics* **2016**, *30*, 69–76. [CrossRef]
49. Entwistle, V.A.; Watt, I.S. Treating Patients as Persons: A Capabilities Approach to Support Delivery of Person-Centered Care. *Am. J. Bioeth.* **2013**, *13*, 29–39. [CrossRef]
50. Epstein, R.M. The ambiguity of personhood. *Am. J. Bioeth.* **2013**, *13*, 42–44. [CrossRef]
51. Ernst, E. Ethics of complementary medicine: Practical issues. *Br. J. Gen. Pract.* **2009**, *59*, 517–519. [CrossRef]
52. Fields, L.M.; Calvert, J.D. Informed consent procedures with cognitively impaired patients: A review of ethics and best practices. *Psychiatry Clin. Neurosci.* **2015**, *69*, 462–471. [CrossRef]
53. Dunn, L.B.; Alici, Y.; Roberts, L.W. Ethical Challenges in the Treatment of Cognitive Impairment in Aging. *Curr. Behav. Neurosci. Rep.* **2015**, *2*, 226–233. [CrossRef]
54. Moye, J.; Karel, M.J.; Azar, A.R.; Gurrera, R.J. Capacity to consent to treatment: Empirical comparison of three instruments in older adults with and without dementia. *Gerontologist* **2004**, *44*, 166–175. [CrossRef]
55. Appelbaum, P.S. Clinical practice. Assessment of patients' competence to consent to treatment. *N. Engl. J. Med.* **2007**, *357*, 1834–1840. [CrossRef]
56. Nissen, N.; Weidenhammer, W.; Schunder-Tatzber, S.; Johannessen, H. Public health ethics for complementary and alternative medicine. *Eur. J. Integr. Med.* **2013**, *5*, 62–67. [CrossRef]
57. Hegyi, G.; Petri, R.P., Jr.; di Sarsina, P.R.; Niemtzow, R.C. Overview of Integrative Medicine Practices and Policies in NATO Participant Countries. *Med. Acupunct.* **2015**, *27*, 318–327. [CrossRef]
58. von Ammon, K.; Cardini, F.; Daig, U.; Dragan, S.; FreiErb, M.; Hegyi, G.; Roberti di Sarsina, P.; Sørensen, J.; Ursoniu, S.; Weidenhammer, W.; et al. Health Technology Assessment (HTA) and a Map of CAM Provision in the EU: Final Report of CAMbrella Work Package No. 5. 2012. Available online: Phaidra.univie.ac.at/detail_object/o:300096 (accessed on 31 August 2019).
59. Ong, C.K.; Petersen, S.; Bodeker, G.C.; Stewart-Brown, S. Health status of people using complementary and alternative medical practitioner services in 4 English counties. *Am. J. Public Health* **2002**, *92*, 1653–1656. [CrossRef]
60. Bodeker, G.; Kronenberg, F. A Public Health Agenda for Traditional, Complementary, and Alternative Medicine. *Am. J. Public Health* **2002**, *92*, 1582–1591. [CrossRef] [PubMed]
61. *White House Commission on Complementary and Alternative Medicine Policy*; Final Report; March 2002. Available online: http://www.whccamp.hhs.gov/finalreport.html (accessed on 31 August 2019).
62. Nahin, R.L.; Nahin, R.; Straus, S.E. Research into complementary and alternative medicine: Problems and potential. *BMJ* **2001**, *322*, 161–164. [CrossRef] [PubMed]
63. Cohen, M.H. *Complementary & Alternative Medicine: Legal Boundaries and Regulatory Perspectives*; Johns Hopkins University Press: Baltimore, MD, USA, 1998.
64. Raposo, V.L. Complementary and alternative medicine, medical liability and the proper standard of care. *Complement. Ther. Clin. Pract.* **2019**, *35*, 183–188. [CrossRef] [PubMed]
65. Moffett, P.; Moore, G. The Standard of Care: Legal History and Definitions: The Bad and Good News. *West. J. Emerg. Med.* **2011**, *12*, 109–112. [PubMed]
66. Gilmour, J.; Harrison, C.; Asadi, L.; Cohen, M.H.; Vohra, S. Complementary and alternative medicine practitioners' standard of care: Responsibilities to patients and parents. *Pediatrics* **2011**, *128* (Suppl. S4), S200–S205. [CrossRef]
67. Cohen, M.H.; Eisenberg, D.M. Potential physician malpractice liability associated with complementary and integrative medical therapies. *Ann. Intern. Med.* **2002**, *136*, 596–603. [CrossRef]

68. Girard, L.; Vohra, S. Ethics of Using Herbal Medicine as Primary or Adjunct Treatment and Issues of Drug–Herb Interaction. Chapter 21. In *Herbal Medicine: Biomolecular and Clinical Aspects*, 2nd ed.; Benzie, I.F.F., Wachtel-Galor, S., Eds.; CRC Press/Taylor & Francis: Boca Raton, FL, USA, 2011.
69. Cohen, M.H. Legal issues in caring for patients with kidney disease by selectively integrating complementary therapies. *Adv. Chronic Kidney Dis.* **2005**, *12*, 300–311. [CrossRef]

International Journal of
Molecular Sciences

MDPI

Review

Ascaroside Pheromones: Chemical Biology and Pleiotropic Neuronal Functions

Jun Young Park [1,2], Hyoe-Jin Joo [2], Saeram Park [2,3] and Young-Ki Paik [1,2,*]

1 Interdisciplinary Program in Integrative Omics for Biomedical Science, Yonsei University, Seoul 03722, Korea
2 Yonsei Proteome Research Center, Yonsei University, Seoul 03722, Korea
3 Department of Chemical Physiology and Dorris Neuroscience Center, The Scripps Research Institute, La Jolla, CA 92037, USA
* Correspondence: paikyk@yonsei.ac.kr or paikyk@gmail.com

Received: 12 July 2019; Accepted: 7 August 2019; Published: 9 August 2019

Abstract: Pheromones are neuronal signals that stimulate conspecific individuals to react to environmental stressors or stimuli. Research on the ascaroside (ascr) pheromones in *Caenorhabditis elegans* and other nematodes has made great progress since ascr#1 was first isolated and biochemically defined in 2005. In this review, we highlight the current research on the structural diversity, biosynthesis, and pleiotropic neuronal functions of ascr pheromones and their implications in animal physiology. Experimental evidence suggests that ascr biosynthesis starts with conjugation of ascarylose to very long-chain fatty acids that are then processed via peroxisomal β-oxidation to yield diverse ascr pheromones. We also discuss the concentration and stage-dependent pleiotropic neuronal functions of ascr pheromones. These functions include dauer induction, lifespan extension, repulsion, aggregation, mating, foraging and detoxification, among others. These roles are carried out in coordination with three G protein-coupled receptors that function as putative pheromone receptors: SRBC-64/66, SRG-36/37, and DAF-37/38. Pheromone sensing is transmitted in sensory neurons via DAF-16-regulated glutamatergic neurotransmitters. Neuronal peroxisomal fatty acid β-oxidation has important cell-autonomous functions in the regulation of neuroendocrine signaling, including neuroprotection. In the future, translation of our knowledge of nematode ascr pheromones to higher animals might be beneficial, as ascr#1 has some anti-inflammatory effects in mice. To this end, we propose the establishment of ***pheromics*** (*pher*omone *omics*) as a new subset of integrated disciplinary research area within chemical ecology for system-wide investigation of animal pheromones.

Keywords: ascaroside pheromone; *C. elegans*; dauer; neuronal signaling; sexual behavior; survival signals; stress response

1. What Are Pheromones?

Pheromones are neuronal signaling molecules synthesized by various organisms and then excreted into the environment, where they typically stimulate individuals of the same species to react to environmental changes (e.g., temperature shifts, biological stimuli, or nutritional changes) [1,2]. It is thought that most organisms, from prokaryotes to higher animals such as humans, can produce and use pheromones for communication between conspecific individuals. In most cases, pheromones trigger neuronal events that are linked to various behavioral responses. The outcomes of such neuronal stimulation are the modulation of developmental and/or physiological programs that can support adaptation to new environments [3]. For example, approximately 1500 insect pheromones have been identified since bombykol was discovered in 1959 [4]. These pheromones mediate common behaviors such as courtship rituals, mating, aggregation, dispersal (e.g., spacing or epideictic pheromones), alarm, recruitment (e.g., trailing pheromones), and maturation [2,4]. In mammals, pheromones are used for marking territories, and for signaling mating and feeding preparedness [5,6].

In humans, there have been numerous reports of putative pheromones; however, their existence has not been experimentally confirmed. For example, a putative human pheromone was proposed to be excreted from the apocrine gland in the male underarm, although its functions have not been characterized [7,8]. Unlike other mammals, humans lack a functional vomeronasal organ (VNO), which processes pheromonal signals in mice and other vertebrates [8–10]. The absence of this key VNO function makes the discovery of human pheromones even more challenging.

The *Caenorhabditis elegans* dauer pheromone, which is part of an important chemical language throughout this nematode's lifespan, has long been known. In 1975, Cassada and Russell first reported the existence of dauer larvae, an alternative developmental stage that prolongs survival under environmental conditions that do not support growth [11]. The observation of dauer larvae might have provoked the search for dauer pheromones. In 1982, the first biological evidence of a nematode pheromone was reported by the Riddle group, who showed that a partially purified *C. elegans* extract could trigger dauer formation in L1/L2 larvae [12]. Indeed, this pioneering work inspired worm biologists to continue to search for pure dauer pheromones.

2. Structural Diversity of Ascaroside (ascr) Pheromones

2.1. Daumone, the First Chemically Characterized Ascr Pheromone

In 2005, the Paik group isolated and chemically characterized the first *C. elegans* pheromone, which they named dauer pheromone, or daumone (now often referred to as **ascr#1**) [13]. Via an activity-guided purification procedure using 300 L of cultured worms, they isolated pure daumone, which has the molecular formula $C_{13}H_{24}O_6$ and an M_r of 276 (Figure 1). Determination of the stereochemical structure of purified daumone, [(2)-(6R)-(3,5-dihydroxy-6-methyltetrahydro-pyran-2-yloxy) heptanoic acid], revealed that it contains one ascarylose (a 3,6-dideoxy sugar also known as rhamnose) linked to the C7 of a methylated short-chain fatty acid (mSCFA) (Figure 1).

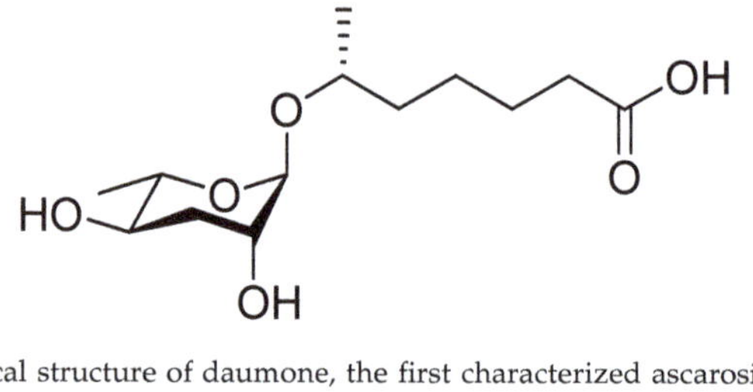

Figure 1. The chemical structure of daumone, the first characterized ascaroside (ascr) pheromone (**ascr#1**), contains an ascarylose sugar and a methylated short-chain fatty acid (mSCFA) linked by an ether bond [(2)-(6R)-(3,5-dihydroxy-6-methyltetrahydropyran-2-yloxy) heptanoic acid] [13].

They also demonstrated that natural and chemically synthesized daumone could equally induce dauer formation in the wild-type *C. elegans* laboratory strain (N2) and in *Caenorhabditis briggsae*. The discovery of daumone, which is indeed a bona fide signaling molecule, not only settled a long-time dispute as to whether the *C. elegans* pheromone acted as a signal or a crowd cue [14], it also opened a new avenue for investigating the chemical biology of ascr pheromones on molecular and system-wide scales. As additional dauer pheromone derivatives (collectively called ascarosides) were identified, daumone was later renamed ascaroside #1 (**ascr#1**) as per Edison's suggestion [15], which was based on the presence of an ascarylose sugar moiety linked to an mSCFA. In this review, we use "ascr pheromones" rather than ascarosides to distinguish between the pheromones and non-pheromonal ascr derivatives or metabolites, consistent with the terminology used for steroid hormones (i.e., steroids vs steroid hormones). This distinction is important, given that more than 200 ascaroside-like compounds with unknown functions have now been identified via metabolomic methods [16,17].

Historically, non-pheromonal ascarosides were first identified among the neutral lipids of parasitic nematodes such as *Ascaris lumbricoides* and *Parascaris equorum* [18,19]. These compounds typically consist of a glycone moiety (one or two ascarylose units, i.e., a 3,6-dideoxy sugar) and an aglycone moiety, a very long chain fatty acid (VLCFA) that contains greater than or equal to 25 carbon atoms [20]. They were mainly recovered from the eggs and reproductive tract tissue of female *A. lumbricoides* nematodes, and they were shown to confer the eggs with chemical resistance against external toxic insults [21]. Therefore, unless otherwise stated, our discussion will be limited to ascr pheromones and their potential neuronal functions.

2.2. Identification of Diverse Ascr Pheromones in Nematodes

In the 15 years since the discovery of the first ascr pheromone, several groups have intensely investigated their chemical biology. For instance, the Clardy group energized the pheromone research community by identifying two additional ascr pheromones (i.e., **ascr#2** and **3**) in cultured worms [22]. **Ascr#2 and 3** contain essentially the same structural backbone as **ascr#1** (**C7**-SCFA) but differ in the number of carbons in the mSCFA moiety linked to the 3,6-dideoxy ascarylose sugar (**ascr#2**: **C6**-mSCFA with a methyl ketone, **ascr#3**: α-β unsaturated **C9**-mSCFA) (Figure 2). To distinguish the ascr pheromone families in this article, we classify them into two groups: the simple ascr pheromones, which contain only ascarylose and mSCFA, and the modular ascr pheromones, which contain modified ascarylose. Other simple and modular ascr pheromones have now been identified and characterized [16,23–27]. In particular, the Schroeder group detected small amounts of several ascarosides (i.e., **ascr#6.1**, **6.2**, **7**, and **8**) among the metabolites of the wild-type N2 strain by comparing its two-dimensional nuclear magnetic resonance spectrum with that of the ascaroside biosynthesis-defective *daf-22(m130)* strain [25]. These ascr pheromones contain an unsaturated seven-carbon mSCFA linked to a *p*-aminobenzoate subunit (i.e., **ascr#8**) or β-glucose (i.e., **glas#10**) [25] (Figure 2).

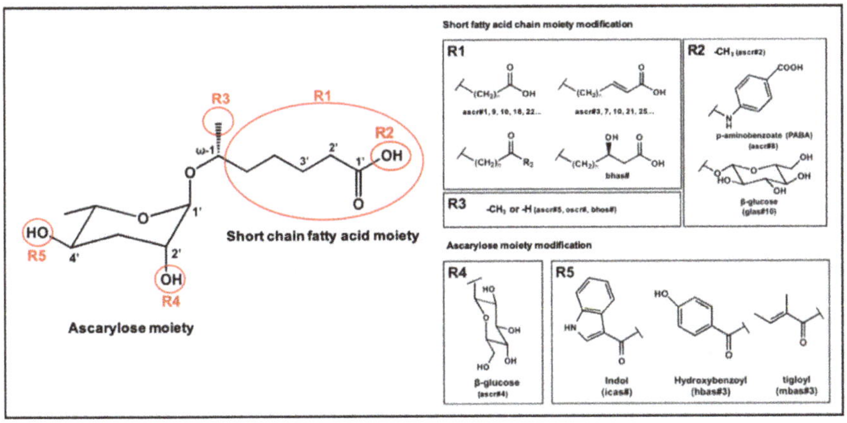

Figure 2. Structural diversity among the known ascr pheromones. Several ascr pheromone analogs modified at various positions (red circles labeled R1–5) have been identified [13,16,17,22–27].

The indole carboxy (IC) ascarosides (icas, e.g., **icas#1**, **3**, **7**, **9**, and **10**) contain a unique indole-3-carbonyl unit attached to the 4′-position of ascarylose [24,27] (Figure 2). Profiling of worm extracts via MS/MS fragmentation and GC-EIMS led to the identification of approximately 200 additional ascr derivatives [16,17]. Notably, structural variations were found in the carbon chain lengths of the mSCFA moiety (e.g., **ascr#18, 21, 22, and 25**). More examples are the presence of hydroxybenzoyl (**hbas#3**) or 2-methyl-2-butenoyl moieties (**mbas#3**) attached to the 4′-position of the ascarylose, ω-linkages at the terminal carbons of the mSCFA moieties, and 2′-hydroxylation of the mSCFA (Figure 2).

Most functionally characterized ascr pheromones are ω-1 linked, i.e., a methyl group is attached to the C1 position at the link between the mSCFA and the ascarylose; however, ω-linked ascr pheromones lacking this linkage have also been reported [16,23] (Figure 2). The structural diversity in the mSCFA is likely generated via the multi-cycled peroxisomal β-oxidation that occurs during ascaroside biosynthesis, although other mechanisms are possible [16,28–36]. Some structural derivatives of ascr pheromones contain other functional groups (e.g., a methyl group, amino acid precursor, glucose, or benzoyl group) linked to the 2′- or 4′-position of the ascarylose moiety or to the 1′-position of the mSCFA moiety, generating a collection of highly diverse ascaroside structures (Figure 2). It is worth noting that ascr pheromone-like derivatives have also been identified in other nematode genera [37–42]. Moreover, the Sternberg group showed that the difference of ascaroside blends between many nematode species was observed with respect to variance of ascr pheromone composition [39]. As most known ascr derivatives share a common structural backbone but differ in their mSCFA moieties or ascarylose modifications (Figure 2), determining their individual functions will be a daunting task.

3. Ascr Pheromone Biosynthesis and Metabolic Regulation

3.1. Ascr Pheromone Biosynthesis

Initially, it was proposed that the ascr pheromone precursors are produced via two distinct reaction pathways, peroxisomal β-oxidation for the SCFA moiety and de novo biosynthesis for the ascarylose moiety (both the simple and modified forms). To produce mature, active ascr pheromone, the SCFA and ascarylose moieties would then be conjugated by UDP (uridine diphosphate)-glucuronosyl transferase (UGT) [29]; however, an alternative pathway has now been proposed. In this alternative pathway, a VLCFA-conjugated ascarylose is first produced and then subsequently subjected to peroxisomal β-oxidation to produce active ascr pheromone [43,44]. This proposal was supported by genetic screens and metabolomic experiments. In *maoc-1*, *dhs-28*, and *daf-22* mutant strains, most of the ascarosides with fatty acid chain lengths of less than nine carbons are not synthesized, whereas non-pheromonal FA-conjugated ascarylose (e.g., VLCFA-, VLCFA-CoA-, and LCFA (long-chain fatty acid linked ascarylose)) accumulates in the worm body [16,17,29,30,43].

Naturally, the source of the ascarylose moiety was an interesting question. The Paik group previously demonstrated that the ascarylose was not derived from the *Escherichia coli* consumed by the worms, but rather that it was de novo synthesized [29]. Sorting out this issue was necessary because ascarylose (a glycoconjugate of ascaroside) is found in the lipopolysaccharide (LPS) of Gram-negative bacteria, and it represents a unique class of sugars with a 3,6-dideoxy sugar structure [45]. In bacteria such as *Yersinia pseudotuberculosis*, ascarylose is produced via a continuous chain of five enzymatic reactions in which CDP-D-glucose is produced from glucose-1-phosphate [46–48]. However, in the course of studying egg shell formation, a gene responsible for ascarylose biosynthesis was found in *C. elegans* [49], supporting the earlier argument in favor of de novo ascarylose biosynthesis [29]. However, additional work is needed to elucidate the detailed mechanism of this step in *C. elegans*.

UGT might be an ideal candidate for catalyzing the conjugation of ascarylose to VLCFAs, as occurs during detoxification reactions in mammals [50,51]. The basis of this prediction is that during detoxification, UGT transfers the monosaccharide glucuronic acid to lipophilic metabolites (e.g., steroids and bile acids) and xenobiotics (e.g., environmental toxins) to render them water-soluble for release [52–54]. However, it remains unknown in *C. elegans* whether an enzyme similar to UGT might catalyze the linkage of fatty acids to ascarylose or cooperate with other enzymes to specifically synthesize ascarosides. For instance, the enzyme encoded by *dgtr-1*, which is involved in egg shell formation, is also thought to be required for ascaroside synthesis because of its homology to the DGAT2 family of acyl-CoA:diacylglycerol acyltransferases, which catalyze the addition of fatty acyl-CoA to diacylglycerol to form triacylglycerol [49].

During the biosynthesis of modular ascarosides (e.g., **icas**, **mbas**, **hbas**, and **osas**), several organic moieties (e.g., amino acid metabolites) are attached to the 4′-position of ascarylose. Using deuterium-labeled tryptophan and axenic in vitro culturing, the Schroeder group found that the indole carbon atom of **icas** is derived from L-tryptophan, while the 4-hydroxybenzoyl group of **hbas** is derived from L-tyrosine or L-phenylalanine. Furthermore, the tigloyl group of **mbas** and the octopamine succinyl group of **osas** are derived from L-isoleucine and L-tyrosine, respectively [16,55]. It has also been suggested that lysosomal ACS-7, an acyl-CoA synthase, catalyzes the linkage of indole-3-carboxy (**icas**) or N-succinyl octopamine groups to ascr [32]. However, the Butcher group showed that ACS-7 appears to transport **icas** to the peroxisomes during the biosynthesis of the short-chain ascaroside **icas** [36]. This dispute on the function and cellular location of ACS-7 remains to be resolved. Based on the findings discussed above, a working model for the biosynthesis of both simple ascr (no attached organic moieties) and modular ascr (various attached organic moieties) in *C. elegans* can be proposed (Figure 3). In this scheme, cytochrome P450 generates (ω-1) or ω-oxygenated VLCFA or LCFA precursors that are then linked to ascarylose to form FA-linked ascarosides (e.g., LCFA). The FA-linked ascarosides then enter the peroxisomal β-oxidation pathway to produce active mSCFA ascr pheromones [16,49].

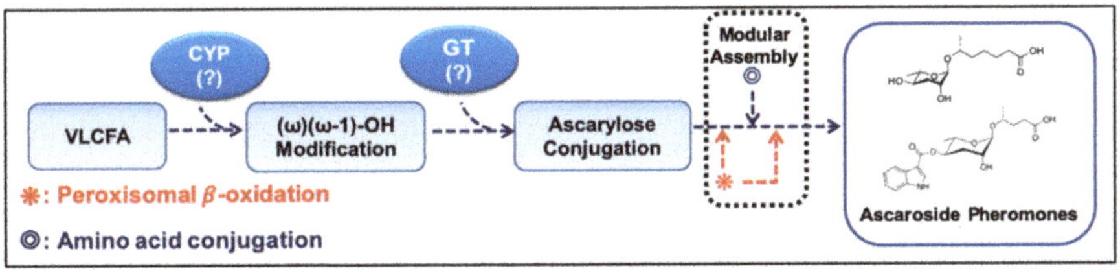

Figure 3. Schematic working model of the ascr pheromone biosynthetic pathway. CYP alters the very long chain fatty acid (VLCFA) produced via elongation of C16 or C18 fatty acids to produce ω-1 or ω-oxygenated VLCFA substrates. Ascarylose is then linked to the ω-1- or ω-oxygenated VLCFAs to form VLCFA-linked ascarosides. Finally, an ascr pheromone containing a shortened fatty acid chain is produced via peroxisomal β-oxidation. In this case, amino acid precursors are linked to specific ascr pheromones. CYP: cytochrome P450, GT: glucuronyltransferase. (?): Names of these enzymes are not known in *C. elegans*.

Peroxisomal β-oxidation is a central metabolic pathway in animals that supplies SCFA components for energy production in mitochondria as well as the main carbon chain precursors for ascr pheromones. The presence of peroxisomes in the intestine and hypodermis of *C. elegans* and the target signals of their peroxisomal proteins have been revealed [56,57]. This topic has been covered in detail by recent publications, and the field is still evolving; therefore, this discussion focuses on important developments related to the production of the mSCFA moieties used in ascr pheromones, as the mSCFAs are a key driver of the structural and functional diversity of ascr pheromones. Research on the ascr biosynthetic pathway has progressed well since the discovery of the nematode acyl-CoA oxidases (ACOX-1 or ACOX-1.1) [30]. ACOXs catalyze the first reaction of peroxisomal β-oxidation by producing enoyl-CoA from acyl-CoA, and they contribute to maintaining the ascr pheromone pool synthesized in response to sudden environmental shifts [30]. Some ascr pheromones (i.e., **ascr#2**, and **3**) are not synthesized by the *acox-1* (*ok2257*) mutant strain [30], whereas the synthesis of others (**ascr#1**, **9**, **10**, **oscr#9**, and **10**) is elevated [16]. These observations suggest that the *acox-1* gene produces multiple ACOX isoforms, which were later found to have different substrate specificities [33–35]. The Butcher group used CRISPR/Cas9 genome editing to elegantly produce various mutant derivatives of the ACOX isoforms and found that the different ACOX isoforms can form various homo- and heterodimers with distinct substrate preferences that produce different ascr pheromones. For example, the ACOX-1.1/ACOX-1.4 heterodimer produces **ascr#1** while the ACOX-1.1/ACOX-1.3 heterodimer produces **ascr#2** [35]. This mechanism for the biosynthesis of such diverse ascr pheromones by the ACOX isoforms is supported by the observation that ACOXs might act on (ω-1)- and ω-oxygenated VLCAs prior to their cyclic stepwise breakdown during peroxisomal β-oxidation. Furthermore, this finding also confirms an earlier report that the ACOXs help to define the ascr pheromone population produced by *C. elegans* [30]. For the second and third reactions of the peroxisomal β-oxidation pathway, MAOC-1 hydrates enoyl-CoA to produce hydroxyacyl-CoA and DHS-28 dehydrogenates hydroxyacyl-CoA to produce 3-ketoacyl-CoA [16,28,29]. Finally, mature mSCFA-containing ascr pheromones are produced via the thiolase activity of DAF-22, a homolog of human SCPx.

3.2. Transcriptional Regulation of Ascr Pheromone Biosynthesis by Environmental Stressors

Although sequence of the biosynthesis of ascr pheromones is known well, it remains unknown how these enzymes are transcriptionally regulated by environmental changes (e.g., temperature increases, nutrition deprivation). To address this question, it was essential to quantify the levels of the approximately ~200 ascr derivatives currently known in *C. elegans*, and to accurately measure the changes in the levels of the ascr pheromones under various physiological states via a standard quantification method [58–60]. The Paik group developed the "PheroQu" method, a multiple reaction monitoring (MRM)-based ascr pheromone quantification method that uses ultra-performance liquid chromatography coupled to mass spectrometry (MS) with only 20 worms. This method enables accurate quantification of the levels of various ascrs in the worm body and in the medium during larval development [59]. With this method, it was found that the biosynthesis of several ascr pheromones (**ascr#1-3**) is robustly influenced by developmental stage, growth condition, and environmental stress (e.g., heat) throughout the life cycle [59].

Upon an increase in ambient temperature, the levels of ascr pheromones increase up to two-fold [30]. It was later found that heat-shock factor 1 (HSF-1) regulates the transcription of ascaroside synthesis genes (e.g., *acox-1*, *dhs-28*, and *daf-22*) in response to external temperature. This finding was supported by chromatin immunoprecipitation assays and increased production of chemically detectable ascarosides (e.g., **ascr#1** and **3**) [31]. Based on this observation, it appears that *C. elegans* requires transcriptional regulation to ensure that a sufficient ascr supply is available upon encountering sudden environmental changes or stress signals, such as poor nutrition or high population density, to prepare for dauer entry. Related to this concept, the Butcher group recently reported that poor nutrition and high temperature can lead to the transformation of one type of ascr (e.g., aggregation-inducing medium-chain **icas**) into another type (e.g., dauer-inducing short-chain **icas**), providing evidence of flexibility in the structure and function of ascr pheromones in response to environmental stress [33,36]. Thus, via combinatorial usage of the products of the *acox* gene family, *C. elegans* has multiple options for adapting to new environments without expending metabolic energy and resources [36].

4. Pleiotropic Neuronal Functions of Ascr Pheromones

4.1. Roles of Ascr Pheromones in Development and Aging

The ascr pheromones influence a variety of functions in the chemosensory neurons that control development, aging, and behaviors in conspecific individuals. Depending on their concentration in the media, they also trigger other important behaviors (e.g., dauer-induction, lifespan extension, mating attraction, repulsion, aggregation, and foraging) that are essential for survival under stressful conditions [26,36,61,62]. Perhaps the best-known function of ascr pheromones is their ability to induce dauer entry, which is a unique system for prolonged survival in *C. elegans*. Reports from several groups showed that there are robust changes in the expression levels of various genes in dauer larvae and dauer entry and exit [63–66]. These findings indicate that ascr pheromones exert their biological functions via some less-characterized signaling pathways involved in neuronal transmission [13,15,26].

By taking advantage of the availability of ascr pheromones, the Paik group characterized the real-time metabolic molecular landscape during dauer formation. These data revealed the metabolic changes underlying the worm's adaptation during the developmental shift to diapause. They measured the genome-wide gene expression changes via DNA microarrays that cover 22,250 unique genes. Their results suggested the presence of a unique adaptive metabolic control mechanism that requires both stage-specific expression of specific genes as well as tight regulation of different modes of fuel metabolite utilization to sustain the energy balance for prolonged survival under adverse conditions [63]. A comprehensive web-based dauer metabolic database for *C. elegans* is available (www.DauerDB.org) for use by the research community and might be broadly useful as a molecular atlas for related nematodes. In addition, using the chemically available pure ascr pheromones, the Lee group routinely produced *C. elegans* dauer larvae and explored that IL2 neurons mediate a phoretic behavior of dauer larvae, called nictation [67]. Furthermore, the same group also characterized nictation as a means of dispersal and survival strategy under harsh conditions through interspecific interaction of *C. elegans* dauer larvae [68].

The clarification of the molecular pathways involved in dauer induction raised questions about the presence of ascr pheromone receptors, which should mediate pheromone sensing to elicit dauer entry. At least three putative pheromone receptors that directly trigger the relevant signaling pathways have been identified in several nematode species. The first ascr pheromone receptor was reported by the Sengupta group, who discovered that the G protein-coupled receptors (GPCRs) SRBC-64 and SRBC-66 are expressed in ASK neurons where they are required for pheromone-induced dauer formation [69]. However, *srbc-64(tm1946)* and *srbc-66(tm2943)* mutant worms failed to form dauer larvae in response to **ascr#1–3** but entered the dauer stage normally in response to **ascr#5** [69]. The decrease in the calcium level in the ASK neurons in response to ascr pheromone observed in adult wild-type worms was not detected in the *srbc-64(tm1946)* and *srbc-66(tm2943)* strains. These mutants did not exhibit long-term responses to pheromones, indicating that other pheromone receptors function via competing signaling cascades depending on the developmental stage [70]. The Bargmann group reported that two other GPCRs, SRG-36 and SRG-37 (which belong to the serpentine receptor class), might act as **ascr#5**-specific ascr pheromone receptors that relay the same dauer entry signals in the ASI neurons [71]. They took advantage of two *C. elegans* strains (LSJ2 and CC1) that had been propagated for long periods of time in liquid axenic media that, unlike the wild-type N2 strain, did not form dauer larvae in response to ascr pheromones (**ascr#1**, **2**, **3**, and **5**). Quantitative trait locus (QTL) mapping and whole-genome sequencing revealed single-nucleotide polymorphisms in *srg-36* and *srg-37* in LSJ2 and CC1, respectively, that specifically prevented the response to **ascr#5**. In *C. briggsae*, another nematode species, the receptor encoded by an *srg* gene paralogous to *srg-36* and *srg-37* responds to **ascr#5** [71]. These results indicate that remodeling of the chemoreceptor repertoire in nematodes allows adaptation to the external environment and that changes in paralogous genes may have common effects across species. In 2012, the Riddle group found that DAF-37 and DAF-38 (also GPCRs) function as a heterodimer to respond to ascr pheromones [72]. DAF-37 responds specifically to **ascr#2**, and its expression in ASI neurons regulates **ascr#2**-mediated dauer formation, whereas its expression in ASK neurons regulates adult behavior. DAF-38, on the other hand, plays a cooperative role in sensing **ascr#2**, **3** and **5** [72]. Other candidate molecules involved in pheromone-induced dauer formation were identified using a forward genetic screen; however, they seem to function in pheromone signaling rather than as pheromone receptors [73,74]. The findings that different pheromone-responsive receptors are expressed in different neurons suggest that additional receptor molecules in other neurons might remain to be identified.

The Scheroder group recently found that **ascr#2**, a ligand of the DAF-37 ascr receptor, mediates an approximately 20% lifespan extension in a sirtuin-dependent manner [75]. This finding revolutionized our thinking on how dauer formation is involved in lifespan extension in *C. elegans*. This new concept, known as ascr-mediated increases of lifespan (AMILS), represents a new paradigm for chemosensation-based non-dauer lifespan extension as it is independent of DAF-16-governed insulin signaling and DAF-12. Given the availability of other ascr pheromones, it would be interesting to investigate whether AMILS is specific to **ascr#2** or whether it exists in other nematode genera or can be regulated by other ascr pheromones.

4.2. Neuronal Effects of Ascr Pheromones on Nematode Social Behaviors

As described above, ascr pheromones have a wide spectrum neuronal functions that not only mediate dauer entry, but also influence adult behaviors and phenotypes, including lifespan extension. For example, very low concentrations (fM–pM) of ascr pheromones attract males, whereas higher concentrations (nM–µM) promote dauer entry [26] (Figure 4). **Ascr#3**, in particular, seems to act as a strong male-attracting pheromone, and various concentrations of **ascr#2–4** appear to exhibit strong synergistic roles in amphid single-ciliated sensory neurons (ADF/ASK) and cephalic companion neurons (CEM) [26,76,77]. **Ascr#8** is also an important male-attracting pheromone at both low and high concentrations [25,76]. These findings clearly confirm that ascr pheromones have neuronal functions that trigger diverse behaviors to ensure prolonged survival in response to environmental changes.

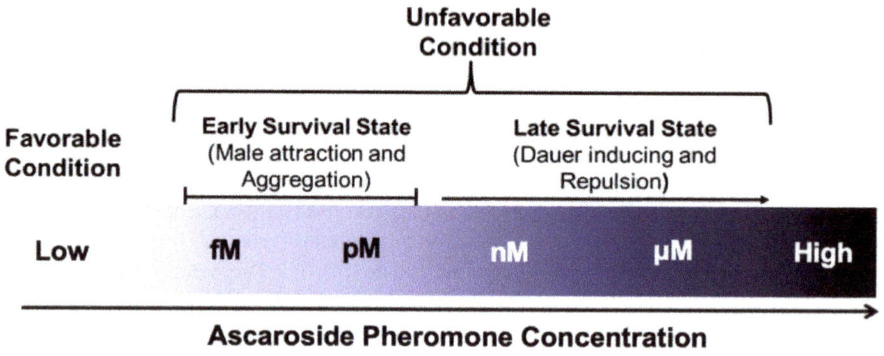

Figure 4. The pleiotropic neuronal functions of major ascr pheromones (e.g., **ascr#1–3**) exerted at their environmental concentrations.

Interestingly, although **ascr#1–3** induce dauer entry of L1 worms at higher concentrations (nM–µM), similar concentrations act as chemorepellents after the L1 stage that stimulate hermaphrodite repulsion [26,62,77–80]. These observations suggest that these pheromones act in a concentration-dependent and stage-specific manner. These repulsive responses appear to be transmitted via the GPA-3-DAF-16/FOXO signaling pathway in sensory neurons, and they affect long-term memory via glutamate signaling regulated by DAF-16 [78]. Note that this behavior is distinct from male attraction behavior because the genetic sex modulates the sensitivity of the ADF neurons to ascr pheromones [77]. The **ascr#3**-dependent avoidance behavior is stimulated by **ascr#3** sensing in the ADL neurons followed by signal propagation to the interneurons, which then regulate the magnitude of the behavioral changes stimulated by pheromone contact in relation to feeding state or early larval development [79,80]. Furthermore, **mbas#3** (an ascaroside linked to a tigloyl group) and **osas#9** (an ascaroside linked to a succinyl octopamine group) also have repulsive effects similar to those of **ascr#3** and **icas#3** [55,81].

At low concentrations (< 10 nM), **ascr#2**, **3**, and **5** can attract hermaphrodites only in specific social strains or strains lacking NPR-1 (e.g., the *npr-1(ad609)* mutant), an important regulator of aggregation behavior [62]. At low concentrations, some IC group-containing ascr pheromones (e.g., **icas#1**, **icas#3**, and **icas#9**) induce aggregation in solitary N2 hermaphrodites as well as in naturally isolated social strains (e.g., CB4856 and RC301), while they induce male attraction at higher concentrations [27]. These responses require the ASK sensory neurons and downstream AIA neurons, but not the RMG neuron required for attraction in *npr-1(ad609)* mutants as previously reported. Like the icas pheromones, **ascr#1**, **2**, **3**, and **5** can act as chemorepellents or aggregation-inducing pheromones, suggesting that their activity is determined by their environmental concentrations. At low concentrations, they induce attraction, whereas at higher concentrations they induce repulsion. One group reported that this behavioral change also depends on the oxygen concentration [82]. In this study, the authors found that

RMG neurons control the oxygen concentration via the URX neurons, resulting in switching between attraction signals in ASK neurons and repulsion signals in ADL neurons. The discovery of the **icas#9** receptors, encoded by *srx-43* and *srx-44*, via QTL mapping and whole-genome sequencing [61,83] revealed that SRX-43 is expressed in ASI neurons, whereas SRX-44 is expressed in ASJ and ADL neurons, and that roaming behavior is determined by the site of their expression [83].

In several asexual species, the rate of sexual reproduction increases in stressful environments, functioning as a survival strategy to generate genetic variation via recombination during outcrossing [84–92]. In *C. elegans*, ascr pheromones induce male mating or aggregation behavior in the early survival state. For example, two naturally occurring strains (CB4856 and JU440) exhibit increased male frequency during the dauer stage that is not observed in the N2 laboratory strain N2. This effect is due to an increase in the male mating rate and increased male survival during the dauer period [93]. The male attraction behavior in response to ascr pheromones is thought to induce an increase in male frequency in dauer-inducing environments [15]; thus, it is likely that larger male populations are beneficial for survival in unfavorable external environments. One study reported that the hermaphrodite reproductive rates of some other naturally isolated strains are regulated by secreted pheromones [94]. In fact, **ascr#3** and **10** are secreted at different rates by males and hermaphrodites [95]. A combination of ascr pheromones secreted by males has been reported to not only affect the hermaphrodite reproductive system, but also to increase heat stress resistance [96]. This male-secreted pheromone also has a male-killing effect, thereby regulating the population size of the species [97]. In sum, the functions and structure of some ascr pheormones are listed in Table 1.

Table 1. The functions and structure of some ascaroside pheromones *.

Name	Chemical Structure	Discovered Receptors	Functions	References
ascr#1	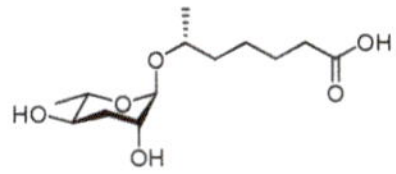	SRBC-64 SRBC-66	Dauer inducing activity Repulsion activity	[13,69,78]
ascr#2	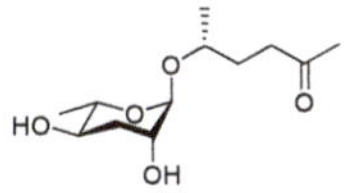	DAF-37 DAF-38 SRBC-64 SRBC-66	Dauer inducing activity Repulsion activity Male attraction activity Foraging activity	[22,26,61,62,69,72,78,82]
ascr#3	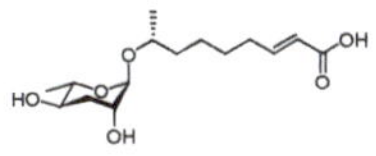	SRBC-64 SRBC-66	Dauer inducing activity Repulsion activity Male attraction activity Foraging activity	[22,26,61,62,69,76–80,82]
ascr#4	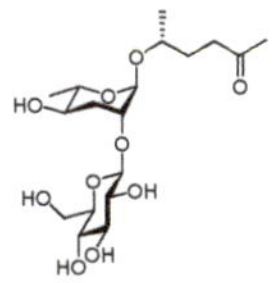	Unknown	Dauer inducing activity Male attraction activity	[26]
ascr#5	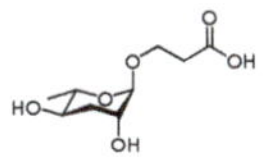	SRG-36 SRG-37	Dauer inducing activity Repulsion activity	[23,62,71,82]
ascr#6.1	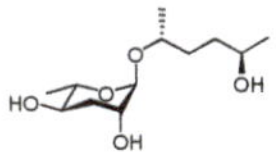	Unknown	Dauer inducing activity	[25]

Table 1. *Cont.*

Name	Chemical Structure	Discovered Receptors	Functions	References
ascr#8	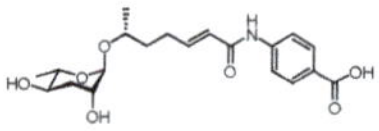	Unknown	Dauer inducing activity Male attraction activity Foraging activity	[25,61,76]
icas#3	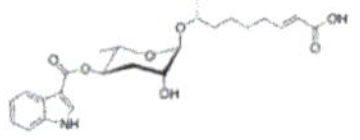	Unknown	Male attraction activity Aggregation activity	[27]
icas#9	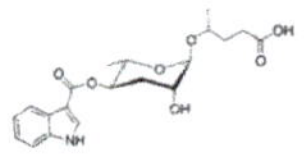	SRX-43SRX-44	Dauer inducing activity Male attraction activity Aggregation activity Foraging activity	[24,27,61,83]
hbas#3	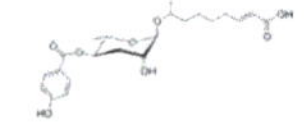	Unknown	Hermaphrodite attraction activity	[16]
mbas#3	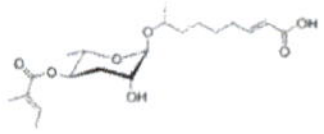	Unknown	Repulsion activity	[16,81]
osas#3	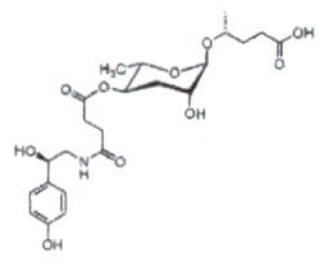	Unknown	Repulsion activity	[55]

* The ascr pheromones listed here were selected based on their identified functions, and citation frequencies.

The concentrations of the ascr pheromones produced by worms and their main functional changes in response to external environmental conditions are outlined in Figure 4. Under favorable conditions, the ascr pheromone concentrations are too low to exert any effects, perhaps due to other environmental factors. However, ascr pheromone synthesis gradually increases as worms encounter unfavorable stress conditions (e.g., high temperature, food limitation, and high population density) [30,59]. It has been hypothesized that ascr pheromones stimulate male mating or aggregation at relatively low concentrations under normal growth conditions, while under stressful conditions that trigger increased ascr pheromone production (and thus higher concentrations), worms may exhibit a repulsive response to ascr pheromone and enter the dauer state. However, the structural basis for the functional differences between ascr pheromones has not yet been clarified.

5. Implications of Ascr Pheromone Metabolism in Neuroprotection

5.1. Implications of Ascr Pheromone Biosynthesis Gene Deficiencies in Neuronal Disorders

Several ascr pheromone biosynthesis defects have been identified in mutant worms deficient for peroxisomal β-oxidation enzymes [16,28–36,98]. The physiological consequences of impaired DAF-22-dependent peroxisomal *β*-oxidation of VLCFAs or fatty acyl-CoAs involved in the production of various aglycone units (mSCFAs with less than nine carbon atoms) required for pheromone biosynthesis indicate that peroxisomal *β*-oxidation of VLCFAs is an essential detoxification process for clearing harmful peroxisomal fatty acids to maintain cellular homoeostasis. This function indicates that ascr pheromones not only regulate stress avoidance, they also maintain cellular homeostasis via the production of excretable FA-ascarylose conjugates (ascarosides) [29]. Here we examine the pleiotropic neuronal functions of ascr pheromones from two different angles, ascr metabolic deficiency and chemotactic responses.

In mammals, it is well known that peroxisomal malfunctions induce developmental defects and neurodevelopmental diseases. These diseases include Zellweger syndrome (ZS) and X-linked adrenoleukodystrophy (X-ALD), which involve severe neurological problems that often lead to death in infants and young children [99–104]. In humans, a single defect in an enzyme involved in peroxisomal fatty acid β-oxidation leads to ZS, which involves abnormal symptoms such as neonatal hypotonia, craniofacial dysmorphia, seizures, and developmental delay [100,103–105]. Mechanistically, it was suggested that the defect in peroxisomal fatty acid β-oxidation results in the accumulation of VLCFAs in the form of triacylglycerols, which are harmful to animals [29,105]. Furthermore, decreased docosahexaenoic acid (DHA; C22:6 (n-3)) levels, plasmalogen depletion, and abnormal neurons myelination (e.g., degenerative loss of myelin (demyelination) or abnormally formed myelin (dysmyelination)) have been suggested to underlie the neuropathologies associated with peroxisomal disorders [106]. In *C. elegans*, ascaroside biosynthesis appears encompass two important physiological roles that affect the worm's quality of life: (1) a social function in which pheromone production affects the behavior and physiology of other individuals, and (2) protection of metabolic homeostasis via the removal of toxic VLCFAs in peroxisomes (Figure 5) [29].

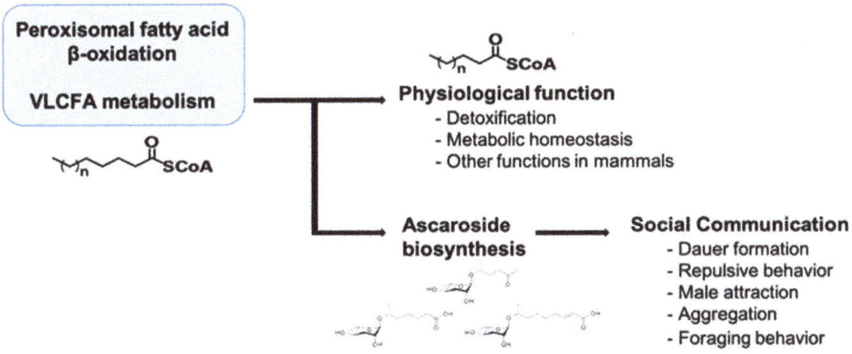

Figure 5. A schematic diagram for the dual role of peroxisomal fatty acid (FA) β-oxidation. By shortening VLCFAs, peroxisomal fatty acid β-oxidation can exert physiological functions such as detoxification and maintenance of metabolic homeostasis. In *Caenorhabditis elegans*, shortened FAs are used to synthesize ascr pheromones, which are important for social communication.

In addition to neurodevelopmental defects, deficiencies in peroxisomal fatty acid β-oxidation seem to be related to other pathologies. In *C. elegans*, animals deficient in peroxisomal fatty acid β-oxidation, such as the *dhs-28(tm2581)* and *daf-22(ok693)* mutant strains, exhibit short lifespans and developmental delays, and are more susceptible to environmental stresses, limiting the worm's survival under harsh conditions [29,107]. In particular, it has recently been suggested that peroxisomal fatty acid β-oxidation has distinct functions in neuronal cells for maintaining normal development and nervous system function [101,106,107]. More interestingly, it was revealed that neuronal peroxisomal fatty acid β-oxidation has an important cell-autonomous function to regulate neuroendocrine signaling activities [107]. The *C. elegans* SCPx gene *daf-22* is expressed in a subset of chemosensory neurons, i.e., the ASK neurons, where its activity is required for exogenous pheromone-induced dauer entry [107]. A deficiency in neuronal peroxisomal fatty acid β-oxidation activates the lipid-induced endoplasmic reticulum (ER) stress response, which then increases the expression of insulin-like peptides in neurons and abnormally enhances insulin/IGF-1 signaling activity to eventually interrupt dauer entry [107]. Meanwhile, ER stress-mediated dauer diapause is also regulated by other sensory neurons, such as the ASI neurons [108]. It has been suggested that the mutated DAF-28 peptide in the *daf-28(sa191)* mutant strain triggers ER stress and activation of the unfolded protein response (UPR) to induce constitutive dauer entry [108–111].

From these studies, it can be inferred that peroxisomal fatty acid β-oxidation is important for neuroprotection via the regulation of metabolic homeostasis (e.g., balance in fatty acid levels), myelination of neuronal cells, and the regulation of cellular signaling; these neuroprotective functions could influence aging, neurodevelopment, and stress resistance. Therefore, it is important to investigate the mechanisms underlying the roles of neuronal peroxisomal fatty acid β-oxidation in neuroprotection and aging in the future. It would also be worthwhile to elucidate the links between neuronal peroxisomal disorders and alterations in neuronal function and neurodevelopment (Figure 6).

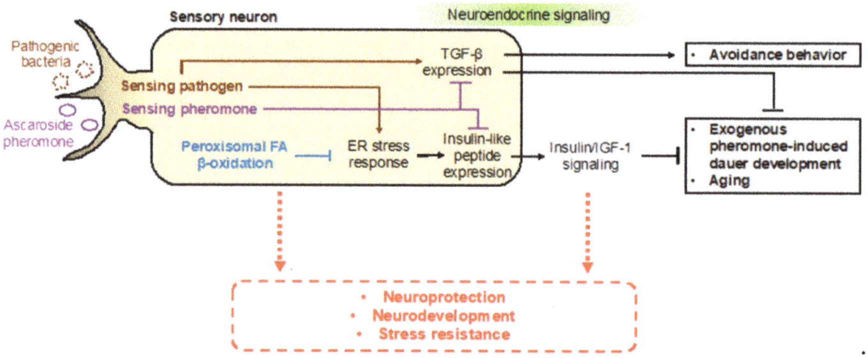

Figure 6. The protective ways in the sensory neurons and their outputs. In *C. elegans*, ascr pheromone sensing affects the expression of neuropeptides, such as insulin and TGF-β. Peroxisomal fatty acid β-oxidation in sensory neurons regulates neuroendocrine signaling (e.g., insulin/IGF-1 signaling) via regulation of insulin-like peptide expression by suppressing the lipid-induced endoplasmic reticulum (ER) stress response. By regulating insulin/IGF-1 signaling, peroxisomal fatty β-oxidation controls both exogenous pheromone-induced dauer entry and aging. Furthermore, pathogens regulate TGF-β expression and ER stress via unfolded protein responses (UPRs). TGF-β expression triggered by pathogens stimulates avoidance behavior. Similarly, such signaling in the nervous system can influence neuroprotection, neurodevelopment, and stress resistance, either directly or via neuroendocrine signaling pathways.

5.2. Implications of Ascr Pheromone Signaling in Chemotactic Responses

Ascr pheromones induce a variety of behaviors [112]; however, these behaviors are controlled not only by the ascr pheromones but also by various other associated factors and environmental conditions. In general, food signals play important roles in determining behaviors and developmental choices in the presence of ascr pheromones in *C. elegans*. For example, calcium/calmodulin-dependent protein kinase I (CMK-1) regulates pheromone-mediated dauer entry in ASI/AWC neurons depending on the feeding state, although not directly via a pheromone-binding receptor [113]. Furthermore, gut-to-neuron signaling induced by feeding conditions affects TGF-β and insulin expression via target of papamycin complex 2 (TORC2), which leads to dauer entry or behavioral changes [114]. Repulsive behavior in response to feeding status is also induced by pheromone-mediated insulin signaling [80]. The combination of these two signals determines the choice between dauer entry or progression to the reproductive state via downstream regulation of DAF-12 and the associated *let-7* microRNA family and hunchback-like-1 (HBL-1) [115]. Ascr pheromones are also involved in chemotactic behavior by regulating endogenous peptide signaling [116]. *C. elegans* exhibits chemotactic attraction toward odorants such as benzaldehyde; however, after prolonged exposure, the chemotactic behavior shifts to a dispersion behavior, and this shift is called olfactory adaptation or food-odor associative learning. The results of the study of Yamada et al. also suggest that NEP-2 (a homolog of the extracellular peptidase neprilysin) and SNET-1 (an NEP-2 suppressor peptide) regulate olfactory adaptation, and that an ascr pheromone that inhibits *snet-1* expression is essential for olfactory adaptation [116].

Factors associated with ascr pheromones and their sensing have also been implicated in other physiological processes, such as aging [117–120]. This change in longevity is not only affected by ascr pheromones, but rather it is also influenced by a combination of other factors, including nutritional state and population density [75,121]. These pheromones act as a kind of warning signal by which *C. elegans* is informed in advance of ongoing changes in growth conditions (e.g., the ratio between food and pheromones). Triggering of this warning signal is also caused by other factors in addition to ascr pheromones. Typically, pathogen-induced avoidance in *C. elegans* has been studied in the context of the innate immune system [122,123]. Interestingly, it appears that the signaling in response to exposure to food bacteria and pathogenic bacteria and the downstream effects are similar, with the difference being the toxicity of the organisms to the worms [124]. Several factors simultaneously play important roles in ascr pheromone-mediated signaling and pathogen avoidance. First, NPR-1, which controls aggregation via ascr pheromones [62], also plays an important role in pathogen avoidance [125–127]. Like pheromones, pathogens are also recognized by sensory neurons [128,129]. Furthermore, the TGF-β ligand and insulin, which also play important roles in dauer entry, also appear to be involved in pathogen avoidance [130–132]. However, DAF-7, a TGF-β ligand, acts in the ASI/ASJ neurons during pathogenic avoidance but primarily in the ASI neurons during ascr pheromone sensing [130,131]. Finally, ER stress or UPR activation in sensory neurons can also be induced by pathogens [133–136]. It is plausible to predict that these physiological effects might involve the same factors to promote the survival of the nematode (Figure 6). Indeed, it has been reported that the use of ascr pheromone in a mammalian system has a therapeutic effect on hepatic inflammation [137,138]. Furthermore, ARTD, a combination of artemisinin and ascr pheromone, can also be used as an effective therapeutic agent in osteoclasts, where it shows a potent cancer inhibitory effect [139]. Thus, this relationship deserves further investigation in the future.

6. Conclusions and Future Directions

In this comprehensive review, we have highlighted some of the major achievements from the past 15 years since the discovery of the first ascr pheromone (**ascr#1**) [13]. The rapid developments in the ascr field have increased the depth of our knowledge with respect to biosynthetic pathways, ascr receptor-mediated neuronal signaling pathways, and potential neuro-physiological effects in animals. We would also like to add a few words on our views of the future of the ascr pheromone field.

(i) Translational research: Given that their biosynthesis has been thoroughly investigated, now is a good time to construct a chemical biology map or database to catalog the structure-function relationships of the more than 200 members of the ascr family. Since some factors involved in ascr biosynthesis also have important neuronal functions in mammals, translation of what we know about nematode ascr pheromones into studies of metabolic diseases might be a promising future step. Some physiological functions of ascr pheromones are also involved in mammalian aging and disease; thus, these pheromones may have implications in human disease. It will also be interesting to unravel the roles of **ascr#1** in disease model animals or mammalian cells [137–139].

(ii) Neuronal pheromone sensing and signaling: Ascr pheromone biosynthesis and their recognition and processing are equally interesting. Previous studies showed that pheromone sensing occurs in sensory neurons, and three receptors specific to some ascr pheromones have been found. However, as the number of newly discovered ascr pheromones increases, how they are sensed and responded to via potential common sensing and signaling pathways remains to be resolved. For example, several GPCRs act as ascr pheromone receptors; however, additional GPCRs have been found in other species [140,141]. Furthermore, several physiological effects induced by ascr pheromones are synergistic, i.e., single pheromones do not always act alone [23,26]. Thus, ascr pheromone sensing and signaling are likely complex and elaborately intertwined and untangling of these knots could provide important clues for understanding neuronal signaling in other species. Given that different ascr pheromones appear to mediate different behaviors across the nematode species depending on environmental conditions, it is reasonable to ask the question, what is the lowest common denominator that underlies the diverse biological functions of ascr pheromones? Fully addressing this question will require additional research on the chemical biology of pheromones in the future.

(iii) Neuronal ascr signaling and behavior: Ascr pheromones were originally found while searching for the factors that influence dauer entry, and they have since been reported to be involved in various behaviors in addition to dauer entry. Interestingly, the effects associated with ascr pheromones are almost exclusively influenced by external environmental cues, many of which involve stress (e.g., poor nutrition, overcrowding, and heat). Therefore, it will be interesting to clarify the biological links between ascr function and stress responses as well as neuroprotection (i.e., the innate immune response, see Section 5.2.).

(iv) Creation of pheromics:In a literature survey of ascr pheromone publications, we noticed many interdisciplinary pheromone research projects and a boom in omics technologies. Examples include, but are not limited to, molecular genetics, chemical biology, metabolomics, proteomics, and genomics. At this juncture, it could be beneficial to create the field of "**pheromics**" (*pheromone omics*) as a new subset of integrated disciplinary research area within chemical ecology with the goal of establishing and supporting a community of researchers involved in the systematic study of the pheromones of living organisms.

Author Contributions: Conceptualization, Y.-K.P., H.-J.J., and J.Y.P.; Supervision, Y.-K.P.; Validation, Y.-K.P.; Visualization, H.-J.J., H.K. and J.Y.P.; Writing—original draft, H.-J.J., J.Y.P. and S.P.; Writing—review and editing, J.Y.P. and Y.-K.P.

Funding: This study was supported by a grant from the National Research Foundation of Korea (2017R1A2B3003200 to Y.-K.P.).

Conflicts of Interest: Authors declare no conflict of interest.

References

1. Karlson, P.; Lüscher, M. 'Pheromones': A New Term for a Class of Biologically Active Substances. *Nature* **1959**, *183*, 55. [CrossRef] [PubMed]
2. Wyatt, T.D. Pheromones. *Curr. Biol.* **2017**, *27*, R739–R743. [CrossRef] [PubMed]
3. Brennan, P.A.; Zufall, F. Pheromonal communication in vertebrates. *Nature* **2006**, *444*, 308. [CrossRef] [PubMed]
4. Tillman, J.A.; Seybold, S.J.; Jurenka, R.A.; Blomquist, G.J. Insect pheromones—an overview of biosynthesis and endocrine regulation. *Insect Biochem. Mol. Biol.* **1999**, *29*, 481–514. [CrossRef]
5. Abel, E.L. Alarm substance emitted by rats in the forced-swim test is a low volatile pheromone. *Physiol. Behav.* **1991**, *50*, 723–727. [CrossRef]
6. McGlone, J.J.; Anderson, D.L. Synthetic maternal pheromone stimulates feeding behavior and weight gain in weaned pigs. *J. Anim. Sci.* **2002**, *80*, 3179–3183. [CrossRef] [PubMed]
7. Wyatt Tristram, D. The search for human pheromones: The lost decades and the necessity of returning to first principles. *Proc. R. Soc. B Biol. Sci.* **2015**, *282*, 20142994. [CrossRef] [PubMed]
8. Wysocki, C.J.; Preti, G. Facts, fallacies, fears, and frustrations with human pheromones. *Anat. Rec. A. Discov. Mol. Cell. Evol. Biol.* **2004**, *281A*, 1201–1211. [CrossRef] [PubMed]
9. McGann, J.P. Poor human olfaction is a 19th-century myth. *Science* **2017**, *356*, eaam7263. [CrossRef]
10. Meredith, M. Human Vomeronasal Organ Function: A Critical Review of Best and Worst Cases. *Chem. Senses* **2001**, *26*, 433–445. [CrossRef]
11. Cassada, R.C.; Russell, R.L. The dauerlarva, a post-embryonic developmental variant of the nematode Caenorhabditis elegans. *Dev. Biol.* **1975**, *46*, 326–342. [CrossRef]
12. Golden, J.W.; Riddle, D.L. A pheromone influences larval development in the nematode Caenorhabditis elegans. *Science* **1982**, *218*, 578–580. [CrossRef] [PubMed]
13. Jeong, P.-Y.; Jung, M.; Yim, Y.-H.; Kim, H.; Park, M.; Hong, E.; Lee, W.; Kim, Y.H.; Kim, K.; Paik, Y.-K. Chemical structure and biological activity of the Caenorhabditis elegans dauer-inducing pheromone. *Nature* **2005**, *433*, 541–545. [CrossRef] [PubMed]
14. Viney, M.E.; Franks, N.R. Is dauer pheromone of Caenorhabditis elegans really a pheromone? *Naturwissenschaften* **2004**, *91*, 123–124. [CrossRef] [PubMed]

15. Edison, A.S. Caenorhabditis elegans pheromones regulate multiple complex behaviors. *Curr. Opin. Neurobiol.* **2009**, *19*, 378–388. [CrossRef]
16. von Reuss, S.H.; Bose, N.; Srinivasan, J.; Yim, J.J.; Judkins, J.C.; Sternberg, P.W.; Schroeder, F.C. Comparative Metabolomics Reveals Biogenesis of Ascarosides, a Modular Library of Small-Molecule Signals in C. elegans. *J. Am. Chem. Soc.* **2012**, *134*, 1817–1824. [CrossRef]
17. von Reuss, S.H.; Dolke, F.; Dong, C. Ascaroside Profiling of Caenorhabditis elegans Using Gas Chromatography–Electron Ionization Mass Spectrometry. *Anal. Chem.* **2017**, *89*, 10570–10577. [CrossRef]
18. Fauré-Frémiet, E. *Le Cycle Erminatif Chez l'Ascaris Megalocephala*; Masson: Issy les Moulineaux, France, 1913.
19. Flury, F. Zur Chemie und Toxikologie der Ascariden. *Arch. Für Exp. Pathol. Pharmakol.* **1912**, *67*, 275–392. [CrossRef]
20. Fouquey, C.; Polonsky, J.; Lederer, E. Chemical structure of ascarylic alcohol isolated from Parascaris equorum. *Bull. Soc. Chim. Biol.* **1957**, *39*, 101–132.
21. Bartley, J.P.; Bennett, E.A.; Darben, P.A. Structure of the Ascarosides from Ascaris suum. *J. Nat. Prod.* **1996**, *59*, 921–926. [CrossRef]
22. Butcher, R.A.; Fujita, M.; Schroeder, F.C.; Clardy, J. Small-molecule pheromones that control dauer development in Caenorhabditis elegans. *Nat. Chem. Biol.* **2007**, *3*, 420–422. [CrossRef] [PubMed]
23. Butcher, R.A.; Ragains, J.R.; Kim, E.; Clardy, J. A potent dauer pheromone component in Caenorhabditis elegans that acts synergistically with other components. *Proc. Natl. Acad. Sci. USA* **2008**, *105*, 14288–14292. [CrossRef] [PubMed]
24. Butcher, R.A.; Ragains, J.R.; Clardy, J. An Indole-Containing Dauer Pheromone Component with Unusual Dauer Inhibitory Activity at Higher Concentrations. *Org. Lett.* **2009**, *11*, 3100–3103. [CrossRef] [PubMed]
25. Pungaliya, C.; Srinivasan, J.; Fox, B.W.; Malik, R.U.; Ludewig, A.H.; Sternberg, P.W.; Schroeder, F.C. A shortcut to identifying small molecule signals that regulate behavior and development in Caenorhabditis elegans. *Proc. Natl. Acad. Sci. USA* **2009**, *106*, 7708–7713. [CrossRef] [PubMed]
26. Srinivasan, J.; Kaplan, F.; Ajredini, R.; Zachariah, C.; Alborn, H.T.; Teal, P.E.A.; Malik, R.U.; Edison, A.S.; Sternberg, P.W.; Schroeder, F.C. A blend of small molecules regulates both mating and development in *Caenorhabditis elegans*. *Nature* **2008**, *454*, 1115–1118. [CrossRef]
27. Srinivasan, J.; von Reuss, S.H.; Bose, N.; Zaslaver, A.; Mahanti, P.; Ho, M.C.; O'Doherty, O.G.; Edison, A.S.; Sternberg, P.W.; Schroeder, F.C. A Modular Library of Small Molecule Signals Regulates Social Behaviors in Caenorhabditis elegans. *PLOS Biol.* **2012**, *10*, e1001237. [CrossRef] [PubMed]
28. Butcher, R.A.; Ragains, J.R.; Li, W.; Ruvkun, G.; Clardy, J.; Mak, H.Y. Biosynthesis of the Caenorhabditis elegans dauer pheromone. *Proc. Natl. Acad. Sci. USA* **2009**, *106*, 1875–1879. [CrossRef]
29. Joo, H.-J.; Yim, Y.-H.; Jeong, P.-Y.; Jin, Y.-X.; Lee, J.-E.; Kim, H.; Jeong, S.-K.; Chitwood, D.J.; Paik, Y.-K. Caenorhabditis elegans utilizes dauer pheromone biosynthesis to dispose of toxic peroxisomal fatty acids for cellular homoeostasis. *Biochem. J.* **2009**, *422*, 61–71. [CrossRef]
30. Joo, H.-J.; Kim, K.-Y.; Yim, Y.-H.; Jin, Y.-X.; Kim, H.; Kim, M.-Y.; Paik, Y.-K. Contribution of the peroxisomal acox gene to the dynamic balance of daumone production in Caenorhabditis elegans. *J. Biol. Chem.* **2010**, *285*, 29319–29325. [CrossRef]
31. Joo, H.-J.; Park, S.; Kim, K.-Y.; Kim, M.-Y.; Kim, H.; Park, D.; Paik, Y.-K. HSF-1 is involved in regulation of ascaroside pheromone biosynthesis by heat stress in Caenorhabditis elegans. *Biochem. J.* **2016**, *473*, 789–796. [CrossRef]
32. Panda, O.; Akagi, A.E.; Artyukhin, A.B.; Judkins, J.C.; Le, H.H.; Mahanti, P.; Cohen, S.M.; Sternberg, P.W.; Schroeder, F.C. Biosynthesis of Modular Ascarosides in C. elegans. *Angew. Chem. Int. Ed.* **2017**, *56*, 4729–4733. [CrossRef] [PubMed]
33. Zhang, X.; Feng, L.; Chinta, S.; Singh, P.; Wang, Y.; Nunnery, J.K.; Butcher, R.A. Acyl-CoA oxidase complexes control the chemical message produced by Caenorhabditis elegans. *Proc. Natl. Acad. Sci. USA* **2015**, *112*, 3955–3960. [CrossRef] [PubMed]
34. Zhang, X.; Li, K.; Jones, R.A.; Bruner, S.D.; Butcher, R.A. Structural characterization of acyl-CoA oxidases reveals a direct link between pheromone biosynthesis and metabolic state in Caenorhabditis elegans. *Proc. Natl. Acad. Sci. USA* **2016**, *113*, 10055–10060. [CrossRef] [PubMed]
35. Zhang, X.; Wang, Y.; Perez, D.H.; Jones Lipinski, R.A.; Butcher, R.A. Acyl-CoA Oxidases Fine-Tune the Production of Ascaroside Pheromones with Specific Side Chain Lengths. *ACS Chem. Biol.* **2018**, *13*, 1048–1056. [CrossRef] [PubMed]

36. Zhou, Y.; Wang, Y.; Zhang, X.; Bhar, S.; Jones Lipinski, R.A.; Han, J.; Feng, L.; Butcher, R.A. Biosynthetic tailoring of existing ascaroside pheromones alters their biological function in C. elegans. *eLife* **2018**, *7*, e33286. [CrossRef]
37. Bose, N.; Ogawa, A.; von Reuss, S.H.; Yim, J.J.; Ragsdale, E.J.; Sommer, R.J.; Schroeder, F.C. Complex Small-Molecule Architectures Regulate Phenotypic Plasticity in a Nematode. *Angew. Chem. Int. Ed.* **2012**, *51*, 12438–12443. [CrossRef] [PubMed]
38. Bose, N.; Meyer, J.M.; Yim, J.J.; Mayer, M.G.; Markov, G.V.; Ogawa, A.; Schroeder, F.C.; Sommer, R.J. Natural Variation in Dauer Pheromone Production and Sensing Supports Intraspecific Competition in Nematodes. *Curr. Biol.* **2014**, *24*, 1536–1541. [CrossRef]
39. Choe, A.; von Reuss, S.H.; Kogan, D.; Gasser, R.B.; Platzer, E.G.; Schroeder, F.C.; Sternberg, P.W. Ascaroside Signaling Is Widely Conserved among Nematodes. *Curr. Biol.* **2012**, *22*, 772–780. [CrossRef]
40. Choe, A.; Chuman, T.; von Reuss, S.H.; Dossey, A.T.; Yim, J.J.; Ajredini, R.; Kolawa, A.A.; Kaplan, F.; Alborn, H.T.; Teal, P.E.A.; et al. Sex-specific mating pheromones in the nematode Panagrellus redivivus. *Proc. Natl. Acad. Sci. USA* **2012**, *109*, 20949–20954. [CrossRef]
41. Dong, C.; Dolke, F.; Reuss, S.H. von Selective MS screening reveals a sex pheromone in Caenorhabditis briggsae and species-specificity in indole ascaroside signalling. *Org. Biomol. Chem.* **2016**, *14*, 7217–7225. [CrossRef]
42. Dong, C.; Reilly, D.K.; Bergame, C.; Dolke, F.; Srinivasan, J.; von Reuss, S.H. Comparative Ascaroside Profiling of Caenorhabditis Exometabolomes Reveals Species-Specific (ω) and (ω–2)-Hydroxylation Downstream of Peroxisomal β-Oxidation. *J. Org. Chem.* **2018**, *83*, 7109–7120. [CrossRef] [PubMed]
43. Izrayelit, Y.; Robinette, S.L.; Bose, N.; von Reuss, S.H.; Schroeder, F.C. 2D NMR-Based Metabolomics Uncovers Interactions between Conserved Biochemical Pathways in the Model Organism Caenorhabditis elegans. *ACS Chem. Biol.* **2013**, *8*, 314–319. [CrossRef] [PubMed]
44. Zagoriy, V.; Matyash, V.; Kurzchalia, T. Long-Chain O-Ascarosyl-alkanediols Are Constitutive Components of Caenorhabditis elegans but Do Not Induce Dauer Larva Formation. *Chem. Biodivers.* **2010**, *7*, 2016–2022. [CrossRef] [PubMed]
45. Griffiths, A.J.; Davies, D.B. Type-specific carbohydrate antigens of pathogenic bacteria. Part 1: Enterobacteriaceae. *Carbohydr. Polym.* **1991**, *14*, 241–279. [CrossRef]
46. Thorson, J.S.; Lo, S.F.; Liu, H.W.; Hutchinson, C.R. Biosynthesis of 3, 6-dideoxyhexoses: New mechanistic reflections upon 2, 6-dideoxy, 4, 6-dideoxy, and amino sugar construction. *J. Am. Chem. Soc.* **1993**, *115*, 6993–6994. [CrossRef]
47. Thorson, J.S.; Lo, S.F.; Ploux, O.; He, X.; Liu, H.W. Studies of the biosynthesis of 3,6-dideoxyhexoses: Molecular cloning and characterization of the asc (ascarylose) region from Yersinia pseudotuberculosis serogroup VA. *J. Bacteriol.* **1994**, *176*, 5483–5493. [CrossRef]
48. Trefzer, A.; Bechthold, A.; Salas, J.A. Genes and enzymes involved in deoxysugar biosynthesis in bacteria. *Nat. Prod. Rep.* **1999**, *16*, 283–299. [CrossRef]
49. Olson, S.K.; Greenan, G.; Desai, A.; Müller-Reichert, T.; Oegema, K. Hierarchical assembly of the eggshell and permeability barrier in C. elegans. *J. Cell Biol* **2012**, *198*, 731–748. [CrossRef]
50. Campbell, J.A.; Davies, G.J.; Bulone, V.; Henrissat, B. A classification of nucleotide-diphospho-sugar glycosyltransferases based on amino acid sequence similarities. *Biochem. J.* **1997**, *326*, 929–939. [CrossRef]
51. Koeller, K.M.; Wong, C.-H. Synthesis of complex carbohydrates and glycoconjugates: Enzyme-based and programmable one-pot strategies. *Chem. Rev.* **2000**, *100*, 4465–4494. [CrossRef]
52. Gems, D.; McElwee, J.J. Broad spectrum detoxification: The major longevity assurance process regulated by insulin/IGF-1 signaling? *Mech. Ageing Dev.* **2005**, *126*, 381–387. [CrossRef] [PubMed]
53. Lindblom, T.H.; Dodd, A.K. Xenobiotic detoxification in the nematode Caenorhabditis elegans. *J. Exp. Zoolog. A Comp. Exp. Biol.* **2006**, *305A*, 720–730. [CrossRef] [PubMed]
54. Tukey, R.H.; Strassburg, C.P. Human UDP-Glucuronosyltransferases: Metabolism, Expression, and Disease. *Annu. Rev. Pharmacol. Toxicol.* **2000**, *40*, 581–616. [CrossRef] [PubMed]
55. Artyukhin, A.B.; Yim, J.J.; Srinivasan, J.; Izrayelit, Y.; Bose, N.; von Reuss, S.H.; Jo, Y.; Jordan, J.M.; Baugh, L.R.; Cheong, M.; et al. Succinylated Octopamine Ascarosides and a New Pathway of Biogenic Amine Metabolism in Caenorhabditis elegans. *J. Biol. Chem.* **2013**, *288*, 18778–18783. [CrossRef] [PubMed]
56. Motley, A.M.; Hettema, E.H.; Ketting, R.; Plasterk, R.; Tabak, H.F. Caenorhabditis elegans has a single pathway to target matrix proteins to peroxisomes. *EMBO Rep.* **2000**, *1*, 40–46. [CrossRef] [PubMed]

57. Yokota, S.; Togo, S.H.; Maebuchi, M.; Bun-ya, M.; Haraguchi, C.M.; Kamiryo, T. Peroxisomes of the nematode Caenorhabditis elegans: Distribution and morphological characteristics. *Histochem. Cell Biol.* **2002**, *118*, 329–336. [CrossRef] [PubMed]
58. Von Reuss, S.H. Exploring Modular Glycolipids Involved in Nematode Chemical Communication. *CHIMIA Int. J. Chem.* **2018**, *72*, 297–303. [CrossRef] [PubMed]
59. Kim, K.-Y.; Joo, H.-J.; Kwon, H.-W.; Kim, H.; Hancock, W.S.; Paik, Y.-K. Development of a method to quantitate nematode pheromone for study of small-molecule metabolism in Caenorhabditis elegans. *Anal. Chem.* **2013**, *85*, 2681–2688. [CrossRef]
60. Kaplan, F.; Srinivasan, J.; Mahanti, P.; Ajredini, R.; Durak, O.; Nimalendran, R.; Sternberg, P.W.; Teal, P.E.A.; Schroeder, F.C.; Edison, A.S.; et al. Ascaroside expression in Caenorhabditis elegans is strongly dependent on diet and developmental stage. *PloS ONE* **2011**, *6*, e17804. [CrossRef]
61. Greene, J.S.; Brown, M.; Dobosiewicz, M.; Ishida, I.G.; Macosko, E.Z.; Zhang, X.; Butcher, R.A.; Cline, D.J.; McGrath, P.T.; Bargmann, C.I. Balancing selection shapes density-dependent foraging behaviour. *Nature* **2016**, *539*, 254–258. [CrossRef]
62. Macosko, E.Z.; Pokala, N.; Feinberg, E.H.; Chalasani, S.H.; Butcher, R.A.; Clardy, J.; Bargmann, C.I. A hub-and-spoke circuit drives pheromone attraction and social behaviour in *C. elegans*. *Nature* **2009**, *458*, 1171–1175. [CrossRef] [PubMed]
63. Jeong, P.-Y.; Kwon, M.-S.; Joo, H.-J.; Paik, Y.-K. Molecular Time-Course and the Metabolic Basis of Entry into Dauer in Caenorhabditis elegans. *PLOS ONE* **2009**, *4*, e4162. [CrossRef]
64. Jones, S.J.M.; Riddle, D.L.; Pouzyrev, A.T.; Velculescu, V.E.; Hillier, L.; Eddy, S.R.; Stricklin, S.L.; Baillie, D.L.; Waterston, R.; Marra, M.A. Changes in Gene Expression Associated with Developmental Arrest and Longevity in Caenorhabditis elegans. *Genome Res.* **2001**, *11*, 1346–1352. [CrossRef] [PubMed]
65. Wadsworth, W.G.; Riddle, D.L. Developmental regulation of energy metabolism in Caenorhabditis elegans. *Dev. Biol.* **1989**, *132*, 167–173. [CrossRef]
66. Wang, J.; Kim, S.K. Global analysis of dauer gene expression in Caenorhabditis elegans. *Development* **2003**, *130*, 1621–1634. [CrossRef] [PubMed]
67. Lee, H.; Choi, M.; Lee, D.; Kim, H.; Hwang, H.; Kim, H.; Park, S.; Paik, Y.; Lee, J. Nictation, a dispersal behavior of the nematode *Caenorhabditis elegans*, is regulated by IL2 neurons. *Nat. Neurosci.* **2012**, *15*, 107–112. [CrossRef]
68. Lee, D.; Yang, H.; Kim, J.; Brady, S.; Zdraljevic, S.; Zamanian, M.; Kim, H.; Paik, Y.; Kruglyak, L.; Andersen, E.C.; et al. The genetic basis of natural variation in a phoretic behavior. *Nat. Commun.* **2017**, *8*, 273. [CrossRef] [PubMed]
69. Kim, K.; Sato, K.; Shibuya, M.; Zeiger, D.M.; Butcher, R.A.; Ragains, J.R.; Clardy, J.; Touhara, K.; Sengupta, P. Two chemoreceptors mediate developmental effects of dauer pheromone in C. elegans. *Science* **2009**, *326*, 994–998. [CrossRef]
70. Sommer, R.J.; Ogawa, A. Hormone Signaling and Phenotypic Plasticity in Nematode Development and Evolution. *Curr. Biol.* **2011**, *21*, R758–R766. [CrossRef]
71. McGrath, P.T.; Xu, Y.; Ailion, M.; Garrison, J.L.; Butcher, R.A.; Bargmann, C.I. Parallel evolution of domesticated Caenorhabditis species targets pheromone receptor genes. *Nature* **2011**, *477*, 321–325. [CrossRef]
72. Park, D.; O'Doherty, I.; Somvanshi, R.K.; Bethke, A.; Schroeder, F.C.; Kumar, U.; Riddle, D.L. Interaction of structure-specific and promiscuous G-protein-coupled receptors mediates small-molecule signaling in Caenorhabditis elegans. *Proc. Natl. Acad. Sci. USA* **2012**, *109*, 9917–9922. [CrossRef]
73. Neal, S.J.; Park, J.; DiTirro, D.; Yoon, J.; Shibuya, M.; Choi, W.; Schroeder, F.C.; Butcher, R.A.; Kim, K.; Sengupta, P. A Forward Genetic Screen for Molecules Involved in Pheromone-Induced Dauer Formation in Caenorhabditis elegans. *G3 Genes Genomes Genet.* **2016**, *6*, 1475–1487. [CrossRef] [PubMed]
74. Park, J.; Choi, W.; Dar, A.R.; Butcher, R.A.; Kim, K. Neuropeptide Signaling Regulates Pheromone-Mediated Gene Expression of a Chemoreceptor Gene in C. elegans. *Mol. Cells* **2018**, *42*, 28–35. [PubMed]
75. Ludewig, A.H.; Izrayelit, Y.; Park, D.; Malik, R.U.; Zimmermann, A.; Mahanti, P.; Fox, B.W.; Bethke, A.; Doering, F.; Riddle, D.L.; et al. Pheromone sensing regulates Caenorhabditis elegans lifespan and stress resistance via the deacetylase SIR-2.1. *Proc. Natl. Acad. Sci. USA* **2013**, *110*, 5522–5527. [CrossRef] [PubMed]

76. Narayan, A.; Venkatachalam, V.; Durak, O.; Reilly, D.K.; Bose, N.; Schroeder, F.C.; Samuel, A.D.T.; Srinivasan, J.; Sternberg, P.W. Contrasting responses within a single neuron class enable sex-specific attraction in Caenorhabditis elegans. *Proc. Natl. Acad. Sci. USA* **2016**, *113*, E1392–E1401. [CrossRef] [PubMed]
77. Fagan, K.A.; Luo, J.; Lagoy, R.C.; Schroeder, F.C.; Albrecht, D.R.; Portman, D.S. A Single-Neuron Chemosensory Switch Determines the Valence of a Sexually Dimorphic Sensory Behavior. *Curr. Biol.* **2018**, *28*, 902–914. [CrossRef]
78. Park, D.; Hahm, J.-H.; Park, S.; Ha, G.; Chang, G.-E.; Jeong, H.; Kim, H.; Kim, S.; Cheong, E.; Paik, Y.-K. A conserved neuronal DAF-16/FoxO plays an important role in conveying pheromone signals to elicit repulsion behavior in Caenorhabditis elegans. *Sci. Rep.* **2017**, *7*, 7260. [CrossRef]
79. Hong, M.; Ryu, L.; Ow, M.C.; Kim, J.; Je, A.R.; Chinta, S.; Huh, Y.H.; Lee, K.J.; Butcher, R.A.; Choi, H.; et al. Early Pheromone Experience Modifies a Synaptic Activity to Influence Adult Pheromone Responses of C. elegans. *Curr. Biol.* **2017**, *27*, 3168–3177. [CrossRef]
80. Ryu, L.; Cheon, Y.; Huh, Y.H.; Pyo, S.; Chinta, S.; Choi, H.; Butcher, R.A.; Kim, K. Feeding state regulates pheromone-mediated avoidance behavior via the insulin signaling pathway in Caenorhabditis elegans. *EMBO J.* **2018**, *37*, e98402. [CrossRef]
81. Zhang, Y.K.; Sanchez-Ayala, M.A.; Sternberg, P.W.; Srinivasan, J.; Schroeder, F.C. Improved Synthesis for Modular Ascarosides Uncovers Biological Activity. *Org. Lett.* **2017**, *19*, 2837–2840. [CrossRef]
82. Fenk, L.A.; Bono, M. de Memory of recent oxygen experience switches pheromone valence in Caenorhabditis elegans. *Proc. Natl. Acad. Sci. USA* **2017**, *114*, 4195–4200. [CrossRef] [PubMed]
83. Greene, J.S.; Dobosiewicz, M.; Butcher, R.A.; McGrath, P.T.; Bargmann, C.I. Regulatory changes in two chemoreceptor genes contribute to a Caenorhabditis elegans QTL for foraging behavior. *eLife* **2016**, *5*, e21454. [CrossRef] [PubMed]
84. Bell, G. *The Masterpiece of Nature: The Evolution and Genetics of Sexuality*; CUP Archive: Cambridge, UK, 1982.
85. Dacks, J.; Roger, A.J. The First Sexual Lineage and the Relevance of Facultative Sex. *J. Mol. Evol.* **1999**, *48*, 779–783. [CrossRef] [PubMed]
86. Dubnau, D. Genetic competence in Bacillus subtilis. *Microbiol. Mol. Biol. Rev.* **1991**, *55*, 395–424.
87. Gemmill, A.W.; Viney, M.E.; Read, A.F. Host Immune Status Determines Sexuality in a Parasitic Nematode. *Evolution* **1997**, *51*, 393–401. [CrossRef] [PubMed]
88. Harris, E.H.; Stern, D.B.; Witman, G.B. *The Chlamydomonas Sourcebook*; Elsevier: San Diego, CA, USA, 2009.
89. Kleiven, O.T.; Larsson, P.; Hobæk, A. Sexual reproduction in Daphnia magna requires three stimuli. *Oikos* **1992**, 197–206. [CrossRef]
90. Lynch, M.; Bürger, R.; Butcher, D.; Gabriel, W. The Mutational Meltdown in Asexual Populations. *J. Hered.* **1993**, *84*, 339–344. [CrossRef] [PubMed]
91. Mai, B.; Breeden, L. CLN1 and Its Repression by Xbp1 Are Important for Efficient Sporulation in Budding Yeast. *Mol. Cell. Biol.* **2000**, *20*, 478–487. [CrossRef]
92. Muller, H.J. The relation of recombination to mutational advance. *Mutat. Res. Mol. Mech. Mutagen.* **1964**, *1*, 2–9. [CrossRef]
93. Morran, L.T.; Cappy, B.J.; Anderson, J.L.; Phillips, P.C. Sexual Partners for the Stressed: Facultative Outcrossing in the Self-Fertilizing Nematode Caenorhabditis Elegans. *Evolution* **2009**, *63*, 1473–1482. [CrossRef]
94. Wharam, B.; Weldon, L.; Viney, M. Pheromone modulates two phenotypically plastic traits—adult reproduction and larval diapause—In the nematode Caenorhabditis elegans. *BMC Evol. Biol.* **2017**, *17*, 197. [CrossRef] [PubMed]
95. Izrayelit, Y.; Srinivasan, J.; Campbell, S.L.; Jo, Y.; von Reuss, S.H.; Genoff, M.C.; Sternberg, P.W.; Schroeder, F.C. Targeted Metabolomics Reveals a Male Pheromone and Sex-Specific Ascaroside Biosynthesis in Caenorhabditis elegans. *ACS Chem. Biol.* **2012**, *7*, 1321–1325. [CrossRef] [PubMed]
96. Aprison, E.Z.; Ruvinsky, I. Sex Pheromones of C. elegans Males Prime the Female Reproductive System and Ameliorate the Effects of Heat Stress. *PLOS Genet.* **2015**, *11*, e1005729. [CrossRef] [PubMed]
97. Shi, C.; Runnels, A.M.; Murphy, C.T. Mating and male pheromone kill Caenorhabditis males through distinct mechanisms. *eLife.* **2017**, *6*, e23493. [CrossRef] [PubMed]
98. Riddle, D.L.; Blumenthal, T.; Meyer, B.J.; Priess, J.R. *C. elegans II*, 2nd ed.; Cold Spring Harbor Laboratory Press: Cold Spring Harbor, NY, USA, 1997.

99. Berger, J.; Dorninger, F.; Forss-Petter, S.; Kunze, M. Peroxisomes in brain development and function. *Biochim. Biophys. Acta BBA—Mol. Cell Res.* **2016**, *1863*, 934–955. [CrossRef]
100. Clayton, P.T. Clinical consequences of defects in peroxisomal β-oxidation. *Biochem. Soc. Trans.* **2001**, *29*, 298–305. [CrossRef] [PubMed]
101. Crane, D.I. Revisiting the neuropathogenesis of Zellweger syndrome. *Neurochem. Int.* **2014**, *69*, 1–8. [CrossRef]
102. De Munter, S.; Verheijden, S.; Régal, L.; Baes, M. Peroxisomal Disorders: A Review on Cerebellar Pathologies. *Brain Pathol.* **2015**, *25*, 663–678. [CrossRef]
103. Wanders, R.J.A. Peroxisomes, lipid metabolism, and peroxisomal disorders. *Mol. Genet. Metab.* **2004**, *83*, 16–27. [CrossRef]
104. Wanders, R.J.A.; Waterham, H.R. Peroxisomal disorders: The single peroxisomal enzyme deficiencies. *Biochim. Biophys. Acta BBA—Mol. Cell Res.* **2006**, *1763*, 1707–1720. [CrossRef]
105. Brites, P.; Mooyer, P.A.W.; el Mrabet, L.; Waterham, H.R.; Wanders, R.J.A. Plasmalogens participate in very-long-chain fatty acid-induced pathology. *Brain* **2009**, *132*, 482–492. [CrossRef] [PubMed]
106. Trompier, D.; Vejux, A.; Zarrouk, A.; Gondcaille, C.; Geillon, F.; Nury, T.; Savary, S.; Lizard, G. Brain peroxisomes. *Biochimie* **2014**, *98*, 102–110. [CrossRef] [PubMed]
107. Park, S.; Paik, Y.-K. Genetic deficiency in neuronal peroxisomal fatty acid β-oxidation causes the interruption of dauer development in Caenorhabditis elegans. *Sci. Rep.* **2017**, *7*, 9358. [CrossRef] [PubMed]
108. Kulalert, W.; Kim, D.H. The Unfolded Protein Response in a Pair of Sensory Neurons Promotes Entry of C. elegans into Dauer Diapause. *Curr. Biol.* **2013**, *23*, 2540–2545. [CrossRef]
109. Malone, E.A.; Thomas, J.H. A screen for nonconditional dauer-constitutive mutations in Caenorhabditis elegans. *Genetics* **1994**, *136*, 879–886. [PubMed]
110. Li, W.; Kennedy, S.G.; Ruvkun, G. daf-28 encodes a C. elegans insulin superfamily member that is regulated by environmental cues and acts in the DAF-2 signaling pathway. *Genes Dev.* **2003**, *17*, 844–858. [CrossRef]
111. Kulalert, W.; Sadeeshkumar, H.; Zhang, Y.K.; Schroeder, F.C.; Kim, D.H. Molecular Determinants of the Regulation of Development and Metabolism by Neuronal eIF2α Phosphorylation in Caenorhabditis elegans. *Genetics* **2017**, *206*, 251–263. [CrossRef]
112. Ludewig, A. Ascaroside signaling in C. elegans. *WormBook* **2013**, 1–22. [CrossRef]
113. Neal, S.J.; Takeishi, A.; O'Donnell, M.P.; Park, J.; Hong, M.; Butcher, R.A.; Kim, K.; Sengupta, P. Feeding state-dependent regulation of developmental plasticity via CaMKI and neuroendocrine signaling. *eLife* **2015**, *4*, e10110. [CrossRef]
114. O'Donnell, M.P.; Chao, P.-H.; Kammenga, J.E.; Sengupta, P. Rictor/TORC2 mediates gut-to-brain signaling in the regulation of phenotypic plasticity in C. elegans. *PLOS Genet.* **2018**, *14*, e1007213. [CrossRef]
115. Ilbay, O.; Ambros, V. Pheromones and Nutritional Signals Regulate the Developmental Reliance on let-7 Family MicroRNAs in C. elegans. *Curr. Biol.* **2019**, *29*, 1735–1745. [CrossRef] [PubMed]
116. Yamada, K.; Hirotsu, T.; Matsuki, M.; Butcher, R.A.; Tomioka, M.; Ishihara, T.; Clardy, J.; Kunitomo, H.; Iino, Y. Olfactory Plasticity Is Regulated by Pheromonal Signaling in Caenorhabditis elegans. *Science* **2010**, *329*, 1647–1650. [CrossRef] [PubMed]
117. Fielenbach, N.; Antebi, A.C. elegans dauer formation and the molecular basis of plasticity. *Genes Dev.* **2008**, *22*, 2149–2165. [CrossRef] [PubMed]
118. Hahm, J.-H.; Kim, S.; Paik, Y.-K. Endogenous cGMP regulates adult longevity via the insulin signaling pathway in Caenorhabditis elegans. *Aging Cell* **2009**, *8*, 473–483. [CrossRef] [PubMed]
119. Artan, M.; Jeong, D.-E.; Lee, D.; Kim, Y.-I.; Son, H.G.; Husain, Z.; Kim, J.; Altintas, O.; Kim, K.; Alcedo, J.; et al. Food-derived sensory cues modulate longevity via distinct neuroendocrine insulin-like peptides. *Genes Dev.* **2016**, *30*, 1047–1057. [CrossRef] [PubMed]
120. Fletcher, M.; Kim, D.H. Age-Dependent Neuroendocrine Signaling from Sensory Neurons Modulates the Effect of Dietary Restriction on Longevity of Caenorhabditis elegans. *PLOS Genet.* **2017**, *13*, e1006544. [CrossRef] [PubMed]
121. Ludewig, A.H.; Gimond, C.; Judkins, J.C.; Thornton, S.; Pulido, D.C.; Micikas, R.J.; Döring, F.; Antebi, A.; Braendle, C.; Schroeder, F.C. Larval crowding accelerates C. elegans development and reduces lifespan. *PLOS Genet.* **2017**, *13*, e1006717. [CrossRef] [PubMed]
122. Schulenburg, H.; Ewbank, J.J. The genetics of pathogen avoidance in Caenorhabditis elegans. *Mol. Microbiol.* **2007**, *66*, 563–570. [CrossRef]

123. Meisel, J.D.; Kim, D.H. Behavioral avoidance of pathogenic bacteria by Caenorhabditis elegans. *Trends Immunol.* **2014**, *35*, 465–470. [CrossRef]
124. Kim, D.H. Bacteria and the Aging and Longevity of Caenorhabditis elegans. *Annu. Rev. Genet.* **2013**, *47*, 233–246. [CrossRef]
125. Reddy, K.C.; Andersen, E.C.; Kruglyak, L.; Kim, D.H. A Polymorphism in npr-1 Is a Behavioral Determinant of Pathogen Susceptibility in C. elegans. *Science* **2009**, *323*, 382–384. [CrossRef] [PubMed]
126. Reddy, K.C.; Hunter, R.C.; Bhatla, N.; Newman, D.K.; Kim, D.H. Caenorhabditis elegans NPR-1–mediated behaviors are suppressed in the presence of mucoid bacteria. *Proc. Natl. Acad. Sci. USA* **2011**, *108*, 12887–12892. [CrossRef] [PubMed]
127. Styer, K.L.; Singh, V.; Macosko, E.; Steele, S.E.; Bargmann, C.I.; Aballay, A. Innate Immunity in Caenorhabditis elegans Is Regulated by Neurons Expressing NPR-1/GPCR. *Science* **2008**, *322*, 460–464. [CrossRef] [PubMed]
128. Chang, H.C.; Paek, J.; Kim, D.H. Natural polymorphisms in *C. elegans* HECW-1 E3 ligase affect pathogen avoidance behaviour. *Nature* **2011**, *480*, 525–529. [CrossRef] [PubMed]
129. Cao, X.; Kajino-Sakamoto, R.; Doss, A.; Aballay, A. Distinct Roles of Sensory Neurons in Mediating Pathogen Avoidance and Neuropeptide-Dependent Immune Regulation. *Cell Rep.* **2017**, *21*, 1442–1451. [CrossRef] [PubMed]
130. Meisel, J.D.; Panda, O.; Mahanti, P.; Schroeder, F.C.; Kim, D.H. Chemosensation of Bacterial Secondary Metabolites Modulates Neuroendocrine Signaling and Behavior of C. elegans. *Cell* **2014**, *159*, 267–280. [CrossRef] [PubMed]
131. Moore, R.S.; Kaletsky, R.; Murphy, C.T. Piwi/PRG-1 Argonaute and TGF-β Mediate Transgenerational Learned Pathogenic Avoidance. *Cell* **2019**, *177*, 1827–1841. [CrossRef] [PubMed]
132. Lee, K.; Mylonakis, E. An Intestine-Derived Neuropeptide Controls Avoidance Behavior in Caenorhabditis elegans. *Cell Rep.* **2017**, *20*, 2501–2512. [CrossRef] [PubMed]
133. Richardson, C.E.; Kooistra, T.; Kim, D.H. An essential role for XBP-1 in host protection against immune activation in *C. elegans*. *Nature* **2010**, *463*, 1092–1095. [CrossRef] [PubMed]
134. Richardson, C.E.; Kinkel, S.; Kim, D.H. Physiological IRE-1-XBP-1 and PEK-1 Signaling in Caenorhabditis elegans Larval Development and Immunity. *PLOS Genet.* **2011**, *7*, e1002391. [CrossRef] [PubMed]
135. Sun, J.; Liu, Y.; Aballay, A. Organismal regulation of XBP-1-mediated unfolded protein response during development and immune activation. *EMBO Rep.* **2012**, *13*, 855–860. [CrossRef] [PubMed]
136. Sun, J.; Singh, V.; Kajino-Sakamoto, R.; Aballay, A. Neuronal GPCR Controls Innate Immunity by Regulating Noncanonical Unfolded Protein Response Genes. *Science* **2011**, *332*, 729–732. [CrossRef] [PubMed]
137. Park, J.H.; Chung, H.Y.; Kim, M.; Lee, J.H.; Jung, M.; Ha, H. Daumone fed late in life improves survival and reduces hepatic inflammation and fibrosis in mice. *Aging Cell* **2014**, *13*, 709–718. [CrossRef] [PubMed]
138. Park, J.H.; Ha, H. Short-term Treatment of Daumone Improves Hepatic Inflammation in Aged Mice. *Korean J. Physiol. Pharmacol.* **2015**, *19*, 269. [CrossRef] [PubMed]
139. Ma, G.T.; Lee, S.K.; Park, K.-K.; Park, J.; Son, S.H.; Jung, M.; Chung, W.-Y. Artemisinin-Daumone Hybrid Inhibits Cancer Cell-Mediated Osteolysis by Targeting Cancer Cells and Osteoclasts. *Cell. Physiol. Biochem.* **2018**, *49*, 1460–1475. [CrossRef] [PubMed]
140. Consortium, T.C. elegans S. Genome Sequence of the Nematode C. elegans: A Platform for Investigating Biology. *Science* **1998**, *282*, 2012–2018.
141. Nagarathnam, B.; Kalaimathy, S.; Balakrishnan, V.; Sowdhamini, R. Cross-Genome Clustering of Human and C. elegans G-Protein Coupled Receptors. *Evol. Bioinforma.* **2012**, *8*, EBO–S9405. [CrossRef] [PubMed]

International Journal of
Molecular Sciences

MDPI

Review

Function of Green Tea Catechins in the Brain: Epigallocatechin Gallate and its Metabolites

Monira Pervin [1,*], Keiko Unno [1,*], Akiko Takagaki [2], Mamoru Isemura [1] and Yoriyuki Nakamura [1]

1 Tea Science Center, Graduate School of Integrated Pharmaceutical and Nutritional Sciences, University of Shizuoka, Shizuoka 422-8526, Japan
2 R&D group, Mitsui Norin Co. Ltd., Shizuoka 426-0133, Japan
* Correspondence: gp1747@u-shizuoka-ken.ac.jp (M.P.); unno@u-shizuoka-ken.ac.jp (K.U.); Tel./Fax: +81-54-264-5822 (M.P. & K.U.)

Received: 3 July 2019; Accepted: 22 July 2019; Published: 25 July 2019

Abstract: Over the last three decades, green tea has been studied for its beneficial effects, including anti-cancer, anti-obesity, anti-diabetes, anti-inflammatory, and neuroprotective effects. At present, a number of studies that have employed animal, human and cell cultures support the potential neuroprotective effects of green tea catechins against neurological disorders. However, the concentration of (−)-epigallocatechin gallate (EGCG) in systemic circulation is very low and EGCG disappears within several hours. EGCG undergoes microbial degradation in the small intestine and later in the large intestine, resulting in the formation of various microbial ring-fission metabolites which are detectable in the plasma and urine as free and conjugated forms. Recently, in vitro experiments suggested that EGCG and its metabolites could reach the brain parenchyma through the blood–brain barrier and induce neuritogenesis. These results suggest that metabolites of EGCG may play an important role, alongside the beneficial activities of EGCG, in reducing neurodegenerative diseases. In this review, we discuss the function of EGCG and its microbial ring-fission metabolites in the brain in suppressing brain dysfunction. Other possible actions of EGCG metabolites will also be discussed.

Keywords: blood–brain barrier; catechin; cognition; epigallocatechin gallate; green tea; microbiota; 5-(3,5-dihydroxyphenyl)-γ-valerolactone

1. Introduction

Tea is derived from the leaves and buds of the plant *Camellia sinensis* L. (Theaceae). Among the different types of tea, such as green tea, black tea, and oolong tea, the health benefits of green tea have been most extensively studied [1,2]. These include anti-cancer [3,4], anti-obesity [5–7], anti-diabetes [8,9], and neuroprotective effects [10–12]. The antioxidant and metal chelating [13,14], anti-carcinogenic [15], anti-apoptotic [16,17], pro-apoptotic, and anti-inflammatory [14,18] properties of catechins are greatly associated with their beneficial health effects, including suppressing neurodegenerative diseases.

Compared to other beverages, green tea is rich in catechins. According to Khokhar et al., 100 mL of green tea (1 g of dry tea leaves brewed for 5 min in 100 mL of hot water) contains on average 67 ± 11 mg of total catechins, including about 30 mg of (−)-epigallocatechin gallate (EGCG), whereas black tea contains 15.4 mg of catechins [19]. In green tea catechins, the main active molecule, EGCG (Figure 1), an ester of (−)-epigallocatechin (EGC) and gallic acid (GA), represents 50–80% of the total catechin content, followed by EGC, (−)-epicatechin gallate (ECG), (−)-epicatechin (EC), and (+)-catechin (C) [20]. Numerous beneficial effects of EGCG have been reported on cognitive function and oxidative damage [21–24]. Several epidemiological studies also showed the association between drinking tea and the beneficial effects on cognitive function [25–28]. For example, a cross-sectional study by Kuriyama et al. showed that daily ingestion of one or two cups of green tea significantly reduced cognitive impairment [25]. In another clinical study by Ide et al., the consumption of green tea (2 g/day) for 3 months significantly improved cognitive function and also reduced the progression of cognitive dysfunction [29].

Male Wistar rats that orally ingested EGCG showed a peak concentration at 1–2 h in systemic circulation, and it remained present in trace amounts after 4 h [30]. Much of orally ingested EGCG undergoes intestinal microbial degradation in the small intestine to EGC and GA, and later in the large intestine, resulting in the formation of various colonic microbial ring-fission metabolites, which are detectable in the plasma and urine [31–34]. These metabolites can exhibit biological activities, and some of them may be attributed to the action of EGCG.

This review discusses the function of EGCG and its metabolites as well as their possible action in the brain in suppressing brain dysfunction. In addition, recent data of other functions of EGCG metabolites are described.

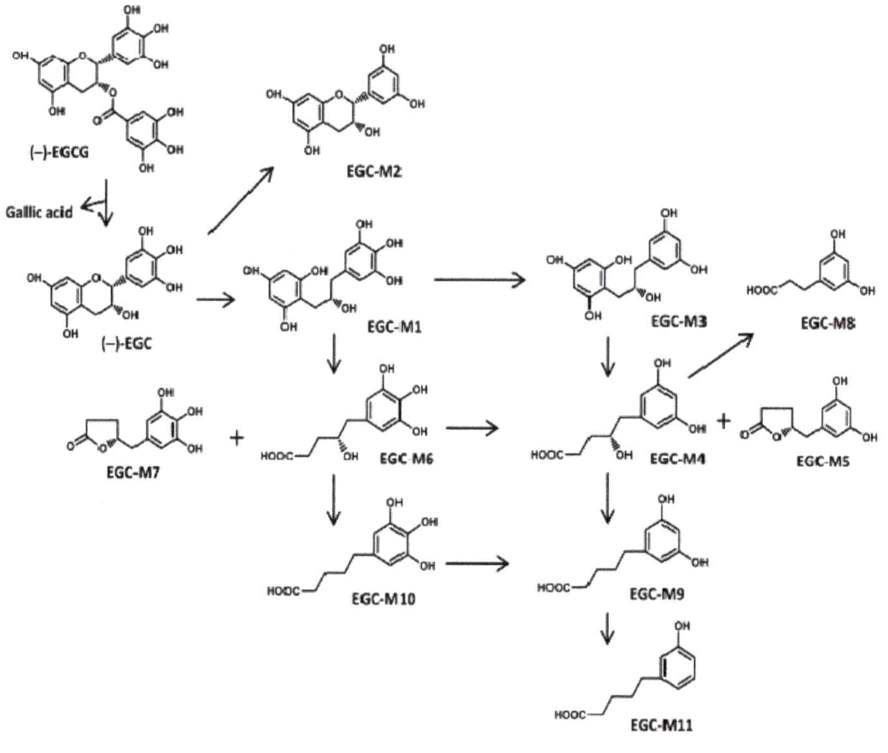

Figure 1. Chemical structures of EGCG metabolites based on data from Takagaki et al. [41].

2. Bioactivity of EGCG and Its Metabolites in the Brain

2.1. Absorption and Bioavailability of EGCG

EGCG is poorly absorbed by the body, it reaches the blood circulation at a very low micromolar concentration, and then it disappears from plasma within several hours [30,35–38]. The oral bioavailability of EGCG is estimated to be about 0.1 to 0.3% in rats and humans [25,26,30,35].

Catechin Ring-Fission Products

EGCG was found to be hydrolyzed by intestinal microbiota to produce EGC and GA. EGC was further degraded to some kinds of ring-fission metabolites in the gut tract. In the large intestine, there are 11 colonic microbial ring-fission metabolites of EGC (EGC-M1–M11) (Table 1, Figure 1) as described by Takagaki et al., i.e. 1-(3,4,5-trihydroxyphenyl) 3-(2,4,6-trihydroxyphenyl)-propan-2-ol (EGC-M1), 4-dehydroxylated epigallocatechin (EGC-M2), 1-(3,5-dihydroxyphenyl)- 3-(2,4,6-trihydroxyphenyl)-propan-2-ol (EGC-M3), 4-hydroxy-5-(3,5-dihydroxyphenyl) valeric acid (EGC-M4), 5-(3,5-dihydroxyphenyl)-γ-valerolactone (EGC-M5), 4-hydroxy-5-(3,4,5-trihydroxyphenyl) valeric acid (EGC-M6), 5-(3,4,5-trihydroxyphenyl)-γ-valerolactone (EGC-M7), 3-(3,5-dihydroxyphenyl) propionic acid (EGC-M8), 5-(3,5-dihydroxyphenyl) valeric acid (EGC-M9), 5-(3,4,5-trihydroxyphenyl) valeric acid (EGC-M10), and 5-(3-hydroxyphenyl) valeric acid (EGC-M11) [39–41]. Among them, EGC-M5 and EGC-M7 were found to be the main metabolites in mice, rat, and human plasma, urine, and bile [42].

Table 1. Microbial ring-fission metabolites of EGCG in rat.

EGCG Metabolites (Microbial Ring-Fission)	Abbreviation
1-(3,4,5-trihydroxyphenyl)-3-(2,4,6-trihydroxyphenyl)-propan-2-ol	(EGC-M1)
4-dehydroxylated epigallocatechin	(EGC-M2)
1-(3,5-dihydroxyphenyl)-3-(2,4,6-trihydroxyphenyl)-propan-2-ol	(EGC-M3)
4-hydroxy-5-(3,5-dihydroxyphenyl) valeric acid	(EGC-M4)
5-(3,5-dihydroxyphenyl)-γ-valerolactone	(EGC-M5)
4-hydroxy-5-(3,4,5-trihydroxyphenyl) valeric acid	(EGC-M6)
5-(3,4,5-trihydroxyphenyl)-γ-valerolactone	(EGC-M7)
3-(3,5-dihydroxyphenyl) propionic acid	(EGC-M8)
5-(3,5-dihydroxyphenyl) valeric acid	(EGC-M9)
5-(3,4,5-trihydroxyphenyl) valeric acid	(EGC-M10)
5-(3-hydroxyphenyl) valeric acid	(EGC-M11)

Adapted from Takagaki et al. [41].

The intestinal microbial ring-fission metabolites of EGCG are present in plasma as free and conjugated forms [31], and in vitro data suggested that they could reach the brain parenchyma through the blood–brain barrier (BBB) and induce neuritogenesis [43], suggesting that they might be important in suppressing neurodegenerative diseases.

The bioavailability of a compound or its metabolites can be determined by quantifying the concentration at the systematic blood flow and at the target organ [44]. It is very important to know the metabolic process and bioavailability of green tea catechins to evaluate their biological activity as well as to understand their beneficial effects on human health. EGCG has much lower bioavailability than other components in catechins [36,45]. For example, after intragastric administration of decaffeinated green tea (200 mg/kg) to male Sprague–Dawley rats, 13.7% of EGC, 31.2% of EC, and 0.1% of EGCG appeared in the blood [36]. The bioavailability of EGCG is significantly different depending on the route of administration, such as intravenous, intragastric, or through peroral ingestion, since intravenously ingested EGCG can equally reach all tissues in a free state (without conjugate) compared to intragastric and peroral administration as a result of the high levels of EGCG in intravenous ingestion. It is much easier for tissues to absorb free EGCG (without conjugate) in intravenous ingestion compared to other routes of administration [38]. On the other hand, the absorption rate of EGCG in plasma was much better in peroral administration [46] compared to intragastric intubation, although the detailed mechanism is not clear [36]. Mice and rats show a difference in bioavailability. For example, in the mice model, there is higher absorption of EGCG (26.5%) [38] than in the rat model (1.6%) [36].

Aglycons (without sugar residues) from plant polyphenols are easily absorbed in the small intestine [47]. However, the majority of polyphenols in plants exist as a form of glycosides, esters, or polymers, and they cannot be absorbed directly from the intestine. Therefore, they are hydrolyzed by intestinal enzymes or gut microbiota. EGCG, the ester of epigallocatechin and GA, is metabolized by intestinal microbiota in rats [39,40,48,49].

In mice, the bioavailability of a single dose of pure EGCG was first reported by Lambert et al. The authors found that after intravenous (21.8 μmol/kg) and intragastric (163.8 μmol/kg) administration of EGCG to male CF-1 mice, the plasma levels of total EGCG reached about 2.7 ± 0.7 and 0.28 ± 0.08 μM, respectively. The levels of free EGCG in the liver, lung, small intestine, and colon were about 3.56, 2.66, 2.40, and 1.20 nmol/g, respectively. The levels of total EGCG in the small intestine and colon were 45.2 and 7.9 nmol/g, but the levels in the liver and lung could not be determined as the concentration was too low [38]. On the other hand, in male Sprague–Dawley rats, the plasma bioavailability of EGCG was 0.1~1.6%, suggesting that the rate of absorption in mice is much higher than in rats [36].

After [4^{-3}H]EGCG (4 mg, 7.4 MBq/kg) was administered to male Wistar rats by intragastric gavage, the absorption, distribution, and excretion in blood, tissues, urine, and feces of EGCG and its metabolites were determined by tracing radioactivity using high-performance liquid chromatography (HPLC) analysis [31]. The results show that the radioactivity of EGCG mostly disappeared in the stomach by 72 h. Peak radioactivity in the small intestine, cecum, and large intestine was detected at 4 h (40.5% of the dose), 8 h (46.4% of the dose), and 8 h (13.2% of the dose), respectively, and the radioactivity was markedly reduced by 24 h and had almost disappeared by 72 h in these tissues. The level of radioactivity in the blood was low at 4 h, began to increase after 8 h, peaked at 24 h, and thereafter decreased. The urinary levels of two major radioactive metabolites, 5-(5-hydroxyphenyl)-γ-valerolactone 3-*O*-β-glucuronide and EGC-M5 were 68% and 16.8% of the ingested radioactivity after 48 h. The authors suggested that intragastrically ingested EGCG is absorbed in the intestine within several hours (<8 h), and thereafter the EGCG metabolites and conjugates are absorbed from the large intestine (>8~48 h), distributed to various tissues via blood circulation, and finally excreted via urine [31]. The degradation of EGCG by gut microbiota could be an important factor in decreasing its bioavailability [50]. When male C57BL/6J mice were given water containing (per mL) ampicillin (1 mg), sulfamethoxazole (1.6 mg), and trimethoprim (0.32 mg) for 11 days and then given a 0.32% Polyphenon E diet containing 643 mg EGCG, 29 mg EGC, 74 mg ECG, 90 mg EC, 45 mg gallocatechin gallate, and 6 mg caffeine per g of Polyphenon E, the levels of EGCG in blood, liver, and urine increased. On the other hand, antibiotic treatment decreased the urinary levels of EGC-M7, the ring-fission metabolites of EGCG, and 5-(3,4-dihydroxyphenyl)-γ-valerolactone, a ring-fission metabolite of EC. This finding suggests that antibiotic treatment eliminated catechin-degrading microbiota in the gut and therefore, increased the levels of EGCG as well as decreased the ring-fission metabolites due to the presence of a low content of microbiota in the gut [50].

In male Sprague–Dawley rats that were given EGCG orally at 150 mg/kg, the plasma and the tissue distribution of EGCG were detected by developed HPLC with electrochemical detection [46]. After 2 h and 5 h of administration of EGCG, the levels of free (without conjugated) and total EGCG (with glucuronides, sulfates, and glucuronides/sulfates) in rat plasma were 0.7, 0.28, 0.82, and 0.5 μM, respectively. The authors also reported unpublished data showing that the plasma level of EGCG in rats 24 h after administration is 0.05 μM, suggesting that the EGCG level was markedly reduced 24 h after administration. The tissue levels of free EGCG in the small intestine and colon were 21.15 and 10.75, as well as 4.75 and 24.41 nmol/g at 2 and 5 h, respectively. They showed that the levels of free EGCG in the kidney, liver, spleen, lung, and brain were 1.02 and 0.54, 1.02 and 0.54, 0.1 and 0.12, 0.4 and 0.14, and 0.19 and 0.18 nmol/g at 2 and 5 h, respectively. These results indicate that the levels of EGCG in plasma and other tissues were high at 2 h and began to decrease 5 h after administration. Moreover, the plasma level of EGCG was very low 24 h after ingestion [46].

A human study by Warden et al. showed that after drinking black tea containing 16.74 mg of EGCG, 15.48 mg of EGC, 36.54 mg of EC, and 31.14 mg of ECG, the plasma concentration of EGCG was at the peak level between 5 and 8 h, but returned to baseline levels by 24 h. After tea ingestion over 6 h, the ingested catechins detected in plasma, urine, and feces were about 0.16%, 1.1%, and 0.42%, respectively, suggesting that level of absorption of catechins in humans is also quite low [51].

Microflora-mediated ring fission metabolites have also been identified in humans. EGCG was found to be hydrolyzed in the small intestine by intestinal microflora to produce EGC and GA and further degraded in the large intestine to produce various kinds of microbial ring fission metabolites [34,52,53]. In a human urinary metabolite profile, the ring-fission metabolites of tea catechins, such as 5-(3, 4-dihydroxyphenyl)-γ-valerolactone, EGC-M5, EGC-M7, and their glucuronide and sulfate conjugates, were found to be the major urinary metabolites at 12–24 h after ingestion of tea (200 mL of reconstituted green tea (from 3 g of tea solids)) in healthy male volunteers [34]. Two catechin ring-fission metabolites, EGC-M7 and 5-(3,4-dihydroxyphenyl)-γ-valerolactone, appeared in urine (4–8 μM) and in plasma (0.1–0.2 μM) approximately 13 h after ingestion of 20 mg/kg of decaffeinated green tea [53]. In addition, the cumulative urinary excretion of these microbial ring-fission metabolites was as high as 8–25 times the levels of ECG and EC [53]. A recent study on colonic ring-fission metabolism in humans identified various urinary metabolites derived from green tea flavan-3-ol (639 μmol of monomeric catechin and 88 μmol of oligomeric catechin), including EGC-M5, EGC-M7, 5-(4,5-dihydroxyphenyl)-γ-valerolactone, and 5-(hydroxyphenyl)-γ-valerolactone, with their glucuronide and sulphate conjugates [54]. The excretion rates of these ring-fission metabolites were as follows: EGC-M5-disulphate (163 μmol), EGC-M5-glucuronide (34.4 μmol), EGC-M7-sulphate (27.7 μmol), EGC-M7-glucuronide (12.1 μmol), methyl-EGC-M7-sulphate (54.7 μmol), methyl-EGC-M7-glucuronide (2.7 μmol), 5-(4,5-dihydroxyphenyl)-γ-valerolactone-disulphate (87.6 μmol), 5-(4,5-dihydroxyphenyl)-γ-valerolactone-glucuronide (16.8 μmol), 5-(hydroxyphenyl)-γ-valerolactone-sulphate (19.7 μmol), and 5-(hydroxyphenyl)-γ-valerolactone-glucuronide (6.6 μmol) [54]. In this study, the bioavailability of green tea flavan-3-ols was about 62% (the ratio between total metabolic excretion and total intake of flavan-3-ols) in 48 h which is higher than that reported previously (39%) in 24 h [52]. This study examined a more complete 48 h metabolic excretion profile and quantified a wider range of colonic microbial metabolites [54].

2.2. Blood–Brain Barrier Permeability of EGCG and Its Metabolites

The BBB is a dynamic system that separates circulating peripheral blood from brain neural tissue in the central nervous system. It is composed of endothelial cells connected through gap junctional proteins, astrocytes, pericytes, and extracellular matrix and works together to regulate the movement of ions, molecules, and cells between the blood and the brain to create a unique microenvironment for proper neuronal function [55]. Therefore, the BBB plays a significant role in transporting intravascular substances into the brain.

After male Sprague–Dawley rats were administrated EGCG at 50 mg/kg, the concentration of EGCG in various brain regions was measured by liquid chromatography tandem mass spectrometry (LC-MS/MS) [56]. The concentration of EGCG in various brain regions was about 5 ng/mL (0.01 μM) and ~4.95% of the orally administered EGCG (100 mg/kg) reached the systemic circulation. However, it was unclear whether EGCG was transferred from blood vessels into the parenchyma [56]. The concentration of EGCG in rat brain tissue (extracted consecutively with ethyl acetate and methanol) was determined to be about 0.5 nmol/g by chemiluminescence-detection HPLC (CL-HPLC) at 60 min after oral administration (500 mg/kg) in male Sprague–Dawley rats [57].

When the blood-to brain distribution ratios of C and EC which were administered (20 mg/kg) to male Sprague–Dawley rats via the femoral vein, which was measured by microdialysis sampling coupled with CL-HPLC, the ratios of C and EC were 0.0726 ± 0.0376 and 0.1065 ± 0.0531, respectively, as determined using the area under the curve for brain and blood [58]. In another study, the transport efficiency of C and EC at 30 mM was determined using two BBB cell lines, RBE-4 (rat brain endothelial cell) and hCMEC/D3 (human brain endothelial cell). Results showed that both C and EC effectively crossed the barrier in a time-dependent manner, and that the percentage of transport efficiency (% in 1 h) of EC (15.4 ± 0.6) was significantly higher than C (7.4 ± 0.7) [59].

Recently, we determined in vitro BBB permeability of EGCG and its metabolites (Table 2) by LC–MS/MS using a BBB kit (RBT-24, PharmaCo-Cell, Nagasaki, Japan) consisting of co-cultures of endothelial cells, pericytes, and astrocytes [43,60]. The in vitro BBB permeability (%, in 0.5 h) of EGCG, EGC, and GA was 4.00 ± 0.17, 4.96 ± 0.55, and 9.42 ± 1.01, respectively (the data from [43] are modified). GA exhibited a higher permeability than EGCG and EGC, perhaps due to the smaller molecular size of GA (MW 170.12) compared to EGCG (MW 458.372) and EGC (MW 306.27). The BBB permeability of EGC was lower than that of EC, and between EC and C. Lower BBB permeability of EGC than that of EC may be due to one more hydroxyl bond of EGC than EC, which affects its permeability. On the other hand, BBB permeability may be influenced by the presence of hydrophobicity of the galloyl bond [43,59,60].

The BBB permeability (%, in 0.5 h) of microbial ring-fission metabolites EGC-M5, and its conjugates, such as glucuronide of EGC-M5 (EGC-M5-GlcUA) and sulfate of EGC-M5 (EGC-M5-Sul), were 5.34 ± 0.23, 3.72 ± 0.01, and 4.34 ± 0.40, respectively. EGC-M5, with a smaller molecular size (MW 208.07), exhibited a slightly higher permeability than its conjugates EGC-M5-GlcUA (MW 384.11) and EGC-M5-Sul (MW 287.02), suggesting that the smaller molecular size of EGC-M5 caused its higher permeability [43].

Table 2. BBB permeability of EGCG metabolites.

Sample	Permeability Coefficient (10^{-6}cm s^{-1})	BBB Permeability (%) (30 min)
EGCG	13.45 ± 0.57	4.00 ± 0.17
EGC	16.70 ± 1.86	4.96 ± 0.55
GA	31.73 ± 3.39	9.42 ± 1.01
EGC-M5	17.99 ± 0.79	5.34 ± 0.23
EGC-M5-GlcUA	12.53 ± 0.02	3.72 ± 0.01
EGC-M5-Sul	14.61 ± 1.35	4.34 ± 0.40
PG	13.79 ± 1.62	4.10 ± 0.48
PG-GlcUA	9.28 ± 1.41	2.76 ± 0.42

Data are expressed as the mean ± SEM (n = 3) [43]. (Data of Ref. 43 are modified).

2.3. Neuritogenic Activity of EGCG and Its Microbial Ring-Fission Metabolites

Since EGCG and its microbial ring-fission metabolites were able to reach brain parenchyma through the BBB, findings on how these bioactive compounds work in the brain and verification of their neuritogenic activity were needed. Human neuroblastoma SH-SY5Y cells (ATCC, CRL-2266) were used to assess neuritogenic activity as they are often used as in vitro models of neuronal function and differentiation [61]. In brief, SH-SY5Y cells were plated as 2.5×10^4 cells/mL in a 24-well plate (500 μL of cell suspension/well). EGCG and its metabolites, which were dissolved in 0.01% DMSO, were added to the culture medium to make a final concentration of 0.01–1.0 μM, and cultured for ~72 h. Neurite length was measured by ImageJ software (Ver. 1.50i) [43,60]. Neurite length was significantly prolonged in cells treated with EGCG and EGC-M5 at 0.05 μM compared to control cells. In addition, SH-SY5Y cell growth was significantly enhanced by 0.05 μM EGCG and its metabolites compared to control cells, but this effect was reduced at higher concentrations (≥ 1.0 μM). Since the data of BBB permeability suggest that 4.0% (0.5 h) of EGCG can pass through blood to brain parenchyma, it may be possible to speculate how much EGCG is needed in the blood for ~0.05 μM EGCG to reach the brain [43,60]. The plasma concentration of EGCG in humans is 0.02 μM after drinking black tea containing 16.74 mg of EGCG [51]. After a few hours of circulation of blood containing 0.02 μM EGCG, its accumulation is ~0.05 μM in the brain. Although EGCG reaches in only trace amounts after 8 h or more of the EGCG intake, EGC-M5, a metabolite of EGCG, can be found in the blood. Whereas the levels of EGCG metabolites such as EGC-M5 and its conjugates in blood have not been determined, they are thought to be circulating in the blood for several hours. Since the BBB permeability of EGC-M5 is slightly higher than that of EGCG and the bioavailability of catechins is reported to be 39% in 24 h [52] and 62% in 48 h [54], EGC-M5 transferred from blood into the brain may also have a role in neuritogenesis. It is necessary to further investigate whether EGCG and its metabolites reach concentrations that cause neuritogenesis in vivo after consuming several cups of green tea per day in humans.

3. Bioactivity of Catechin Ring-Fission Metabolites

Catechin metabolites show several biological activities, including anti-oxidative, anti-inflammatory, anti-cancer, immunomodulatory, anti-thrombotic, and blood pressure-lowering activities (Table 3).

Table 3. Bioactivity of catechin metabolites.

Catechin Metabolites	Bioactivity	Reference
5-(3,4-dihydroxyphenyl)-γ-valerolactone	Anti-oxidative	[63]
5-(3,4-dihydroxyphenyl)-γ-valerolactone	Anti-oxidative	[65]
5-(3-hydroxyphenyl)-γ-valerolactone	Anti-oxidative	[63]
(EGC-M1)	Anti-cancer	[62]
(EGC-M4)	Anti-oxidative	[63]
(EGC-M5)	Antidiabetic effect	[41]
(EGC-M5)	Neuritogenic activity	[43]
(EGC-M5)	Blood–brain barrier penetrating activity	[43]
(EGC-M5)	Anti-oxidative	[63]
(EGC-M5)	Immunomodulatory activity	[66]
(EGC-M5)	Blood pressure lowering activity	[67]
(EGC-M6)	Antidiabetic effect	[41]
(EGC-M6)	Anti-cancer	[62]
(EGC-M7)	Antidiabetic effect	[41]
(EGC-M7)	Anti-cancer	[64]
(EGC-M7)	Anti-inflammatory	[64]
(EGC-M7)	Blood pressure lowering activity	[67]
(EGC-M9)	Anti-oxidative	[63]
(EGC-M10)	Anti-oxidative	[63]
(EGC-M10)	Anti-cancer	[62]
(EGC-M11)	Antidiabetic effect	[41]
(EGC-M11)	Anti-oxidative	[63]

Hara-Terawaki et al. evaluated anti-cancer effects of catechin metabolites against human cervical cancer cells (HeLa cells) [62]. The authors screened the inhibitory activities of 11 kinds of metabolites (EGC-M1-M11) produced from EGCG by intestinal microbiota on proliferation of HeLa cells. Among the 11 metabolites, EGC-M1, EGC-M6, and EGC-M10 inhibited the proliferation of HeLa cells at a final concentration of 50 μg/mL [62]. Another study by Takagaki et al. investigated the anti-oxidative activity of catechin metabolites by flow injection analysis coupled to an on-line antioxidant detection system with the 2, 20-azinobis (3-ethylbenzothiazoline-6-sulfonic acid) radical cation. The radical scavenging abilities of EGCG metabolites, such as EGC-M4, EGC-M5, EGC-M9, EGC-M10, and EGC-M11, as well as 5-(3, 4 dihydroxyphenyl)-γ-valerolactone, and 5-(3-hydroxyphenyl)-γ-valerolactone), which are ring-fission metabolites produced from EC or ECG, were found to be stronger than those of parental catechins [63]. Two ring-fission metabolites of tea catechins were tested for their anti-cancer and anti-inflammatory activities against a panel of immortalized and malignant human cell lines [64]. EGC-M7 had significantly strong inhibitory activity at 15–73 μM than 5-(3,4-dihydroxyphenyl)-γ-valerolactone at 50 μM against human colon cancer cells (HT-29 and HCT-116), human esophageal squamous cell carcinoma (KYSE150), human normal immortalized intestinal cells (INT-407), and rat intestinal epithelial cells (IEC-6). EGC-M7 also showed anti-inflammatory activity at 20 μM by inhibiting nitric oxide production (50%) in lipopolysaccharide (LPS)-stimulated murine macrophage (RAW264.7) cells [64]. The anti-oxidant activity of a ring-fission metabolite 5-(3,4-dihydroxyphenyl)-γ-valerolactone from (−)-epicatechin was described by Unno et al. [65]. In another study, EGC-M5 was found to have immunomodulatory activity by enhancing the activity of $CD4^+$ T cells and the cytotoxic activity of natural killer cells in BALB/c mice [66]. EGCG microbial metabolites were found to have blood pressure lowering activity in rats. A single oral intake of EGCG metabolites, EGC-M5 and EGC-M7, was examined to observe systolic blood pressure (SBP) using spontaneously hypertensive rats. There was a significant decrease in SBP 2 h after administration (150 mg/kg) of EGC-M7 and 4 h after administration (200 mg/kg) of EGC-M5, compared to the control group [67]. More recently, EGCG microbial metabolites were found to have antidiabetic effects in vitro and in vivo [41]. Glucose uptake ability of EGCG metabolites was measured with differentiated rat L6 myoblast cells by using 2-deoxyglucose. The treatment with EGC-M5, EGC-M6, EGC-M7, and EGC-M11 at 3 μM for 15 min significantly increased glucose uptake by 164.2%, 165.2%, 167.6%, and 146.3%, respectively, compared to control cells [41]. Moreover, oral administration of EGC-M5 at 32 mg/kg of body weight significantly suppressed postprandial hyperglycemia at 15 min (150.5 ± 13.6 mg/dL) and 30 min (108.5 ± 17.2 mg/dL) after oral glucose loading, compared to the saline control group [41].

The above studies indicate an important contribution of intestinal microflora-derived ring fission metabolites of catechins on protection against various diseases, including neurodegenerative diseases.

4. Conclusions and Future Expectation

Several studies including animal, human, and cell cultures support the potential neuroprotective activities of green tea catechins against neurological disorders. Very recently, EGCG was found to be safe and potential in improving cognition using both preclinical (mice) and clinical (human) studies [68]. The concentrations of EGCG, which is the main and the most active component among catechins, are very low in human and rat plasma and EGCG disappears within several hours from systemic circulation (<8 h) due to fast and extensive metabolism (methylation, glucuronidation, and sulfation) and microbial metabolism and degradation, resulting in the formation of various microbial ring-fission metabolites, which are detectable (>8 h) in the plasma and urine [30,31,33]. These microbial ring-fission metabolites show much higher bioavailability [52,55]. Intact EGCG and its metabolites reached the brain parenchyma through the BBB and induced neuritogenesis at a low concentration (0.05 μM) [43,60].

Based on our and other findings, we propose a possible action of EGCG and its metabolites in the brain as follows. When humans drink green tea, intact EGCG at a very low micromolar level reaches the brain parenchyma through the BBB and may induce neurite outgrowth, and after EGCG disappears, metabolized EGCG may promote neurite outgrowth, resulting in the prevention of cognitive dysfunction [43,60]. On the other hand, EGCG and its metabolites that reached the brain may reduce oxidative damage, since the levels of lipid peroxidation were significantly reduced in the brain of senescence-accelerated mouse prone 10 (SAMP10) that ingested EGCG [60]. In addition, EGCG metabolites have anti-oxidant activity [63,65]. Thus, microbial ring-fission metabolites may play an important role in suppressing brain dysfunction. However, differences in intestinal microbiota may have great importance on the variability of metabolites as well as the absorption rate among humans [52–54,69]. To date, there are no findings on the neuroprotective action of microbial ring-fission metabolites of EGCG in vivo. It is becoming epidemiologically clear that intake of green tea suppresses cognitive decline [11,70,71]. In the future it will be necessary to examine not only the relationship between green tea intake and brain function but also the relationship between brain function and the concentrations of EGCG and its metabolites in the blood.

Author Contributions: M.P.—Corresponding author; Outline of whole manuscript; Literature search; Text preparation; K.U.—Corresponding author; Project design; Outline of whole manuscript; Text preparation; A.T.—Literature search; Text preparation; Preparation of Figure 1.; M.I.—Literature search; Outline of whole manuscript; Text preparation; Y.N.—Project design; Text preparation. All authors read and approved the final manuscript.

Funding: This research received no external funding.

Conflicts of Interest: The authors declare no conflict of interest.

References

1. Miyoshi, N.; Pervin, M.; Suzuki, T.; Unno, K.; Isemura, M.; Nakamura, Y. Green tea catechins for well-being and therapy: Prospects and opportunities. *Bot. Targets Ther.* **2015**, *5*, 85–96.
2. Suzuki, T.; Miyoshi, N.; Hayakawa, S.; Imai, S.; Isemura, M.; Nakamura, Y. Health Benefits of Tea Consumption. In *Beverage Impacts on Health and Nutrition*, 2nd ed.; Wilson, T., Templ, N.J., Eds.; Springer International Publishing: Cham, Switzerland, 2016; pp. 49–67, ISBN 978-3-319-23672-8.
3. Carlson, J.R.; Bauer, B.A.; Vincent, A.; Limburg, P.J.; Wilson, T. Reading the tea leaves: Anticarcinogenic properties of (-)-epigallocatechin-3-gallate. *Mayo Clin. Proc.* **2007**, *82*, 725–732. [CrossRef]
4. Yang, C.S.; Wang, H. Cancer preventive activities of tea catechins. *Molecules* **2016**, *21*, 1679. [CrossRef] [PubMed]
5. Suzuki, T.; Pervin, M.; Goto, S.; Isemura, M.; Nakamura, Y. Beneficial effects of tea and the green tea catechin epigallocatechin-3-gallate on obesity. *Molecules* **2016**, *21*, 1305. [CrossRef] [PubMed]
6. Friedrich, M.; Petzke, K.J.; Raederstorff, D.; Wolfram, S.; Klaus, S. Acute effects of epigallocatechin gallate from green tea on oxidation and tissue incorporation of dietary lipids in mice fed a high-fat diet. *Int. J. Obes.* **2012**, *36*, 735–743. [CrossRef]
7. Lee, M.S.; Kim, C.T.; Kim, Y. Green tea (-)-epigallocatechin-3-gallate reduces body weight with regulation of multiple genes expression in adipose tissue of diet-induced obese mice. *Ann. Nutr. Metab.* **2009**, *54*, 151–157. [CrossRef]
8. Lombo, C.; Morgado, C.; Tavares, I.; Neves, D. Effects of prolonged ingestion of epigallocatechin gallate on diabetes type 1-induced vascular modifications in the erectile tissue of rats. *Int. J. Impot. Res.* **2016**, *28*, 133–138. [CrossRef]
9. Othman, A.I.; El-Sawi, M.R.; El-Missiry, M.A.; Abukhalil, M.H. Epigallocatechin-3-gallate protects against diabetic cardiomyopathy through modulating the cardiometabolic risk factors, oxidative stress, inflammation, cell death and fibrosis in streptozotocin-nicotinamide-induced diabetic rats. *Biomed. Pharmacother.* **2017**, *94*, 362–373. [CrossRef]
10. Yokogoshi, H. Green tea in the protection against neurodegeneration. In *Health Benefits of Green Tea: An Evidence-Based Approach*, 1st ed.; Hara, Y., Yang, C.S., Isemura, M., Tomita, I., Eds.; CABI International: Oxfordshire, UK, 2016; pp. 185–229, ISBN 978-178639-239-8.
11. Pervin, M.; Unno, K.; Ohishi, T.; Tanabe, H.; Miyoshi, N.; Nakamura, Y. Beneficial effects of green tea catechins on neurodegenerative diseases. *Molecules* **2018**, *23*, 1297. [CrossRef]
12. Xicota, L.; Rodriguez-Morato, J.; Dierssen, M.; de la Torre, R. Potential Role of (-)-Epigallocatechin-3-Gallate (EGCG) in the Secondary Prevention of Alzheimer Disease. *Curr. Drug Targets* **2017**, *2*, 174–195. [CrossRef]
13. Yang, Y.; Qin, Y.J.; Yip, Y.W.Y.; Chan, K.P.; Chu, K.O.; Chu, W.K.; Ng, T.K.; Pang, C.P.; Chan, S.O. Green tea catechins are potent anti-oxidants that ameliorate sodium iodate-induced retinal degeneration in rats. *Sci. Rep.* **2016**, *6*, 29546. [CrossRef] [PubMed]
14. Cavet, M.E.; Harrington, K.L.; Vollmer, T.R.; Ward, K.W.; Zhang, J.Z. Anti-inflammatory and anti-oxidative effects of the green tea polyphenol epigallocatechin gallate in human corneal epithelial cells. *Mol. Vis.* **2011**, *17*, 533–542. [PubMed]
15. Xiang, L.P.; Wang, A.; Ye, J.H.; Zheng, X.Q.; Polito, C.A.; Lu, J.L.; Li, Q.S.; Liang, Y.R. Suppressive effects of tea catechins on breast cancer. *Nutrients* **2016**, *8*, 458. [CrossRef]
16. Ding, M.L.; Ma, H.; Man, Y.G.; Lv, H.Y. Protective effects of a green tea polyphenol, epigallocatechin-3-gallate, against sevoflurane-induced neuronal apoptosis involve regulation of CREB/BDNF/TrkB and PI3K/Akt/mTOR signalling pathways in neonatal mice. *Can. J. Physiol. Pharmacol.* **2017**, *95*, 1396–1405. [CrossRef] [PubMed]
17. He, Y.; Tan, D.; Bai, B.; Wu, Z.; Ji, S. Epigallocatechin-3-gallate attenuates acrylamide-induced apoptosis and astrogliosis in rat cerebral cortex. *Toxicol. Mech. Methods* **2017**, *27*, 298–306. [CrossRef] [PubMed]
18. Liu, D.; Perkins, J.T.; Hennig, B. EGCG prevents PCB-126-induced endothelial cell inflammation via epigenetic modifications of NF-κB target genes in human endothelial cells. *J. Nutr. Biochem.* **2016**, *28*, 164–170. [CrossRef]
19. Khokhar, S.; Magnusdottir, S.G. Total phenol, catechin, and caffeine contents of teas commonly consumed in the United Kingdom. *J. Agric. Food Chem.* **2002**, *50*, 565–570. [CrossRef]
20. Khan, N.; Mukhtar, H. Tea polyphenols for health promotion. *Life Sci.* **2007**, *81*, 519–533. [CrossRef]

21. Chang, X.; Rong, C.; Chen, Y.; Yang, C.; Hu, Q.; Mo, Y.; Zhang, C.; Gu, X.; Zhang, L.; He, W.; et al. (-)-Epigallocatechin-3-gallate attenuates cognitive deterioration in Alzheimer's disease model mice by upregulating neprilysin expression. *Exp. Cell. Res.* **2015**, *334*, 136–145. [CrossRef]
22. Unno, K.; Takabayashi, F.; Kishido, T.; Oku, N. Suppressive effect of green tea catechins on morphologic and functional regression of the brain in aged mice with accelerated senescence (SAMP10). *Exp. Gerontol.* **2004**, *39*, 1027–1034. [CrossRef]
23. Unno, K.; Takabayashi, F.; Yoshida, H.; Choba, D.; Fukutomi, R.; Kikunaga, N.; Kishido, T.; Oku, N.; Hoshino, M. Daily consumption of green tea catechin delays memory regression in aged mice. *Biogerontology* **2007**, *8*, 89–95. [CrossRef] [PubMed]
24. Unno, K.; Ishikawa, Y.; Takabayashi, F.; Sasaki, T.; Takamori, N.; Iguchi, K.; Hoshino, M. Daily ingestion of green tea catechins from adulthood suppressed brain dysfunction in aged mice. *Biofactors* **2008**, *34*, 263–271. [CrossRef]
25. Kuriyama, S.; Hozawa, A.; Ohmori, K.; Shimazu, T.; Matsui, T.; Ebihara, S.; Awata, S.; Nagatomi, R.; Arai, H.; Tsuji, I. Green tea consumption and cognitive function: A cross-sectional study from the Tsurugaya Project 1. *Am. J. Clin. Nutr.* **2006**, *83*, 355. [CrossRef] [PubMed]
26. Noguchi-Shinohara, M.; Yuki, S.; Dohmoto, C.; Ikeda, Y.; Samuraki, M.; Iwasa, K.; Yokogawa, M.; Asai, K.; Komai, K.; Nakamura, H.; et al. Consumption of green tea, but not black tea or coffee, is associated with reduced risk of cognitive decline. *PLoS ONE* **2014**, *9*, e96013. [CrossRef] [PubMed]
27. Feng, L.; Gwee, X.; Kua, E.H.; Ng, T.P. Cognitive function and tea consumption in community dwelling older Chinese in Singapore. *J. Nutr. Health Aging* **2010**, *14*, 433–438. [CrossRef] [PubMed]
28. Gu, Y.J.; He, C.H.; Li, S.; Zhang, S.Y.; Duan, S.Y.; Sun, H.P.; Shen, Y.P.; Xu, Y.; Yin, J.Y.; Pan, C.W. Tea consumption is associated with cognitive impairment in older Chinese adults. *Aging Ment. Health* **2018**, *22*, 1232–1238. [CrossRef] [PubMed]
29. Ide, K.; Yamada, H.; Takuma, N.; Park, M.; Wakamiya, N.; Nakase, J.; Ukawa, Y.; Sagesaka, Y.M. Green tea consumption affects cognitive dysfunction in the elderly: A pilot study. *Nutrients* **2014**, *6*, 4032–4042. [CrossRef]
30. Unno, T.; Takeo, T. Absorption of (-)-epigallocatechin gallate into the circulation system of rats. *Biosci. Biotechnol. Biochem.* **1995**, *59*, 1558–1559. [CrossRef]
31. Kohri, T.; Matsumoto, N.; Yamakawa, M.; Suzuki, M.; Nanjo, F.; Hara, Y.; Oku, N. Metabolic fate of (-)-[4-(3)H] epigallocatechin gallate in rats after oral administration. *J. Agric. Food Chem.* **2001**, *49*, 4102–4112. [CrossRef]
32. Clifford, M.N.; van der Hooft, J.J.; Crozier, A. Human studies on the absorption, distribution, metabolism, and excretion of tea polyphenols. *Am. J. Clin. Nutr.* **2013**, *98*, 1619S–1630S. [CrossRef]
33. Stalmach, A.; Troufflard, S.; Serafini, M.; Crozier, A. Absorption, metabolism and excretion of Choladi green tea flavan-3-ols by humans. *Mol. Nutr. Food Res.* **2009**, *53*, S44–S53. [CrossRef] [PubMed]
34. Sang, S.; Lee, M.J.; Yang, I.; Buckley, B.; Yang, C.S. Human urinary metabolite profile of tea polyphenols analyzed by liquid chromatography/electrospray ionization tandem mass spectrometry with data-dependent acquisition. *Rapid Commun. Mass. Spectrom.* **2008**, *22*, 1567–1578. [CrossRef] [PubMed]
35. Nakagawa, K.; Miyazawa, T. Chemiluminescence-high performance liquid chromatographic determination of tea catechin, (-)-epigallocatechin 3-gallate, at picomole levels in rat and human plasma. *Anal. Biochem.* **1997**, *248*, 41–49. [CrossRef]
36. Chen, L.; Lee, M.J.; Li, H.; Yang, C.S. Absorption, distribution, and elimination of tea polyphenols in rats. *Drug Metab. Dispos.* **1997**, *25*, 1045–1050. [PubMed]
37. Zhu, M.; Chen, Y.; Li, R.C. Oral absorption and bioavailability of tea catechins. *Planta Med.* **2000**, *66*, 444–447. [CrossRef] [PubMed]
38. Lambert, J.D.; Lee, M.J.; Lu, H.; Meng, X.; Hong, J.J.; Seril, D.N.; Sturgill, M.G.; Yang, C.S. Epigallocatechin-3-gallate is absorbed but extensively glucuronidated following oral administration to mice. *J. Nutr.* **2003**, *133*, 4172–4177. [CrossRef] [PubMed]
39. Takagaki, A.; Nanjo, F. Metabolism of (−)-epigallocatechin gallate by rat intestinal flora. *J. Agric. Food Chem.* **2010**, *58*, 1313–1321. [CrossRef]
40. Takagaki, A.; Kato, Y.; Nanjo, F. Isolation and characterization of rat intestinal bacteria involved in biotransformation of (−)-epigallocatechin. *Arch. Microbiol.* **2014**, *196*, 681–695. [CrossRef]

41. Takagaki, A.; Yoshioka, Y.; Yamashita, Y.; Nagano, T.; Ikeda, M.; Hara-Terawaki, A.; Seto, R.; Ashida, H. Effects of microbial metabolites of (-)-epigallocatechin gallate on glucose uptake in L6 skeletal muscle cell and glucose tolerance in ICR mice. *Biol. Pharm. Bull.* **2019**, *42*, 212–221. [CrossRef]
42. Feng, W.Y. Metabolism of green tea catechins: An overview. *Curr. Drug Metab.* **2006**, *7*, 755–809. [CrossRef]
43. Unno, K.; Pervin, M.; Nakagawa, A.; Iguchi, K.; Hara, A.; Takagaki, A.; Nanjo, F.; Minami, A.; Nakamura, Y. Blood-brain barrier permeability of green tea catechin metabolites and their neuritogenic activity in human neuroblastoma SH-SY5Y cells. *Mol. Nutr. Food Res.* **2017**, *61*, 1700294. [CrossRef] [PubMed]
44. Peterson, B.; Weyers, M.; Steenekamp, J.H.; Steyn, J.D.; Gouws, C.; Hamman, J.H. Drug bioavailability enhancing agents of natural origin (bioenhancers) that modulate drug membrane permeation and pre-systemic metabolism. *Pharmaceutics* **2019**, *11*, 33. [CrossRef] [PubMed]
45. Kim, S.; Lee, M.J.; Hong, J.; Li, C.; Smith, T.J.; Yang, G.Y.; Seril, D.N.; Yang, C.S. Plasma and tissue levels of tea catechins in rats and mice during chronic consumption of green tea polyphenols. *Nutr. Cancer* **2000**, *37*, 41–48. [CrossRef] [PubMed]
46. Raneva, V.G.; Shimizu, Y.; Shimasaki, H. Antioxidant activity in plasma and tissues distribution of (-)-epigallocatechin gallate after oral administration to rats. *J. Oleo Sci.* **2005**, *54*, 289–298. [CrossRef]
47. Pandey, K.B.; Rizvi, S.I. Plant polyphenols as dietary antioxidants in human health and disease. *Oxid. Med. Cell. Longev.* **2009**, *2*, 270–278. [CrossRef] [PubMed]
48. Takagaki, A.; Nanjo, F. Catabolism of (+)-catechin and (−)-epicatechin by rat intestinal microbiota. *J. Agric. Food Chem.* **2013**, *61*, 4927–4935. [CrossRef] [PubMed]
49. Takagaki, A.; Nanjo, F. Bioconversion of (−)-epicatechin, (+)-epicatechin, (−)-catechin, and (+)-catechin by (−)-epigallocatechinmetabolizing bacteria. *Biol. Pharm. Bull.* **2015**, *38*, 789–794. [CrossRef]
50. Liu, A.B.; Tao, S.; Lee, M.J.; Hu, Q.; Meng, X.; Lin, Y.; Yang, C.S. Effects of gut microbiota and time of treatment on tissue levels of green tea polyphenols in mice. *Biofactors* **2018**, *44*, 348–360. [CrossRef]
51. Warden, B.A.; Smith, L.S.; Beecher, G.R.; Balentine, D.A.; Clevidence, B.A. Catechins are bioavailable in men and women drinking black tea throughout the day. *J. Nutr.* **2001**, *131*, 1731–1737. [CrossRef]
52. Del Rio, D.; Calani, L.; Cordero, C.; Salvantore, S.; Pellegrini, N.; Brighenti, F. Bioavailability and catabolism of green tea flavan-3-ols in humans. *Nutrition* **2010**, *26*, 1110–1116. [CrossRef]
53. Li, C.; Lee, M.J.; Sheng, S.; Meng, X.; Prabhu, S.; Winnik, B.; Huang, B.; Chung, J.Y.; Yan, S.; Ho, C.T.; et al. Structural identification of two metabolites of catechins and their kinetics in human urine and blood after tea ingestion. *Chem. Res. Toxicol.* **2000**, *13*, 177–184. [CrossRef] [PubMed]
54. Calani, L.; Del Rio, D.; Luisa Callegari, M.; Morelli, L.; Brighenti, F. Updated bioavailability and 48 h excretion profile of flavan-3-ols from green tea in humans. *Int. J. Food Sci. Nutr.* **2012**, *63*, 513–521. [CrossRef] [PubMed]
55. Daneman, R.; Prat, A. The blood-brain barrier. *Cold. Spring Harb. Perspect. Biol.* **2015**, *7*, a020412. [CrossRef] [PubMed]
56. Lin, L.C.; Wang, M.N.; Tseng, T.Y.; Sung, J.S.; Tsai, T.H. Pharmacokinetics of (-)-epigallocatechin-3-gallate in conscious and freely moving rats and its brain regional distribution. *J. Agric. Food Chem.* **2007**, *55*, 1517–1524. [CrossRef] [PubMed]
57. Nakagawa, K.; Miyazawa, T. Absorption and distribution of tea catechin, (-)-epigallocatechin-3-gallate, in the rat. *J. Nutr. Sci. Vitaminol.* **1997**, *43*, 679–684. [CrossRef] [PubMed]
58. Wu, L.; Zhang, Q.L.; Zhang, X.Y.; Lv, C.; Li, J.; Yuan, Y.; Yin, F.X. Pharmacokinetics and blood-brain barrier penetration of (+)-catechin and (-)-epicatechin in rats by microdialysis sampling coupled to high-performance liquid chromatography with chemiluminescence detection. *J. Agric. Food Chem.* **2012**, *60*, 9377–9383. [CrossRef]
59. Faria, A.; Pestana, D.; Teixeira, D.; Couraud, P.O.; Romero, I.; Weksler, B.; de Freitas, V.; Mateus, N.; Calhau, C. Insights into the putative catechin and epicatechin transport across blood-brain barrier. *Food Funct.* **2011**, *2*, 39–44. [CrossRef]
60. Pervin, M.; Unno, K.; Nakagawa, A.; Takahashi, Y.; Iguchi, K.; Yamamoto, H.; Hoshino, M.; Hara, A.; Takagaki, A.; Nanjo, F.; et al. Blood brain barrier permeability of (-)-epigallocatechin gallate, its proliferation-enhancing activity of human neuroblastoma SH-SY5Y cells, and its preventive effect on age-related cognitive dysfunction in mice. *Biochem. Biophys. Rep.* **2017**, *9*, 180–186. [CrossRef]
61. Price, R.D.; Oe, T.; Yamaji, T.; Matsuoka, N. A simple, flexible, nonfluorescent system for the automated screening of neurite outgrowth. *J. Biomol. Screen.* **2006**, *11*, 155–164. [CrossRef]

62. Hara-Terawaki, A.; Takagaki, A.; Kobayashi, H.; Nanjo, F. Inhibitory activity of catechin metabolites produced by intestinal microbiota on proliferation of HeLa cells. *Biol. Pharm. Bull.* **2017**, *40*, 1331–1335. [CrossRef]
63. Takagaki, A.; Otani, S.; Nanjo, F. Antioxidative activity of microbial metabolites of (-)-epigallocatechin gallate produced in rat intestines. *Biosci. Biotechnol. Biochem.* **2011**, *75*, 582–585. [CrossRef] [PubMed]
64. Lambert, J.D.; Rice, J.E.; Hong, J.; Hou, Z.; Yang, C.S. Synthesis and biological activity of the tea catechin metabolites, M4 and M6 and their methoxy-derivatives. *Bioorg. Med. Chem. Lett.* **2005**, *15*, 873–876. [CrossRef] [PubMed]
65. Unno, T.; Tamemoto, K.; Yayabe, F.; Kakuda, T. Urinary excretion of 5-(3′,4′-dihydroxyphenyl)-gamma-valerolactone, a ring-fission metabolite of (-)-epicatechin, in rats and its in vitro antioxidant activity. *J. Agric. Food Chem.* **2003**, *51*, 6893–6898. [CrossRef] [PubMed]
66. Kim, Y.H.; Won, Y.S.; Yang, X.; Kumazoe, M.; Yamashita, S.; Hara, A.; Takagaki, A.; Goto, K.; Nanjo, F.; Tachibana, H. Green tea catechin metabolites exert immunoregulatory effects on CD4 $^{(+)}$ T cell and natural killer cell activities. *J. Agric. Food Chem.* **2016**, *64*, 3591–3597. [CrossRef] [PubMed]
67. Takagaki, A.; Nanjo, F. Effects of metabolites produced from (-)-epigallocatechin gallate by rat intestinal bacteria on angiotensin I-converting enzyme activity and blood pressure in spontaneously hypertensive rats. *J. Agric. Food Chem.* **2015**, *63*, 8262–8266. [CrossRef] [PubMed]
68. de la Torre, R.; de Sola, S.; Farré, M.; Xicota, L.; Cuenca-Royo, A.; Rodriguez, J.; León, A.; Langohr, K.; Gomis-González, M.; Hernandez, G.; et al. A phase 1, randomized double-blind, placebo controlled trial to evaluate safety and efficacy of epigallocatechin-3-gallate and cognitive training in adults with Fragile X syndrome. *Clin. Nutr.* **2019**. [CrossRef] [PubMed]
69. Lee, H.C.; Jenner, A.M.; Low, C.S.; Lee, Y.K. Effect of tea phenolics and their aromatic fecal bacterial metabolites on intestinal microbiota. *Res. Microbiol.* **2006**, *157*, 876–884. [CrossRef]
70. Kakutani, S.; Watanabe, H.; Murayama, N. Green tea intake and risks for dementia, Alzheimer's disease, mild cognitive impairment, and cognitive impairment: A systematic review. *Nutrients* **2019**, *11*, 1165. [CrossRef]
71. Mancini, E.; Beglinger, C.; Drewe, J.; Zanchi, D.; Lang, U.E.; Borgwardt, S. Green tea effects on cognition, mood and human brain function: A systematic review. *Phytomedicine* **2017**, *34*, 26–37. [CrossRef]

International Journal of
Molecular Sciences

MDPI

Review

Anti-Inflammatory Activities of Marine Algae in Neurodegenerative Diseases

Maria Cristina Barbalace [1], Marco Malaguti [1], Laura Giusti [2], Antonio Lucacchini [3], Silvana Hrelia [1,*] and Cristina Angeloni [2,*]

1 Department for Life Quality Studies, Alma Mater Studiorum-University of Bologna, 40126 Bologna, Italy; maria.barbalace2@unibo.it (M.C.B.); marco.malaguti@unibo.it (M.M.)
2 School of Pharmacy, University of Camerino, 62032 Camerino, Italy; laura.giusti@unicam.it
3 Department of Clinical and Experimental Medicine, University of Pisa, Pisa 56126, Italy; antonio.lucacchini@gmail.com
* Correspondence: silvana.hrelia@unibo.it (S.H.); cristina.angeloni@unicam.it (C.A.)

Received: 31 May 2019; Accepted: 19 June 2019; Published: 22 June 2019

Abstract: Neuroinflammation is one of the main contributors to the onset and progression of neurodegenerative diseases such as Alzheimer's and Parkinson's diseases. Microglial and astrocyte activation is a brain defense mechanism to counteract harmful pathogens and damaged tissues, while their prolonged activation induces neuroinflammation that can trigger or exacerbate neurodegeneration. Unfortunately, to date there are no pharmacological therapies able to slow down or stop the progression of neurodegeneration. For this reason, research is turning to the identification of natural compounds with protective action against these diseases. Considering the important role of neuroinflammation in the onset and development of neurodegenerative pathologies, natural compounds with anti-inflammatory activity could be good candidates for developing effective therapeutic strategies. Marine organisms represent a huge source of natural compounds, and among them, algae are appreciated sources of important bioactive components such as antioxidants, proteins, vitamins, minerals, soluble dietary fibers, polyunsaturated fatty acids, polysaccharides, sterols, carotenoids, tocopherols, terpenes, phycobilins, phycocolloids, and phycocyanins. Recently, numerous anti-inflammatory compounds have been isolated from marine algae with potential protective efficacy against neuroinflammation. This review highlights the key inflammatory processes involved in neurodegeneration and the potential of specific compounds from marine algae to counteract neuroinflammation in the CNS.

Keywords: neuroinflammation; neurodegeneration; algae; seaweeds; neurodegenerative diseases

1. Introduction

Neurodegeneration refers to a progressive and permanent loss of neurons in specified regions of the brain and spinal cord. It is the pathological condition that characterizes many neurodegenerative diseases, including Alzheimer's disease (AD), Parkinson's disease (PD), multiple sclerosis (MS), Huntington's disease (HD), amyotrophic lateral sclerosis (ALS) [1], and traumatic brain injury (TBI) [2]. The main cellular and molecular events that trigger neurodegeneration are oxidative stress, abnormal protein deposition, damaged mitochondrial function, induction of apoptosis, impairment of proteostasis, and neuroinflammation [3]. Since the first identification of the main neurodegenerative disorders, research on the molecular mechanisms underlying these pathologies has focused on major anatomical changes such as neuronal loss and protein aggregation [4]. In recent years, more and more studies have highlighted the key role of the immune system in the initiation and progression of neurodegeneration [5,6] due to changes in cytokine signaling, immune cell proliferation and migration, altered phagocytosis, and reactive gliosis as common features of neurodegeneration [4].

Neuroinflammation, or, more specifically, the activation of the neuroimmune cells microglia and astrocytes into proinflammatory states, is an effective endogenous defense that protects the central nervous system (CNS) against microorganisms and injuries. It is usually a positive mechanism that aims to eliminate threats and restore homeostasis [7]. However, prolonged neuroinflammatory events can lead to a series of events that conclude with progressive neuronal damage that characterizes many neurodegenerative disorders [8]. The glial cells, microglia and astrocytes, have a pro- and anti-inflammatory role and are involved in different functions under physiological and disease conditions, such as phagocytosis, steroid release, free radical reduction, and cellular repair [9]. Glial cells exert a proinflammatory action through the production of cytokines and reactive oxygen species (ROS) that lead to synaptic dysfunction, loss of synapses, and neuronal death resulting in CNS injury. Until now, most research has been focused on microglial cells as key actors of neuroinflammation in neurodegeneration, but recently new scientific evidence has shown the important contribution of astrocytes to the inflammation that characterizes neurodegenerative diseases [10–12]. Unfortunately, to date there are no pharmacological therapies able to slow down or stop the progression of these devastating pathologies. For this reason, research is turning to the identification of natural compounds with protective action against these diseases. Considering the important role of neuroinflammation in the onset and development of neurodegenerative pathologies, natural compounds with anti-inflammatory activity could be good candidates to develop effective therapeutic strategies. Marine organisms represent a huge source of natural compounds, some of which have different structural characteristics from those of terrestrial origin. Marine-derived natural compounds could produce different pharmacological effects, like anti-diabetic [13], anti-inflammatory [14], antioxidant [15], anticancer [16], and anti-obesity [17] activities, and open the way for the development of new drugs [18]. Of note, seven marine-derived natural compounds have been approved for clinical use [19].

Among marine organisms, algae are one of the most valuable resources of the sea. Epidemiological studies comparing Japanese and Western diets show an association between algae consumption and a lower incidence of chronic degenerative diseases [20]. Algae are appreciated sources of important bioactive components such as antioxidants, proteins, vitamins, minerals, soluble dietary fibers, polyunsaturated fatty acids, polysaccharides, sterols, carotenoids, tocopherols, terpenes, phycobilins, phycocolloids, and phycocyanins [20]. Recently, Fernando et al. [21] summarized the latest knowledge about the potential anti-inflammatory activity of marine algae derivatives, evidencing their potential protective efficacy against neuroinflammation too. In particular, marine algae have been shown to counteract neuroinflammation by acting at different cellular levels: inhibiting pro-inflammatory enzymes such as COX-2 and iNOS [22], modulating MAPK pathways [23], and NK-kB activation [24], among others. Currently there are no clinical trials on the effects of marine algae against neuroinflammation but, given their important biological activities, as demonstrated by in vitro and animal studies, we believe that they will be carried out in the near future. Moreover, as anti-inflammatory drugs can trigger complications and important side effects [25,26], identifying novel anti-inflammatory agents from marine algae could be a valid solution to overcome this problem. In fact, anti-inflammatory natural compounds have been demonstrated to be safe thanks to their long use in folk medicine [27].

This review highlights the key inflammatory processes involved in neurodegeneration and the potential of marine algae and specific compounds from marine algae to counteract neuroinflammation in the CNS. The most recent and relevant results on the promising anti-inflammatory activities of marine algae related to neuroprotection have been selected.

2. Methods

A PubMed search was conducted. The combinations of terms that we used for this search were "marine algae and neuroinflammation," "marine algae and clinical studies," "marine algae and inflammation," "marine algae and toxicity," and "marine algae." We also combined the terms marine algae and neuroinflammation with fucosterol, phlorotannins, astaxanthin, polysaccharides, glycoprotein, chlorophyll, lutein, zeaxanthin, violaxanthin, neoxanthin, or β-carotene. No restrictions were placed on the date of the articles or the language of publication. Studies with a clearly described methodology were included.

3. Molecular Mechanisms of Neuroinflammation

Neuroinflammation is a defense process aimed to protect both the brain and the spinal cord from tissue damage or pathogen invasion [8]. Generally, inflammatory processes involve numerous cellular types and mediators with the aim of separating, via the formation of a glial scar, damaged tissue from healthy tissue [28]. When an insult occurs at brain level, the immune response is mediated through cross-talk between the CNS and the periphery. In fact, due to inflammation, blood-brain barrier (BBB) permeability is increased and leucocytes can infiltrate into the CNS [9].

At the brain level, microglia, astrocytes, and oligodendrocytes constitute the neuroglial cells [29]. Microglia have been demonstrated to be derived from primitive macrophages [30] and are now considered the resident immune system of the brain [31]. In non-activated conditions, microglia contribute to brain homeostasis [32] by modulating neuronal survival and maintenance thanks to the ability to release neurotrophic factors such as basic fibroblast growth factor and nerve growth factor (NGF) [33]. Acting as immune cells, microglial cells are also responsible for the phagocytosis of cell debris and contribute to the apoptosis of defective cells [34,35]. More recently, astrocytes, which are known to be involved in CNS homeostasis by sustaining synapse plasticity, have also been demonstrated to participate in protective signaling pathways such as those modulated by glycoprotein gp130, which is crucial for glial cells' survival [36], and by the transforming growth factor beta (TGFβ), whose signaling has been shown to exert immunosuppressive effects and to inhibit nuclear factor κB (NF-κB) nuclear translocation [37].

Beside their neuroprotective properties, the microglia supervise the brain environment by modulating the immune functions in response to tissue damage, degeneration, and pathogen infections [38]. Their activation can be triggered by different stimuli such as lipopolysaccharide (LPS), a well-known toll-like receptor (TLRs) ligand [39], and they represent the first line of defense against infections [40]. Microglia activation results in both morphological and biochemical changes: cells lose their shape and begin to secrete inflammatory biomarkers such as cytokines, eicosanoids, nitric oxide, and ROS [41,42].

Even though neuroinflammation does not usually trigger neurodegenerative diseases, it is directly involved in neuronal dysfunctions and contributes to neuronal death and to neurodegenerative disease progression [43]. In fact, diseases such as PD, AD, ALS, and MS, as well as ischemia and TBI, are associated with chronic inflammation and long-lasting microglia activation [44]. Such chronic inflammatory states result in an abnormal increased cytokine levels [45], the production of neurotoxic mediators, and oxidative stress that triggers a pro-inflammatory cycle [46] and amplifies degenerative processes such as abnormal protein deposition, mitochondrial dysfunction, and BBB permeability impairment [44,47,48].

Chronic inflammation in neurodegenerative diseases is sustained by TLRs activation at the glial level [49]. Among TLRs, TLR4 is the most expressed in microglia [50]; its activation has been demonstrated to be responsible for chronic inflammation in AD, where Aβ-oligomers interact with TLR4 and increase its expression [51,52], and in PD, where TLR4 protein expression is also increased in both in vitro and in vivo model systems [53]. Moreover, TLR4 has been found to be responsible for inflammation in spinal cord injury and stroke [54]. TLR4 activation triggers two different downstream proinflammatory signaling pathways, leading to cytokine expression. Among these pathways, the phosphoinositide 3-kinase/protein kinase B (PI3K/Akt) pathway, mammalian target of rapamycin (mTOR) activation, and mitogen-activated protein kinases cascades (MAPKs) are the main ones involved and lead to NF-κB activation [7,55,56]. Once activated, PI3K triggers Akt phosphorylation, which in turn activates mTOR. The mTOR pathway plays a pivotal role in the regulation of NF-κB and inflammation [57]. NF-κB signaling is considered particularly important in every neuroinflammation-related disease. After initial TLR4 activation, the sequence of events that leads to the translocation of NF-κB to the nucleus includes the activation of the protein IκB kinase, phosphorylation of the IκB inhibitory protein, and the consequent release of active NF-κB [58]. As a dimer, NF-κB translocates to the nucleus, where it activates the transcription of its target genes such as inducible nitric oxide synthase (iNOS), cyclooxygenase (COX2), tumor necrosis factor alpha (TNF-α), interleukin (IL)-6, and IL-1β by binding to p65-responsive element [56]. During neuroinflammation, NF-κB signaling is also stimulated in astrocytes [59], where its translocation to the nucleus and the subsequent cytokines expression is triggered by IL-17-receptor [60] and lactosyl ceramide, a lipid mediator produced by astrocytes [61]. Astrocytes' contribution to neuroinflammation and neurotoxicity has, thus, been demonstrated in models of different neurodegenerative diseases such as brain injury [62] and spinal cord and nerve injury [63,64], where NF-κB inactivation resulted in positive outcomes.

MAPKs are proteins involved in the regulation of multiple cellular functions. In particular, they are involved in the regulation of apoptosis, cell differentiation, and proliferation.

In activated microglia, increased signaling of p38 MAPK and c-Jun N-terminal kinases (JNK) has been described [65]. These MAPKs induce, through the transcription factor activating protein-1 (AP-1), the transcription of proinflammatory genes such as COX2, TNF-α, and IL-6. The involvement of p38 and JNK signaling in the LPS-activated MG6 microglial cell line has recently been confirmed, showing that LPS treatment strongly induces phospho-p38/p38 and phospho-JNK/JNK ratio, the AP-1 translocation to the nucleus, iNOS protein expression, and NO production [65].

PI3K/Akt and MAPK are not the only pathways involved in neuroinflammation; the Janus Kinase/Signal Transducers and Activators of Transcription (JAKs/STATs) signaling pathway represents a further pathway able to trigger inflammation in the CNS [66]. Several cytokines trigger this pathway by binding their specific receptors and promoting JAK kinase activity, both in microglia and astrocytes. Once activated, JAK phosphorylates STAT, which dimerizes and translocates to the nucleus, where it promotes the expression of cytokine-responsive genes. At least four JAK and seven STAT proteins have been identified [67]. Specific combinations of JAKs and STATs are involved in the response to different cytokines, allowing each cytokine to transduce its own message [66]. JAKs/STATs are involved in the inflammatory response occurring in most neurodegenerative diseases. In MS, endoplasmic reticulum stress induces astrocyte activation through JAK1/STAT3 signaling [68]. IL-6 and IFN-γ, two major activators of JAKs/STATs signaling, are elevated in PD [69]; moreover, in primary microglial cell culture it has been demonstrated that the inhibition of JAK 1/2 prevents the release of NO, TNF-α, and IL-1β induced by α-synuclein treatment [70,71].

Besides classical inflammatory pathways, non-classical pathways, such as the Hippo pathway, have been related to neuroinflammation and in particular to astrocyte activation [72]. In its typical sequence of events, the Hippo pathway involves numerous kinases such as Mst 1/2, Sav1, and Last 1/2. Last 1/2 phosphorylates and thus inactivates by proteasomal degradation or cytoplasmic retention, two transcription factors: YAP and TAZ. When dephosphorylated YAP and TAZ migrate to the nucleus, where they promote the expression of downstream genes [73]. YAP has been found to be highly expressed in astrocytes and its deletion induced astrocytic activation in both cell cultures and in vivo studies [72]. In astrocytes, IFNβ induced YAP activation, which, in turn, promoted the expression of the suppressor of cytokine signaling 3 (SOCS3), a negative regulator of JAK-STAT. In fact, YAP(-/-) astrocytes showed hyperactivation of the JAK-STAT pathway and astrocyte activation [72].

Neuroinflammation represents a crucial aspect of neurodegenerative disease progression. Targeting neuroinflammatory pathways seems to be a promising strategy to counteract neurodegenerative diseases. As different pathways are involved in the onset of neuroinflammation, compounds with different molecular targets are the best candidates to fight this condition. On these bases, beside drug development, the study of natural bioactive compounds, thanks to their varied and complex structures, can help with the identification of effective anti-inflammatory agents.

4. Marine Algae

Algae are photosynthetic eukaryotic organisms that present a complex and controversial taxonomy. More than 20,000 species of algae have been identified, and on the basis of their size they are divided in macroalgae (seaweeds) and microalgae. Macroalgae are multicellular marine plants, while microalgae are small unicellular or simple multicellular species [74]. Marine macroalgae can be classified into three classes according to their pigments: Brown (Phaeophyta) Green (Chlorophyta), and Red (Rhodophyta). The pigments responsible for the algae's color are: fucoxanthin (Phaephyta); chlorophyll a, b, lutein, zeaxanthin violaxanthin neoxanthin, and β-carotene (Chlorophyta); phycobilliproteins and lutein, zeaxanthin, and β-carotene (Rhodophyta). The classification of microalgae is extremely complex considering the thousands of species present even in small areas of water.

Microalgae are classified into groups based on different characteristics: pigment composition, morphological variations (rounded, oval, cylindrical, and fusiform cells), the presence of thorns, cilia, flagella etc. In addition, they can be classified based on their sizes: picoplankton (0.2–2 μm), nanoplankton (2–20 μm), and microplankton (20–200 μm). Recently, Corrêa et al., at the 16th IEEE International Conference on Machine Learning and Applications in 2017, proposed a deep learning technique to solve the problem by using as input low-resolution images [75].

Marine algae are composed of various substances: carbohydrates, lipids, proteins, amino acids, vitamins, minerals, and secondary metabolites such as phytosterols and polyphenols [76]. The chemical composition of macroalgae is considerably different between species and dependent on the season (sunlight), habitat (salinity, depth in the sea), and environmental conditions.

4.1. Carbohydrates

Among the various components, carbohydrates are the most abundant constituents of marine algae. Moreover, polysaccharides are usually the major component of red, green, and brown algae [77,78], and monosaccharides and oligosaccharides are also present. The storage polysaccharide is laminarin in brown algae and floridean in starch (more branched than amylopectin) in green and red algae. Algae cell walls are characterized by the presence of uncommon polysaccharides that can be sulfated, acetylated, etc. Marine algae carbohydrates are promising compounds in various fields, such as food, pharmaceutical, and biomedical. Noteworthy therapeutic applications are due to their antiviral, antibacterial, and antitumoral activities, antioxidant, antilipidemic, and antiglycemic properties, and anti-inflammatory and immunomodulatory characteristics. In particular, alginate-derived oligosaccharides inhibit neuroinflammation [79]. Laminarin (a polysaccharide composed of (1,3)-β-D-glucan with β(1,6) branching), particularly abundant in *Laminaria* species, has been demonstrated to possess antibacterial and chemopreventive activities, together with prebiotic activity [80], important in modulating gut microbiota, which in turn can regulate neuroinflammation [81]. Algae polysaccharides have been also utilized in the cosmeceutical industries due to their chemical and physical properties exhibiting potential benefits for skin [82].

Table 1 shows the different carbohydrates of brown, green, and red macroalgae. The oligosaccharides derived from polysaccharides are also important. They are produced by chemical or enzymatic hydrolysis and present numerous activities such as antioxidant, anti-inflammatory, and anti-melanogenic [83–87]. Microalgae also produce polysaccharides, and release in particular sulfated polysaccharides (carrageenan, ulvan, and fucoidan) [88–90]. Polysaccharides found in the cell wall vary among microalgae genera and species. Microalgae present an advantage with respect to macroalgae because they are easy to grow and culture and do not depend on the climate or season.

Table 1. Carbohydrates in marine algae.

Carbohydrates	Brown Macroalgae	Red Macroalgae	Green Macroalgae
monosaccharides	glucose, galactose, xylose, fucose, uronic acid, glucuronic acid mannuronic acid, guluronic acid	glucose, galactose, mannose	glucose, mannose, xylose, rhamnose, glucuronic acid, uronic acid
polysaccharides	laminarin alginate, fucoidan (sulphated), cellulose, mannitol	carrageenans (sulfated), agar (sulfated), floridean starch, cellulose, lignin, funoran	ulvan (sulfated), mannan, galactans (sulfated), xylans, floridean starch, cellulose, lignin

To date the ability of algae-derived polysaccharides to counteract neuroinflammation has not yet been fully explored.

4.2. Lipids

Algae contain different types of lipid phospholipids, non-polar glycerolipids, glycolipids, betaine lipids, and some unusual lipids, e.g., sulfolipid (sulfoquinovosyldiacylglycerol) sterols [91].

Marine macroalgae have a low lipid content but the proportion of long-chained polyunsaturated fatty acids (PUFA) is relatively high. In macroalgae, PUFAs are represented by omega-3 and omega-6 fatty acids. The content of PUFAs is generally higher in those living in cold water. Eicosapentaenoic acid (EPA) is the principal fatty acid. PUFAs have health benefits: they regulate blood clotting and blood pressure and develop functions of the brain and nervous systems [92,93]. They also decrease the risk of many chronic diseases such as arthritis, diabetes, and obesity [94,95], and regulate the signaling of microglia, mostly in the context of neuroinflammation and behavior [93].

Sterols. Among macroalgae, cholesterol is the most representative sterol in all the red algae; fucosterol, which has anti-inflammatory activity, is the chief sterol in brown algae [96], and in green algae the dominant sterol is isofucosterol clionasterol. Microalgae are characterized by the presence of unusual dihydroxysterols, pavlovols, crinosterols, and stigmasterols. It has been proposed that sterols, due to their ability to cross the blood-brain barrier, can prevent neuroinflammation [97,98], but there are few reports of the neuroprotective activities of algae-derived phytosterols.

4.3. Proteins and Amino Acids

Macroalgae and microalgae have been used as a source of human nutrition for thousands of years by some indigenous populations. This is due to their significant protein content, which is even greater than some ground plant sources. Algae proteins are rich in aspartic and glutamic acid, the latter contributing to the typical taste (umami). Green macroalgae, and especially red macroalgae, have a higher protein content than brown macroalgae. Macroalgae also contain a number of bioactive amino acids and peptides (e.g., taurine, carnosine, and glutathione and mycosporine-like) [99] that have been demonstrated to exert antioxidant and antiapoptotic effects in the rat brain [100]. Lectins are a group of glycoproteins isolated from algae [101] that present several properties including anti-inflammatory [102,103] antibiotic, cytotoxic, mitogenic, antinociceptive, and anti-viral due to their ability to bind to specific glycan structures [104]. Marine algae, with their high protein content, are now considered a precious source of bioactive peptides, obtained after enzymatic digestion, with considerable health potential. These biopeptides have been demonstrated to exhibit antioxidant, anticancer, antihypertensive, antiatherosclerotic, and immunomodulatory activities [105]. In the future it is desirable that research address the potential neuroprotective role of these biopeptides, elucidating their mechanism of action.

4.4. Phenols

Phenolic compounds are a class of chemical compounds characterized by hydroxyl groups directly attached to aromatic hydrocarbon rings. The simplest is composed of one aromatic ring and is called phenol. Phenolic compounds can be single phenols or polyphenols, depending on the number of phenol units in the molecule.

Phenols are largely represented in all the organisms belonging to the Plant kingdom; however, the phenols present in marine algae are different to those produced by terrestrial plants [104].

The best known polyphenols in marine algae are phloroglucinols and phlorotannins. Phlorotannins can be classified into subclasses: eckols, fuhalols, fucophlorethols, phlorethols, fucols, and ishofuhalols.

The largest proportion of phenolic compounds is in green and red algae (bromophenols, phenolic acids, and flavonoids). Phlorotannins are found only in marine brown algae [106,107].

Phenols and polyphenols from marine algae have attracted much attention for their anticancer, antioxidant, antimicrobial, and anti-inflammatory activities [108]. To, date several mechanisms behind microglial activation have been reported (see Section 3), and research is moving towards the discovery of alternative anti-inflammatory compounds from natural renewable sources that could potentially counteract neuroinflammation and, therefore, neuronal injury in neurodegenerative diseases, characterized by complex and deeply related phenomena. Marine algae rich in phenols are good candidates for potential application in the nutraceutical sector.

4.5. Isoprenoids

Carotenoids and terpenoids are two important classes of isoprenoids belonging to the marine algae. Carotenoids contains eight isoprene units, while terpenoids contain five isoprene units.

The carotenoids that consist of only hydrocarbons are carotenes, while those with oxo, hydroxyl, or epoxy groups are called xanthophylls. The most diffuse carotenoids in marine algae are: β-carotene, fucoxanthin, astaxanthin, canthaxanthin, and lutein. Fucostantin is mostly present in brown algae and in planktonic microalgae, while β-carotene is predominant in green microalgae [109,110].

The potential health-promoting effects of these carotenoids are: antioxidant activity, anti-inflammatory effects, anticancer activity anti-obese effect, antidiabetic activity, hepatoprotective effect, antiangiogenic effect, and cerebrovascular protective effect [111–113]. In particular, fucoxanthin has been demonstrated to decrease inflammation and oxidative damage [114] and astaxanthin has been demonstrated to decrease the expression of IL-6 in activated microglial cells [115], all factors implicated in the pathogenesis of neurodegenerative diseases.

Brown macroalgae are considered one of the principal source of biologically and ecologically relevant terpenoids, mainly diterpenes and merodilterpenes [116]. In *Sargassum*, meroterpenoids prevail, in particular sargachromenol, which presents anti-inflammatory and neuroprotective effects. Also, green algae are a source of terpenes, in particular the genus *Caulerpa*, which is represented by about 60 species living in tropical and subtropical waters that biosynthesize acyclic and monocyclic sesqui- and diterpenes [117] with neuroprotective activities.

5. Marine Algae and Neuroinflammation

As previously mentioned, activated microglia are a critical modulator of the neuroinflammation process, triggering a self-feeding loop with the neighboring astrocytes through the release of pro-inflammatory cytokines, including TNF-α and IL-1β [118]. In this context, a persistent and unrestrained neuroinflammatory loop harms neuronal cells and can promote neurodegenerative diseases [119]. Recent years have been characterized by a huge boost in nutritional research to discover natural compounds with anti-inflammatory properties and potential neuroprotective capacity. Marine algae have been part of a healthy diet in East Asia for centuries and represent a rich reservoir of structurally different bioactive compounds with great potential for pharmaceutical applications.

Increasingly, reports have shown the anti-inflammatory action of marine algae [120], as well as of their major components such as phlorotannins and pigments [121–123].

The methanol extract of *Ulva conglobata,* a green alga consumed as a marine vegetable, has been demonstrated to possess anti-inflammatory potential [22]. In particular, the extract was tested in hippocampal neuronal HT22 cells and microglial BV2 cells. In HT22 cells, 40 and 50 μg/mL *Ulva conglobata* extract was able to significantly restore cellular viability compared to glutamate-treated cells. Moreover, *Ulva conglobata* extract effectively suppressed IFN-γ-induced microglial activation, and 50 μg/mL inhibited NO release and reduced the expression of iNOS and COX-2 enzymes. Kim et al. [124] found that the hexane fraction of brown seaweed *Myagropsis myagroides* ethanolic extract exhibits the highest anti-inflammatory activity among different solvent fractions. In LPS-stimulated BV-2 cells, 25 μg/mL *Myagropsis myagroides* extract had the potential to revert the induction of pro-inflammatory mediators such as NO, PGE_2, and the cytokines IL-6 and TNF-α through the prevention of NF-κB nuclear translocation and MAPKs phosphorylation. Surprisingly, they did not identify the active compound responsible for these effects. Meanwhile, another report from the same authors suggested that the anti-inflammatory activity of *Myagropsis myagroides* ethanolic extract in LPS-stimulated BV-2 cells could be completely ascribed to the presence of sargachromenol [125]. A study assessed the anti-neuroinflammatory capacity of three extracts obtained from Malaysian seaweed: *Padina australis, Sargassum polycystum,* and *Caulerpa racemosa* [126]. All the extracts reduced the elevation of inflammatory mediators like NO, TNF-α, IL-6, and IL-1β, with the brown seaweeds (*Padina, Sargassum*) showing stronger inhibitory activity compared to the green seaweed (*Caulerpa*).

The so-called "cholinergic hypothesis" suggests a correlation between memory impairment in AD and the reduction of neurotransmitter acetylcholine [127]. The preservation of acetylcholine levels could be useful in view of a multitarget therapy. Fucosterol, a sterol mainly found in brown algae including *Padina australis,* was isolated to investigate its cholinesterase and inflammatory inhibitory properties [128]. It was observed that fucosterol inhibits acetylcholinesterase (AChE) and butyrylcholinesterase (BChE), both responsible for acetylcholine hydrolysis, and significantly prevents the production of pro-inflammatory mediators in LPS-induced C8-B4 microglial cells and in Aβ-induced BV-2 microglial cells.

Ecklonia cava, an edible brown alga used for the production of food ingredients, animal feed, and fertilizers, has been shown to possess anti-inflammatory activity [129,130].

Three of the major phlorotannins that can be found in *Ecklonia cava* eckol, dieckol, and 8,8′-bieckol, were investigated for their protective effects against $A\beta_{25\text{-}35}$-induced neuroinflammatory damage in PC12 cells [130]. The results indicated that all phlorotannins tested possess antioxidant and protective effects against Aβ damage, while dieckol has the strongest ability to combat apoptosis and Ca^{2+} overload and more effectively inhibits the increase of inflammatory markers and the protein levels of p65, the best studied NF-κB subunit. Therefore, the neuroprotective property of dieckol with a diphenyl ether linkage was greater than that of 8,8′-bieckol with a biaryl linkage, although these two compounds are both dimers of eckol.

These data were further confirmed by Jung et al. [129], who isolated dieckol from *Ecklonia cava* extract, reporting its potential as an anti-inflammatory agent by reducing the release and stimulation of pro-inflammatory cytokines and enzymes together with an intracellular scavenging activity. Also, a component from *Ecklonia stolonifera,* phlorofucofuroeckol B, was identified as a potent suppressor of inflammation, inhibiting IκB-α/NF-κB and Akt/ERK/JNK pathways [23]. A study conducted by Kim et al. [131] demonstrated, for the first time, that floridoside, a natural glycerol galactoside from the red alga *Laurencia undulata,* possesses the potential to counteract the neuronal damage induced by neuroinflammation in vitro, preventing ROS and NO overload due to iNOS and COX-2 overexpression. Among algae pigments, fucoxanthin is one of the main carotenoids found in brown algae [132]. In an $A\beta_{42}$ -induced microglial activation model, fucoxanthin significantly reduced the rates of inflammatory and oxidative damage, protecting DNA from oxidation and attenuating the increasing of inflammatory enzymes [114]. Astaxanthin, a red carotenoid pigment, occurs naturally in plants and marine seaweeds, but also in shellfish and crustaceans [133]. It has been shown to possess a variety of pharmacological effects, including anti-inflammatory and antioxidative activity [133–136].

Increasing evidence correlates a neuronal inflammation status with the development of depression [137,138]. In a rat model of LPS-induced depressive-like behaviors, 80 mg/kg astaxanthin had an antidepressant-like effect due to the restoration of LPS-induced alterations of brain inflammatory markers (i.e., IL-1β, IL-6, and TNF-α), as well as iNOS, nNOS, and COX-2 expression via the modulation of NF-κB activation [24].

In addition, Zhang et al. [139] found that astaxanthin administration could alleviate early brain injury via suppressing the inflammation damage induced by subarachnoid hemorrhage. In particular, 75 mg/kg astaxanthin significantly reduced the elevated cortical levels of inflammatory mediators, together with the degree of neutrophil infiltration.

A food supplement approved by the U.S. Food and Drug Administration (FDA), named Aquamin, is a natural multi-mineral derived from the marine red seaweed *Lithothamnion corallioides*. Aquamin was evaluated for its anti-neuroinflammatory potential, and in cortical glial-enriched cells was able to suppress the release of LPS-induced TNF-α and IL-1β. Recently, several authors suggested that anti-inflammatory and antioxidative agents could prevent the deposition of Aβ and the subsequent brain damage [140,141]. Indeed, in the promoter of neuronal beta-secretase 1 (BACE1), the enzyme involved in Aβ buildup, NF-κB DNA consensus sequences are present [142]. So, it could be beneficial in treating AD to reduce microglia-mediated neuroinflammation and increase microglia scavenger activity for toxic Aβ aggregates [143]. The ethanol extract of *Nannochloropsis oceanica* demonstrated anti-inflammatory, antioxidative, and anti-amyloidogenesis activities in a mouse model of LPS-induced AD [141]. The authors recently found that the main component of *Nannochloropsis oceanica* is eicosapentaenoic acid (EPA), suggesting that it could be responsible for the neuroprotective effects. The depolymerization of the polysaccharide alginate, found in many marine brown algae, produces alginate-derived oligosaccharide with various biological activities depending on the degradation method used [79]. The alginate-derived oligosaccharide produced by enzymatic depolymerization showed anti-inflammatory activity by repressing the LPS and Aβ-induced production of inflammatory cytokines and mediators in microglial cells. These effects have been associated with the inactivation of the TLR4/NF-κB axis [79]. Interestingly, the interaction between this oligosaccharide and TLR4 promotes the uptake of toxic Aβ aggregates. Regarding the possibility of alginate-derived oligosaccharide crossing the BBB, the authors declared an average molecular weight of 1500 Da and previous works demonstrated that oligosaccharides produced by enzymatic depolymerization are able to pass through the BBB easily [25,144]. Differently, Bi et al. [13] synthesized a seleno-polysaccharide from alginate-derived polymannuronate. Using in vitro/in vivo models of microglia and astrocyte activation, the pre-treatment with seleno-polymannuronate reduced the overgeneration of proinflammatory mediators, including NO, PGE_2, TNF-α, IL-6, and IL-1β as well as iNOS and COX-2, by suppressing the MAPK/NF-κB signaling pathway. Cui et al. [145] assessed whether fucoidan, a class of fucose-enriched sulfated polysaccharides isolated from *Laminaria japonica*, protects dopaminergic neurons from inflammation-mediated damage in a PD inflammatory rat model induced by an intranigral injection of LPS. Fucoidan was able to improve behavioral deficits in mice by protecting them from the loss of dopaminergic neurons. Other important anti-AD and anti-inflammatory effects have been manifested by the glycoproteins purified from brown alga *Undaria pinnatifida* [146]. *Undaria pinnatifida* displayed dose-responsive inhibition for AChE and BChE with an IC_{50} of 63.56 and 99.03 μg/mL, respectively, and has been shown to inhibit BACE1, acting on the neurotransmitter acetylcholine and on the formation and accumulation of Aβ aggregates. Moreover, *Undaria pinnatifida* promotes cell survival and neurite extension, preventing inflammation status.

Epidemiological studies demonstrate a negative correlation between the use of non-steroidal anti-inflammatory drugs (NSAIDs) and the incidence of inflammation in the nervous system, which in turn participates in the development of neurodegenerative diseases [120]. The NSAIDs' mechanism of action involves the inhibition of the inflammatory mediator release. Marine algae can control the inflammatory process in microglia, suggesting their potential role as neuroprotective agents. Moreover, the signaling pathways involved in the neuroprotective activity of algae are multiple. The complexity of neurodegenerative diseases makes them difficult to counteract with single-target molecules. In this context, marine algae, with their pleiotropic effects, have a great potential for application as anti-neuroinflammatory agents. However, further studies are needed, along with clinical trials to confirm marine algae's anti-neuroinflammatory activity.

6. Conclusions

The wide range of biological and bioactive molecules found in marine algae represents a challenge for researchers involved in the study of neuroinflammation/neurodegeneration processes. Marine algae extracts and many marine algae constituents belonging to different chemical classes have been demonstrated to exert preventive/protective effects against neuro-inflammation (Table 2). In particular, they have been demonstrated to be effective in reducing inflammatory mediators like NO, TNF-α, IL-6, and IL-1β, in downregulating inflammatory enzymes like iNOS and COX-2, and in modulating the signaling pathways that lead to NF-κB activation. Moreover, most of the compounds isolated from marine algae have also shown antioxidant activity. Oxidative stress represents a hallmark of neuroinflammation and its counteraction could be a successful strategy in the prevention of neurodegeneration. ROS production is strictly related to neuro-inflammation, and marine algae compounds with both antioxidant and anti-inflammatory activities are good candidates to counteract neurodegeneration thanks to their pleiotropic activity. A better knowledge of these molecules should be associated with an implementation in the extraction and purification procedures in order to obtain marine algae extracts with standardized concentrations to be applied in in vitro studies. In fact, the choice of an appropriate extraction method can deeply influence the presence and concentration of the bioactive compounds. Moreover, the ability of marine algae constituents to cross the blood-brain barrier has not been investigated, which calls into question the possibility of developing them as neuroprotective agents. Also, studies on potential adverse effects are lacking. Although still in their infancy, studies on the anti-neuroinflammatory effects of marine algae compounds should be corroborated by clinical trials. Currently there is a paucity of information reported in the literature, which only contains studies on in vitro or animal models. Human studies could strengthen the choice of marine algae products as potential nutraceutical compounds for the prevention of neuro-inflammation.

Table 2. Studies showing anti-neuroinflammatory activities of marine algae.

Marine Algae Extract/Bioactive Compound	Treatment Conc.	Experimental Model	Key Findings
Ulva conglobata methanol extract	10-50 μg/mL	mouse hippocampal HT-22 cells; mouse microglial BV-2 cells	Restoration of cellular viability in HT-22 cells; downregulation of COX-2 and iNOS in BV-2 cells [22]
Exane fraction of *Myagropsis myagroides* ethanolic extract	5-25 μg/mL	mouse microglial BV-2 cells	Decreased release of inflammatory cytokines, inactivation of NF-κB and reduced mRNA and protein levels of iNOS and COX-2 [124]
Myagropsis myagroides ethanolic extract	5–25 μg/mL	mouse microglial BV-2 cells	Reduction in NO, PGE_2, IL-6, IL-1β and TNF-α release; inhibition of ERKs-JNKs/NF-κB axis [125]
Padina australis, *Sargassum polycystum* and *Caulerpa racemosa* extracts	0.05–0.4 mg/mL	mouse microglial C8-B4 cells	Decreased release of pro-inflammatory mediators (NO, PGE_2, IL-6, IL-1β and TNF-α) [126]

Table 2. *Cont.*

Marine Algae Extract/Bioactive Compound	Treatment Conc.	Experimental Model	Key Findings
Fucosterol from *Padina australis*	0.004–192 μM	mouse microglial C8-B4 and BV-2 cells	Inhibition of AChE and BChE; reduction in release of NO, PGE_2, IL-6, IL-1β and TNF-α in LPS-stimulated C8-B4 cells; prevented production of NO, IL-6 and TNF-α in $A\beta_{42}$-stimulated BV-2 cells [128]
Eckol, dieckol and 8,8′-bieckol from *Ecklonia cava*	1–50 μM	rat neuronal PC12 cells	Antioxidant activity; anti-apoptotic effects; decrease in key inflammatory proteins (COX-2, iNOS, IL-1β and TNF-α) [130]
Dieckol from *Ecklonia cava*	50–300 μg/mL	mouse microglial BV-2 cells	Inhibition of LPS-induced iNOS and COX-2 protein and mRNA expression; suppression of p-38/ NF-κB pathway; ROS scavenging activity [129]
Phlorofucofuroeckol B from *Ecklonia stolonifera*	10–40 μM	mouse microglial BV-2 cells	Inhibition of IκB-α/NF-κB and Akt/ERK/JNK pathways [23]
Floridoside from *Laurencia undulata*	1–50 μM	mouse microglial BV-2 cells	Inhibition of LPS-induced NO and ROS production; downregulation of COX-2 and iNOS mRNA and protein levels by reducing p38 and ERK phosphorylation [131]
Fucoxanthin	5–50 μM	mouse microglial BV-2 cells	Attenuation of $A\beta_{42}$-induced cytokines release (NO, PGE_2, IL-6, IL-1β and TNF-α) and enzymes upregulation (COX-2, iNOS) by suppressing MAPKs phosphorylation; protection from H_2O_2-induced ROS release and DNA damage by recovering antioxidant enzymes [114]
Astaxanthin	20–80 mg/Kg	male ICR mice	Reversed LPS-induced depressive-like behaviors; attenuation of cytokines level (IL-6, IL-1β and TNF-α) and antagonization of iNOS, nNOS and COX-2 expression in the hippocampus and prefrontal cortex [24]
Astaxanthin	75 mg/Kg	male Sprague-Dawley rats	Amelioration in cerebral edema, blood-brain barrier disruption, neurological dysfunction and neuronal degeneration after the induction of subarachnoid hemorrhage; downregulation of NF-κB activity, and intercellular adhesion molecule-1, IL-1β and TNF-α expression [139]

Table 2. *Cont.*

Marine Algae Extract/Bioactive Compound	Treatment Conc.	Experimental Model	Key Findings
Aquamin™	0.05–2 mg/mL	cortical glial-enriched cultures from Sprague-Dawley rat pups	Attenuation of LPS-induced IL-1β and TNF-α secretion [147]
Nannochloropsis oceanica ethanol extract	50–100 mg/Kg	male ICR mice	Decrease of ROS and malondialdehyde levels; improvement of LPS-induced memory impairment; suppression of $A\beta_{42}$ generation by downregulating APP and BACE1 expression [141]
Alginate-derived oligosaccharide	50–500 μg/mL	mouse microglial BV-2 cells	Inhibition of LPS/ $A\beta_{42}$-induced NO and PGE_2 production, COX-2 and iNOS expression, and cytokines secretion; attenuation of TLR4 and NF-κB overexpression; promotion of Aβ phagocytosis [79]
Seleno-polymannuronate	0.5 mg/mL, 0.8 mg/mL	primary microglia and astrocytes from BALB/c mouse pups; female BALB/c mice	In LPS-activated primary cells, attenuation of NF-κB and MAPK signaling with the reduction of NO, PGE_2 production, downregulation of COX-2 and iNOS expression, and IL-6, IL-1β and TNF-α secretion; decrease of Iba1- and GFAP-positive cells in the brain of a mouse model of LPS-induced inflammation [13]
Fucoidan	7.5 mg/Kg, 15 mg/Kg; 31.25–125 μg/mL	male Sprague-Dawley rats; primary microglia from neonatal Sprague–Dawley rats	Improvement of behavioral deficits and prevention of dopaminergic neuron loss; inhibition of ROS and TNF-α release [145]
Glycoprotein from *Undaria pinnatifida*	5–45 μg/mL	primary hippocampal cells from embryonal Sprague–Dawley rats	Inhibition of AChE, BChE and BACE1; promotion of cell survival and neurite extension [146]

Funding: This work was supported by MIUR-PRIN 2015 (N. 20152HKF3Z).

Conflicts of Interest: The authors declare no conflict of interest.

Abbreviations

AChE	Acetylcholinesterase
AD	Alzheimer's disease
ALS	Amyotrophic lateral sclerosis
AP-1	Activating protein 1
A-	Amyloid beta
BACE1	Beta-secretase 1
BBB	Blood-brain barrier
BChE	Butyrylcholinesterase
CNS	Central nervous system
COX-2	Cyclooxygenase-2
EPA	Eicosapentaenoic acid
HD	Huntington's disease
IL	Interleukin
iNOS	Inducible nitric oxide synthase
JNK	c-Jun N-terminal kinases
LPS	Lipopolysaccharide
MAPKs	Mitogen-activated protein kinases cascade
MS	Multiple sclerosis
NF-κB	Nuclear factor κB
NGF	Nerve growth factor
nNOS	Neuronal nitric oxide synthase
NO	Nitric oxide
NSAIDs	Non-steroidal anti-inflammatory drugs
PD	Parkinson's disease
PGE_2	Prostaglandin E2
PUFA	Polyunsaturated fatty acids
TBI	Traumatic brain injury
TGFβ	Transforming growth factor beta
TLRs	Toll-like receptor
TNF-α	Tumor necrosis factor alpha

References

1. Höglund, K.; Salter, H. Molecular biomarkers of neurodegeneration. *Expert Rev. Mol. Diagn.* **2013**, *13*, 845–861. [CrossRef] [PubMed]
2. Angeloni, C.; Prata, C.; Dalla Sega, F.V.; Piperno, R.; Hrelia, S. Traumatic brain injury and NADPH oxidase: A deep relationship. *Oxid. Med. Cell Longev.* **2015**, *2015*, 370312. [CrossRef] [PubMed]
3. Tarozzi, A.; Angeloni, C.; Malaguti, M.; Morroni, F.; Hrelia, S.; Hrelia, P. Sulforaphane as a potential protective phytochemical against neurodegenerative diseases. *Oxid. Med. Cell Longev.* **2013**, *2013*, 415078. [CrossRef] [PubMed]
4. Hammond, T.R.; Marsh, S.E.; Stevens, B. Immune Signaling in Neurodegeneration. *Immunity* **2019**, *50*, 955–974. [CrossRef] [PubMed]
5. Frank-Cannon, T.C.; Alto, L.T.; McAlpine, F.E.; Tansey, M.G. Does neuroinflammation fan the flame in neurodegenerative diseases? *Mol. Neurodegener* **2009**, *4*, 47. [CrossRef] [PubMed]
6. Morales, I.; Guzmán-Martínez, L.; Cerda-Troncoso, C.; Farías, G.A.; Maccioni, R.B. Neuroinflammation in the pathogenesis of Alzheimer's disease. A rational framework for the search of novel therapeutic approaches. *Front. Cell Neurosci.* **2014**, *8*, 112. [CrossRef] [PubMed]
7. Glass, C.K.; Saijo, K.; Winner, B.; Marchetto, M.C.; Gage, F.H. Mechanisms underlying inflammation in neurodegeneration. *Cell* **2010**, *140*, 918–934. [CrossRef]
8. Spencer, J.P.; Vafeiadou, K.; Williams, R.J.; Vauzour, D. Neuroinflammation: modulation by flavonoids and mechanisms of action. *Mol. Asp. Med.* **2012**, *33*, 83–97. [CrossRef]
9. Schain, M.; Kreisl, W.C. Neuroinflammation in Neurodegenerative Disorders-a Review. *Curr. Neurol. Neurosci. Rep.* **2017**, *17*, 25. [CrossRef]

10. Sofroniew, M.V. Multiple roles for astrocytes as effectors of cytokines and inflammatory mediators. *Neuroscientist* **2014**, *20*, 160–172. [CrossRef]
11. Neal, M.; Richardson, J.R. Epigenetic regulation of astrocyte function in neuroinflammation and neurodegeneration. *Biochim Biophys Acta. Mol. Basis Dis.* **2018**, *1864*, 432–443. [CrossRef] [PubMed]
12. Motori, E.; Puyal, J.; Toni, N.; Ghanem, A.; Angeloni, C.; Malaguti, M.; Cantelli-Forti, G.; Berninger, B.; Conzelmann, K.K.; Götz, M.; et al. Inflammation-induced alteration of astrocyte mitochondrial dynamics requires autophagy for mitochondrial network maintenance. *Cell Metab* **2013**, *18*, 844–859. [CrossRef] [PubMed]
13. Decheng, B.; Qiuxian, L.; Qingguo, H.; Nan, C.; Hanxing, H.; Weishan, F.; Jiang, Y.; Xiaofan, L.; Hong, X.; Xiuting, L.; et al. Seleno-polymannuronate attenuates neuroinflammation by suppressing microglial and astrocytic activation. *J. Funct. Foods* **2018**, *51*, 113–120.
14. Ning, C.; Wang, H.D.; Gao, R.; Chang, Y.C.; Hu, F.; Meng, X.; Huang, S.Y. Marine-derived protein kinase inhibitors for neuroinflammatory diseases. *Biomed. Eng. Online* **2018**, *17*, 46. [CrossRef] [PubMed]
15. Park, E.J.; Pezzuto, J.M. Antioxidant marine products in cancer chemoprevention. *Antioxid Redox Signal.* **2013**, *19*, 115–138. [CrossRef] [PubMed]
16. Pejin, B.; Jovanovic, K.K.; Savic, A.G. New antitumour natural products from marine red algae: Covering the period from 2003 to 2012. *Mini. Rev. Med. Chem.* **2015**, *15*, 720–730. [CrossRef] [PubMed]
17. Jin, Q.; Yu, H.; Li, P. The Evaluation and Utilization of Marine-derived Bioactive Compounds with Anti-obesity Effect. *Curr. Med. Chem.* **2018**, *25*, 861–878. [CrossRef] [PubMed]
18. Huang, C.; Zhang, Z.; Cui, W. Marine-Derived Natural Compounds for the Treatment of Parkinson's Disease. *Mar. Drugs* **2019**, *17*, 221. [CrossRef]
19. Newman, D.J.; Cragg, G.M. Advanced preclinical and clinical trials of natural products and related compounds from marine sources. *Curr. Med. Chem.* **2004**, *11*, 1693–1713. [CrossRef] [PubMed]
20. Brown, E.S.; Allsopp, P.J.; Magee, P.J.; Gill, C.I.; Nitecki, S.; Strain, C.R.; McSorley, E.M. Seaweed and human health. *Nutr. Rev.* **2014**, *72*, 205–216. [CrossRef] [PubMed]
21. Fernando, I.P.S.; Nah, J.W.; Jeon, Y.J. Potential anti-inflammatory natural products from marine algae. *Env. Toxicol Pharm.* **2016**, *48*, 22–30. [CrossRef] [PubMed]
22. Jin, D.Q.; Lim, C.S.; Sung, J.Y.; Choi, H.G.; Ha, I.; Han, J.S. Ulva conglobata, a marine algae, has neuroprotective and anti-inflammatory effects in murine hippocampal and microglial cells. *Neurosci. Lett.* **2006**, *402*, 154–158. [CrossRef] [PubMed]
23. Yu, D.K.; Lee, B.; Kwon, M.; Yoon, N.; Shin, T.; Kim, N.G.; Choi, J.S.; Kim, H.R. Phlorofucofuroeckol B suppresses inflammatory responses by down-regulating nuclear factor κB activation via Akt, ERK, and JNK in LPS-stimulated microglial cells. *Int. Immunopharmacol* **2015**, *28*, 1068–1075. [CrossRef]
24. Jiang, X.; Chen, L.; Shen, L.; Chen, Z.; Xu, L.; Zhang, J.; Yu, X. Trans-astaxanthin attenuates lipopolysaccharide-induced neuroinflammation and depressive-like behavior in mice. *Brain Res.* **2016**, *1649*, 30–37. [CrossRef] [PubMed]
25. Fan, Y.; Hu, J.; Li, J.; Yang, Z.; Xin, X.; Wang, J.; Ding, J.; Geng, M. Effect of acidic oligosaccharide sugar chain on scopolamine-induced memory impairment in rats and its related mechanisms. *Neurosci. Lett.* **2005**, *374*, 222–226. [CrossRef]
26. Scarpignato, C.; Dolak, W.; Lanas, A.; Matzneller, P.; Renzulli, C.; Grimaldi, M.; Zeitlinger, M.; Bjarnason, I. Rifaximin Reduces the Number and Severity of Intestinal Lesions Associated With Use of Nonsteroidal Anti-Inflammatory Drugs in Humans. *Gastroenterology* **2017**, *152*, 980–982.e983. [CrossRef]
27. Yuan, G.; Wahlqvist, M.L.; He, G.; Yang, M.; Li, D. Natural products and anti-inflammatory activity. *Asia Pac. J. Clin. Nutr.* **2006**, *15*, 143–152.
28. Adams, K.L.; Gallo, V. The diversity and disparity of the glial scar. *Nat. Neurosci.* **2018**, *21*, 9–15. [CrossRef]
29. Nayak, D.; Roth, T.L.; McGavern, D.B. Microglia development and function. *Annu. Rev. Immunol.* **2014**, *32*, 367–402. [CrossRef]
30. Ginhoux, F.; Greter, M.; Leboeuf, M.; Nandi, S.; See, P.; Gokhan, S.; Mehler, M.F.; Conway, S.J.; Ng, L.G.; Stanley, E.R.; et al. Fate mapping analysis reveals that adult microglia derive from primitive macrophages. *Science* **2010**, *330*, 841–845. [CrossRef]
31. Subhramanyam, C.S.; Wang, C.; Hu, Q.; Dheen, S.T. Microglia-mediated neuroinflammation in neurodegenerative diseases. *Semin Cell Dev. Biol.* **2019**. [CrossRef] [PubMed]

32. Schwartz, M.; Kipnis, J.; Rivest, S.; Prat, A. How do immune cells support and shape the brain in health, disease, and aging? *J. Neurosci.* **2013**, *33*, 17587–17596. [CrossRef] [PubMed]
33. Ueno, M.; Fujita, Y.; Tanaka, T.; Nakamura, Y.; Kikuta, J.; Ishii, M.; Yamashita, T. Layer V cortical neurons require microglial support for survival during postnatal development. *Nat. Neurosci.* **2013**, *16*, 543–551. [CrossRef] [PubMed]
34. Takahashi, K.; Rochford, C.D.; Neumann, H. Clearance of apoptotic neurons without inflammation by microglial triggering receptor expressed on myeloid cells-2. *J. Exp. Med.* **2005**, *201*, 647–657. [CrossRef] [PubMed]
35. Liao, H.; Bu, W.Y.; Wang, T.H.; Ahmed, S.; Xiao, Z.C. Tenascin-R plays a role in neuroprotection via its distinct domains that coordinate to modulate the microglia function. *J. Biol. Chem.* **2005**, *280*, 8316–8323. [CrossRef] [PubMed]
36. Drögemüller, K.; Helmuth, U.; Brunn, A.; Sakowicz-Burkiewicz, M.; Gutmann, D.H.; Mueller, W.; Deckert, M.; Schlüter, D. Astrocyte gp130 expression is critical for the control of Toxoplasma encephalitis. *J. Immunol.* **2008**, *181*, 2683–2693. [CrossRef] [PubMed]
37. Cho, M.L.; Min, S.Y.; Chang, S.H.; Kim, K.W.; Heo, S.B.; Lee, S.H.; Park, S.H.; Cho, C.S.; Kim, H.Y. Transforming growth factor beta 1(TGF-beta1) down-regulates TNFalpha-induced RANTES production in rheumatoid synovial fibroblasts through NF-kappaB-mediated transcriptional repression. *Immunol. Lett.* **2006**, *105*, 159–166. [CrossRef]
38. Wyss-Coray, T.; Mucke, L. Inflammation in neurodegenerative disease—A double-edged sword. *Neuron* **2002**, *35*, 419–432. [CrossRef]
39. Hoshino, K.; Takeuchi, O.; Kawai, T.; Sanjo, H.; Ogawa, T.; Takeda, Y.; Takeda, K.; Akira, S. Cutting edge: Toll-like receptor 4 (TLR4)-deficient mice are hyporesponsive to lipopolysaccharide: evidence for TLR4 as the Lps gene product. *J. Immunol.* **1999**, *162*, 3749–3752.
40. Tang, Y.; Le, W. Differential Roles of M1 and M2 Microglia in Neurodegenerative Diseases. *Mol. Neurobiol.* **2016**, *53*, 1181–1194. [CrossRef]
41. Colton, C.A.; Gilbert, D.L. Production of superoxide anions by a CNS macrophage, the microglia. *FEBS Lett.* **1987**, *223*, 284–288. [CrossRef]
42. Graeber, M.B.; Streit, W.J.; Kreutzberg, G.W. Axotomy of the rat facial nerve leads to increased CR3 complement receptor expression by activated microglial cells. *J. Neurosci. Res.* **1988**, *21*, 18–24. [CrossRef] [PubMed]
43. Chen, W.W.; Zhang, X.; Huang, W.J. Role of neuroinflammation in neurodegenerative diseases. *Mol. Med. Rep.* **2016**, *13*, 3391–3396. [CrossRef] [PubMed]
44. Lull, M.E.; Block, M.L. Microglial activation and chronic neurodegeneration. *Neurotherapeutics* **2010**, *7*, 354–365. [CrossRef] [PubMed]
45. Licastro, F.; Hrelia, S.; Porcellini, E.; Malaguti, M.; Di Stefano, C.; Angeloni, C.; Carbone, I.; Simoncini, L.; Piperno, R. Peripheral Inflammatory Markers and Antioxidant Response during the Post-Acute and Chronic Phase after Severe Traumatic Brain Injury. *Front. Neurol.* **2016**, *7*, 189. [CrossRef] [PubMed]
46. Tansey, M.G.; McCoy, M.K.; Frank-Cannon, T.C. Neuroinflammatory mechanisms in Parkinson's disease: potential environmental triggers, pathways, and targets for early therapeutic intervention. *Exp. Neurol.* **2007**, *208*, 1–25. [CrossRef] [PubMed]
47. Das Sarma, J. Microglia-mediated neuroinflammation is an amplifier of virus-induced neuropathology. *J. Neurovirol.* **2014**, *20*, 122–136. [CrossRef]
48. Chen, H.; Chan, D.C. Mitochondrial dynamics–fusion, fission, movement, and mitophagy–in neurodegenerative diseases. *Hum. Mol. Genet.* **2009**, *18*, R169–R176. [CrossRef]
49. Xiang, W.; Chao, Z.Y.; Feng, D.Y. Role of Toll-like receptor/MYD88 signaling in neurodegenerative diseases. *Rev. Neurosci.* **2015**, *26*, 407–414. [CrossRef]
50. Lehnardt, S.; Massillon, L.; Follett, P.; Jensen, F.E.; Ratan, R.; Rosenberg, P.A.; Volpe, J.J.; Vartanian, T. Activation of innate immunity in the CNS triggers neurodegeneration through a Toll-like receptor 4-dependent pathway. *Proc. Natl. Acad. Sci USA* **2003**, *100*, 8514–8519. [CrossRef]
51. Balducci, C.; Frasca, A.; Zotti, M.; La Vitola, P.; Mhillaj, E.; Grigoli, E.; Iacobellis, M.; Grandi, F.; Messa, M.; Colombo, L.; et al. Toll-like receptor 4-dependent glial cell activation mediates the impairment in memory establishment induced by β-amyloid oligomers in an acute mouse model of Alzheimer's disease. *Brain Behav. Immun.* **2017**, *60*, 188–197. [CrossRef] [PubMed]

52. Capiralla, H.; Vingtdeux, V.; Zhao, H.; Sankowski, R.; Al-Abed, Y.; Davies, P.; Marambaud, P. Resveratrol mitigates lipopolysaccharide- and Aβ-mediated microglial inflammation by inhibiting the TLR4/NF-κB/STAT signaling cascade. *J. Neurochem.* **2012**, *120*, 461–472. [CrossRef] [PubMed]
53. Lv, R.; Du, L.; Liu, X.; Zhou, F.; Zhang, Z.; Zhang, L. Rosmarinic acid attenuates inflammatory responses through inhibiting HMGB1/TLR4/NF-κB signaling pathway in a mouse model of Parkinson's disease. *Life Sci.* **2019**, *223*, 158–165. [CrossRef] [PubMed]
54. Caso, J.R.; Pradillo, J.M.; Hurtado, O.; Lorenzo, P.; Moro, M.A.; Lizasoain, I. Toll-like receptor 4 is involved in brain damage and inflammation after experimental stroke. *Circulation* **2007**, *115*, 1599–1608. [CrossRef] [PubMed]
55. Lu, Y.C.; Yeh, W.C.; Ohashi, P.S. LPS/TLR4 signal transduction pathway. *Cytokine* **2008**, *42*, 145–151. [CrossRef] [PubMed]
56. Hanada, T.; Yoshimura, A. Regulation of cytokine signaling and inflammation. *Cytokine Growth Factor Rev.* **2002**, *13*, 413–421. [CrossRef]
57. Shabab, T.; Khanabdali, R.; Moghadamtousi, S.Z.; Kadir, H.A.; Mohan, G. Neuroinflammation pathways: a general review. *Int J. Neurosci* **2017**, *127*, 624–633. [CrossRef]
58. Kawai, T.; Akira, S. Signaling to NF-kappaB by Toll-like receptors. *Trends Mol. Med.* **2007**, *13*, 460–469. [CrossRef]
59. Qian, Y.; Liu, C.; Hartupee, J.; Altuntas, C.Z.; Gulen, M.F.; Jane-Wit, D.; Xiao, J.; Lu, Y.; Giltiay, N.; Liu, J.; et al. The adaptor Act1 is required for interleukin 17-dependent signaling associated with autoimmune and inflammatory disease. *Nat. Immunol* **2007**, *8*, 247–256. [CrossRef]
60. Colombo, E.; Di Dario, M.; Capitolo, E.; Chaabane, L.; Newcombe, J.; Martino, G.; Farina, C. Fingolimod may support neuroprotection via blockade of astrocyte nitric oxide. *Ann. Neurol.* **2014**, *76*, 325–337. [CrossRef]
61. Mayo, L.; Trauger, S.A.; Blain, M.; Nadeau, M.; Patel, B.; Alvarez, J.I.; Mascanfroni, I.D.; Yeste, A.; Kivisäkk, P.; Kallas, K.; et al. Regulation of astrocyte activation by glycolipids drives chronic CNS inflammation. *Nat. Med.* **2014**, *20*, 1147–1156. [CrossRef] [PubMed]
62. Füchtbauer, L.; Groth-Rasmussen, M.; Holm, T.H.; Løbner, M.; Toft-Hansen, H.; Khorooshi, R.; Owens, T. Angiotensin II Type 1 receptor (AT1) signaling in astrocytes regulates synaptic degeneration-induced leukocyte entry to the central nervous system. *Brain Behav. Immun.* **2011**, *25*, 897–904. [CrossRef] [PubMed]
63. Brambilla, R.; Bracchi-Ricard, V.; Hu, W.H.; Frydel, B.; Bramwell, A.; Karmally, S.; Green, E.J.; Bethea, J.R. Inhibition of astroglial nuclear factor kappaB reduces inflammation and improves functional recovery after spinal cord injury. *J. Exp. Med.* **2005**, *202*, 145–156. [CrossRef] [PubMed]
64. Fu, E.S.; Zhang, Y.P.; Sagen, J.; Candiotti, K.A.; Morton, P.D.; Liebl, D.J.; Bethea, J.R.; Brambilla, R. Transgenic inhibition of glial NF-kappa B reduces pain behavior and inflammation after peripheral nerve injury. *Pain* **2010**, *148*, 509–518. [CrossRef] [PubMed]
65. Youssef, M.; Ibrahim, A.; Akashi, K.; Hossain, M.S. PUFA-Plasmalogens Attenuate the LPS-Induced Nitric Oxide Production by Inhibiting the NF-kB, p38 MAPK and JNK Pathways in Microglial Cells. *Neuroscience* **2019**, *397*, 18–30. [CrossRef] [PubMed]
66. Yan, Z.; Gibson, S.A.; Buckley, J.A.; Qin, H.; Benveniste, E.N. Role of the JAK/STAT signaling pathway in regulation of innate immunity in neuroinflammatory diseases. *Clin. Immunol.* **2018**, *189*, 4–13. [CrossRef] [PubMed]
67. Yeung, Y.T.; Aziz, F.; Guerrero-Castilla, A.; Arguelles, S. Signaling Pathways in Inflammation and Anti-inflammatory Therapies. *Curr. Pharm Des.* **2018**, *24*, 1449–1484. [CrossRef]
68. Meares, G.P.; Liu, Y.; Rajbhandari, R.; Qin, H.; Nozell, S.E.; Mobley, J.A.; Corbett, J.A.; Benveniste, E.N. PERK-dependent activation of JAK1 and STAT3 contributes to endoplasmic reticulum stress-induced inflammation. *Mol. Cell Biol.* **2014**, *34*, 3911–3925. [CrossRef]
69. Chen, H.; O'Reilly, E.J.; Schwarzschild, M.A.; Ascherio, A. Peripheral inflammatory biomarkers and risk of Parkinson's disease. *Am. J. Epidemiol.* **2008**, *167*, 90–95. [CrossRef]
70. Lee, E.J.; Woo, M.S.; Moon, P.G.; Baek, M.C.; Choi, I.Y.; Kim, W.K.; Junn, E.; Kim, H.S. Alpha-synuclein activates microglia by inducing the expressions of matrix metalloproteinases and the subsequent activation of protease-activated receptor-1. *J. Immunol.* **2010**, *185*, 615–623. [CrossRef]
71. Qin, H.; Buckley, J.A.; Li, X.; Liu, Y.; Fox, T.H.; Meares, G.P.; Yu, H.; Yan, Z.; Harms, A.S.; Li, Y.; et al. Inhibition of the JAK/STAT Pathway Protects Against α-Synuclein-Induced Neuroinflammation and Dopaminergic Neurodegeneration. *J. Neurosci.* **2016**, *36*, 5144–5159. [CrossRef] [PubMed]

72. Huang, Z.; Wang, Y.; Hu, G.; Zhou, J.; Mei, L.; Xiong, W.C. YAP Is a Critical Inducer of SOCS3, Preventing Reactive Astrogliosis. *Cereb. Cortex* **2016**, *26*, 2299–2310. [CrossRef] [PubMed]
73. Piccolo, S.; Dupont, S.; Cordenonsi, M. The biology of YAP/TAZ: hippo signaling and beyond. *Physiol. Rev.* **2014**, *94*, 1287–1312. [CrossRef] [PubMed]
74. Blunt, J.W.; Carroll, A.R.; Copp, B.R.; Davis, R.A.; Keyzers, R.A.; Prinsep, M.R. Marine natural products. *Nat. Prod. Rep.* **2018**, *35*, 8–53. [CrossRef]
75. Correa, I.; Drews, P.; Botelho, S.; de Souza, M.S.; Tavano, V.M. Deep Learning for Microalgae Classification. In Proceedings of the 16th IEEE International Conference on Machine Learning and Applications (ICMLA), Cancun, Mexico, 18–21 December 2017. [CrossRef]
76. Peng, Y.; Hu, J.; Yang, B.; Lin, X.-P.; Zhou, X.-F.; Yang, X.-W.; Liu, Y. *Chemical composition of seaweeds, Seaweed Sustainability: Food and Non-Food Applications*; Academic Press: Salt Lake City, USA, 2015; pp. 79–124.
77. Goo, B.G.; Baek, G.; Choi, D.J.; Park, Y.I.; Synytsya, A.; Bleha, R.; Seong, D.H.; Lee, C.G.; Park, J.K. Characterization of a renewable extracellular polysaccharide from defatted microalgae Dunaliella tertiolecta. *Bioresour. Technol.* **2013**, *129*, 343–350. [CrossRef] [PubMed]
78. Kurniawati, H.A.; Ismadji, S.; Liu, J.C. Microalgae harvesting by flotation using natural saponin and chitosan. *Bioresour. Technol.* **2014**, *166*, 429–434. [CrossRef]
79. Zhou, R.; Shi, X.Y.; Bi, D.C.; Fang, W.S.; Wei, G.B.; Xu, X. Alginate-Derived Oligosaccharide Inhibits Neuroinflammation and Promotes Microglial Phagocytosis of β-Amyloid. *Mar. Drugs* **2015**, *13*, 5828–5846. [CrossRef]
80. O'Sullivan, L.; Murphy, B.; McLoughlin, P.; Duggan, P.; Lawlor, P.G.; Hughes, H.; Gardiner, G.E. Prebiotics from marine macroalgae for human and animal health applications. *Mar. Drugs* **2010**, *8*, 2038–2064. [CrossRef]
81. Rea, K.; Dinan, T.G.; Cryan, J.F. The microbiome: A key regulator of stress and neuroinflammation. *Neurobiol. Stress* **2016**, *4*, 23–33. [CrossRef]
82. Wang, H.D.; Chen, C.C.; Huynh, P.; Chang, J.S. Exploring the potential of using algae in cosmetics. *Bioresour. Technol.* **2015**, *184*, 355–362. [CrossRef]
83. Yun, E.J.; Choi, I.G.; Kim, K.H. Red macroalgae as a sustainable resource for bio-based products. *Trends Biotechnol.* **2015**, *33*, 247–249. [CrossRef] [PubMed]
84. Chen, H.; Yan, X.; Zhu, P.; Lin, J. Antioxidant activity and hepatoprotective potential of agaro-oligosaccharides in vitro and in vivo. *Nutr. J.* **2006**, *5*, 31. [CrossRef] [PubMed]
85. Enoki, T.; Okuda, S.; Kudo, Y.; Takashima, F.; Sagawa, H.; Kato, I. Oligosaccharides from agar inhibit pro-inflammatory mediator release by inducing heme oxygenase 1. *Biosci. Biotechnol. Biochem.* **2010**, *74*, 766–770. [CrossRef] [PubMed]
86. Bin, B.H.; Kim, S.T.; Bhin, J.; Lee, T.R.; Cho, E.G. The Development of Sugar-Based Anti-Melanogenic Agents. *Int. J. Mol. Sci* **2016**, *17*, 583. [CrossRef] [PubMed]
87. Fernando, I.P.S.; Sanjeewa, K.K.A.; Samarakoon, K.W.; Lee, W.W.; Kim, H.S.; Kang, N.; Ranasinghe, P.; Lee, H.S.; Jeon, Y.J. A fucoidan fraction purified from Chnoospora minima; a potential inhibitor of LPS-induced inflammatory responses. *Int. J. Biol Macromol.* **2017**, *104*, 1185–1193. [CrossRef] [PubMed]
88. Raposo, M.F.; de Morais, R.M.; Bernardo de Morais, A.M. Bioactivity and applications of sulphated polysaccharides from marine microalgae. *Mar. Drugs* **2013**, *11*, 233–252. [CrossRef] [PubMed]
89. Ho, S.H.; Chen, C.Y.; Chang, J.S. Effect of light intensity and nitrogen starvation on CO2 fixation and lipid/carbohydrate production of an indigenous microalga Scenedesmus obliquus CNW-N. *Bioresour. Technol.* **2012**, *113*, 244–252. [CrossRef]
90. Cheng, Y.-S.; Zheng, Y.; Labavitch, J.M.; VanderGheynst, J.S. The impact of cell wall carbohydrate composition on the chitosan flocculation of Chlorella. *Process. Biochem.* **2011**, *46*, 1927–1933. [CrossRef]
91. Kumari, P.; Kumar, M.; Reddy, C.R.K. *Algal lipids, fatty acids and sterols. Functional ingredients from algae for foods and nutraceuticals*; Woodhead Publishing: Cambridge, UK, 2013; pp. 87–134.
92. Manuelli, M.; Della Guardia, L.; Cena, H. Enriching Diet with n-3 PUFAs to Help Prevent Cardiovascular Diseases in Healthy Adults: Results from Clinical Trials. *Int. J. Mol. Sci* **2017**, *18*, 1552. [CrossRef]
93. Layé, S.; Nadjar, A.; Joffre, C.; Bazinet, R.P. Anti-Inflammatory Effects of Omega-3 Fatty Acids in the Brain: Physiological Mechanisms and Relevance to Pharmacology. *Pharm. Rev.* **2018**, *70*, 12–38. [CrossRef]
94. Van Ginneken, V.J.; Helsper, J.P.; de Visser, W.; van Keulen, H.; Brandenburg, W.A. Polyunsaturated fatty acids in various macroalgal species from North Atlantic and tropical seas. *Lipids Health Dis.* **2011**, *10*, 104. [CrossRef] [PubMed]

95. Santos, M.A.Z.; Colepicolo, P.; Pupo, D.; Fujii, M.T.; de Pereira, C.M.P.; Mesko, M.F. Antarctic red macroalgae: a source of polyunsaturated fatty acids. *J. Appl. Phycol.* **2017**, *29*, 759–767. [CrossRef]
96. Jung, H.A.; Jin, S.E.; Ahn, B.R.; Lee, C.M.; Choi, J.S. Anti-inflammatory activity of edible brown alga Eisenia bicyclis and its constituents fucosterol and phlorotannins in LPS-stimulated RAW264.7 macrophages. *Food Chem. Toxicol.* **2013**, *59*, 199–206. [CrossRef] [PubMed]
97. Sun, Y.; Lin, Y.; Cao, X.; Xiang, L.; Qi, J. Sterols from Mytilidae show anti-aging and neuroprotective effects via anti-oxidative activity. *Int. J. Mol. Sci* **2014**, *15*, 21660–21673. [CrossRef] [PubMed]
98. Chen, H.; Han, C.; Wu, J.; Liu, X.; Zhan, Y.; Chen, J.; Chen, Y.; Gu, R.; Zhang, L.; Chen, S.; et al. Accessible Method for the Development of Novel Sterol Analogues with Dipeptide-like Side Chains That Act as Neuroinflammation Inhibitors. *ACS Chem. Neurosci.* **2016**, *7*, 305–315. [CrossRef] [PubMed]
99. Harnedy, P.A.; FitzGerald, R.J. BIOACTIVE PROTEINS, PEPTIDES, AND AMINO ACIDS FROM MACROALGAE(1). *J. Phycol.* **2011**, *47*, 218–232. [CrossRef]
100. Aydın, A.F.; Çoban, J.; Doğan-Ekici, I.; Betül-Kalaz, E.; Doğru-Abbasoğlu, S.; Uysal, M. Carnosine and taurine treatments diminished brain oxidative stress and apoptosis in D-galactose aging model. *Metab Brain Dis.* **2016**, *31*, 337–345. [CrossRef]
101. Korhonen, H.; Pihlanto, A. Bioactive peptides: production and functionality. *Int. Dairy J.* **2006**, *16*, 945–960. [CrossRef]
102. Vanderlei, E.S.; Patoilo, K.K.; Lima, N.A.; Lima, A.P.; Rodrigues, J.A.; Silva, L.M.; Lima, M.E.; Lima, V.; Benevides, N.M. Antinociceptive and anti-inflammatory activities of lectin from the marine green alga Caulerpa cupressoides. *Int. Immunopharmacol.* **2010**, *10*, 1113–1118. [CrossRef]
103. Silva, L.M.; Lima, V.; Holanda, M.L.; Pinheiro, P.G.; Rodrigues, J.A.; Lima, M.E.; Benevides, N.M. Antinociceptive and anti-inflammatory activities of lectin from marine red alga Pterocladiella capillacea. *Biol. Pharm Bull.* **2010**, *33*, 830–835. [CrossRef]
104. Coelho, L.C.; Silva, P.M.; Lima, V.L.; Pontual, E.V.; Paiva, P.M.; Napoleão, T.H.; Correia, M.T. Lectins, Interconnecting Proteins with Biotechnological/Pharmacological and Therapeutic Applications. *Evid Based Complement. Altern. Med.* **2017**, *2017*, 1594074. [CrossRef]
105. Fan, X.; Bai, L.; Zhu, L.; Yang, L.; Zhang, X. Marine algae-derived bioactive peptides for human nutrition and health. *J. Agric. Food Chem.* **2014**, *62*, 9211–9222. [CrossRef] [PubMed]
106. Eom, S.H.; Kim, Y.M.; Kim, S.K. Antimicrobial effect of phlorotannins from marine brown algae. *Food Chem. Toxicol* **2012**, *50*, 3251–3255. [CrossRef] [PubMed]
107. Corona, G.; Coman, M.M.; Guo, Y.; Hotchkiss, S.; Gill, C.; Yaqoob, P.; Spencer, J.P.E.; Rowland, I. Effect of simulated gastrointestinal digestion and fermentation on polyphenolic content and bioactivity of brown seaweed phlorotannin-rich extracts. *Mol. Nutr. Food Res.* **2017**, *61*. [CrossRef] [PubMed]
108. Montero, L.; del Pilar Sánchez-Camargo, A.; Ibáñez, E.; Gilbert-López, B. Phenolic Compounds from Edible Algae: Bioactivity and Health Benefits. *Curr. Med. Chem.* **2018**, *25*, 4808–4826. [CrossRef] [PubMed]
109. Sivagnanam, S.P.; Yin, S.; Choi, J.H.; Park, Y.B.; Woo, H.C.; Chun, B.S. Biological Properties of Fucoxanthin in Oil Recovered from Two Brown Seaweeds Using Supercritical CO_2 Extraction. *Mar. Drugs* **2015**, *13*, 3422–3442. [CrossRef] [PubMed]
110. Mikami, K.; Hosokawa, M. Biosynthetic pathway and health benefits of fucoxanthin, an algae-specific xanthophyll in brown seaweeds. *Int. J. Mol. Sci.* **2013**, *14*, 13763–13781. [CrossRef] [PubMed]
111. Peng, J.; Yuan, J.P.; Wu, C.F.; Wang, J.H. Fucoxanthin, a marine carotenoid present in brown seaweeds and diatoms: metabolism and bioactivities relevant to human health. *Mar. Drugs* **2011**, *9*, 1806–1828. [CrossRef] [PubMed]
112. Gateau, H.; Solymosi, K.; Marchand, J.; Schoefs, B. Carotenoids of Microalgae Used in Food Industry and Medicine. *Mini. Rev. Med. Chem.* **2017**, *17*, 1140–1172. [CrossRef]
113. Eggersdorfer, M.; Wyss, A. Carotenoids in human nutrition and health. *Arch. Biochem. Biophys* **2018**, *652*, 18–26. [CrossRef]
114. Pangestuti, R.; Vo, T.S.; Ngo, D.H.; Kim, S.K. Fucoxanthin ameliorates inflammation and oxidative reponses in microglia. *J. Agric. Food Chem.* **2013**, *61*, 3876–3883. [CrossRef]
115. Kim, Y.H.; Koh, H.K.; Kim, D.S. Down-regulation of IL-6 production by astaxanthin via ERK-, MSK-, and NF-κB-mediated signals in activated microglia. *Int. Immunopharmacol.* **2010**, *10*, 1560–1572. [CrossRef] [PubMed]

116. Gaysinski, M.; Ortalo-Magné, A.; Thomas, O.P.; Culioli, G. Extraction, Purification, and NMR Analysis of Terpenes from Brown Algae. *Methods Mol. Biol.* **2015**, *1308*, 207–223. [CrossRef] [PubMed]
117. Yang, P.; Liu, D.Q.; Liang, T.J.; Li, J.; Zhang, H.Y.; Liu, A.H.; Guo, Y.W.; Mao, S.C. Bioactive constituents from the green alga Caulerpa racemosa. *Bioorg. Med. Chem.* **2015**, *23*, 38–45. [CrossRef] [PubMed]
118. Kirkley, K.S.; Popichak, K.A.; Afzali, M.F.; Legare, M.E.; Tjalkens, R.B. Microglia amplify inflammatory activation of astrocytes in manganese neurotoxicity. *J. Neuroinflammation* **2017**, *14*, 99. [CrossRef] [PubMed]
119. Heneka, M.T.; Carson, M.J.; El Khoury, J.; Landreth, G.E.; Brosseron, F.; Feinstein, D.L.; Jacobs, A.H.; Wyss-Coray, T.; Vitorica, J.; Ransohoff, R.M.; et al. Neuroinflammation in Alzheimer's disease. *Lancet Neurol.* **2015**, *14*, 388–405. [CrossRef]
120. Pangestuti, R.; Kim, S.K. Neuroprotective effects of marine algae. *Mar. Drugs* **2011**, *9*, 803–818. [CrossRef]
121. Lee, J.Y.; Lee, M.S.; Choi, H.J.; Choi, J.W.; Shin, T.; Woo, H.C.; Kim, J.I.; Kim, H.R. Hexane fraction from Laminaria japonica exerts anti-inflammatory effects on lipopolysaccharide-stimulated RAW 264.7 macrophages via inhibiting NF-kappaB pathway. *Eur. J. Nutr.* **2013**, *52*, 409–421. [CrossRef]
122. Khan, M.N.; Lee, M.C.; Kang, J.Y.; Park, N.G.; Fujii, H.; Hong, Y.K. Effects of the brown seaweed Undaria pinnatifida on erythematous inflammation assessed using digital photo analysis. *Phytother. Res.* **2008**, *22*, 634–639. [CrossRef]
123. Kang, J.Y.; Khan, M.N.; Park, N.H.; Cho, J.Y.; Lee, M.C.; Fujii, H.; Hong, Y.K. Antipyretic, analgesic, and anti-inflammatory activities of the seaweed Sargassum fulvellum and Sargassum thunbergii in mice. *J. Ethnopharmacol.* **2008**, *116*, 187–190. [CrossRef]
124. Kim, S.; Kim, J.I.; Choi, J.W.; Kim, M.; Yoon, N.Y.; Choi, C.G.; Choi, J.S.; Kim, H.R. Anti-inflammatory effect of hexane fraction from Myagropsis myagroides ethanolic extract in lipopolysaccharide-stimulated BV-2 microglial cells. *J. Pharm. Pharm.* **2013**, *65*, 895–906. [CrossRef] [PubMed]
125. Kim, S.; Lee, M.S.; Lee, B.; Gwon, W.G.; Joung, E.J.; Yoon, N.Y.; Kim, H.R. Anti-inflammatory effects of sargachromenol-rich ethanolic extract of Myagropsis myagroides on lipopolysaccharide-stimulated BV-2 cells. *Bmc Complement. Altern Med.* **2014**, *14*, 231. [CrossRef]
126. Aisya, G.S.; Ching, T.S.; Yee, G.S. Antioxidative, Anticholinesterase and Anti-Neuroinflammatory Properties of Malaysian Brown and Green Seaweeds. *Int. J. Ind. Manuf. Eng.* **2014**, *8*, 895–906.
127. Craig, L.A.; Hong, N.S.; McDonald, R.J. Revisiting the cholinergic hypothesis in the development of Alzheimer's disease. *Neurosci. Biobehav. Rev.* **2011**, *35*, 1397–1409. [CrossRef] [PubMed]
128. Hoong, W.C.; John, I.; Yee, G.S.; Ling, C.E.W.; Ching, T.S.; Moi, P.S.; Aisya, G.S.; Tiong, Y.; Irvine, G.A. Fucosterol inhibits the cholinesterase activities and reduces the release of pro-inflammatory mediators in lipopolysaccharide and amyloid-induced microglial cells. *J. Appl. Phycol.* **2018**, *30*, 3261–3270.
129. Jung, W.K.; Heo, S.J.; Jeon, Y.J.; Lee, C.M.; Park, Y.M.; Byun, H.G.; Choi, Y.H.; Park, S.G.; Choi, I.W. Inhibitory effects and molecular mechanism of dieckol isolated from marine brown alga on COX-2 and iNOS in microglial cells. *J. Agric. Food Chem.* **2009**, *57*, 4439–4446. [CrossRef] [PubMed]
130. Lee, S.; Youn, K.; Kim, D.H.; Ahn, M.R.; Yoon, E.; Kim, O.Y.; Jun, M. Anti-Neuroinflammatory Property of Phlorotannins from Ecklonia cava on Aβ25-35-Induced Damage in PC12 Cells. *Mar. Drugs* **2019**, *17*, 7. [CrossRef] [PubMed]
131. Kim, M.; Li, Y.X.; Dewapriya, P.; Ryu, B.; Kim, S.K. Floridoside suppresses pro-inflammatory responses by blocking MAPK signaling in activated microglia. *BMB Rep.* **2013**, *46*, 398–403. [CrossRef]
132. Heo, S.J.; Yoon, W.J.; Kim, K.N.; Ahn, G.N.; Kang, S.M.; Kang, D.H.; Affan, A.; Oh, C.; Jung, W.K.; Jeon, Y.J. Evaluation of anti-inflammatory effect of fucoxanthin isolated from brown algae in lipopolysaccharide-stimulated RAW 264.7 macrophages. *Food Chem. Toxicol.* **2010**, *48*, 2045–2051. [CrossRef]
133. Lin, T.Y.; Lu, C.W.; Wang, S.J. Astaxanthin inhibits glutamate release in rat cerebral cortex nerve terminals via suppression of voltage-dependent Ca^{2+} entry and mitogen-activated protein kinase signaling pathway. *J. Agric. Food Chem.* **2010**, *58*, 8271–8278. [CrossRef]
134. Wu, H.; Niu, H.; Shao, A.; Wu, C.; Dixon, B.J.; Zhang, J.; Yang, S.; Wang, Y. Astaxanthin as a Potential Neuroprotective Agent for Neurological Diseases. *Mar. Drugs* **2015**, *13*, 5750–5766. [CrossRef] [PubMed]
135. Wu, D.; Xu, H.; Chen, J.; Zhang, L. Effects of Astaxanthin Supplementation on Oxidative Stress. *Int. J. Vitam. Nutr. Res.* **2019**, 1–16. [CrossRef] [PubMed]
136. Balietti, M.; Giannubilo, S.R.; Giorgetti, B.; Solazzi, M.; Turi, A.; Casoli, T.; Ciavattini, A.; Fattorettia, P. The effect of astaxanthin on the aging rat brain: gender-related differences in modulating inflammation. *J. Sci. Food Agric.* **2016**, *96*, 4295. [CrossRef] [PubMed]

137. Krogh, J.; Benros, M.E.; Jørgensen, M.B.; Vesterager, L.; Elfving, B.; Nordentoft, M. The association between depressive symptoms, cognitive function, and inflammation in major depression. *Brain Behav. Immun.* **2014**, *35*, 70–76. [CrossRef] [PubMed]
138. Miller, A.H.; Raison, C.L. The role of inflammation in depression: from evolutionary imperative to modern treatment target. *Nat. Rev. Immunol.* **2015**, *16*, 22–34. [CrossRef] [PubMed]
139. Zhang, X.S.; Zhang, X.; Wu, Q.; Li, W.; Wang, C.X.; Xie, G.B.; Zhou, X.M.; Shi, J.X.; Zhou, M.L. Astaxanthin offers neuroprotection and reduces neuroinflammation in experimental subarachnoid hemorrhage. *J. Surg. Res.* **2014**, *192*, 206–213. [CrossRef] [PubMed]
140. Lee, J.C.; Hou, M.F.; Huang, H.W.; Chang, F.R.; Yeh, C.C.; Tang, J.Y.; Chang, H.W. Marine algal natural products with anti-oxidative, anti-inflammatory, and anti-cancer properties. *Cancer Cell Int.* **2013**, *13*, 55. [CrossRef]
141. Choi, J.Y.; Hwang, C.J.; Lee, H.P.; Kim, H.S.; Han, S.B.; Hong, J.T. Inhibitory effect of ethanol extract of Nannochloropsis oceanica on lipopolysaccharide-induced neuroinflammation, oxidative stress, amyloidogenesis and memory impairment. *Oncotarget* **2017**, *8*, 45517–45530. [CrossRef]
142. Xiang, Y.; Meng, S.; Wang, J.; Li, S.; Liu, J.; Li, H.; Li, T.; Song, W.; Zhou, W. Two novel DNA motifs are essential for BACE1 gene transcription. *Sci Rep.* **2014**, *4*, 6864. [CrossRef]
143. Smith, A.M.; Gibbons, H.M.; Dragunow, M. Valproic acid enhances microglial phagocytosis of amyloid-beta(1-42). *Neuroscience* **2010**, *169*, 505–515. [CrossRef]
144. Guo, X.; Xin, X.; Gan, L.; Nie, Q.; Geng, M. Determination of the accessibility of acidic oligosaccharide sugar chain to blood-brain barrier using surface plasmon resonance. *Biol. Pharm. Bull.* **2006**, *29*, 60–63. [CrossRef] [PubMed]
145. Cui, Y.Q.; Jia, Y.J.; Zhang, T.; Zhang, Q.B.; Wang, X.M. Fucoidan protects against lipopolysaccharide-induced rat neuronal damage and inhibits the production of proinflammatory mediators in primary microglia. *CNS. Neurosci.* **2012**, *18*, 827–833. [CrossRef] [PubMed]
146. Rafiquzzaman, S.M.; Kim, E.Y.; Lee, J.M.; Mohibbullah, M.; Alam, M.B.; Moon, I.S.; Kim, J.-M.; Kong, I.-S. Anti-Alzheimers and anti-inflammatory activities of a glycoprotein purified from the edible brown alga Undaria pinnatifida. *Food Res. Int.* **2015**, *77*, 118–124.
147. Ryan, S.; O'Gorman, D.M.; Nolan, Y.M. Evidence that the marine-derived multi-mineral Aquamin has anti-inflammatory effects on cortical glial-enriched cultures. *Phytother. Res.* **2011**, *25*, 765–767. [CrossRef] [PubMed]

International Journal of
Molecular Sciences

MDPI

Review

Considerations for the Use of Polyphenols as Therapies in Neurodegenerative Diseases

Justine Renaud [1] and Maria-Grazia Martinoli [1,2,*]

1 Cellular Neurobiology, Department of Medical Biology, Université du Québec, Trois-Rivières, Québec, QC G9A5H7, Canada; justine.renaud@uqtr.ca
2 Department of Psychiatry & Neuroscience, Université Laval and CHU Research Center, Ste-Foy, QC G1V 4G2, Canada
* Correspondence: maria-grazia.martinoli@uqtr.ca; Tel.: +1-819-376-5011 (ext. 3994)

Received: 19 March 2019; Accepted: 12 April 2019; Published: 16 April 2019

Abstract: Over the last two decades, the increase in the incidence of neurodegenerative diseases due to the increasingly ageing population has resulted in a major social and economic burden. At present, a large body of literature supports the potential use of functional nutrients, which exhibit potential neuroprotective properties to mitigate these diseases. Among the most studied dietary molecules, polyphenols stand out because of their multiple and often overlapping reported modes of action. However, ambiguity still exists as to the significance of their influence on human health. This review discusses the characteristics and functions of polyphenols that shape their potential therapeutic actions in neurodegenerative diseases while the less-explored gaps in knowledge of these nutrients will also be highlighted.

Keywords: neurodegeneration; neuroprotection; nutraceuticals; bioavailability; stress response

1. Introduction

It is widely acknowledged that nutrition plays a key role in the occurrence and progression of non-communicable diseases. A body of epidemiological evidence shows that a diet rich in fruit and vegetables reduces the incidence of cardiovascular diseases [1–4], type 2 diabetes [5,6], stroke [7,8] and numerous cancers [9–11]. Other studies find an inverse association between the consumption of green tea and cognitive decline [12,13]. These observed health benefits are thought to be at least partly attributable to a class of non-essential nutrients named polyphenols, found abundantly in fruits and vegetables [14,15].

Together with cancer and cardiovascular diseases, neurodegenerative disorders constitute a potential application for the benefits of polyphenols [16,17]. This includes Parkinson's and Alzheimer's diseases which lack clear etiopathogenetic origins and arise from the interaction between aging, environment and genetic risk factors. The etiology of these diseases is further complicated by a number of proposed causative mechanisms, including oxidative stress, neuroinflammation, protein aggregation, iron toxicity and mitochondrial dysfunction. Polyphenols are reported to improve many of these factors at a cellular level, which makes their use in complex neurodegenerative disorders compelling. In this review, the properties that may influence the functionality and bioavailability of dietary polyphenols in the central nervous system (CNS) are discussed with a particular focus on therapeutic applications and limitations.

2. Chemico-Structural Characteristics

2.1. Classification

Plant polyphenols were originally classified in the early literature as "vegetable tannins" owing to their tanning action on animal skins [18]. The first comprehensive description, referred to as the White–Bate-Smith–Swain–Haslam (WBSSH) definition, recommended that the term polyphenol be exclusively used to describe water-soluble phenolic compounds having a molecular mass ranging between 500 to 4000 Da, possessing at least 12 phenolic hydroxyl groups and 5 to 7 aromatic rings per 1000 Da [19]. A less restrictive interpretation was proposed offering a broader view of the WBSSH definition to include simpler phenolic compounds with potential biological activities others than tanning [20]:

"The term "polyphenol" should be used to define compounds exclusively derived from the shikimate/phenylpropanoid and/or the polyketide pathway, featuring more than one phenolic unit and deprived of nitrogen-based functions. This definition lets out all monophenolic structures as well as all their naturally occurring derivatives such as phenyl esters, methyl phenyl ethers and O-phenyl glycosides."

A majority of plant polyphenols originate from phenylalanine which is deaminated to cinnamic acid, which then enters the phenylpropanoid pathway [21]. Plant metabolism utilizes the phenylpropanoid unit C6-C3, a phenol ring with a 3-carbon side chain, as a building block to construct polyphenols. Classification of the resulting molecules is dictated by the number of phenol rings (C6) they contain and the structural elements binding these rings to one another. The main subclasses, varying in complexity, are phenolic acids (C6-C3 and C6-C1), flavonoids (C6-C3-C6), stilbenes (C6-C2-C6) and lignans (C6-C3-C3-C6). Within these subclasses, hydroxylations and O-glycosylations at various positions as well as *cis-trans* isomerization give rise to the thousands of polyphenols (estimated to be >8000) identified to date, resulting in a complex range of molecules with potential pharmacological values. Details of these polyphenols alongside their occurrence in various food products are available on databases such as Phenol-Explorer managed by the Institut National de la Recherche Agronomique (www.phenol-explorer.eu).

2.2. Structure versus Biofunctionality in Neuroprotection

The structural properties shared by polyphenols are important to their potential therapeutic applications, particularly in neuroprotection. These include the presence of phenol rings, variable hydroxylation patterns and conjugated double bonds all of which grant polyphenols metal-chelating, fibril-destabilizing, estrogen-like, enzyme-binding and antioxidative properties. These modes of action allow polyphenols to provide a defense against many pathophysiological aspects of neurodegenerative diseases, namely oxidative stress, neuroinflammation, protein aggregation, iron toxicity and mitochondrial dysfunction. These are detailed below:

The redox properties of divalent metals, such as copper, zinc and iron, are essential for cellular homeostasis. When in excess, however, these metals generate surplus reactive oxygen species. This excess can be reversed by chelation with polyphenols that possess at least one galloyl or catechol group (hydroxyl groups in the *ortho*-position) which are powerful bidentate chelators of divalent metals [22], whereas polyphenols having only a phenol substitution (one hydroxyl function) or possessing a resorcinol group (*meta*-position hydroxyl pair) are less potent monodentate chelators [23,24]. For chelation to occur, a deprotonation step of the phenolic group is necessary and has been shown to be possible at physiological pH [23].

Self-assembly of amyloidogenic fibrils including tau, beta amyloid (Aβ) and α-synuclein all neuropathologically relevant proteins involves interactions between aromatic residues [25]. Using similar aromatic interactions, as described above, phenol moieties in polyphenols can interfere with fibril assembly [26], possibly by weakening cross-β structures. This interference seems to arise from hydrophobic and π stacking interactions [27], although the formation of covalent bonds through

Schiff base reactions has also been proposed for the green tea polyphenol epigallocatechin-3-gallate (EGCG) [28,29]. Analysis of binding energies between polyphenols and protein fibrils has also shown favorable entropic and enthalpic dynamics that suggest the stabilization of H-bonds [30].

Polyphenols, referred to as phytoestrogens, have the ability to bind estrogen receptors (ERs), usually with a greater affinity for ERβ [31,32]. Depending on structure, dose, cell type and estrogen response element (ERE) sequence, different polyphenols have a weak or strong antagonistic or agonistic effect on ERs, resulting in a wide spectrum of activities in cells [33–36]. To enable binding to ERs, a structure should be composed of a phenolic ring with a configuration resembling that of estradiol, as found in flavonoid isoflavones or the stilbene resveratrol, for instance. Also, a specific hydroxylation pattern and an adequate distance between substituted hydroxyl groups are necessary to bind ERs.

Polyphenols can also share structural similarities with endogenous ligands, such as cyclic adenosine monophosphate (cAMP) or nucleoside triphosphates, endowing them with the aptitude to activate or inhibit key enzymes [37,38]. To date, the modulatory effects of several polyphenols on enzymes have been confirmed in cellular or animal models, these include resveratrol on cAMP phosphodiesterases [39], theaflavins on the adenosine triphosphate (ATP) synthase and respiratory chain [40] and curcumin on glyoxalase 1 [41]. The presence of appropriately spaced ketone and hydroxyl groups in a planar configuration, bestow some polyphenols, such as curcumin, with the ability to mimic an enediolate intermediate in physiological conditions [42] is an example of structural elements that make enzyme binding possible.

Apart from the functions described above which result from the unique chemical structures of polyphenols, the most vastly studied characteristic of this class of chemicals is their antioxidative action. Polyphenols are thought to exert their antioxidative action directly, by scavenging free radical species firsthand, and/or indirectly, by activating endogenous antioxidative pathways. Direct antioxidative effects usually occur through H-atom transfer from polyphenols' (ArOH) hydroxyl (OH) groups to the free radicals (R•):

$$\mathrm{ArOH} + \mathrm{R}\bullet \rightarrow \mathrm{ArO}\bullet + \mathrm{RH} \qquad (1)$$

The existence of multiple conjugated double bonds in polyphenols allows unpaired electron to be delocalized over the aromatic ring, yielding a much more stable and much less reactive, polyphenolic radical (ArO•) (Equation (1)). Some polyphenols also exert indirect antioxidative effects through the Kelch-like ECH-associated protein 1/nuclear factor erythroid 2-related factor 2/antioxidant response elements (Keap1/Nrf2/ARE) regulatory pathway made possible by the presence of electrophilic functions (α,β-unsaturated carbonyl group, 1,2- and 1,4-quinones or other groups) that alkylate thiol sensors in the cysteine pocket of Keap1 [43,44]. Others, like stilbenes, engage their resorcinol hydroxyl functions in hydrogen bonds with the Kelch pocket of Keap1 [45]. Both these events lead to the disruption of the Keap1/Nrf2 complex, allowing Nrf2 to translocate to the nucleus where it can trigger the expression of antioxidant proteins like heme oxygenase-1 via binding of adenylate and uridylate (AU)-rich elements (AREs). This cysteine-modifying function of polyphenols may also have implications for the activity of various other enzymes [44].

3. Factors Influencing Pharmacokinetics and Bioavailability

To be effective in the prevention or amelioration of neurodegenerative diseases, polyphenols must be bioavailable. Extensive reports on the bioavailability of the most common dietary polyphenols can be found elsewhere [46–48]. In this review, we will first discuss the obstacles that hinder polyphenol bioavailability and address CNS permeability in particular.

3.1. Food Matrix or Vehicle

Oral administration is the most usual route if polyphenols are given pharmacologically but this often conflicts with bioavailability. Particular factors include interaction with vehicle, transformations by digestive and microbial enzymes and absorption by the gastrointestinal tract [49].

Food matrices are central to the efficacy of polyphenols [50]. Few studies have been conducted and inconsistent results have been obtained, demonstrating either a negligible [51,52] or a significant [53–56] contribution of the food matrix to polyphenol absorption. Indeed, peculiar factors such as the type of lipid matrix used may mediate in the release of polyphenols in the gastrointestinal tract [57,58]. Ethanol may also play a role in polyphenols absorption with studies showing improved bioavailability of quercetin in rats when administered in 30% ethanol, an alcohol content that is unsustainable in the diet [59]. In humans administered normal or dealcoholized red wine there was no differences in plasma levels of catechin but increased catechin excretion with red wine probably due to a diuretic effect of alcohol [60]. However, matrix effects are too peculiar to be fully reviewed here.

3.2. Gastrointestinal Transformations and Absorption

Absorption and metabolism of polyphenols have extensively studied (see for review, References [61,62]). Whereas aglycones are normally well absorbed by the small intestine, nutritional polyphenols are more commonly present as glycosides, esters and polymers, which cannot be efficiently assimilated in the upper portion of the gut.

Molecules not absorbed in the upper gastrointestinal tract continue to the colon to become substrates for the gut microbiota, responsible for a very wide array of reactions, some of which yield monomers or aglycones from glycosylated polyphenols (see for review [63]). Smaller, better-absorbed phenolic acids may also be produced by the gut microbiota. For example, microbiotic degradation of quercetin mainly generates 3,4-dihydroxyphenylacetic, 3-methoxy-4-hydroxyphenylacetic (homovanillic acid) and 3-hydroxyphenylacetic acid [64]. In volunteers challenged with 75 mg of rutin, a quercetin glycoside, the total urinary excretion of microbial metabolites accounted for as much as 50% of the ingested dose [65]. Importantly, the sum of these gastrointestinal transformations and food matrix interactions can either increase or decrease the absorption of the resulting metabolites in the bloodstream.

3.3. Plasma Bioavailability, Transformations and Cellular Uptake

Once in the blood stream, enzymes in the liver and kidneys further modify polyphenols into various conjugated forms, a process that serves to detoxify potentially harmful substances. Molecules are rendered more hydrophilic in order to facilitate their urinary elimination, which usually lowers bioavailability [66,67]. While metabolites usually constitute the greatest fraction of circulating polyphenolic species, some forms undergo enterohepatic recirculation via biliary secretion, followed by deconjugation into free polyphenols by the gut microbiota and reabsorption in the colon [68–70]. Additional hepatic reactions may also occur which revert circulating metabolites back to the free form [71–73], as is the case for the conversion of resveratrol sulphate to bioactive resveratrol by sulphatases in humans [73]. Moreover, glucuronide and sulphate metabolites retain some of their beneficial effects in vitro [74,75]. Thus, chronic administration of polyphenols may be an efficient strategy to increase plasma bioavailability in humans, as reported for epigallocatechin-3-gallate (EGCG) [76].

The final step in the action of polyphenols is cellular uptake, which depends not only on how they have been metabolized but also on their interaction with circulating proteins, fatty acids and lipoproteins [77] with the bioefficacy of therapeutic agents heavily relying on binding to such serum transporters [78]. Resveratrol for example, is lipophilic which requires transformation into a more hydrophilic form, by sulphation, glucuronidation or binding to proteins enabling circulation in appropriate concentrations [79]. The formation of complexes between resveratrol and transporter proteins, principally albumin [80–82] and lipoproteins [83–86], impedes its uptake by cells [79]. Fatty acids are also known to improve the ability of resveratrol to bind transporter proteins [87].

While the binding by transporter proteins diminishes the availability of the free form of the polyphenol, it is thought to provide a polyphenol reservoir, important in the systemic distribution of bound species [77]. Some studies have proposed that these complexes are retained at the cell membrane by albumin and lipoprotein receptors, offering a carrier-mediated mechanism by which polyphenols may gain entry to cells [77] in addition to passive diffusion [79]. There is also the possibility that polyphenols need not enter cells to have an effect, as when free resveratrol binds integrin $\alpha V\beta 3$ [88] to produce an angiosuppressive effect (Belleri et al., 2008) and when it triggers p53-dependent apoptosis of breast cancer cells [89].

3.4. Accumulation in the Brain Parenchyma

Drugs targeting the brain must ultimately be able to accumulate in the brain parenchyma, in a biologically active form and in sufficient concentrations. Three important obstacles stand in the way of this: the blood-brain barrier (BBB), efflux transporters and multidrug resistance-associated proteins [90,91]. Youdim and colleagues were the first to demonstrate polyphenols crossing the BBB in an in vitro model, describing superior penetration of lipophilic (methylated conjugates) in comparison to hydrophilic molecules (sulphated or glucuronidated) [92,93]. Another study identified a stereoselective process in the passage of flavonoid catechins across the BBB [94]. Yet, the exact mechanisms polyphenols use by to traverse the BBB in vivo, either via diffusion or via transporters, remains to be elucidated.

Although information on transport of polyphenols into the brain is limited compared to the measurement of plasma levels, an increasing number of studies have measured polyphenols and metabolites in the brains of rodents and pigs [95], as reviewed elsewhere [90,96,97]. Entry into the CNS of the most commonly studied polyphenols has been reported several times, for resveratrol [67,98–101], EGCG [102,103] and quercetin [93,104–106].

However, differences in uptake are reported depending on the route of administration and the methods used for measurement. For example, in one study, orally administered tritiated resveratrol in rats (50 mg/kg b.w.) was reported to reach 1.7% of the ingested dose in the plasma and below 0.1% in the brain after 2 h [67]. Interestingly, 18 h after administration, the CNS retained 43% of the resveratrol measured at 2 h, mainly in the free form. Despite this retention in the brain, resveratrol levels, measured by high-performance liquid chromatography (HPLC) are lower than in the liver, kidney, testes and lungs [99]. However, another study was unable to detect brain resveratrol or metabolites in rats fed a 0.2% resveratrol diet for 45 days using HPLC with a detection limit of 0.5 pmol/mL/mg [107]. Other studies have also used chromatographic methods to measure resveratrol in rat brains using different protocols. In one study, 15 mg/kg b.w. of resveratrol were administered intravenously (i.v.), a relatively high dose, with brain tissue concentrations reaching ~0.17 nmol/g after 90 min [99]. Another study administered escalating oral doses of resveratrol (100–400 mg/kg b.w.) for 3 days and detected ~1.7 nmol/g in the brain by liquid chromatography-mass spectrometry [100].

Some polyphenols are extensively transformed before they reach the brain, which may dampen their bioavailability, as discussed above. As an example, curcumin is highly lipophilic and, in theory, should easily gain entry to the brain [108]. However, before reaching the BBB, the free form of curcumin is rapidly conjugated, rendering it only sparingly bioavailable to the CNS [109]. Conversely, catechins efficiently cross the BBB after oral administration but are found in glucuronidated and 3′-*O*-methyl glucuronidated forms in the brain [102,110]. To date, it remains unclear whether conjugation occurs before or after entry into the brain. Nevertheless, strategies exist to boost CNS concentrations of the aglycone form, for example by continuous administration aimed at promoting tissue accumulation [103]. Following 24 h of continuous intragastric administration, EGCG levels in the CNS reached 5–10% of concentrations measured in the plasma [103]. These results imply, however, that a very high plasma concentration is needed for EGCG to accumulate in therapeutically reasonable concentrations in the brain. The necessity of maintaining high circulating concentrations may raise questions regarding the safety and tolerability of polyphenols.

3.5. Synergistic Effects

Some polyphenols interact beneficially when administered in combination. Synergistic pharmacokinetics are at the basis of emerging multi-drug therapies [111–113] developed to surmount problems of low efficacy, acquired resistance and undesirable side effects in standalone treatments. Polyphenols synergize via multiple mechanisms, extensively reviewed elsewhere [114–116]. Although synergistic chemosensitization properties of polyphenols are well known, for example EGCG-induced downregulation of endoplasmic reticulum stress response elements rendering temozolomide treatments more efficient in a mouse model of glioma [117], what follows will concentrate solely on neuroprotective mechanisms.

Underlying the efficacy of herb and plant extracts, different polyphenols may concurrently regulate the same or separate targets in cells, resulting in a concerted agonistic effect. For instance, combinations of resveratrol and quercetin [118,119] or epicatechin and quercetin [120] synergize to protect against amyloid-like aggregation, oxidative stress and oxygen-glucose deprivation in vitro. An earlier report of synergy between polyphenols showed that treatment of neuronal PC12 cells with suboptimal doses of resveratrol in combination with catechin conferred greater protection against Aβ toxicity than the sum of their individual actions [121]. However, when measuring their free radical scavenging activities, the authors found their combined antioxidative effect to be merely additive, suggesting that their synergistic neuroprotective competences at combined subliminal doses may depend on other cellular mechanisms [121]. Very few studies have addressed neuroprotective synergy in vivo though a combination of polyphenols was found to synergistically rescue photoreceptors in an animal model of retinal degeneration [122].

Synergy can also occur between polyphenols, drugs and hormones. Many in vitro reports support this, as is the case for the potentiation of neurite outgrowth by a subeffective dose of brain-derived neurotrophic factor (BDNF) in conjunction with green tea catechins [123,124], as well as the protection of primary neurons and astrocytes by a cocktail of suboptimal doses of resveratrol and melatonin via upregulation of heme oxygenase-1 [125]. One of the first reports of polyphenol-drug synergy in rodents showed EGCG favorably interacting with rasagiline, an irreversible inhibitor of dopamine-metabolizing monoamine oxidase B (MAO-B) for the treatment of Parkinson's disease [126,127]. When administered alone in suboptimal doses, neither EGCG nor rasagiline were capable of rescuing nigrostriatal neurons in a 1,2,3,6-tetrahydropyridine (MPTP)-injured mouse model of Parkinson's disease [128]. However, in combination these agents in low doses promoted the survival of the dopaminergic nigrostriatal pathway, demonstrating their synergistic effect. Interestingly, the ability of rasagiline to promote the expression of BDNF in concert with EGCG-induced induction of protein kinase C produced a sum agonistic effect converging at their downstream effector Akt/protein kinase B, thought to account for their neuroprotective action. Other examples of polyphenol-drug synergies exist for valproate and resveratrol in ischemic stroke [129] as well as for glatiramer acetate and EGCG in experimental autoimmune encephalomyelitis [130].

Many polyphenols readily regulate absorption in the gastrointestinal tract, clearance at the level of the kidneys and detoxification in the liver by modulating the activity of transport proteins or metabolic enzymes, which may improve their own oral availability. This property has potential for use in Parkinson's disease by minimizing levodopa methylation in the liver by inhibiting human catechol-*O*-methyl transferase (COMT), thereby enhancing bioavailability of the drug [131]. Flavonoids are also known to be potent inhibitors of cytochrome P450 (CYP) enzymes [132,133] whose activity reduces polyphenol bioavailability. This potential to enhance bioavailability of metabolism-sensitive drugs constitutes a clear example of polyphenol synergy that may be relevant in human treatment.

4. Safety and Tolerability

In addition to favorable pharmacokinetics, polyphenols must be safe and well-tolerated in humans. Several investigations have already addressed safety and tolerability issues (see for review [134–137]). What follows is a summary of these findings.

4.1. *Side Effects from Dosage and Chronicity*

Virtually all investigations performed in humans using a wide array of polyphenol preparations found that they are safe and tolerable in the short- [138,139], medium- [46,140] and long-term [141–143]. Generally, side effects are uncommon and are mild and transient and include minor gastrointestinal problems and, more rarely, headaches, dizziness and rashes. In a phase II trial, 24 Alzheimer's patients were administered 2 or 4 g of curcuminoids daily for 48 weeks and 3 withdrew due to minor gastrointestinal issues [143]. A study using a single 5 g/70 kg b.w. intake of resveratrol, representing 1/40 of the nephrotoxic dose and 1/4 of the highest dose reported to be safe in rats [144], did not show any serious adverse effects [138]. A great number of investigations have also addressed the safety of specific diets enriched in polyphenol-rich foods. Of particular interest, black cohosh, soy and red clover regimens aimed at reducing menopausal symptoms in women have proven to be safe, with occasional mild gastrointestinal issues, musculoskeletal and connective tissue troubles, as well as weight gain (see for review, Reference [134]).

4.2. *Adverse Pharmacological Interactions*

While a consensus has been reached on the safety and tolerability of polyphenols in most individuals, certain contexts preclude their use. Grapefruit juice is an example of the possible effects of polyphenols under specific conditions. Apigenin, naringenin, nobiletin and hesperetin in grapefruit juice potently inhibit the detoxifying enzymes, members of the CYP family, responsible for the metabolism of several prescription drugs [132,145–148]. Interestingly, enzymatic inhibition is apparently irreversible following the ingestion of 200–300 mL of juice, leading to increased drug bioavailability and toxicity for up to 24 h after intake. Medical professionals are now mindful of the risks of consuming grapefruit juice in individuals already taking antidepressants such as buspirone (Buspar) and sertraline (Zoloft), beta-blockers, anti-cancer agents, fexofenadine (Allegra) or certain statins (atorvastatin) among other drugs [149–152]. Several other adverse interactions exist between polyphenols and drugs [153,154] and have been extensively discussed elsewhere [136].

4.3. *Tumorigenicity*

As previously discussed, certain polyphenols, termed phytoestrogens, are biofunctional due to their resemblance to steroid hormones. Members of the flavonoid and stilbene subclasses indeed possess the capacity to bind ERs [155] and testosterone receptors [156], albeit with much lower affinities than endogenous ligands. Many studies find phytoestrogens to be safe with respect to incidences of cancers [157,158] and support their role in inhibiting aberrant cell proliferation [159–165]. Nevertheless, a few publications draw attention to the possible carcinogenic actions of some phytoestrogens that should not be ignored [166]. In particular, soy genistein and daidzein (0.001–10 μM) may stimulate the growth of malignant breast tumors, both in vitro and in vivo [166,167].

In the case of the stilbene resveratrol, studies confirm its ability to bind both ERs [168], however with 7000 times less affinity than estradiol [33]. Interestingly, its effects are apparent for select EREs regulated by ERα but not for EREs dependent on ERβ activation. Unlike other ERα agonists, resveratrol does not appear to provoke mammary or uterine tissue proliferation in rats [169] and even promotes neuronal differentiation in vitro [170]. In light of this, resveratrol's favorable effects may in fact partially hinge on tissue-specific expression profiles of ERα and ERβ [171]. More recently, a study delineated the discriminatory ability of resveratrol to impede inflammation without promoting cell proliferation through pathway-selective ERα activation [172]. Crystallographic studies of the ligand-binding domain revealed resveratrol to bind in the opposite orientation to estradiol, which may be at the core of its pathway selectivity and its proven safety in humans [135], particularly with regard to carcinogenesis.

5. Clinical Progress

The therapeutic potential of polyphenols is clear from the overwhelming body of literature supporting their beneficial effects in countless preclinical disease settings (see for review [16,17]). Notwithstanding the weight of epidemiological, anecdotal and fundamental evidence, translation from bench-to-bedside has proven challenging despite relentless efforts to test polyphenols in human trials (see [90] for a review). Currently, only a single trial looking at polyphenols in neurodegenerative disease has reached phase III clinical testing [173]. In this randomized, double-blind, placebo-controlled parallel group study, disease progression will be assessed after 48 weeks of daily oral EGCG treatments in multiple system atrophy patients.

The example of a standardized Ginkgo biloba extract, rich in flavonoids, yielded particularly disappointing results with numerous failed phase I trials [106,174–176]. These studies addressed dementia prevention in large cohorts of healthy or mildly cognitively impaired elderly individuals administered oral Ginkgo biloba twice daily for several years [177] but no reduction in the incidence of cognitive decline or Alzheimer's disease was found [178–182]. Other phase I and II clinical attempts have also been unsuccessful in confirming the putative positive effects of curcumin in Alzheimer's disease patients [143,183]. The reasons behind these results may be due to preclinical models failing to fulfill their predictive purpose or clinical trials may simply be incapable of detecting the beneficial effects of polyphenols due to a flawed approach. What is important to keep in mind is that successful clinical trials are not common, on account of the inherent difficulty of translating applications between rodents and humans.

To address this, the required recruitment profile for testing Ginkgo biloba extracts was re-evaluated, yielding positive results in a new round of clinical trials, this time performed in full-blown Alzheimer's disease and vascular dementia. These trials successfully uncovered the benefits of several months of a daily Ginkgo biloba treatment on cognition and neuropsychiatric symptoms [141,142]. Changing the endpoints and focusing on prefrontal dopaminergic functions in elderly humans with self-reported mild cognitive decline was another fruitful strategy to reveal the beneficial effects of Ginkgo biloba [184]. Nevertheless, the cholinesterase inhibitor rivastigmine, commercially known as Exelon, has been shown to be more efficient than Ginkgo biloba in treating Alzheimer's disease and remains the drug of choice to ameliorate cognitive impairment in mild to moderate forms of the disease [185].

Several other phase I trials have been successful in confirming small positive effects in healthy individuals. A variety of polyphenols, including resveratrol, were found to increase cerebral blood flow without, however, improving cognitive performances in young adults, whether administered in a single dose [186–188] or chronically over 28 days [189]. However, other groups found that longer chronic interventions in elderly humans using either cocoa flavanols or resveratrol enhanced dentate gyrus-related cognitive functions [190] and hippocampal-related memory functions [191], respectively. In Alzheimer's disease patients, resveratrol reached phase II trials on the basis of its modulatory role on neuroinflammation, cognitive decline and cerebrospinal fluid (CSF) levels of Aβ40 [192,193]. Following a twice-daily oral regime for one year, resveratrol and its metabolites were present in the CSF, validating its ability to cross the BBB in humans [192]. Despite its relatively low bioavailability, resveratrol remains a candidate for potential use in human neurodegenerative diseases.

6. Future Strategies for Pharmaceutical Development for Neuroprotection

Polyphenols have interesting properties that justify efforts to translate their potential neuroprotective effects into treatment for human neurodegenerative diseases. However, their questionable bioavailability, modest effects in humans and the impossibility of applying patent protection on natural molecules detracts from the appeal of polyphenols for pharmaceutical use. Nevertheless, several strategies have been used by drug development in recent years to tackle these issues.

6.1. Alternative Preparations and Prodrug Approches

The engineering of novel structural analogues inspired by existing polyphenols or formulating specific preparations of polyphenols, such as the well-defined Ginkgo Biloba extract 761, may be patentable options. Among the latest innovations, chemical engineering of pro-drug polyphenolic structures has shown promising results. For instance, acetylation of EGCG or resveratrol via esterification of their hydroxyl moieties yields stable pro-drugs in vivo whose acetyl groups can be hydrolyzed intracellularly by esterases to release the free polyphenol within the cell [194–196]. This strategy minimizes polyphenol auto-oxidation and allows better lipophilicity-dependent cellular uptake [197–199]. Production of conjugates with improved bioefficacy has also been a good approach to promote polyphenols absorption and activity. For example, the glutamoyl diester of curcumin is a more potent neuroprotective agent than curcumin [200] and similar approaches have been deployed for resveratrol [201,202]. More importantly, prodrugs of resveratrol are promising as recently reviewed in Biasutto et al. [203], for delivery to the brain parenchyma.

6.2. Alternative Drug Delivery Systems

Another favorable approach is the development of novel encapsulation technologies. Progress in vehicle formulation has allowed polyphenols to be contained in lipid nanocapsules [204–206], nanoparticles [206,207], exosomes [208], nanocomposites [209], emulsified formulations [206,210,211] or in gel form [212]. Several reports demonstrate increased bioavailability for encapsulated polyphenols in rodents [213,214]. Another unusual approach is the administration of biologically compatible carbon nanotubes [215] grafted with polyphenols, such as gallic acid [216]. This method was shown to enhance the antioxidative properties of grafted agents [216] and to improve their ability to traverse biological barriers [215,217], although the application of such conjugates is still not common, and the outcomes have not been sufficiently addressed. Possible health concerns of using carbon nanotubes also warrant further investigations [215,218]. Another simple tactic consists in improving solubility of polyphenols in circulation, such as for the lipophilic resveratrol [219], via coupling to cyclodextrins, which have the capacity to form inclusion complexes and this approach has already been exploited in other drug delivery strategies [220]. Overall, each of these methods has advantages and disadvantages but brain accessibility is generally augmented owing to improved BBB infiltration by lipophilic vehicles, brain targeting by encapsulation and blocking the metabolism of polyphenols [221].

6.3. Alternative Administration Routes

In order to target the human brain more efficiently, the route of administration is another variable that can be altered. The most promising of these is intranasal administration, usually paired with one of the previously described encapsulation techniques, which has proved successful for brain-targeted drugs in humans, at least for increased bioavailability and the avoidance peripheral side effects [222,223]. Notable examples are the administration of insulin for the treatment of Alzheimer's disease [224] and apomorphine for the treatment of Parkinson's disease [225]. The mechanisms by which drugs can be delivered to the brain parenchyma are only beginning to be explored. It would appear that drugs administered nasally either enter the brain through retrograde axonal transport at the level of the olfactory sensory cells or by penetration into the CSF across the nasal epithelium [226]. Although studies with polyphenols are scarce in preclinical models [227–229], intranasal curcumin administration has gained attention (see for review [230]) due to its very poor oral bioavailability [231] but promising neuroprotective actions. Curcumin is highly lipophilic and may easily cross the BBB [108] if it is delivered into the bloodstream and protected from enzymatic modifications [232]. While it is generally recognized as a safe route, intranasal administration sometimes leads to minor adverse effects, principally nasal irritation, constituting a potential problem in the development of intranasal polyphenol administration [225,233]. More unusual administration systems for polyphenols include rectal suppositories for efficient systemic distribution, bone-marrow administration for immunomodulatory effects and controlled-release implant strategies for targeting tumors. Intrathecal administration for direct distribution in the CSF of curcumin remains a favorable yet invasive option for brain targeting (see for review, Reference [230]).

7. On the Topic of Dose-Response

To prove that polyphenols can accumulate in high-enough concentrations in target tissues as the brain is linked to the antioxidative properties of polyphenols in vitro [234,235].

More recently, the physiological significance of the direct antioxidative actions of polyphenols is met with skepticism, particularly with regard to the action in the brain, due to limited gastrointestinal absorption, propensity to undergo biotransformation and rapid excretion by the kidneys [97,236]. On the one hand, H-atom transfer must always occur faster than at least one of the reactions of free-radical-production cascades (e.g., the limiting step in lipid peroxidation) and this is improbable [237]. On the other hand, polyphenol concentrations, which rarely exceed micromolar concentrations in plasma or tissues [238] are substantially inferior to those of endogenous antioxidants such as ascorbate (30–100 μM) and urate (140–200 μM) [239]. Consequently, it is argued that their contribution to the total antioxidative capacity of the plasma never exceeds 2% and may therefore be irrelevant in a physiological context [236,240]. In fact, direct antioxidative effects of polyphenols have not been measured in the brain [97]. Also, studies demonstrating the anti-inflammatory properties of polyphenol analogues, other than direct antioxidative actions, challenges the idea that their health effects stem from their ability to hamper oxidative stress [201,202].

Nowadays, it is acknowledged that high circulating concentrations of polyphenols may not be required to achieve certain clinical endpoints. Indeed, by interacting with various enzymatic targets, for instance Keap1, very small doses of polyphenols may benefit from the cascades of events that ensue in cells. Despite this, efforts continue to focus on enhancing bioavailability rather than on identifying an adequate dose-response framework that could predict the behavior of this class of molecules. This oversight may partly account for the apparent difficulty of translating preclinical findings into actual positive outcomes in humans. Where disappointingly modest clinical benefits have been shown, is increasing the dose always a judicious strategy? The answer may not be as obvious as once thought.

Explanations have been proposed to explain the bioefficacy of polyphenols at very low doses. One of these is that polyphenols exert their biological effects in a non-linear fashion by exhibiting a biphasic dose-response profile. One such model predicts J or inverted U dose-response curves depending on the endpoint [241,242]. The biphasic theory stipulates low-dose stimulatory and high-dose inhibitory effects [243,244]. It direct stimulatory effects at low concentrations followed by biological overcompensation at higher doses [245]. In neuroprotection, hormesis predicts very low doses as beneficial and higher doses as potentially harmful. The application of this theory is thus intimately linked with whether polyphenols are indeed stressors that induce a defense response in cells. This has yet to be confirmed for polyphenols.

At present, the biphasic hypothesis explaining the bioefficacy of polyphenols at very low doses is gaining momentum, resveratrol constituting the best example. A wealth of reports support the hormetic action of resveratrol in various applications, ranging from cancer to neuroscience, extensively reviewed elsewhere [246]. In some instances, resveratrol stimulates cancer cell proliferation at very low doses but inhibits carcinogenesis in higher concentrations [247]. Other reports show resveratrol inducing atherosclerotic lesions at high doses, while it remains cardioprotective at lower concentrations [248]. In neurons, resveratrol promotes survival at very low concentrations but is neurotoxic at higher doses [121,249]. One study performed in mice and primary cortical neurons proposed a mechanism possibly underlying the biphasic response of energy-depleted neurons to resveratrol, showing protection at low doses and toxicity at higher doses [250]. The authors explained resveratrol's bimodal effects via its stimulatory action on silent mating type information regulation 2 homolog 1 (SIRT1), whose low-grade activity can suppress oxidative stress [251]. However, when stimulated by greater doses of resveratrol, SIRT1 expends too much-reduced nicotinamide adenine dinucleotide (NAD+) where neurons are already energetically depleted, causing energy failure. During an ischemic event, resveratrol administration could be either beneficial or detrimental, depending on dosage and timing and the bioenergetic status of neurons.

At present, these studies are usually performed in pre-clinical models and do not necessarily reflect what could occur in humans. The best-documented evidence of biphasic dose-responses in humans is for radiation, for instance in cancer treatments or in atomic bomb survivors [252,253]. However, reservations remain on the significance of such a dose-response relationship in the human brain, as it is highly unlikely that polyphenols could ever increase bioavailability in the parenchyma beyond low concentrations. This means that the observed bioefficacy of polyphenols may already be optimal where modest benefits are found in trials. Indeed, one distinct feature of the biphasic hypothesis provides that beneficial effects at low doses stem from cellular overcompensation mechanisms in response to the polyphenol-induced stress [254]. Beyond the optimal concentration at which maximal benefits are seen this compensation reaction is slowly overwhelmed by the increasing stress polyphenols directly exert on the cell. Even at the optimal concentration, these beneficial effects are thus thought to be at best partial. If this theory holds true, this could explain the results of clinical trials to date, even upon increasing dosages.

8. Concluding Remarks

The chemical structure of polyphenols confers them metal-chelating, fibril-destabilizing, estrogen-like, enzyme-binding and indirect antioxidative effects supporting their usefulness in neurodegenerative diseases. Epidemiological evidence shows a strong association between polyphenol consumption and reduced occurrence of various neurodegenerative diseases. Preclinical models lend them neuroprotective properties. Some clinical trials have even been successful in revealing small but measurable improvements in human health and have confirmed their safety in various settings. Nevertheless, the limited bioavailability of polyphenols together with their apparent bioefficacy remains under-explored. Investigators must demonstrate that polyphenols exert significant health benefits. However, in neurodegenerative diseases, polyphenol trials consistently fail in early clinical testing. To overcome this, researchers must optimize the design of their trials, subjects (disease stage, participant profile, cohort age and medical history), polyphenol administration (polyphenol formulation, route, dosage, frequency and duration) and endpoints (motor symptoms, cognitive decline, neuroinflammation, neuron integrity, CNS vascular health, etc.). As reviewed here, polyphenols are sensitive to a great number of physiological conditions that impinge on their bioavailability and biofunctionality, which may account for the markedly high inter individual variation observed in clinical investigations, which cannot be explained by biphasic dose-response theories.

Despite a large amount of information from many pre-clinical disease models and applications, a working theoretical framework that could aid in predicting outcomes in humans cannot be agreed. A priority would consist of determining the maximal health benefits that could be achieved from polyphenol monotherapies as they most usually stand alone in trials. Can we really expect standalone treatments to fulfill hard-to-reach clinical endpoints? If epidemiological evidence is strong for the protective effects of consuming complex mixtures of polyphenols in food, it may be unjustified to expect single molecules to be as effective. Perhaps concentrating on the concerted effects between polyphenols with each other or with other drugs that show partial benefits, such as the MAO-B inhibitor rasagiline [127] or levodopa [131], may overcome the as yet modest effects in humans. Evaluating polyphenols in preventive clinical paradigms may also constitute a more realistic strategy.

Besides, recent nutrigenomics data show that the interaction between genes and food bioactive compounds can positively or negatively influence an individual's health and possibly will aid with the prescription of customized diets according to an individual's genotype. Thus, the next approaches to clinical research with polyphenols should consider that dietary bioactive compounds such as polyphenols can be attributed to epigenetic mechanisms such as the regulation of histone deacetylases

(HDAC) and histone acetyltransferase (HAT) activities and acetylation of histones and non-histone chromatin proteins [255,256].

Author Contributions: Conceptualization, J.R. and M.-G.M.; Supervision, M.-G.M.; Validation, M.-G.M.; Visualization, J.R.; Writing—original draft, J.R.; Writing—review & editing, M.-G.M.

Funding: This research was funded by the Natural Sciences and Engineering Research Council (NSERC) of Canada (no. 04321) to M.-G.M. J.R. is recipient of a Vanier Graduate Scholarship from the NSERC and a Doctoral Training Scholarship from the Fonds de recherche en santé du Québec.

Acknowledgments: The authors would like to thank Lynda M. Williams (Rowett Institute, University of Aberdeen, Scotland, UK) for revising the manuscript and helpful suggestions.

Conflicts of Interest: The authors declare no conflict of interest.

Abbreviations

Aβ	beta amyloid
AREs	uridylate (AU)-rich elements
ATP	adenosine triphosphate
BBB	blood-brain barrier
BDNF	brain-derived neurotrophic factor
COMT	catechol-*O*-methyl transferase
CNS	central nervous system
CSF	cerebrospinal fluid
cAMP	cyclic adenosine monophosphate
CYP	cytochrome P450
EGCG	epigallocatechin-3-gallate
ERs	estrogen receptors
ERα	estrogen receptor alpha
ERβ	estrogen receptor beta
ERE	estrogen response element
HPLC	high-performance liquid chromatography
Keap1/Nrf2/ARE	Kelch-like ECH-associated protein 1/nuclear factor erythroid 2-related factor 2/antioxidant response elements
MAO-B	monoamine oxidase B
NAD+	nicotinamide adenine dinucleotide
ArO•	polyphenolic radical
SIRT1	silent mating type information regulation 2 homolog 1
WBSSH	White–Bate-Smith–Swain–Haslam
MPTP	1,2,3,6-tetrahydropyridine

References

1. Estruch, R.; Ros, E.; Salas-Salvadó, J.; Covas, M.I.; Corella, D.; Arós, F.; Gómez-Gracia, E.; Ruiz-Gutiérrez, V.; Fiol, M.; Lapetra, J.; et al. Primary prevention of cardiovascular disease with a Mediterranean diet. *N. Engl. J. Med.* **2013**, *368*, 1279–1290. [CrossRef] [PubMed]
2. Hertog, M.G.; Feskens, E.J.; Hollman, P.C.; Katan, M.B.; Kromhout, D. Dietary antioxidant flavonoids and risk of coronary heart disease: The Zutphen Elderly Study. *Lancet* **1993**, *342*, 1007–1011. [CrossRef]
3. Joshipura, K.J.; Hu, F.B.; Manson, J.E.; Stampfer, M.J.; Rimm, E.B.; Speizer, F.E.; Colditz, G.; Ascherio, A.; Rosner, B.; Spiegelman, D.; et al. The effect of fruit and vegetable intake on risk for coronary heart disease. *Ann. Intern. Med.* **2001**, *134*, 1106–1114. [CrossRef]
4. Von Ruesten, A.; Feller, S.; Bergmann, M.M.; Boeing, H. Diet and risk of chronic diseases: Results from the first 8 years of follow-up in the EPIC-Potsdam study. *Eur. J. Clin. Nutr.* **2013**, *67*, 412–419. [CrossRef] [PubMed]
5. Carter, P.; Gray, L.J.; Troughton, J.; Khunti, K.; Davies, M.J. Fruit and vegetable intake and incidence of type 2 diabetes mellitus: Systematic review and meta-analysis. *BMJ* **2010**, *341*, c4229. [CrossRef]

6. Sargeant, L.A.; Khaw, K.T.; Bingham, S.; Day, N.E.; Luben, R.N.; Oakes, S.; Welch, A.; Wareham, N.J. Fruit and vegetable intake and population glycosylated haemoglobin levels: The EPIC-Norfolk Study. *Eur. J. Clin. Nutr.* **2001**, *55*, 342–348. [CrossRef] [PubMed]
7. He, F.J.; Nowson, C.A.; MacGregor, G.A. Fruit and vegetable consumption and stroke: Meta-analysis of cohort studies. *Lancet* **2006**, *367*, 320–326. [CrossRef]
8. Ness, A.R.; Powles, J.W. Fruit and vegetables and cardiovascular disease: A review. *Int. J. Epidemiol.* **1997**, *26*, 1–13. [CrossRef]
9. Lunet, N.; Lacerda-Vieira, A.; Barros, H. Fruit and vegetables consumption and gastric cancer: A systematic review and meta-analysis of cohort studies. *Nutr. Cancer* **2005**, *53*, 1–10. [CrossRef]
10. Masala, G.; Assedi, M.; Bendinelli, B.; Ermini, I.; Sieri, S.; Grioni, S.; Sacerdote, C.; Ricceri, F.; Panico, S.; Mattiello, A.; et al. Fruit and vegetables consumption and breast cancer risk: The EPIC Italy study. *Breast Cancer Res. Treat.* **2012**, *132*, 1127–1136. [CrossRef]
11. Steinmetz, K.A.; Potter, J.D. Vegetables, fruit and cancer prevention: A review. *J. Am. Diet. Assoc.* **1996**, *96*, 1027–1039. [CrossRef]
12. Feng, L.; Gwee, X.; Kua, E.H.; Ng, T.P. Cognitive function and tea consumption in community dwelling older Chinese in Singapore. *J. Nutr. Health Aging* **2010**, *14*, 433–438. [CrossRef] [PubMed]
13. Kuriyama, S.; Hozawa, A.; Ohmori, K.; Shimazu, T.; Matsui, T.; Ebihara, S.; Awata, S.; Nagatomi, R.; Arai, H.; Tsuji, I. Green tea consumption and cognitive function: A cross-sectional study from the Tsurugaya Project 1. *Am. J. Clin. Nutr.* **2006**, *83*, 355–361. [CrossRef]
14. Manach, C.; Scalbert, A.; Morand, C.; Rémésy, C.; Jiménez, L. Polyphenols: Food sources and bioavailability. *Am. J. Clin. Nutr.* **2004**, *79*, 727–747. [CrossRef]
15. Scalbert, A.; Williamson, G. Dietary intake and bioavailability of polyphenols. *J. Nutr.* **2000**, *130*, 2073S–2085S. [CrossRef] [PubMed]
16. Bhullar, K.S.; Rupasinghe, H.P. Polyphenols: Multipotent therapeutic agents in neurodegenerative diseases. *Oxid. Med. Cell. Longev.* **2013**, *2013*, 891748. [CrossRef]
17. Ebrahimi, A.; Schluesener, H. Natural polyphenols against neurodegenerative disorders: Potentials and pitfalls. *Ageing Res. Rev.* **2012**, *11*, 329–345. [CrossRef]
18. Bate-Smith, E.C.; Swain, T. *Comparative Biochemistry*; Florkin, M., Mason, H.S., Eds.; Academic Press: Cambridge, MA, USA, 1962; Volume 3A, pp. 705–809.
19. Haslam, E.; Cai, Y. Plant polyphenols (vegetable tannins): Gallic acid metabolism. *Nat. Prod. Rep.* **1994**, *11*, 41–66. [CrossRef]
20. Quideau, S. Why Bother with Polyphenols? Groupe Polyphenols. 2011. Available online: http://www.groupepolyphenols.com/the-society/why-bother-with-polyphenols/ (accessed on 5 October 2014).
21. Parr, A.J.; Bolwell, G.P. Phenols in the Plant and in Man. The Potential for Possible Nutritional Enhancement of the Diet by Modifying the Phenols Content or Profile. *J. Sci. Food Agric.* **2000**, *80*, 985–1012. [CrossRef]
22. Petry, N.; Egli, I.; Zeder, C.; Walczyk, T.; Hurrel, R. Polyphenols and phytic acid contribute to the low iron bioavailability from common beans in young women. *J. Nutr.* **2010**, *140*, 1977–1982. [CrossRef]
23. Hider, R.C.; Liu, Z.D.; Khodr, H.H. Metal chelation of polyphenols. *Methods Enzymol.* **2001**, *335*, 190–203. [PubMed]
24. Purawatt, S.; Siripinyanond, A.; Shiowatana, J. Flow field-flow fractionation-inductively coupled optical emission spectrometric investigation of the size-based distribution of iron complexed to phytic and tannic acids in food suspension: Implications for iron availability. *Anal. Bioanal. Chem.* **2007**, *289*, 733–742. [CrossRef] [PubMed]
25. Gazit, E. A possible role for π-stacking in self-assembly of amyloid fibrils. *FASEB J.* **2002**, *16*, 77–83. [CrossRef]
26. Porat, Y.; Abramowitz, A.; Gazit, E. Inhibition of amyloid fibril formation by polyphenols: Structural similarity and aromatic interactions as a common inhibition mechanism. *Chem. Biol. Drug Des.* **2006**, *67*, 27–37. [CrossRef] [PubMed]
27. Bieschke, J. Natural compounds may open new routes to treatment of amyloid diseases. *Neurotherapeutics* **2013**, *10*, 429–439. [CrossRef]
28. Ishii, T.; Ichikawa, T.; Minoda, K.; Kusaka, K.; Ito, S.; Suzuki, Y.; Akagawa, M.; Mochizuki, K.; Goda, T.; Nakayama, T. Human serum albumin as an antioxidant in the oxidation of (−)-epigallocatechin gallate: Participation of reversible covalent binding for interaction and stabilization. *Biosci. Biotechnol. Biochem.* **2011**, *75*, 100–106. [CrossRef]

29. Palhano, F.L.; Lee, J.; Grimster, N.P.; Kelly, J.W. Toward the molecular mechanism(s) by which EGCG treatment remodels mature amyloid fibrils. *J. Am. Chem. Soc.* **2013**, *135*, 7503–7510. [CrossRef]
30. Maiti, T.K.; Ghosh, K.S.; Dasgupta, S. Interaction of (−)-epigallocatechin-3-gallate with human serum albumin: Fluorescence, fourier transform infrared, circular dichroism and docking studies. *Proteins* **2006**, *64*, 355–362. [CrossRef] [PubMed]
31. Yildiz, F. *Phytoestrogens in Functional Foods*; Taylor & Francis Ltd.: Abingdon, UK, 2005; Volume 210–211, pp. 3–5.
32. Turner, J.V.; Agatonovic-Kustrin, S.; Glass, B.D. Molecular aspects of phytoestrogen selective binding at estrogen receptors. *J. Pharm. Sci.* **2007**, *96*, 1879–1885. [CrossRef] [PubMed]
33. Bowers, J.L.; Tyulmenkov, V.V.; Jernigan, S.C.; Klinge, C.M. Resveratrol acts as a mixed agonist/antagonist for estrogen receptors alpha and beta. *Endocrinology* **2000**, *141*, 3657–3667. [CrossRef]
34. Gagné, B.; Gélinas, S.; Bureau, G.; Lagacé, B.; Ramassamy, C.; Chiasson, K.; Valastro, B.; Martinoli, M.G. Effects of estradiol, phytoestrogens and Ginkgo biloba extracts against 1-methyl-4-phenyl-pyridine-induced oxidative stress. *Endocrine* **2003**, *21*, 89–95. [CrossRef]
35. Cossette, L.J.; Gaumond, I.; Martinoli, M.G. Combined effect of xenoestrogens and growth factors in two estrogen-responsive cell lines. *Endocrine* **2002**, *18*, 303–308. [CrossRef]
36. Gélinas, S.; Martinoli, M.G. Neuroprotective effect of estradiol and phytoestrogens on MPP+-induced cytotoxicity in neuronal PC12 cells. *J. Neurosci. Res.* **2002**, *70*, 90–96. [CrossRef] [PubMed]
37. Ferrel, J.E.; Chang Sing PD, G.; Loew, G.; King, R.; Marnsour, J.M.; Mansour, T.E. Structure/activity studies of flavonoids as inhibitors of cyclic AMP phosphodiesterase and relationship to quantum chemical indices. *Mol. Pharmacol.* **1979**, *16*, 556–568.
38. Cochet, C.; Feige, J.J.; Pirollet, F.; Keramidas, M.; Chambaz, E.M. Selective inhibition of a cyclic nucleotide independent protein kinase (G type casein kinase) by quercetin and related polyphenols. *Biochem. Pharmacol.* **1982**, *31*, 1357–1361. [CrossRef]
39. Park, S.J.; Ahmad, F.; Philp, A.; Baar, K.; Williams, T.; Luo, H.; Ke, H.; Rehmann, H.; Taussig, R.; Brown, A.L.; et al. Resveratrol ameliorates aging-related metabolic phenotypes by inhibiting cAMP phosphodiesterases. *Cell* **2012**, *148*, 421–433. [CrossRef]
40. Li, B.; Vik, S.B.; Tu, Y. Theaflavins inhibit the ATP synthase and the respiratory chain without increasing superoxide production. *J. Nutr. Biochem.* **2011**, *23*, 953–960. [CrossRef] [PubMed]
41. Santel, T.; Pflug, G.; Hemdan, N.Y.; Schäfer, A.; Hollenbach, M.; Buchold, M.; Hintersdorf, A.; Lindner, I.; Otto, A.; Bigl, M. Curcumin inhibits glyoxalase 1: A possible link to its anti-inflammatory and anti-tumor activity. *PLoS ONE* **2008**, *3*, e3508. [CrossRef] [PubMed]
42. Takasawa, R.; Takahashi, S.; Saeki, K.; Sunaga, S.; Yoshimori, A.; Tanuma, S. Structure-activity relationship of human GLO I inhibitory natural flavonoids and their growth inhibitory effects. *Bioorg. Med. Chem.* **2008**, *16*, 3969–3975. [CrossRef]
43. Foresti, R.; Bains, S.K.; Pitchumony, T.S.; de Castro Brás, L.E.; Drago, F.; Dubois-Randé, J.L.; Bucolo, C.; Motterlini, R. Small molecule activators of the Nrf2-HO-1 antioxidant axis modulate heme metabolism and inflammation in BV2 microglia cells. *Pharmacol. Res.* **2013**, *76*, 132–148. [CrossRef]
44. Ishii, T.; Ishikawa, M.; Miyoshi, N.; Yasunaga, M.; Akagawa, M.; Uchida, K.; Nakamura, Y. Catechol type polyphenol is a potential modifier of protein sulfhydryls: Development and application of a new probe for understanding the dietary polyphenol actions. *Chem. Res. Toxicol.* **2009**, *22*, 1689–1698. [CrossRef] [PubMed]
45. Bhakkiyalakshmi, E.; Dineshkumar, K.; Karthik, S.; Sireesh, D.; Hopper, W.; Paulmurugan, R.; Ramkumar, K.M. Pterostilbene-mediated Nrf2 activation: Mechanistic insights on Keap1:Nrf2 interface. *Bioorg. Med. Chem.* **2016**, *24*, 3378–3386. [CrossRef] [PubMed]
46. Chow, H.H.; Cai, Y.; Hakim, I.A.; Crowell, J.A.; Shahi, F.; Brooks, C.A.; Dorr, R.T.; Hara, Y.; Alberts, D.S. Pharmacokinetics and safety of green tea polyphenols after multiple-dose administration of epigallocatechin gallate and polyphenon E in healthy individuals. *Clin. Cancer Res.* **2003**, *9*, 3312–3319. [PubMed]
47. Graefe, E.U.; Wittig, J.; Mueller, S.; Riethling, A.K.; Uehleke, B.; Drewelow, B.; Pforte, H.; Jacobasch, G.; Derendorf, H.; Veit, M. Pharmacokinetics and bioavailability of quercetin glycosides in humans. *J. Clin. Pharmacol.* **2001**, *41*, 492–499. [CrossRef] [PubMed]
48. Manach, C.; Williamson, G.; Morand, C.; Scalbert, A.; Rémésy, C. Bioavailability and bioefficacy of polyphenols in humans. I. Review of 97 bioavailability studies. *Am. J. Clin. Nutr.* **2005**, *81*, 230S–242S. [CrossRef] [PubMed]

49. Huang, W.; Lee, S.L.; Lawrence, X.Y. Mechanistic approaches to predicting oral drug absorption. *Aaps J.* **2009**, *11*, 217–224. [CrossRef] [PubMed]
50. Scholz, S.; Williamson, G. Interactions affecting the bioavailability of dietary polyphenols in vivo. *Int. J. Vitam. Nutr. Res.* **2007**, *77*, 224–235. [CrossRef]
51. Van het Hof, K.H.; Kivits, G.A.; Weststrate, J.A.; Tijburg, L.B. Bioavailability of catechins from tea: The effect of milk. *Eur. J. Clin. Nutr.* **1998**, *52*, 356–359. [CrossRef]
52. Hollman, P.C.; van Het Hof, K.H.; Tijburg, L.B.; Katan, M.B. Addition of milk does not affect the absorption of flavonols from tea in man. *Free Radic. Res.* **2001**, *34*, 297–300. [CrossRef]
53. Yamashita, S.; Sakane, T.; Harada, M.; Sugiura, N.; Koda, H.; Kiso, Y.; Sezaki, H. Absorption and metabolism of antioxidative polyphenolic compounds in red wine. *Ann. N. Y. Acad. Sci.* **2002**, *957*, 325–328. [CrossRef]
54. Hollman, P.C.; van Trijp, J.M.; Buysman, M.N.; van der Gaag, M.S.; Mengelers, M.J.; de Vries, J.H.; Katan, M.B. Relative bioavailability of the antioxidant flavonoid quercetin from various foods in man. *FEBS Lett.* **1997**, *418*, 152–156. [CrossRef]
55. Olthof, M.R.; Hollman PC, H.; Vree, T.B.; Katan, M.B. Bioavailabilities of quercetin-3-glucoside and quercetin-4′-glucoside do not differ in humans. *J. Nutr.* **2000**, *130*, 1200–1203. [CrossRef] [PubMed]
56. Manach, C.; Morand, C.; Crespy, V.; Demingné, C.; Texier, O.; Régérat, F.; Rémésy, C. Quercetin is recovered in human plasma as conjugated derivatives which retain antioxidant properties. *FEBS Lett.* **1998**, *426*, 331–336. [CrossRef]
57. Corte-real, J.; Richling, E.; Hoffmann, L.; Bohn, T. Selective factors governing in vitro b-carotene bioaccessibility: Negative influence of low filtration cutoffs and alterations by emulsifiers and food matrices. *Nutr. Res.* **2014**, *34*, 1101–1110. [CrossRef] [PubMed]
58. Mandalari, G.; Vardaku, M.; Faulks, R.; Bisignano, C.; Martorana, M.; Smeriglio, A.; Trombetta, D. Food matrix effects of polyphenols accessibility from almond skin during simulated human digestion. *Nutrients* **2016**, *8*, 568. [CrossRef] [PubMed]
59. Azuma, K.; Ippoushi, K.; Ito, H.; Higashio, H.; Terao, J. Combination of lipids and emulsifiers enhances the absorption of orally administered quercetin in rats. *J. Agric. Food Chem.* **2002**, *50*, 1706–1712. [CrossRef]
60. Donovan, J.L.; Kasim-Karakas, S.; German, J.B.; Waterhouse, A.L. Urinary excretion of catechin metabolites by human subjects after red wine consumption. *Br. J. Nutr.* **2002**, *87*, 31–37. [CrossRef] [PubMed]
61. Williams, G. The role of polyphenols in modern nutrition. *Nutr. Bull.* **2017**, *42*, 226–235. [CrossRef]
62. Murota, K.; Nakamura, Y.; Uehara, M. Flavonoid metabolism: The interaction of metabolites and gut microbiota. *Biosci. Biotechnol. Biochem.* **2018**, *82*, 600–610. [CrossRef] [PubMed]
63. Shortt, C.; Hasselwander, O.; Meynier, A.; Nauta, A.; Fernández, E.N.; Putz, P.; Rowland, I.; Swann, J.; Türk, J.; Vermeiren, J.; et al. Systematic review of the effects of the intestinal microbiota on selected nutrients and non-nutrients. *Eur. J. Nutr.* **2018**, *57*, 25–49. [CrossRef]
64. Aura, A.M.; O'Leary, K.A.; Williamson, G.; Ojala, M.; Bailey, M.; Puupponen-Pimiä, R.; Nuutila, A.M.; Oksman-Caldentey, K.M.; Poutanen, K. Quercetin derivatives are deconjugated and converted to hydroxyphenylacetic acids but not methylated by human fecal flora in vitro. *J. Agric. Food Chem.* **2002**, *50*, 1725–1730. [CrossRef]
65. Sawai, Y.; Kohsaka, K.; Nishiyama, Y.; Ando, K. Serum concentrations of rutoside metabolites after oral administration of a rutoside formulation to humans. *Arzneimittelforschung* **1987**, *37*, 729–732.
66. Walle, T. Bioavailability of resveratrol. *Ann. N. Y. Acad. Sci.* **2011**, *1215*, 9–15. [CrossRef]
67. Abd El-Mohsen, M.; Bayele, H.; Kuhnle, G.; Gibson, G.; Debnam, E.; Kaila Srai, S.; Rice-Evans, C.; Spencer, J.P. Distribution of [^{3}H]*trans*-resveratrol in rat tissues following oral administration. *Br. J. Nutr.* **2006**, *96*, 62–70. [CrossRef]
68. Olivares-Vicente, M.; Barrajon-Catalan, E.; Herranz-Lopez, M.; Segura-Carretero, A.; Joven, J.; Encinar, J.A.; Micol, V. Plant derived polyphenols in human health: Biological activity, matabolites and putative molecular targets. *Curr. Drug Matab.* **2018**, *19*, 351–369. [CrossRef]
69. Maier-Salamon, A.; Böhmdorfer, M.; Riha, J.; Thalhammer, T.; Szekeres, T.; Jaeger, W. Interplay between metabolism and transport of resveratrol. *Ann. N. Y. Acad. Sci.* **2013**, *1290*, 98–106. [CrossRef]
70. Marier, J.F.; Vachon, P.; Gritsas, A.; Zhang, J.; Moreau, J.P.; Ducharme, M.P. Metabolism and disposition of resveratrol in rats: Extent of absorption, glucuronidation and enterohepatic recirculation evidenced by a linked-rat model. *J. Pharmacol. Exp. Ther.* **2002**, *302*, 369–373. [CrossRef]

71. Wenzel, E.; Somoza, V. Metabolism and bioavailability of *trans*-resveratrol. *Mol. Nutr. Food Res.* **2005**, *49*, 472–481. [CrossRef]
72. Vitrac, X.; Desmoulière, A.; Brouillaud, B.; Krisa, S.; Deffieux, G.; Barthe, N.; Rosenbaum, J.; Mérillon, J.M. Distribution of [^{14}C]-*trans*-resveratrol, a cancer chemopreventive polyphenol, in mouse tissues after oral administration. *Life Sci.* **2003**, *72*, 2219–2233. [CrossRef]
73. Andreadi, C.; Britton, R.G.; Patel, K.R.; Brown, K. Resveratrol sulfates provide an intracellular reservoir for generation of parent resveratrol, which induces autophagy in cancer cells. *Autophagy* **2014**, *10*, 524–525. [CrossRef]
74. Marchiani, A.; Mammi, S.; Siligardi, G.; Hussain, R.; Tessari, I.; Bubacco, L.; Delogu, G.; Fabbri, D.; Dettori, M.A.; Sanna, D.; et al. Small molecules interacting with α-synuclein: Antiaggregating and cytoprotective properties. *Amino Acids* **2013**, *45*, 327–338. [CrossRef] [PubMed]
75. Zunino, S.J.; Storms, D.H. Resveratrol-3-*O*-glucuronide and resveratrol-4′-*O*-glucuronide reduce DNA strand breakage but not apoptosis in Jurkat T cells treated with camptothecin. *Oncol. Lett.* **2017**, *14*, 2517–2522. [CrossRef] [PubMed]
76. Henning, S.M.; Niu, Y.; Liu, Y.; Lee, N.H.; Hara, Y.; Thames, G.D.; Minutti, R.R.; Carpenter, C.L.; Wang, H.; Heber, D. Bioavailability and antioxidant effect of epigallocatechin gallate administered in purified form versus as green tea extract in healthy individuals. *J. Nutr. Biochem.* **2005**, *16*, 610–616. [CrossRef] [PubMed]
77. Delmas, D.; Aires, V.; Limagne, E.; Dutarte, P.; Mazué, F.; Ghiringhelli, F.; Latruffe, N. Transport, stability and biological activity of RESV. *Ann. N. Y. Acad. Sci.* **2011**, *1215*, 48–59. [CrossRef] [PubMed]
78. Khan, M.A.; Muzammil, S.; Musarrat, J. Differential binding of tetracyclines with serum albumin and induced structural alterations in drug-bound protein. *Int. J. Biol. Macromol.* **2002**, *30*, 243–249. [CrossRef]
79. Lancon, A.; Delmas, D.; Osman, H.; Thénot, J.P.; Jannin, B.; Latruffe, N. Human hepatic cell uptake of resveratrol: Involvement of both passive diffusion and carrier-mediated process. *Biochem. Biophys. Res. Commun.* **2004**, *316*, 1132–1137. [CrossRef]
80. Bourassa, P.; Kanakis, C.D.; Tarantilis, P.; Pollissiou, M.G.; Tajmir-Riahi, H.A. Resveratrol, genistein and curcumin bind bovine serum albumin. *J. Phys. Chem. B* **2010**, *114*, 3348–3354. [CrossRef]
81. Lu, Z.; Zhang, Y.; Liu, H.; Yuan, J.; Zheng, Z.; Zou, G. Transport of a cancer chemopreventive polyphenol, resveratrol: Interaction with serum albumin and hemoglobin. *J. Fluoresc.* **2007**, *17*, 580–587. [CrossRef]
82. Yang, F.; Bian, C.; Zhu, L.; Zhao, G.; Huang, Z.; Huang, M. Effect of human serum albumin on drug metabolism: Structural evidence of esterase activity of human serum albumin. *J. Struct. Biol.* **2007**, *157*, 348–355. [CrossRef]
83. Blache, D.; Rustan, I.; Durand, P.; Lesgards, G.; Loreau, N. Gas chromatographic analysis of resveratrol in plasma, lipoproteins and cells after in vitro incubations. *J. Chromatogr. B Biomed. Sci. Appl.* **1997**, *702*, 103–110. [CrossRef]
84. Belguendouz, L.; Fremont, L.; Gozzelino, M.T. Interaction of transresveratrol with plasma lipoproteins. *Biochem. Pharmacol.* **1998**, *55*, 811–816. [CrossRef]
85. Burkon, A.; Somoza, V. Quantification of free and protein-bound *trans*-resveratrol metabolites and identification of *trans*-resveratrol-C/O-conjugated diglucuronides – two novel resveratrol metabolites in human plasma. *Mol. Nutr. Food Res.* **2008**, *52*, 549–557. [CrossRef] [PubMed]
86. Urpi-Sarda, M.; Zamora-Ros, R.; Lamuela-Raventos, R.; Cherubini, A.; Jauregui, O.; de la Torre, R.; Covas, M.I.; Estruch, R.; Jaeger, W.; Andres-Lacueva, C. HPLC-tandem mass spectrometric method to characterize resveratrol metabolism in humans. *Clin. Chem.* **2007**, *53*, 292–299. [CrossRef] [PubMed]
87. Jannin, B.; Menzel, M.; Berlot, J.P.; Delmas, D.; Lançon, A.; Latruffe, N. Transport of resveratrol, a cancer chemopreventive agent, to cellular targets: Plasmatic protein binding and cell uptake. *Biochem. Pharmacol.* **2004**, *68*, 1113–1118. [CrossRef]
88. Lin, H.Y.; Lansing, L.; Merillon, J.M.; Davis, F.B.; Thang, H.Y.; Shih, A.; Vitrac, X.; Krisa, S.; Keating, T.; Cao, H.J.; et al. Integrin alphaVbeta3 contains a receptor site for resveratrol. *FASEB J.* **2006**, *20*, 1742–1744. [CrossRef] [PubMed]
89. Lin, H.Y.; Thang, H.Y.; Keating, T.; Wu, Y.H.; Shih, A.; Hammond, D.; Sun, M.; Hercbergs, A.; Davis, F.B.; Davis, P.J. Resveratrol is pro-apoptotic and thyroid hormone is anti-apoptotic in glioma cells: Both actions are integrin and ERK mediated. *Carcinogenesis* **2008**, *29*, 62–69. [CrossRef] [PubMed]
90. Figueira, I.; Meneses, R.; Macedo, D.; Costa, I.; Dos Santos, C.N. Polyphenols Beyond Barriers: A Glimpse into the Brain. *Curr. Neuropharmacol.* **2017**, *15*, 562–594. [CrossRef] [PubMed]

91. Liu, Z.; Hu, M. Natural polyphenol disposition via coupled metabolic pathways. *Expert Opin. Drug Metab. Toxicol.* **2007**, *3*, 389–406. [CrossRef]
92. Youdim, K.A.; Dobbie, M.S.; Kuhnle, G.; Proteggente, A.R.; Abbott, N.J.; Rice-Evans, C. Interaction between flavonoids and the blood-brain barrier: In vitro studies. *J. Neurochem.* **2003**, *85*, 180–192. [CrossRef]
93. Youdim, K.A.; Qaiser, M.Z.; Begley, D.J.; Rice-Evans, C.A.; Abbott, N.J. Flavonoid permeability across an in situ model of the blood-brain barrier. *Free Radic. Biol. Med.* **2004**, *36*, 592–604. [CrossRef] [PubMed]
94. Faria, A.; Pestana, D.; Teixeira, D.; Couraud, P.O.; Romero, I.; Weksler, B.; de Freitas, V.; Mateus, N.; Calhau, C. Insights into the putative catechin and epicatechin transport across blood-brain barrier. *Food Funct.* **2011**, *2*, 39–44. [CrossRef] [PubMed]
95. Milbury, P.E.; Kalt, W. Xenobiotic metabolism and berry flavonoid transport across the blood-brain barrier. *J. Agric. Food Chem.* **2010**, *58*, 3950–3956. [CrossRef] [PubMed]
96. Campos-Bedolla, P.; Walter, F.R.; Veszelka, S.; Deli, M.A. Role of the blood-brain barrier in the nutrition of the central nervous system. *Arch. Med. Res.* **2014**, *45*, 610–638. [CrossRef]
97. Schaffer, S.; Halliwell, B. Do polyphenols enter the brain and does it matter? Some theoretical and practical considerations. *Genes Nutr.* **2012**, *7*, 99–109. [CrossRef] [PubMed]
98. Asensi, M.; Medina, I.; Ortega, A.; Carretero, J.; Bano, M.C.; Obrador, E.; Estrela, J.M. Inhibition of cancer growth by resveratrol is related to its low bioavailability. *Free Radic. Biol. Med.* **2002**, *33*, 387–398. [CrossRef]
99. Juan, M.E.; Maijo, M.; Planas, J.M. Quantification of *trans*-resveratrol and its metabolites in rat plasma and tissues by HPLC. *J. Pharmaceut. Biomed. Anal.* **2010**, *51*, 391–398. [CrossRef]
100. Vingtdeux, V.; Giliberto, L.; Zhao, H.; Chandakkar, P.; Wu, Q.; Simon, J.E.; Janle, E.M.; Lobo, J.; Ferruzzi, M.G.; Davies, P.; et al. AMP-activated protein kinase signalling activation by resveratrol modulates amyloid-beta peptide metabolism. *J. Biol. Chem.* **2010**, *285*, 9100–9113. [CrossRef] [PubMed]
101. Wang, Q.; Xu, J.; Rottinghaus, G.E.; Simonyi, A.; Lubahn, D.; Sun, G.Y.; Sun, A.Y. Resveratrol protects against global cerebral ischemic injury in gerbils. *Brain Res.* **2002**, *958*, 439–447. [CrossRef]
102. Abd El-Mohsen, M.M.; Kuhnle, G.; Rechner, A.R.; Schroeter, H.; Rose, S.; Jenner, P.; Rice-Evans, C.A. Uptake and metabolism of epicatechin and its access to the brain after oral ingestion. *Free Radic. Biol. Med.* **2002**, *33*, 1693–1702. [CrossRef]
103. Suganuma, M.; Okabe, S.; Oniyama, M.; Tada, Y.; Ito, H.; Fujiki, H. Wide distribution of [3H](-)-epigallocatechin gallate, a cancer preventive tea polyphenol, in mouse tissue. *Carcinogenesis* **1998**, *19*, 1771–1776. [CrossRef]
104. Dajas, F.; Rivera, F.; Blasina, F.; Arredondo, F.; Echeverry, C.; Lafon, L.; Morquio, A.; Heinzen, H. Cell culture protection and in vivo neuroprotective capacity of flavonoids. *Neurotox. Res.* **2003**, *5*, 425–432. [CrossRef] [PubMed]
105. Mitsunaga, Y.; Takanaga, H.; Matsuo, H.; Naito, M.; Tsuruo, T.; Ohtani, H.; Sawada, Y. Effect of bioflavonoids on vincristine transport across blood-brain barrier. *Eur. J. Pharmacol.* **2000**, *395*, 193–201. [CrossRef]
106. Watanabe, C.M.; Wolffram, S.; Ader, P.; Rimbach, G.; Packer, L.; Maguire, J.J.; Schultz, P.G.; Gohil, K. The in vivo neuromodulatory effects of the herbal medicine ginkgo biloba. *Proc. Natl. Acad. Sci. USA* **2001**, *98*, 6577–6580. [CrossRef]
107. Karuppagounder, S.S.; Pinto, J.T.; Xu, H.; Chen, H.L.; Beal, M.F.; Gibson, G.E. Dietary supplementation with resveratrol reduces plaque pathology in a transgenic model of Alzheimer's disease. *Neurochem. Int.* **2009**, *54*, 111–118. [CrossRef] [PubMed]
108. Yang, F.; Lim, G.P.; Begum, A.N.; Ubeda, O.J.; Simmons, M.R.; Ambegaokar, S.S.; Chen, P.P.; Kayed, R.; Glabe, C.G.; Frautschy, S.A.; et al. Curcumin inhibits formation of amyloid beta oligomers and fibrils, binds plaques and reduces amyloid in vivo. *J. Biol. Chem.* **2005**, *280*, 5892–5901. [CrossRef]
109. Kelloff, G.J.; Boone, C.W.; Crowell, J.A.; Steele, V.E.; Lubet, R.A.; Doody, L.A.; Malone, W.F.; Hawk, E.T.; Sigman, C.C. New agents for cancer chemoprevention. *J. Cell. Biochem.* **1996**, *26*, 1–28. [CrossRef]
110. Mandel, S.; Weinreb, O.; Reznichenko, L.; Kalfon, L.; Amit, T. Green tea catechins as brain-permeable, non toxic iron chelators to "iron out iron" from the brain. *J. Neural Transm. Suppl.* **2006**, *2006*, 249–257.
111. Degenhardt, L.; Mathers, B.; Vickerman, P.; Rhodes, T.; Latkin, C.; Hickman, M. Prevention of HIV infection for people who inject drugs: Why individual, structural and combination approaches are needed. *Lancet* **2010**, *376*, 285–301. [CrossRef]

112. Kerbel, R.S.; Yu, J.; Tran, J.; Man, S.; Viloria-Petit, A.; Klement, G.; Coomber, B.L.; Rak, J. Possible mechanisms of acquired resistance to anti-angiogenic drugs: Implications for the use of combination therapy approaches. *Cancer Metastasis Rev.* **2001**, *20*, 79–86. [CrossRef]
113. Rodríguez, A.; Mendia, A.; Sirvent, J.M.; Barcenilla, F.; de la Torre-Prados, M.V.; Solé-Violán, J.; Rello, J.; CAPUCI Study Group. Combination antibiotic therapy improves survival in patients with community-acquired pneumonia and shock. *Crit. Care Med.* **2007**, *35*, 1493–1498. [CrossRef]
114. Yang, Y.; Zhang, Z.; Li, S.; Ye, X.; Li, X.; He, K. Synergy effects of herb extracts: Pharmacokinetics and pharmacodynamic basis. *Fitoterapia* **2014**, *92*, 133–147. [CrossRef]
115. Wagner, H.; Ulrich-Merzenich, G. Synergy research: Approaching a new generation of phytopharmaceuticals. *Phytomedicine* **2009**, *16*, 97–110. [CrossRef]
116. Ulrich-Merzenich, G.; Panek, D.; Zeitler, H.; Vetter, H.; Wagner, H. Drug development from natural products: Exploiting synergistic effects. *Indian J. Exp. Biol.* **2010**, *48*, 208–219. [PubMed]
117. Chen, T.C.; Wang, W.; Golden, E.B.; Thomas, S.; Sivakumar, W.; Hofman, F.M.; Louie, S.G.; Schönthal, A.H. Green tea epigallocatechin gallate enhances therapeutic efficacy of temozolomide in orthotopic mouse glioblastoma models. *Cancer Lett.* **2011**, *302*, 100–108. [CrossRef]
118. Gazova, Z.; Siposova, K.; Kurin, E.; Mučaji, P.; Nagy, M. Amyloid aggregation of lysozyme: The synergy study of red wine polyphenols. *Proteins* **2013**, *81*, 994–1004. [CrossRef] [PubMed]
119. Kurin, E.; Mučaji, P.P.; Nagy, M. In vitro antioxidant activities of three red wine polyphenols and their mixtures: An interaction study. *Molecules* **2012**, *17*, 14336–14348. [CrossRef] [PubMed]
120. Nichols, M.; Zhang, J.; Polster, B.M.; Elustondo, P.A.; Thirumaran, A.; Pavlov, E.V.; Robertson, G.S. Synergistic neuroprotection by epicatechin and quercetin: Activation of convergent mitochondrial signaling pathways. *Neuroscience* **2015**, *308*, 75–94. [CrossRef]
121. Conte, A.; Pellegrini, S.; Tagliazucchi, D. Synergistic protection of PC12 cells from beta-amyloid toxicity by resveratrol and catechin. *Brain Res. Bull.* **2003**, *62*, 29–38. [CrossRef]
122. Sanz, M.M.; Johnson, L.E.; Ahuja, S.; Ekström, P.A.; Romero, J.; van Veen, T. Significant photoreceptor rescue by treatment with a combination of antioxidants in an animal model for retinal degeneration. *Neuroscience* **2007**, *145*, 1120–1129. [CrossRef]
123. Gundimeda, U.; McNeill, T.H.; Barseghian, B.A.; Tzeng, W.S.; Rayudu, D.V.; Cadenas, E.; Gopalakrishna, R. Polyphenols from green tea prevent antineuritogenic action of Nogo-A via 67-kDa laminin receptor and hydrogen peroxide. *J. Neurochem.* **2015**, *132*, 70–84. [CrossRef]
124. Gundimeda, U.; McNeill, T.H.; Fan, T.K.; Deng, R.; Rayudu, D.; Chen, Z.; Cadenas, E.; Gopalakrishna, R. Green tea catechins potentiate the neuritogenic action of brain-derived neurotrophic factor: Role of 67-kDa laminin receptor and hydrogen peroxide. *Biochem. Biophys. Res. Commun.* **2014**, *445*, 218–224. [CrossRef]
125. Kwon, K.J.; Kim, J.N.; Kim, M.K.; Lee, J.; Ignarro, L.J.; Kim, H.J.; Shin, C.Y.; Han, S.H. Melatonin synergistically increases resveratrol-induced heme oxygenase-1 expression through the inhibition of ubiquitin-dependent proteasome pathway: A possible role in neuroprotection. *J. Pineal. Res.* **2011**, *50*, 110–123. [CrossRef]
126. Olanow, C.W.; Rascol, O.; Hauser, R.; Feigin, P.D.; Jankovic, J.; Lang, A.; Langston, W.; Melamed, E.; Poewe, W.; Stocchi, F.; et al. A doubleblind, delayed-start trial of rasagiline in Parkinson's disease. *N. Engl. J. Med.* **2009**, *361*, 1268–1278. [CrossRef]
127. Masellis, M.; Collinson, S.; Freeman, N.; Tampakeras, M.; Levy, J.; Tchelet, A.; Eyal, E.; Berkovich, E.; Eliaz, R.E.; Abler, V.; et al. D2 receptor gene variants and response to rasagiline in early Parkinson's disease: A pharmacogenetic study. *Brain* **2016**, *139*, 2050–2062. [CrossRef]
128. Reznichenko, L.; Kalfon, L.; Amit, T.; Youdim, M.B.; Mandel, S.A. Low dosage of rasagiline and epigallocatechin gallate synergistically restored the nigrostriatal axis in MPTP-induced parkinsonism. *Neurodegener. Dis.* **2010**, *7*, 219–231. [CrossRef]
129. Faggi, L.; Pignataro, G.; Parrella, E.; Porrini, V.; Vinciguerra, A.; Cepparulo, P.; Cuomo, O.; Lanzillotta, A.; Mota, M.; Benarese, M.; et al. Association of Valproate and Resveratrol Reduces Brain Injury in Ischemic Stroke. *Int. J. Mol. Sci.* **2018**, *19*, 172. [CrossRef]
130. Herges, K.; Millward, J.M.; Hentschel, N.; Infante-Duarte, C.; Aktas, O.; Zipp, F. Neuroprotective effect of combination therapy of glatiramer acetate and epigallocatechin-3-gallate in neuroinflammation. *PLoS ONE* **2011**, *6*, e25456. [CrossRef]

131. Kang, K.S.; Wen, Y.; Yamabe, N.; Fukui, M.; Bishop, S.C.; Zhu, B.T. Dual beneficial effects of (-)-epigallocatechin-3-gallate on levodopa methylation and hippocampal neurodegeneration: In vitro and in vivo studies. *PLoS ONE* **2010**, *5*, e11951. [CrossRef]
132. Doostdar, H.; Burke, M.D.; Mayer, R.T. Bioflavonoids: Selective substrates and inhibitors for cytochrome P450CYP1A and CYP1B1. *Toxicology* **2000**, *144*, 31–38. [CrossRef]
133. Chang, T.K.; Chen, J.; Yeung, E.Y. Effect of Ginkgo biloba extract on procarcinogen-bioactivating human CYP1 enzymes: Identification of isorhamnetin, kaempferol and quercetin as potent inhibitors of CYP1B1. *Toxicol. Appl. Pharmacol.* **2006**, *213*, 18–26. [CrossRef]
134. Czuczwar, P.; Paszkowski, T.; Lisiecki, M.; Woźniak, S.; Stępniak, A. The safety and tolerance of phytotherapies in menopausal medicine—A review of the literature. *Prz Menopauzalny* **2017**, *16*, 8–11. [CrossRef]
135. Cottart, C.H.; Nivet-Antoine, V.; Laguillier-Morizot, C.; Beaudeux, J.L. Resveratrol bioavailability and toxicity in humans. *Mol. Nutr. Food Res.* **2010**, *54*, 7–16. [CrossRef]
136. Margină, D.; Ilie, M.; Grădinaru, D.; Androutsopoulos, V.P.; Kouretas, D.; Tsatsakis, A.M. Natural products-friends or foes? *Toxicol. Lett.* **2015**, *236*, 154–167. [CrossRef]
137. Galati, G.; O'Brien, P.J. Potential toxicity of flavonoids and other dietary phenolics: Significance for their chemopreventive and anticancer properties. *Free Radic. Biol. Med.* **2004**, *37*, 287–303. [CrossRef]
138. Boocock, D.J.; Faust, G.E.; Patel, K.R.; Schinas, A.M.; Brown, V.A.; Ducharme, M.P.; Booth, T.D.; Crowell, J.A.; Perloff, M.; Gescher, A.J.; et al. Phase I dose escalation pharmacokinetic study in healthy volunteers of resveratrol, a potential cancer chemopreventive agent. *Cancer Epidemiol. Biomark. Prev.* **2007**, *16*, 1246–1252. [CrossRef]
139. Vaz-da-Silva, M.; Loureiro, A.I.; Falcao, A.; Nunes, T.; Rocha, J.F.; Fernandes-Lopes, C.; Soares, E.; Wright, L.; Almeida, L.; Soares-da-Silva, P. Effect of food on the pharmacokinetic profile of *trans*-resveratrol. *Int. J. Clin. Pharmacol. Ther.* **2008**, *46*, 564–570. [CrossRef]
140. Almeida, L.; Vaz-da-Silva, M.; Falcao, A.; Soares, E.; Costa, R.; Loureiro, A.I.; Fernandes-Lopes, C.; Rocha, J.F.; Nunes, T.; Wright, L.; et al. Pharmacokinetic and safety profile of *trans*-resveratrol in a rising multiple-dose study in healthy volunteers. *Mol. Nutr. Food Res.* **2009**, *53*, 7–15. [CrossRef]
141. Herrschaft, H.; Nacu, A.; Likhachev, S.; Sholomov, I.; Hoerr, R.; Schlaefke, S. Ginkgo biloba extract EGb 761® in dementia with neuropsychiatric features: A randomised, placebo-controlled trial to confirm the efficacy and safety of a daily dose of 240 mg. *J. Psychiatr. Res.* **2012**, *46*, 716–723. [CrossRef]
142. Ihl, R.; Tribanek, M.; Bachinskaya, N.; GOTADAY Study Group. Efficacy and tolerability of a once daily formulation of Ginkgo biloba extract EGb 761® in Alzheimer's disease and vascular dementia: Results from a randomised controlled trial. *Pharmacopsychiatry* **2012**, *45*, 41–46. [CrossRef]
143. Ringman, J.M.; Frautschy, S.A.; Teng, E.; Begum, A.N.; Bardens, J.; Beigi, M.; Gylys, K.H.; Badmaev, V.; Heath, D.D.; Apostolova, L.G.; et al. Oral curcumin for Alzheimer's disease: Tolerability and efficacy in a 24-week randomized, double blind, placebo-controlled study. *Alzheimers Res. Ther.* **2012**, *4*, 43. [CrossRef]
144. Crowell, J.A.; Korytko, P.J.; Morrissey, R.L.; Booth, T.D.; Levine, B.S. Resveratrol-associated renal toxicity. *Toxicol. Sci.* **2004**, *82*, 614–619. [CrossRef]
145. Bailey, D.G.; Malcolm, J.; Arnold, O.; Spence, J.D. Grapefruit juice drug interactions. *Br. J. Clin. Pharmacol.* **1998**, *46*, 101–110. [CrossRef]
146. Hodek, P.; Trefil, P.; Stiborová, M. Flavonoids—Potent and versatile biologically active compounds interacting with cytochromes P450. *Chem. Biol. Interact.* **2002**, *139*, 1–21. [CrossRef]
147. Uno, T.; Yasui-Furukori, N. Effect of grapefruit in relation to human pharmacokinetic study. *Curr. Clin. Pharmacol.* **2006**, *1*, 157–161.
148. Koga, N.; Ohta, C.; Kato, Y.; Haragushi, K.; Endo, T.; Ogawa, K.; Ohta, H.; Yano, H. In vitro metabolism of nobiletin, a polymethoxy-flavonoid, by human liver microsomes and cytochrome P450. *Xenobiotica* **2011**, *41*, 927–933. [CrossRef]
149. Ofer, M.; Wolffram, S.; Koggel, A.; Spahn-Langguth, H.; Langguth, P. Modulation of drug transport by selected flavonoids: Involvement of P-gp and OCT? *Eur. J. Pharm. Sci.* **2005**, *25*, 263–271. [CrossRef]
150. Bailey, D.G. Fruit juice inhibition of uptake transport: A new type of food–drug interaction. *Br. J. Clin. Pharmacol.* **2010**, *70*, 645–655. [CrossRef]
151. Owira, P.M.; Ojewole, J.A. The grapefruit: An old wine in a new glass? Metabolic and cardiovascular perspectives Cardiovasc. *J. Afr.* **2010**, *21*, 280–285.

152. Jáuregui-Garrido, B.; Jáuregui-Lobera, I. Interactions between antihypertensive drugs and food. *Nutr. Hosp.* **2012**, *27*, 1866–1875.
153. Eaton, E.A.; Walle, U.K.; Lewis, A.J.; Hudson, T.; Wilson, A.A.; Walle, T. Flavonoids, potent inhibitors of the human P-form phenolsulfotransferase: Potential role in drug metabolism and chemoprevention. *Drug Metab. Dispos.* **1996**, *24*, 232–237.
154. Von Moltke, L.L.; Weemhoff, J.L.; Bedir, E.; Khan, I.A.; Harmatz, J.S.; Goldman, P.; Greenlatt, D.J. Inhibition of human cytochrome P450 by components of Ginkgo biloba. *J. Pharm. Pharmacol.* **2004**, *56*, 1039–1044. [CrossRef]
155. Martin, P.M.; Horwitz, K.B.; Ryan, D.S.; McGuire, W.L. Phytoestrogen interaction with estrogen receptors in human breast cancer cells. *Endocrinology* **1978**, *103*, 1860–1867. [CrossRef]
156. Nifli, A.P.; Bosson-Kouame, A.; Papadopoulou, N.; Kogia, C.; Kampa, M.; Castagnino, C.; Stournaras, C.; Vercauteren, J.; Castanas, E. Monomeric and oligomeric flavanols are agonists of membrane androgen receptors. *Exp. Cell. Res.* **2005**, *309*, 329–339. [CrossRef]
157. Stubert, J.; Gerber, B. Isoflavones—Mechanism of Action and Impact on Breast Cancer Risk. *Breast Care* **2009**, *4*, 22–29. [CrossRef]
158. Zhang, G.Q.; Chen, J.L.; Liu, Q.; Zhang, Y.; Zeng, H.; Zhao, Y. Soy Intake Is Associated with Lower Endometrial Cancer Risk: A Systematic Review and Meta-Analysis of Observational Studies. *Medicine'(Baltimore)* **2015**, *94*, e2281. [CrossRef]
159. Mgbonyebi, O.P.; Russo, J.; Russo, I.H. Antiproliferative effect of synthetic resveratrol on human breast epithelial cells. *Int. J. Oncol.* **1998**, *12*, 865–869. [CrossRef]
160. Della Ragione, F.; Cucciola, V.; Borriello, A.; Della Pietra, V.; Racioppi, L.; Soldati, G.; Manna, C.; Galletti, P.; Zappia, V. Resveratrol arrests the cell division cycle at S/G2 phase transition. *Biochem. Biophys. Res. Commun.* **1998**, *250*, 53–58. [CrossRef]
161. Hsieh, T.C.; Burfeind, P.; Laud, K.; Backer, J.M.; Traganos, F.; Darzynkiewicz, Z.; Wu, J.M. Cell cycle effects and control of gene expression by resveratrol in human breast carcinoma cell lines with different metastatic potentials. *Int. J. Oncol.* **1999**, *15*, 245–252. [CrossRef] [PubMed]
162. Kaushik, S.; Shyam, H.; Sharma, R.; Balapure, A.K. Genistein synergizes centchroman action in human breast cancer cells. *Indian J. Pharmacol.* **2016**, *48*, 637–642.
163. Choi, E.J.; Jung, J.Y.; Kim, G.H. Genistein inhibits the proliferation and differentiation of MCF-7 and 3T3-L1 cells via the regulation of ERα expression and induction of apoptosis. *Exp. Ther. Med.* **2014**, *8*, 454–458. [CrossRef]
164. Gehm, B.D.; McAndrews, J.M.; Chien, P.Y.; Jameson, J.L. Resveratrol, a polyphenolic compound found in grapes and wine, is an agonist for the estrogen receptor. *Proc. Natl. Acad. Sci. USA* **1997**, *94*, 14138–14143. [CrossRef] [PubMed]
165. Lu, R.; Serrero, G. Resveratrol, a natural product derived from grape, exhibits antiestrogenic activity and inhibits the growth of human breast cancer cells. *J. Cell. Physiol.* **1999**, *179*, 297–304. [CrossRef]
166. De Lemos, M.L. Effects of soy phytoestrogens genistein and daidzein on breast cancer growth. *Ann. Pharmacother.* **2001**, *35*, 1118–1121. [CrossRef]
167. Shike, M.; Doane, A.S.; Russo, L.; Cabal, R.; Reis-Filho, J.S.; Gerald, W.; Cody, H.; Khanin, R.; Bromberg, J.; Norton, L. The effects of soy supplementation on gene expression in breast cancer: A randomized placebo-controlled study. *J. Natl. Cancer Inst.* **2014**, *106*. [CrossRef]
168. Saleh, M.C.; Connell, B.J.; Saleh, T.M. Resveratrol induced neuroprotection is mediated via both estrogen receptor subtypes, ER(α) and ER(β). *Neurosci. Lett.* **2013**, *548*, 217–221. [CrossRef]
169. Turner, R.T.; Evans, G.L.; Zhang, M.; Maran, A.; Sibonga, J.D. Is resveratrol an estrogen agonist in growing rats? *Endocrinology* **1999**, *140*, 50–54. [CrossRef] [PubMed]
170. Lecomte, S.; Lelong, M.; Bourgine, G.; Efstathiou, T.; Saligaut, C.; Pakdel, F. Assessment of the potential activity of major dietary compounds as selective estrogen receptormodulators in two distinct cell models for proliferation and differentiation. *Toxicol. Appl. Pharmacol.* **2017**, *325*, 61–70. [CrossRef] [PubMed]
171. Bookout, A.L.; Jeong, Y.; Downes, M.; Yu, R.T.; Evans, R.M.; Mangelsdorf, D.J. Anatomical profiling of nuclear receptor expression reveals a hierarchical transcriptional network. *Cell* **2006**, *126*, 789–799. [CrossRef] [PubMed]

172. Nwachukwu, J.C.; Srinivasan, S.; Bruno, N.E.; Parent, A.A.; Hughes, T.S.; Pollock, J.A.; Gjyshi, O.; Cavett, V.; Nowak, J.; Garcia-Ordonez, R.D.; et al. Resveratrol modulates the inflammatory response via an estrogen receptor-signal integration network. *Elife* **2014**, *3*, e02057. [CrossRef] [PubMed]
173. Levin, J.; Maaß, S.; Schuberth, M.; Respondek, G.; Paul, F.; Mansmann, U.; Oertel, W.H.; Lorenzl, S.; Krismer, F.; Seppi, K.; Poewe, W.; et al. The PROMESA-protocol: Progression rate of multiple system atrophy under EGCG supplementation as anti-aggregation-approach. *J. Neural Transm. (Vienna)* **2016**, *123*, 439–445. [CrossRef]
174. Maurer, K.; Ihl, R.; Dierks, T.; Frölich, L. Clinical efficacy of Ginkgo biloba special extract EGb 761 in dementia of the Alzheimer type. *J. Psychiatr. Res.* **1997**, *31*, 645–655. [CrossRef]
175. Montes, P.; Ruiz-Sanchez, E.; Rojas, C.; Rojas, P. Ginkgo biloba Extract 761: A Review of Basic Studies and Potential Clinical Use in Psychiatric Disorders. *Cns Neurol. Disord. Drug Targets* **2015**, *14*, 132–149. [CrossRef] [PubMed]
176. Schulz, V. Ginkgo extract or cholinesterase inhibitors in patients with dementia: What clinical trials and guidelines fail to consider. *Phytomedicine* **2003**, *10*, 74–79. [CrossRef] [PubMed]
177. Andrieu, S.; Ousset, P.J.; Coley, N.; Ouzid, M.; Mathiex-Fortunet, H.; Vellas, B.; GuidAge study GROUP. GuidAge study: A 5-year double blind, randomised trial of EGb 761 for the prevention of Alzheimer's disease in elderly subjects with memory complaints. i. rationale, design and baseline data. *Curr. Alzheimer Res.* **2008**, *5*, 406–415. [CrossRef] [PubMed]
178. Vellas, B.; Coley, N.; Ousset, P.J.; Berrut, G.; Dartigues, J.F.; Dubois, B.; Grandjean, H.; Pasquier, F.; Piette, F.; Robert, P.; et al. Long-term use of standardised Ginkgo biloba extract for the prevention of Alzheimer's disease (GuidAge): A randomised placebo-controlled trial. *Lancet Neurol.* **2012**, *11*, 851–859. [CrossRef]
179. DeKosky, S.T.; Williamson, J.D.; Fitzpatrick, A.L.; Kronmal, R.A.; Ives, D.G.; Saxton, J.A.; Lopez, O.L.; Burke, G.; Carlson, M.C.; Fried, L.P.; et al. Ginkgo biloba for prevention of dementia: A randomized controlled trial. *JAMA* **2008**, *300*, 2253–2262. [CrossRef] [PubMed]
180. Dodge, H.H.; Zitzelberger, T.; Oken, B.S.; Howieson, D.; Kaye, J. A randomized placebo-controlled trial of Ginkgo biloba for the prevention of cognitive decline. *Neurology* **2008**, *70*, 1809–1817. [CrossRef]
181. Snitz, B.E.; O'Meara, E.S.; Carlson, M.C.; Arnold, A.M.; Ives, D.G.; Rapp, S.R.; Saxton, J.; Lopez, O.L.; Dunn, L.O.; Sink, K.M.; et al. Ginkgo biloba for preventing cognitive decline in older adults: A randomized trial. *JAMA* **2009**, *302*, 2663–2670. [CrossRef]
182. Scherrer, B.; Andrieu, S.; Ousset, P.J.; Berrut, G.; Dartigues, J.F.; Dubois, B.; Pasquier, F.; Piette, F.; Robert, P.; Touchon, J.; et al. Analysing Time to Event Data in Dementia Prevention Trials: The Example of the GuidAge Study of EGb761. *J. Nutr. Health Aging* **2015**, *19*, 1009–1011. [CrossRef]
183. Baum, L.; Lam, C.W.; Cheung, S.K. Six-month randomized, placebo-controlled, double-blind, pilot clinical trial of curcumin in patients with Alzheimer disease. *J. Clin. Psychopharmacol.* **2008**, *28*, 110–113. [CrossRef]
184. Beck, S.M.; Ruge, H.; Schindler, C.; Burkart, M.; Miller, R.; Kirschbaum, C.; Goschke, T. Effects of Ginkgo biloba extract EGb 761® on cognitive control functions, mental activity of the prefrontal cortex and stress reactivity in elderly adults with subjective memory impairment—A randomized double-blind placebo-controlled trial. *Hum. Psychopharmacol.* **2016**, *31*, 227–242. [CrossRef]
185. Nasab, N.M.; Bahrammi, M.A.; Nikpour, M.R.; Rahim, F.; Naghibis, S.N. Efficacy of rivastigmine in comparison to ginkgo for treating Alzheimer's dementia. *J. Pak. Med. Assoc.* **2012**, *62*, 677–680.
186. Kennedy, D.O.; Wightman, E.L.; Reay, J.L.; Lietz, G.; Okello, E.J.; Wilde, A.; Haskell, C.F. Am Effects of resveratrol on cerebral blood flow variables and cognitive performance in humans: A double-blind, placebo-controlled, crossover investigation. *J. Clin. Nutr.* **2010**, *91*, 1590–1597. [CrossRef]
187. Wightman, E.L.; Haskell, C.F.; Forster, J.S.; Veasey, R.C.; Kennedy, D.O. Epigallocatechin gallate, cerebral blood flow parameters, cognitive performance and mood in healthy humans: A double-blind, placebo-controlled, crossover investigation. *Hum. Psychopharmacol.* **2012**, *27*, 177–186. [CrossRef] [PubMed]
188. Wightman, E.L.; Reay, J.L.; Haskell, C.F.; Williamson, G.; Dew, T.P.; Kennedy, D.O. Effects of resveratrol alone or in combination with piperine on cerebral blood flow parameters and cognitive performance in human subjects: A randomised, double-blind, placebo-controlled, cross-over investigation. *Br. J. Nutr.* **2014**, *112*, 203–213. [CrossRef]
189. Wightman, E.L.; Haskell-Ramsay, C.F.; Reay, J.L.; Williamson, G.; Dew, T.; Zhang, W.; Kennedy, D.O. The effects of chronic *trans*-resveratrol supplementation on aspects of cognitive function, mood, sleep, health and cerebral blood flow in healthy, young humans. *Br. J. Nutr.* **2015**, *114*, 1427–1437. [CrossRef] [PubMed]

190. Brickman, A.M.; Khan, U.A.; Provenzano, F.A.; Yeung, L.K.; Suzuki, W.; Schroeter, H.; Wall, M.; Sloan, R.P.; Small, S.A. Enhancing dentate gyrus function with dietary flavanols improves cognition in older adults. *Nat. Neurosci.* **2014**, *17*, 1798–1803. [CrossRef]
191. Witte, A.V.; Kerti, L.; Margulies, D.S.; Flöel, A. Effects of resveratrol on memory performance, hippocampal functional connectivity and glucose metabolism in healthy older adults. *J. Neurosci.* **2014**, *34*, 7862–7870. [CrossRef]
192. Moussa, C.; Hebron, M.; Huang, X.; Ahn, J.; Rissman, R.A.; Aisen, P.S.; Turner, R.S. Resveratrol regulates neuro-inflammation and induces adaptive immunity in Alzheimer's disease. *J. Neuroinflammation* **2017**, *14*, 1. [CrossRef] [PubMed]
193. Turner, R.S.; Thomas, R.G.; Craft, S.; van Dyck, C.H.; Mintzer, J.; Reynolds, B.A.; Brewer, J.B.; Rissman, R.A.; Raman, R.; Aisen, P.S. A randomized, double-blind, placebo-controlled trial of resveratrol for Alzheimer disease. Alzheimer's Disease Cooperative Study. *Neurology* **2015**, *85*, 1383–1391. [CrossRef]
194. Riva, S.; Monti, D.; Luisetti, M.; Danieli, B. Enzymatic modification of natural compounds with pharmacological properties. *Ann. N. Y. Acad. Sci.* **1998**, *864*, 70–80. [CrossRef]
195. Colin, D.; Gimazane, A.; Lizard, G.; Izard, J.C.; Solary, E.; Latruffe, N.; Delmas, D. Effects of resveratrol analogs on cell cycle progression, cell cycle associated proteins and 5fluoro-uracil sensitivity in human derived colon cancer cells. *Int. J. Cancer* **2009**, *124*, 2780–2788. [CrossRef]
196. Marel, A.K.; Lizard, G.; Izard, J.C.; Latruffe, N.; Delmas, D. Inhibitory effects of *trans*-resveratrol analogs molecules on the proliferation and the cell cycle progression of human colon tumoral cells. *Mol. Nutr. Food Res.* **2008**, *52*, 538–548. [CrossRef]
197. Chao, J.; Li, H.; Cheng, K.W.; Yu, M.S.; Chang, R.C.; Wang, M. Protective effects of pinostilbene, a resveratrol methylated derivative, against 6-hydroxydopamine-induced neurotoxicity in SH-SY5Y cells. *J. Nutr. Biochem.* **2010**, *21*, 482–489. [CrossRef]
198. Huo, C.; Wan, S.B.; Lam, W.H.; Li, L.; Wang, Z.; Landis-Piwowar, K.R.; Chen, D.; Dou, Q.P.; Chan, T.H. The challenge of developing green tea polyphenols as therapeutic agents. *Inflammopharmacology* **2008**, *16*, 248–252. [CrossRef]
199. Lambert, M.; Olsen, L.; Jaroszewski, J.W. Stereoelectronic effects on 1H nuclear magnetic resonance chemical shifts in methoxybenzenes. *J. Org. Chem.* **2006**, *71*, 9449–9457. [CrossRef]
200. Mythri, R.B.; Harish, G.; Dubey, S.K.; Misra, K.; Bharath, M.M. Glutamoyl diester of the dietary polyphenol curcumin offers improved protection against peroxynitrite-mediated nitrosative stress and damage of brain mitochondria in vitro: Implications for Parkinson's disease. *Mol. Cell. Biochem.* **2011**, *347*, 135–143. [CrossRef]
201. Solberg, N.O.; Chamberlin, R.; Vigil, J.R.; Deck, L.M.; Heidrich, J.E.; Brown, D.C.; Brady, C.I.; Vander, T.A.; Garwood, M.; Bisoffi, M.; et al. Optical and SPION-Enhanced MR Imaging Shows that *trans*-Stilbene Inhibitors of NF-κB Concomitantly Lower Alzheimer's Disease Plaque Formation and Microglial Activation in AβPP/PS-1 Transgenic Mouse Brain. *J. Alzheimers Dis.* **2014**, *40*, 191–212. [CrossRef]
202. Hauss, F.; Liu, J.; Michelucci, A.; Coowar, D.; Morga, E.; Heuschling, P.; Luu, B. Dual bioactivity of resveratrol fatty alcohols: Differentiation of neural stem cells and modulation of neuroinflammation. *Bioorg. Med. Chem. Lett.* **2007**, *17*, 4218–4222. [CrossRef]
203. Biasutti, L.; Mattarei, A.; Azzolini, M.; La Spina, M.; Sassi, N.; Romio, M.; Paradisi, C.; Zoratti, M. Resveratol derivatives as pharmacological tools. *Ann. N. Y. Acad. Sci.* **2017**, *1403*, 27–37. [CrossRef]
204. Frozza, R.L.; Bernardi, A.; Hoppe, J.B.; Meneghetti, A.B.; Battastini, A.M.; Pohlmann, A.R.; Guterres, S.S.; Salbego, C. Lipid-core nanocapsules improve the effects of resveratrol against Abeta-induced neuroinflammation. *J. Biomed. Nanotechnol.* **2013**, *9*, 2086–2104. [CrossRef] [PubMed]
205. Neves, A.R.; Lúcio, M.; Martins, S.; Lima, J.L.; Reis, S. Novel resveratrol nanodelivery systems based on lipid nanoparticles to enhance its oral bioavailability. *Int. J. Nanomed.* **2013**, *8*, 177–187.
206. Souto, E.B.; Severino, P.; Basso, R.; Santana, M.H. Encapsulation of antioxidants in gastrointestinal-resistant nanoparticulate carriers. *Methods Mol. Biol.* **2013**, *1028*, 37–46. [PubMed]
207. Augustin, M.A.; Sanguansri, L.; Lockett, T. Nano- and micro-encapsulated systems for enhancing the delivery of resveratrol. *Ann. N. Y. Acad. Sci.* **2013**, *1290*, 107–112. [CrossRef]
208. Tsilioni, I.; Panagiotidou, S.; Theoharides, T.C. Exosomes in neurologic and psychiatric disorders. *Clin. Ther.* **2014**, *36*, 882–888. [CrossRef] [PubMed]

209. Siddique, Y.H.; Khan, W.; Singh, B.R.; Naqvi, A.H. Synthesis of alginate-curcumin nanocomposite and its protective role in transgenic Drosophila model of Parkinson's disease. *ISRN Pharmacol.* **2013**, *2013*, 794582. [CrossRef]
210. Amiot, M.J.; Romier, B.; Anh Dao, T.M.; Fanciullino, R.; Ciccolini, J.; Burcelin, R.; Pechere, L.; Emond, C.; Savouret, J.F.; Seree, E. Optimization of *trans*-resveratrol bioavailability for human therapy. *Biochimie* **2013**, *95*, 1233–1238. [CrossRef]
211. Xiao, Y.; Chen, X.; Yang, L.; Zhu, X.; Zou, L.; Meng, F.; Ping, Q. Preparation and oral bioavailability study of curcuminoid-loaded microemulsion. *J. Agric. Food Chem.* **2013**, *61*, 3654–3660. [CrossRef]
212. Ahirrao, M.; Shrotriya, S. In vitro and in vivo evaluation of cubosomal in situ nasal gel containing resveratrol for brain targeting. *Drug Dev. Ind. Pharm.* **2017**, *43*, 1686–1693. [CrossRef]
213. Fang, Z.; Bhandari, B. Encapsulation of polyphenols—A review. *Trends Food Sci. Technol.* **2010**, *21*, 510–523. [CrossRef]
214. Yang, Q.Q.; Wei, X.L.; Fang, R.Y.; Wang, M.; Ge, Y.T.; Zhang, D.; Cheng, L.Z.; Corke, H. Nanochemoprevention with therapeutic benefits: An updates review focused on epigallocatechin gallate delivery. *Crit. Rev. Food Nutr.* **2019**, *23*, 1–22. [CrossRef] [PubMed]
215. Cirillo, G.; Hampel, S.; Spizzirri, U.G.; Parisi, O.I.; Picci, N.; Iemma, F. Carbon nanotubes hybrid hydrogels in drug delivery: A perspective review. *BioMed Res. Int.* **2014**, *2014*, 825017. [CrossRef]
216. Cirillo, G.; Hampel, S.; Klingeler, R.; Puoci, F.; Iemma, F.; Curcio, M.; Parisi, O.I.; Spizzirri, U.G.; Picci, N.; Leonhardt, A.; et al. Antioxidant multi-walled carbon nanotubes by free radical grafting of gallic acid: New materials for biomedical applications. *J. Pharm. Pharmacol.* **2011**, *63*, 179–188. [CrossRef]
217. Kostarelos, K.; Lacerda, L.; Pastorin, G.; Wu, W.; Wieckowski, S.; Luandsilivay, J.; Godefroy, S.; Pantarotto, D.; Briand, J.P.; Muller, S.; et al. Cellular uptake of functionalized carbon nanotubes is independent of functional group and cell type. *Nat. Nanotechnol.* **2007**, *2*, 108–113. [CrossRef]
218. Colvin, V.L. The potential environmental impact of engineered nanomaterials. *Nat. Biotechnol.* **2003**, *21*, 1166–1170. [CrossRef]
219. Lopez-Nicolas, J.M.; Núñez-Delicado, E.; Pérez-López, A.J.; Barrachina, A.C.; Cuadra-Crespo, P. Determination of stoichiometric coefficients and apparent formation constants for beta-cyclodextrin complexes of *trans*-resveratrol using reversed-phase liquid chromatography. *J. Chromatogr. A* **2006**, *1135*, 158–165. [CrossRef]
220. Laza-Knoerr, A.L.; Gref, R.; Couvreur, P. Cyclodextrins for drug delivery. *J. Drug Target* **2010**, *18*, 645–656. [CrossRef]
221. Tapeinos, C.; Battaglini, M.; Ciofani, G. Advances in the design of solid lipid nanoparticles and nanostructured lipid carriers for targeting brain diseases. *J. Control. Release* **2017**, *264*, 306–332. [CrossRef]
222. Hanson, L.R.; Frey, W.H., 2nd. Intranasal delivery bypasses the blood-brain barrier to target therapeutic agents to the central nervous system and treat neurodegenerative disease. *BMC Neurosci.* **2008**, *9*, 5. [CrossRef] [PubMed]
223. Chapman, C.D.; Frey, W.H., 2nd; Craft, S.; Danielyan, L.; Hallschmid, M.; Schiöth, H.B.; Benedict, C. Intranasal treatment of central nervous system dysfunction in humans. *Pharm. Res.* **2013**, *30*, 2475–2484. [CrossRef] [PubMed]
224. Rosenbloom, M.H.; Barclay, T.R.; Pyle, M.; Owens, B.L.; Cagan, A.B.; Anderson, C.P.; Frey, W.H., 2nd; Hanson, L.R. A single-dose pilot trial of intranasal rapid-acting insulin in apolipoprotein E4 carriers with mild-moderate Alzheimer's disease. *CNS Drugs* **2014**, *28*, 1185–1189. [CrossRef]
225. Dewey, R.B., Jr.; Maraganore, D.M.; Ahlskog, J.E.; Matsumoto, J.Y. A double-blind, placebo-controlled study of intranasal apomorphine spray as a rescue agent for off-states in Parkinson's disease. *Mov. Disord.* **1998**, *13*, 782–787. [CrossRef]
226. Crowe, T.P.; Greenlee, M.H.W.; Kanthasamy, A.G.; Hsu, W.H. Mechanism of intranasal drug delivery directly to the brain. *Life Sci.* **2018**, *195*, 44–52. [CrossRef]
227. Desai, P.P.; Patravale, V.B. Curcumin Cocrystal Micelles-Multifunctional Nanocomposites for Management of Neurodegenerative Ailments. *J. Pharm. Sci.* **2018**, *107*, 1143–1156. [CrossRef]
228. Lungare, S.; Hallam, K.; Badhan, R.K. Phytochemical-loaded mesoporous silica nanoparticles for nose-to-brain olfactory drug delivery. *Int. J. Pharm.* **2016**, *513*, 280–293. [CrossRef]
229. Nasr, M. Development of an optimized hyaluronic acid-based lipidic nanoemulsion co-encapsulating two polyphenols for nose to brain delivery. *Drug Deliv.* **2016**, *23*, 1444–1452. [CrossRef]

230. Helson, L. Curcumin (diferuloylmethane) delivery methods: A review. *Biofactors* **2013**, *39*, 21–26. [CrossRef]
231. Sharma, R.A.; McLelland, H.R.; Hill, K.A.; Ireson, C.R.; Euden, S.A.; Manson, M.M.; Pirmohamed, M.; Marnett, L.J.; Gescher, A.J.; Steward, W.P. Pharmacodynamic and pharmacokinetic study of oral Curcuma extract in patients with colorectal cancer. *Clin. Cancer Res.* **2001**, *7*, 1894–1900.
232. DiMauro, T.M. Intranasally Administering Curcumin to the Brain to Treat Alzheimer's Disease. U.S. Patent US20080075671A1, 22 September 2006.
233. Dhuria, S.V.; Hanson, L.R.; Frey, W.H., 2nd. Intranasal delivery to the central nervous system: Mechanisms and experimental considerations. *J. Pharm. Sci.* **2010**, *99*, 1654–1673. [CrossRef]
234. Rice-Evans, C.A.; Miller, N.J.; Bolwell, P.G.; Bramley, P.M.; Pridham, J.B. The relative antioxidant activities of plant-derived polyphenolic flavonoids. *Free Radic. Res.* **1995**, *22*, 375–383. [CrossRef]
235. Pannala, A.S.; Rice-Evans, C.A.; Halliwell, B.; Singh, S. Inhibition of peroxynitrite-mediated tyrosine nitration by catechin polyphenols. *Biochem. Biophys. Res. Commun.* **1997**, *232*, 164–168. [CrossRef]
236. Hollman, P.C.; Cassidy, A.; Comte, B.; Heinonen, M.; Richelle, M.; Richling, E.; Serafini, M.; Scalbert, A.; Sies, H.; Vidry, S. The biological relevance of direct antioxidant effects of polyphenols for cardiovascular health in humans is not established. *J. Nutr.* **2011**, *141*, 989S–1009S. [CrossRef]
237. Di Meo, F.; Lemaur, V.; Cornil, J.; Lazzaroni, R.; Duroux, J.L.; Olivier, Y.; Trouillas, P. Free radical scavenging by natural polyphenols: Atom versus electron transfer. *J. Phys. Chem. A* **2013**, *117*, 2082–2092. [CrossRef]
238. Del Rio, D.; Rodriguez-Mateos, A.; Spencer, J.P.; Tognolini, M.; Borges, G.; Crozier, A. Dietary (poly)phenolics in human health: Structures, bioavailability and evidence of protective effects against chronic diseases. *Antioxid. Redox Signal.* **2013**, *18*, 1818–1892. [CrossRef]
239. Hollman, P.C. Unravelling of the health effects of polyphenols is a complex puzzle complicated by metabolism. *Arch. Biochem. Biophys.* **2014**, *559*, 100–105. [CrossRef]
240. Benzie, I.F.; Szeto, Y.T.; Strain, J.J.; Tomlinson, B. Consumption of green tea causes rapid increase in plasma antioxidant power in humans. *Nutr. Cancer.* **1999**, *34*, 83–87. [CrossRef]
241. Calabrese, E.J. Neuroscience and hormesis: Overview and general findings. *Crit. Rev. Toxicol.* **2008**, *38*, 249–252. [CrossRef]
242. Calabrese, E.J.; Baldwin, L.A. The frequency of U-shaped dose responses in the toxicological literature. *Toxicol. Sci.* **2001**, *62*, 330–338. [CrossRef]
243. Calabrese, E.J.; Baldwin, L.A. The hormetic dose-response model is more common than the threshold model in toxicology. *Toxicol. Sci.* **2003**, *71*, 246–250. [CrossRef]
244. Calabrese, E.J.; Baldwin, L.A. Toxicology rethinks its central belief. *Nature* **2003**, *421*, 691–692. [CrossRef]
245. Peper, A. Aspects of the relationship between drug dose and drug effect. *Dose Response* **2009**, *7*, 172–192. [CrossRef] [PubMed]
246. Calabrese, E.J.; Mattson, M.P.; Calabrese, V. Resveratrol commonly displays hormesis: Occurrence and biomedical significance. *Hum. Exp. Toxicol.* **2010**, *29*, 980–1015. [CrossRef] [PubMed]
247. Szende, B.; Tyihák, E.; Király-Véghely, Z. Dose-dependent effect of resveratrol on proliferation and apoptosis in endothelial and tumor cell cultures. *Exp. Mol. Med.* **2000**, *32*, 88–92. [CrossRef] [PubMed]
248. Dudley, J.; Das, S.; Mukherjee, S.; Das, D.K. Resveratrol, a unique phytoalexin present in red wine, delivers either survival signal or death signal to the ischemic myocardium depending on dose. *J. Nutr. Biochem.* **2009**, *20*, 443–452. [CrossRef] [PubMed]
249. Conte, A.; Pellegrini, S.; Tagliazucchi, D. Effect of resveratrol and catechin on PC12 tyrosine kinase activities and their synergistic protection from beta-amyloid toxicity. *Drugs Exp. Clin. Res.* **2003**, *29*, 243–255. [PubMed]
250. Liu, D.; Gharavi, R.; Pitta, M.; Gleichmann, M.; Mattson, M.P. Nicotinamide prevents NAD+ depletion and protects neurons against excitotoxicity and cerebral ischemia: NAD+ consumption by SIRT1 may endanger energetically compromised neurons. *Neuromol. Med.* **2009**, *11*, 28–42. [CrossRef]
251. Brunet, A.; Sweeney, L.B.; Sturgill, J.F.; Chua, K.F.; Greer, P.L.; Lin, Y.; Tran, H.; Ross, S.E.; Mostoslavsky, R.; Cohen, H.Y.; et al. Stress-dependent regulation of FOXO transcription factors by the SIRT1 deacetylase. *Science* **2004**, *303*, 2011–2015. [CrossRef] [PubMed]
252. Doss, M. Evidence supporting radiation hormesis in atomic bomb survivor cancer mortality data. *Dose Response* **2012**, *10*, 584–592. [CrossRef] [PubMed]
253. Tang, F.R.; Loke, W.K. Molecular mechanisms of low dose ionizing radiation-induced hormesis, adaptive responses, radioresistance, bystander effects and genomic instability. *Int. J. Radiat. Biol.* **2015**, *91*, 13–27. [CrossRef]

254. Calabrese, E.J. Biphasic dose responses in biology, toxicology and medicine: Accounting for their generalizability and quantitative features. *Environ. Pollut.* **2013**, *182*, 452–460. [CrossRef]
255. Martins, I.J. Increased risk for obesity and diabetes with neurodegeneration in developing countries. *J. Mol. Genet. Med.* **2013**, *S1:001*, 1–8. [CrossRef]
256. Faggi, L.; Porrini, V.; Lanzillotta, A.; Benarese, M.; Mota, M.; Tsoukalas, D.; Parrella, E.; Pizzi, M. A polyphenol-enriched supplement exerts potent epigenetic-protective activity in a cell-based model of brain ischemia. *Nutrients* **2019**, *11*, 345. [CrossRef] [PubMed]

International Journal of
Molecular Sciences

MDPI

Review

Curcumin and Heme Oxygenase: Neuroprotection and Beyond

Emanuela Mhillaj [1,2,†], Andrea Tarozzi [3,†], Letizia Pruccoli [3], Vincenzo Cuomo [4], Luigia Trabace [5] and Cesare Mancuso [1,2,*]

1 Fondazione Policlinico Universitario A. Gemelli IRCCS, 00168 Rome, Italy; mhillaj.emanuela@gmail.com
2 Università Cattolica del Sacro Cuore, 00168 Rome, Italy
3 Department for Life Quality Studies, Alma Mater Studiorum, University of Bologna, 47900 Rimini, Italy; andrea.tarozzi@unibo.it (A.T.); letizia.pruccoli2@unibo.it (L.P.)
4 Department of Physiology and Pharmacology "V. Erspamer", Sapienza University of Rome, 00185 Rome, Italy; vincenzo.cuomo@uniroma1.it
5 Department of Clinical and Experimental Medicine, University of Foggia, 71122 Foggia, Italy; luigia.trabace@unifg.it
* Correspondence: cesare.mancuso@unicatt.it; Tel.: +39-06-3015-4367; Fax: +39-06-3050-159
† These authors contributed equally to this work.

Received: 10 April 2019; Accepted: 14 May 2019; Published: 16 May 2019

Abstract: Curcumin is a natural polyphenol component of *Curcuma longa Linn*, which is currently considered one of the most effective nutritional antioxidants for counteracting free radical-related diseases. Several experimental data have highlighted the pleiotropic neuroprotective effects of curcumin, due to its activity in multiple antioxidant and anti-inflammatory pathways involved in neurodegeneration. Although its poor systemic bioavailability after oral administration and low plasma concentrations represent restrictive factors for curcumin therapeutic efficacy, innovative delivery formulations have been developed in order to overwhelm these limitations. This review provides a summary of the main findings involving the heme oxygenase/biliverdin reductase system as a valid target in mediating the potential neuroprotective properties of curcumin. Furthermore, pharmacokinetic properties and concerns about curcumin's safety profile have been addressed.

Keywords: curcumin; free radicals; heme oxygenase; neuroprotection; safety profile

1. Introduction

Curcumin (1,7-bis[4-hydroxy 3-methoxy phenyl]-1,6-heptadiene-3,5-dione) is a polyphenol compound contained in the rhizome of *Curcuma longa Linn*. Indeed, turmeric contains several polyphenols, the most abundant being curcumin (~77%), demethoxycurcumin (~15%), and *bis*-demethoxycurcumin (~3%) [1]. Considering that curcumin prevails over the other congeners, most of the literature in this field has explored the beneficial effects of this compound, although a few papers have studied the physical and biological properties of related curcuminoids [2,3].

In addition to the culinary use due to its spicy and pleasant taste, curcumin has been considered for thousands of years, by traditional Indian medicine, as an effective remedy in the treatment of several diseases [4–6]. Chemically speaking, the curcumin structure presents two aromatic rings holding *o*-methoxy phenolic groups, linked by an α,β-unsaturated β-diketone moiety (Figure 1) [7].

These three reactive functional sites are responsible for the multiple different biological effects of curcumin. Indeed, literature data have reported that the antioxidant activity of curcumin as a free radical scavenger is mediated primarily by the phenolic groups, which undergo oxidation through electron transfer and hydrogen abstraction mechanisms (reviewed in [8]). On the other hand, many studies have demonstrated that curcumin exerts beneficial effects by enhancing the cell stress response in several experimental models, thus supporting the adjuvant role proposed for this dietary supplement in free radical-derived disorders, mainly neurodegenerative diseases [6,9]. In this light, several research studies underlined the pivotal role played by the heme oxygenase/biliverdin reductase system (HO/BVR) as a determinant of curcumin's neuroprotective effects (see below). Unfortunately, despite the huge amount of preclinical studies confirming the pleiotropic effects of curcumin due to HO modulation, the clinical evidence is not strong enough to include chronic curcumin supplementation as an effective strategy to prevent or contrast neurodegeneration. One of the reasons behind the dichotomy between preclinical and clinical results has been identified in curcumin pharmacokinetics in humans; first of all, the poor bioavailability after ingestion and the effective concentrations reached in tissues. However, several efforts have been made over recent years to overcome these limitations, with encouraging results.

The aim of this review is to summarize the preclinical and clinical outcomes which have appeared in the scientific literature, supporting or contrasting the claimed therapeutic efficacy of curcumin in neurodegeneration. The reason why the focus has been on the HO/BVR system depends on the several lines of evidence highlighting its role as a determinant of curcumin neuroprotection. Finally, some safety issues related to curcumin supplementation have been also reported.

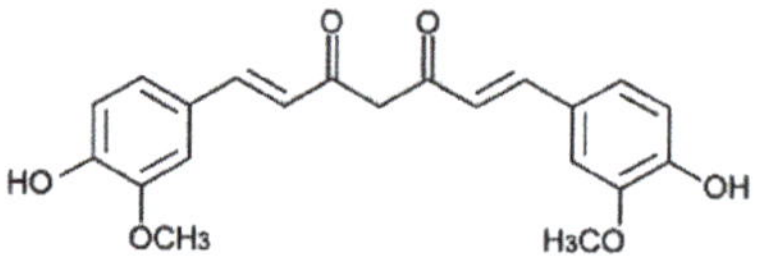

Figure 1. Chemical structure of curcumin.

2. The Heme Oxygenase/Biliverdin Reductase Pathway

Heme oxygenase catalyzes the oxygen- and NADPH-dependent oxidation of hemoproteins' heme moieties at the alpha-meso carbon bridge, yielding equimolar amounts of ferrous iron, carbon monoxide (CO), and biliverdin (BV), the latter being further reduced into bilirubin (BR) by biliverdin reductase [10,11]. Heme oxygenase exists as two main isoforms, named HO-1 and HO-2. Although these isozymes share the same mechanism of action, their regulation and distribution are quite different. Heme oxygenase-1 is the inducible isoform and both its gene transcription and protein levels increase in response to free radicals, e.g., reactive oxygen species and reactive nitrogen species (ROS and RNS, respectively) [11]. Furthermore, HO-1 is the major isoform detected in both the liver and spleen, even if it is expressed, at lower levels, in some brain areas, such as the hippocampus and hypothalamus [11,12]. Conversely, the constitutive isoform HO-2 is involved in the physiological turnover of heme and is mainly detectable in neurons and testes [13,14].

The cytoprotective effects of the HO/BVR system depend on several factors: (i) the degradation of heme, which may become toxic under unbalanced redox conditions; (ii) the generation of CO, which improves mitochondrial biogenesis, counteracts NADPH oxidase-induced ROS generation, activates pro-survival systems (e.g., the protein kinase B/Akt and extracellular signal-related kinase (ERK)/p38 mitogen-activated protein kinase (MAPK) signaling pathways), modulates the release of neuroinflammatory mediators (e.g., interleukin-1β and prostaglandins), dilates cerebral and peripheral vessels, and inhibits platelet aggregation; (iii) the antioxidant and antiviral activities of BR [14–20]. Interestingly, the modulation of both mitochondrial respiratory chains and NADPH oxidase accounts for CO's antiproliferative effects [21].

Under oxidative stress and inflammatory conditions, several transcription factors, including nuclear factor erythroid 2-related factor 2 (Nrf2), nuclear factor k-light-chain-enhancer of activated B cells (NF-kB), and hypoxia-inducible factor 1 (HIF1), are established as pivotal regulators of HO-1 induction in the brain [22,23]. Among these transcription factors, Nrf2 plays the conservative role of a positive regulator of HO-1 induction in the development and progression of many diseases [24].

Conversely, a few negative regulators, such as Keap-1 and Bach1, can modulate the crosstalk between the Nrf2 and HO-1 [25,26].

3. Curcumin, Neuroprotection, and the HO/BVR Pathway

Over the last 15 years, many papers have appeared in the scientific literature dealing with the cytoprotective effects of curcumin through the up-regulation of HO-1 (see Table 1).

Table 1. Contribution of HO-1 up-regulation to the biological effects of curcumin in preclinical in vitro and in vivo models.

Preclinical Model	Curcumin (Concentration or Dose)	Effect(s)	Reference(s)
Endothelial cells	2–30 μM	Enhancement of cellular resistance against oxidative damage. Alleviation of vasodilator dysfunction	[27–30]
Renal tubule cells	1–50 μM	Cytoprotection. Inhibition of fibrosis.	[31–33]
Anti-Thy 1 glomerulonephritis rats Nephrectomized rats	100 mg/kg i.p. 75 mg/kg per os	Reduction of renal fibrosis and proteinuria. Inhibition of lipid peroxidation, inflammation and renal fibrosis. Amelioration of renal function.	[34,35]
Hepatocytes	1–50 μM	Cytoprotection against cold/rewarming- or ethanol-induced damages.	[36–38]
Monocytes	1–20 μM	Activation of ARE-modulated genes via PKCδ. Inhibition of inflammation.	[39,40]
Macrophages	0.5–50 μM	Inhibition of inflammation.	[41–43]
Cardiac myoblasts	5–30 μM	Inhibition of apoptosis. Cytoprotection against cold-storage damage.	[44,45]
Smooth muscle cells	1–20 μM	Inhibition of proliferation.	[46]
LPS-treated mice	30 mg/kg i.p.	Prevention of pulmonary sequestration of neutrophils.	[47]
Pancreatic islets	6–10 μM	Inhibition of islet damage during cryopreservation. Improvement of insulin secretion.	[48,49]
Rat testicular injury	200 mg/kg i.v. 200 mg/kg per os for 30 days before and 45 days after injury.	Inhibition of lipid peroxidation and increase in testicular spermatogenesis. Reduced lipid peroxidation; improvement of serum testosterone level.	[50,51]
Fibroblasts	5–25 μM	Induction of apoptosis and modulation of pathological scar formation.	[52]
High-fat-diet-fed mice	50 mg/kg per os	Improvement in muscular oxidative stress and glucose tolerance.	[53]
Bladder cancer cells	10 μM	Modulation of cancer cell proliferation.	[54]
Breast cancer cells	5–20 μM	Inhibition of tumor invasion.	[55]
Hepatoma cells expressing HCV	5–25 μM	Inhibition of HCV replication.	[56]
Lung cancer cells expressing influenza virus	0.1–10 μM	Inhibition of virus-induced lung injury.	[57]
Keratinocytes	1–30 μM	Anti-inflammatory activity.	[58]
Metabolic syndrome in rats	5 mg/kg i.p. for 6 weeks	Prevention of hyperinsulinemia and amelioration of endothelial-dependent relaxation.	[59]

ARE, antioxidant responsive element; HCV, hepatitis C virus; i.p., intraperitoneal route of administration; i.v., intravenous route of administration; PKC, protein kinase C.

The following are the main studies supporting the neuroprotective effects of curcumin via the modulation of the HO/BVR pathway.

Scapagnini et al. [60] have shown how curcumin (5–25 μM) up-regulates HO-1 in cultured rat hippocampal neurons and, thus, the polyphenol enhances the cell stress response against glucose oxidase-mediated oxidative damage. Shin et al. [61] reported that curcumin (200 mg/kg by intraperitoneal route (i.p.)) reduced kainic acid-induced seizures in mice through the increased expression of HO-1 and endothelial nitric oxide synthase (eNOS) in hippocampal astrocytes, whereas Park and Chun [62] demonstrated that curcumin (0.1–10 μM) reduces oxidative stress, apoptosis, and mitochondrial damage through the direct involvement of HO-1 in BV-2 microglial cells.

These early studies were followed by several others describing the neuroprotective effects of curcumin in neurovascular disorders. Curcumin (100 mg/kg i.p. or 5–30 μM), via HO-1 over-expression, was neuroprotective in a rat model of focal ischemia [63] and in rat cerebellar granule neurons exposed to hemin [64]. In an experimental system of rat hypoxic-ischemic brain injury, curcumin (150 mg/kg per os for three days) overexpressed HO-1 with a mechanism related to Nrf2 nuclear translocation [65]. In addition, curcumin (1–100 μM) has been shown to up-regulate HO-1 and, through this mechanism, it prevents oxygen glucose deprivation-induced damage in rat brain microvascular endothelial cells, a model mimicking the blood–brain barrier (BBB) function [66].

With regard to neurodegenerative diseases, in a rodent model of Alzheimer's disease (AD), e.g., the SAMP8 mouse, 500 mg/kg of curcumin in a five month diet increased *HO-1* gene expression, together with regulators of mitochondrial function, e.g., the translocator protein (TSPO) [67]. Similarly, by up-regulating HO-1, curcumin (1.25–20 μM) inhibited programmed cell death and prevented the loss of mitochondrial function in SH-SY5Y neuroblastoma cells transfected with appoptosin, a pro-apoptotic protein overexpressed in AD [68]. Concerning neurodegenerative diseases, curcumin (100 mg/kg twice a day for 50 days intragastrically) contrasted extrapyramidal symptoms and increased HO-1 expression, through Akt/Nrf2 phosphorylation, in the substantia nigra pars compacta of rats treated with rotenone, a pharmacological tool able to destroy dopaminergic neurons and, therefore, used to induce experimental Parkinson's disease (PD) [69]. It is no longer a hypothesis that the cytoprotective effects of curcumin against neuroinflammation depend on the inhibition, HO-1-mediated, of cytokine release and iNOS overexpression in rat microglia [70,71].

Finally, curcumin (15 μM or 200 mg/kg for four days) has been shown to counteract both hydrogen peroxide-induced damage in human retinal pigment cells [72] and cisplatin-induced ototoxicity in outer hair cells [73].

As far as the modulation of HO-2 by curcumin and the potential neuroprotective features, only limited evidence is available. As shown by Yin et al. [74], curcumin (5 μM) up-regulated HO-1 but down-regulated HO-2 in *APPswe* transfected SH-SY5Y. In the same experimental system, curcumin was able to activate phosphoinositide 3-kinase (PI3K) and Akt [74]. By keeping this in mind, it is necessary to draw the conclusion that in selected experimental settings, the neuroprotective outcomes of curcumin strictly depend on the fine-tuning of the HO-1/HO-2 balance, in concert with the modulation of other pro-survival systems, such as PI3K and Akt.

An accurate analysis of both previous paragraphs and Table 1 has drawn attention to the fact that the concentrations of curcumin responsible for protective effects on various organs and tissues, primarily on the brain, were obtained with polyphenol concentrations in the micromolar size range. That said, curcumin, per os, has about a 60% bioavailability, due to a marked first-pass metabolism [9,75]. This implies a low concentration of curcumin in both blood and tissues, even at high doses. Curcumin plasma levels up to 0.16 μM have been detected in humans treated with polyphenol at supra-maximal doses (10–12 g/day), whereas at the lowest doses, curcumin (450–3600 mg/day for one week) reached the plasma concentration of about 0.003 μM [76,77]. In chronic administrations, curcumin (1–4 g/day for six months) exhibited plasma concentrations in the range of 0.06–0.27 μM [78]. With regard to tissue levels, the available data are quite limited. In patients suffering from colorectal cancer and treated with curcumin (1.8 to 3.6 g/day for seven days), concentrations of polyphenol in colorectal tumor tissue and normal tissue were about 7 nmol/g and 20 nmol/g, respectively [79]. These data lead to the conclusion that the plasma concentrations of curcumin that can be reached in the plasma, even after high dose chronic supplementation, are at least two–three orders of magnitude lower than those at which the polyphenol has shown therapeutic effects in in vitro preclinical models. The calculation of the concentrations of curcumin in the tissues is more difficult and may appear less accurate. In the brain, which is protected by BBB, the achievable curcumin concentrations are even lower than those detected in the blood and other tissues. These analytical data have important consequences also from a functional point of view. In subjects with AD, supplementation with curcumin (1–4 g/day for six months) reduced neither peripheral biomarkers of inflammation (e.g., isoprostanes) nor amyloid-β-peptide (Aβ) serum levels; importantly, curcumin did not improve cognitive functions—evaluated through the mini-mental status examination test—in AD patients [78]. Concerning the contribution of the HO/BVR system to the cytoprotective effects of curcumin, the study by Klickovic et al. [80] is significant, showing how 10 healthy male subjects treated with 12 g curcumin *per* os, did not have any significant induction of *HO-1* gene and protein in peripheral blood mononuclear cells up to 48 h from treatment.

In order to overcome limitations due to the poor bioavailability after ingestion and the low plasma concentrations, new formulations of curcumin complexed with liposoluble matrices have been developed (for an extensive review on this topic see [81]) (Table 2).

Table 2. The main pharmacokinetic parameters of curcumin and some of its novel formulations (adapted from [82]).

Formulation	AUC	C_{max}	T_{max}	$T_{1/2}$
Curcumin	~312 ng/mL·h [a]	~ 245 nM [a]	0.5 h [a]	~1.0 h [a]
Curcumin-PLGA	~3224 ng/mL·h [b]	~ 710 nM [b]	2.0 h [b]	
Curcumin-TMC	~12,760 ng/mL·h [c]	~3.3 μM [c]	2.0 h [c]	~12 h [c]
Curcumin-SLN	~42,000 ng/mL·h [d]	~38 μM [d]	0.5 h [d]	

[a] Male Sprague-Dawley rats treated with 250 mg/kg curcumin per os; [b] male Sprague-Dawley rats treated with 100 mg/kg curcumin-PLGA per os; [c] Balb/c mice treated with 50 mg/kg curcumin-TMC per os; [d] male Wistar rats treated with 50 mg/kg curcumin-SLN per os; AUC, area under the curve; C_{max}, peak plasma concentration; PLGA, poly(lactic-co-glycolic) acid; SLN, solid lipid nanoparticles; T_{max}, time necessary to reach the C_{max}; $T_{1/2}$, half-life; TMC, N-trimethyl chitosan.

Among the matrices complexed with curcumin, the ones that are better characterized, from a pharmacokinetic viewpoint, are poly(lactic-co-glycolic) acid (PLGA) derivatives, solid lipid nanoparticles (SLN), and N-trimethyl-chitosan (TMC) [82,83]. Preclinical studies in rodents (Table 2) have shown how the complexation of curcumin with these different carriers increases the C_{max} of both SLN and TMC (155 times and 13 times greater than curcumin, respectively) markedly, suggesting a more effective absorption of the active ingredient [82]. Furthermore, the increase in the area under the curve demonstrates how the presence of SLN or TMC can improve curcumin bioavailability by about 135 times and 41 times, respectively [82]. Finally, an approximately 10-fold increase in the half-life ($T_{1/2}$) of curcumin in the case of formulations based on SLN and TMC implies an extension of the time of persistence of the active agent in the body and, therefore, a more prolonged pharmacological action [82]. Unfortunately, no studies are available in the literature on the interaction of such novel curcumin liposoluble formulations and HO. Indeed, few studies which have been carried out using novel gelatin-based water-soluble formulations of curcumin and remarkable results have been reported. The oral administration of water-soluble curcumin (2–10 mg/kg per os for 45 days) increased plasma insulin levels and improved glucose absorption in diabetic rats by up-regulating HO-1 expression in the pancreas and liver [84]. The same authors supported the beneficial effects of water-soluble curcumin (2–10 mg/kg per os up to one week) in an experimental model of erectile dysfunction. At a dose of 10 mg/kg, water-soluble curcumin over-expressed HO-1 and soluble guanylyl cyclase (sGC) as early as 1 h after treatment, with a concomitant increase in intracavernosal pressure. These effects were maintained over one week from treatment [85].

Although not strictly related to any modulation of the HO system, it is worth mentioning a novel formulation of curcumin complexed with exosomes; these latter are extracellular microvesicles (diameter ranging from 30 to 100 nm) able to carry several types of agents, thus enhancing their bioavailability [86]. Interestingly, curcumin-exosome has been shown to improve cognitive function in a preclinical model of AD, through the inhibition of tau hyperphosphorylation via Akt activation [87].

4. Curcumin's Safety Profile

In any case, regardless of whether it is pure curcumin or new liposoluble or water-soluble formulations, it is worth considering the possibility that the administration of high doses of curcumin causes toxic effects. An organic extract, called turmeric oleoresin, containing a high percentage of curcumin (79–85%), at the concentration of 50,000 ppm (equivalent to 2600 mg/kg and 2800 mg/kg in male and female rats, respectively) has been shown to increase the incidence of ulcers, hyperplasia, and inflammation in the forestomach, cecum, and colon of male and female rats supplemented for two years [88]. Increased evidence of small intestine carcinomas in male mice supplemented with curcumin (0.2 mg/kg) has also been described [88]. Furthermore, curcumin (0.5–2% with the diet for either 2 or 12 weeks) exhibited iron-chelating activity in mice, thus suggesting its involvement in the onset of hypochromic anemia [88]. Finally, curcumin (1 g or 4 g per os for one or six months) modestly increased cholesterol plasma levels in Chinese subjects aged 50 years or older [89]. Regarding the interaction with drug-metabolizing enzymes, curcumin has been shown to inhibit not only several subtypes of cytochrome P450 (CYP), such as CYP1A2, CYP2A6, CYP2B6, CYP2C9, CYP2D6, and CYP3A4, but also uridine dinucleotide phosphate glucuronosyltransterases (UGT), sulfotransferase, glutathione-S-transferase, and organic anion transporting polypeptides (OATP) [9,75,90]. Among the drugs metabolized by these enzymes, whose blood levels may be altered by curcumin and for which further research is needed to assess the effects in cases of chronic supplementation, there are midazolam, talinolol, nifedipine, rosuvastatin, docetaxel, warfarin, clopidogrel, and norfloxacin ([90] and references therein).

In April 2017, the European Food Scientific Agency (EFSA) pointed out that there is no scientific evidence strong enough to justify the use of curcumin in inflammatory diseases, such as osteoarthritis and rheumatoid arthritis [91].

5. Conclusions

In this review, we have summarized the conflicting preclinical and clinical results on the neuroprotective effects of curcumin. Furthermore, we have made our best efforts to provide a critical analysis of the pharmacological issues responsible for this divergence, which have precluded the full development of curcumin supplementation as a useful strategy in neurodegenerative diseases. The intriguing results, in terms of improved absorption and bioavailability, obtained with lipid- and water-soluble curcumin formulations, should prompt researchers to transfer this technology to clinical studies, with the hope of overwhelming the pharmacokinetic limitations experienced with standard curcumin. The contribution of pharmaceutical companies to scale up and transpose into clinics these encouraging preclinical results is more than welcome.

Author Contributions: Conceptualization, C.M., A.T., L.T.; Data Curation, E.M., L.P.; Writing—Original Draft Preparation, C.M., E.M., A.T.; Writing—Review & Editing, E.M., A.T., L.P., V.C., L.T., C.M.; Supervision, C.M., L.T., V.C.

Funding: This work was supported by Catholic University grant "Fondi Ateneo" to C.M.

Conflicts of Interest: The authors declare no conflict of interest.

References

1. Den Haan, J.; Morrema, T.H.J.; Rozemuller, A.J.; Bouwman, F.H.; Hoozemans, J.J.M. Different Curcumin Forms Selectively Bind Fibrillar Amyloid Beta In Post Mortem Alzheimer's Disease Brains: Implications for In-Vivo Diagnostics. *Acta Neuropathol. Commun.* **2018**, *6*, 75. [CrossRef] [PubMed]
2. Lestari, M.L.; Indrayanto, G. Profiles Drug Subst. *Excip. Relat. Methodol.* **2014**, *39*, 113–204.
3. Jitoe-Masuda, A.; Fujimoto, A.; Masuda, T. Curcumin: From Chemistry to Chemistry-Based Functions. *Curr. Pharm. Des.* **2013**, *19*, 2084–2092.
4. Mantzorou, M.; Pavlidou, E.; Vasios, G.; Tsagalioti, E.; Giaginis, C. Effects of Curcumin Consumption on Human Chronic Diseases: A Narrative Review of The Most Recent Clinical Data. *Phytother. Res.* **2018**, *32*, 957–975. [CrossRef]
5. Marchiani, A.; Rozzo, C.; Fadda, A.; Delogu, G.; Ruzza, P. Curcumin and Curcumin-Like Molecules: From Spice to Drugs. *Curr. Med. Chem.* **2014**, *21*, 204–222. [CrossRef]
6. Calabrese, V.; Bates, T.E.; Mancuso, C.; Cornelius, C.; Ventimiglia, B.; Cambria, M.T.; Di Renzo, L.; de Lorenzo, A.; Dinkova-Kostova, A.T. Curcumin and The Cellular Stress Response in Free Radical-Related Diseases. *Mol. Nutr. Food Res.* **2008**, *52*, 1062–1073. [CrossRef] [PubMed]
7. Priyadarsini, K.I. The Chemistry of Curcumin: From Extraction to Therapeutic Agent. *Molecules* **2014**, *19*, 20091–20112. [CrossRef] [PubMed]
8. Del Prado-Audelo, M.L.; Caballero-Floran, I.H.; Meza-Toledo, J.A.; Mendoza-Munoz, N.; Gonzalez-Torres, M.; Floran, B.; Cortes, H.; Leyva-Gomez, G. Formulations of Curcumin Nanoparticles for Brain Diseases. *Biomolecules* **2019**, *9*, 56. [CrossRef] [PubMed]
9. Mancuso, C.; Siciliano, R.; Barone, E.; Preziosi, P. Natural Substances and Alzheimer's Disease: From Preclinical Studies to Evidence Based Medicine. *Biochim. Biophys. Acta* **2012**, *1822*, 616–624. [CrossRef] [PubMed]
10. Mancuso, C. Bilirubin and Brain: A Pharmacological Approach. *Neuropharmacology* **2017**, *118*, 113–123. [CrossRef] [PubMed]
11. Maines, M.D. The Heme Oxygenase System: A Regulator of Second Messenger Gases. *Annu. Rev. Pharmacol. Toxicol.* **1997**, *37*, 517–554. [CrossRef] [PubMed]
12. Mancuso, C. Heme Oxygenase and Its Products in The Nervous System. *Antioxid. Redox Signal.* **2004**, *6*, 878–887. [PubMed]
13. Maines, M.D. The Heme Oxygenase System: Update 2005. *Antioxid. Redox Signal.* **2005**, *7*, 1761–1766. [CrossRef] [PubMed]
14. Mancuso, C.; Santangelo, R.; Calabrese, V. The Heme Oxygenase/Biliverdin Reductase System: A Potential Drug Target in Alzheimers Disease. *J. Biol. Regul. Homeost. Agents* **2013**, *27*, 75–87. [PubMed]
15. Mancuso, C.; Barone, E.; Guido, P.; Miceli, F.; Di Domenico, F.; Perluigi, M.; Santangelo, R.; Preziosi, P. Inhibition of Lipid Peroxidation and Protein Oxidation By Endogenous and Exogenous Antioxidants in Rat Brain Microsomes In Vitro. *Neurosci. Lett.* **2012**, *518*, 101–105. [CrossRef]
16. Mancuso, C.; Tringali, G.; Grossman, A.; Preziosi, P.; Navarra, P. The Generation of Nitric Oxide and Carbon Monoxide Produces Opposite Effects on The Release of Immunoreactive Interleukin-1β From the Rat Hypothalamus In Vitro: Evidence for The Involvement of Different Signaling Pathways. *Endocrinology* **1998**, *139*, 1031–1037. [CrossRef] [PubMed]
17. Mancuso, C.; Perluigi, M.; Cini, C.; de Marco, C.; Giuffrida Stella, A.M.; Calabrese, V. Heme Oxygenase and Cyclooxygenase in The Central Nervous System: A Functional Interplay. *J. Neurosci. Res.* **2006**, *84*, 1385–1391. [CrossRef]
18. Suliman, H.B.; Piantadosi, C.A. Mitochondrial Biogenesis: Regulation by Endogenous Gases during Inflammation and Organ Stress. *Curr. Pharm. Des.* **2014**, *20*, 5653–5662. [CrossRef]
19. Basuroy, S.; Tcheranova, D.; Bhattacharya, S.; Leffler, C.W.; Parfenova, H. Nox4 Nadph Oxidase-Derived Reactive Oxygen Species, Via Endogenous Carbon Monoxide, Promote Survival of Brain Endothelial Cells During Tnf-α-Induced Apoptosis. *Am. J. Physiol. Cell Physiol.* **2011**, *300*, C256–C265. [CrossRef]
20. Santangelo, R.; Mancuso, C.; Marchetti, S.; di Stasio, E.; Pani, G.; Fadda, G. Bilirubin: An Endogenous Molecule with Antiviral Activity In Vitro. *Front. Pharmacol.* **2012**, *3*, 36. [CrossRef]

21. Taille, C.; El-Benna, J.; Lanone, S.; Boczkowski, J.; Motterlini, R. Mitochondrial Respiratory Chain and Nad(P)H Oxidase Are Targets for The Antiproliferative Effect of Carbon Monoxide in Human Airway Smooth Muscle. *J. Biol. Chem.* **2005**, *280*, 25350–25360. [CrossRef]
22. Jazwa, A.; Cuadrado, A. Targeting Heme Oxygenase-1 for Neuroprotection and Neuroinflammation in Neurodegenerative Diseases. *Curr. Drug Targets* **2010**, *11*, 1517–1531. [CrossRef]
23. Wardyn, J.D.; Ponsford, A.H.; Sanderson, C.M. Dissecting Molecular Cross-Talk Between Nrf2 And Nf-κb Response Pathways. *Biochem. Soc. Trans.* **2015**, *43*, 621–626. [CrossRef]
24. Loboda, A.; Damulewicz, M.; Pyza, E.; Jozkowicz, A.; Dulak, J. Role of Nrf2/Ho-1 System in Development, Oxidative Stress Response and Diseases: An Evolutionarily Conserved Mechanism. *Cell Mol. Life Sci.* **2016**, *73*, 3221–3247. [CrossRef]
25. Sun, J.; Hoshino, H.; Takaku, K.; Nakajima, O.; Muto, A.; Suzuki, H.; Tashiro, S.; Takahashi, S.; Shibahara, S.; Alam, J.; et al. Hemoprotein Bach1 Regulates Enhancer Availability of Heme Oxygenase-1 Gene. *Embo. J.* **2002**, *21*, 5216–5224. [CrossRef]
26. Kitamuro, T.; Takahashi, K.; Ogawa, K.; Udono-Fujimori, R.; Takeda, K.; Furuyama, K.; Nakayama, M.; Sun, J.; Fujita, H.; Hida, W.; et al. Bach1 Functions as A Hypoxia-Inducible Repressor for The Heme Oxygenase-1 Gene in Human Cells. *J. Biol. Chem.* **2003**, *278*, 9125–9133. [CrossRef]
27. Motterlini, R.; Foresti, R.; Bassi, R.; Green, C.J. Curcumin, An Antioxidant and Anti-Inflammatory Agent, Induces Heme Oxygenase-1 And Protects Endothelial Cells Against Oxidative Stress. *Free Radic. Biol. Med.* **2000**, *28*, 1303–1312. [CrossRef]
28. Scapagnini, G.; Foresti, R.; Calabrese, V.; Giuffrida Stella, A.M.; Green, C.J.; Motterlini, R. Caffeic Acid Phenethyl Ester and Curcumin: A Novel Class of Heme Oxygenase-1 Inducers. *Mol. Pharmacol.* **2002**, *61*, 554–561. [CrossRef]
29. El-Bassossy, H.M.; El-Maraghy, N.N.; El-Fayoumi, H.M.; Watson, M.L. Haem Oxygenase-1 Induction Protects Against Tumour Necrosis Factor Alpha Impairment of Endothelial-Dependent Relaxation in Rat Isolated Pulmonary Artery. *Br. J. Pharmacol.* **2009**, *158*, 1527–1535. [CrossRef]
30. Fang, X.D.; Yang, F.; Zhu, L.; Shen, Y.L.; Wang, L.L.; Chen, Y.Y. Curcumin Ameliorates High Glucose-Induced Acute Vascular Endothelial Dysfunction in Rat Thoracic Aorta. *Clin. Exp. Pharmacol. Physiol.* **2009**, *36*, 1177–1182. [CrossRef]
31. Hill-Kapturczak, N.; Thamilselvan, V.; Liu, F.; Nick, H.S.; Agarwal, A. Mechanism of Heme Oxygenase-1 Gene Induction by Curcumin In Human Renal Proximal Tubule Cells. *Am. J. Physiol. Renal. Physiol.* **2001**, *281*, F851–F859. [CrossRef]
32. Balogun, E.; Hoque, M.; Gong, P.; Killeen, E.; Green, C.J.; Foresti, R.; Alam, J.; Motterlini, R. Curcumin Activates the Haem Oxygenase-1 Gene Via Regulation of Nrf2 and The Antioxidant-Responsive Element. *Biochem. J.* **2003**, *371*, 887–895. [CrossRef]
33. Zhang, L.; Fang, Y.; Xu, Y.; Lian, Y.; Xie, N.; Wu, T.; Zhang, H.; Sun, L.; Zhang, R.; Wang, Z. Curcumin Improves Amyloid β-Peptide (1-42) Induced Spatial Memory Deficits Through Bdnf-Erk Signaling Pathway. *PLoS ONE* **2015**, *10*, E0131525. [CrossRef]
34. Gaedeke, J.; Noble, N.A.; Border, W.A. Curcumin Blocks Fibrosis in Anti-Thy 1 Glomerulonephritis Through Up-Regulation of Heme Oxygenase 1. *Kidney Int.* **2005**, *68*, 2042–2049. [CrossRef] [PubMed]
35. Soetikno, V.; Sari, F.R.; Lakshmanan, A.P.; Arumugam, S.; Harima, M.; Suzuki, K.; Kawachi, H.; Watanabe, K. Curcumin Alleviates Oxidative Stress, Inflammation, And Renal Fibrosis in Remnant Kidney Through the Nrf2-Keap1 Pathway. *Mol. Nutr. Food Res.* **2013**, *57*, 1649–1659. [CrossRef]
36. Mcnally, S.J.; Harrison, E.M.; Ross, J.A.; Garden, O.J.; Wigmore, S.J. Curcumin Induces Heme Oxygenase 1 Through Generation of Reactive Oxygen Species, P38 Activation and Phosphatase Inhibition. *Int. J. Mol. Med.* **2007**, *19*, 165–172. [CrossRef] [PubMed]
37. Mcnally, S.J.; Harrison, E.M.; Ross, J.A.; Garden, O.J.; Wigmore, S.J. Curcumin Induces Heme Oxygenase-1 In Hepatocytes and Is Protective in Simulated Cold Preservation and Warm Reperfusion Injury. *Transplantation* **2006**, *81*, 623–626. [CrossRef] [PubMed]
38. Bao, W.; Li, K.; Rong, S.; Yao, P.; Hao, L.; Ying, C.; Zhang, X.; Nussler, A.; Liu, L. Curcumin Alleviates Ethanol-Induced Hepatocytes Oxidative Damage Involving Heme Oxygenase-1 Induction. *J. Ethnopharmacol.* **2010**, *128*, 549–553. [CrossRef]

39. Rushworth, S.A.; Ogborne, R.M.; Charalambos, C.A.; O'connell, M.A. Role of Protein Kinase C Delta in Curcumin-Induced Antioxidant Response Element-Mediated Gene Expression in Human Monocytes. *Biochem. Biophys. Res. Commun.* **2006**, *341*, 1007–1016. [CrossRef] [PubMed]
40. Hsu, H.Y.; Chu, L.C.; Hua, K.F.; Chao, L.K. Heme Oxygenase-1 Mediates the Anti-Inflammatory Effect of Curcumin Within Lps-Stimulated Human Monocytes. *J. Cell Physiol.* **2008**, *215*, 603–612. [CrossRef]
41. Kim, K.M.; Pae, H.O.; Zhung, M.; Ha, H.Y.; Ha, Y.A.; Chai, K.Y.; Cheong, Y.K.; Kim, J.M.; Chung, H.T. Involvement of Anti-Inflammatory Heme Oxygenase-1 In the Inhibitory Effect of Curcumin on The Expression of Pro-Inflammatory Inducible Nitric Oxide Synthase in Raw264.7 Macrophages. *Biomed. Pharmacother.* **2008**, *62*, 630–636. [CrossRef]
42. Zhong, Y.; Liu, T.; Lai, W.; Tan, Y.; Tian, D.; Guo, Z. Heme Oxygenase-1-Mediated Reactive Oxygen Species Reduction Is Involved in The Inhibitory Effect of Curcumin on Lipopolysaccharide-Induced Monocyte Chemoattractant Protein-1 Production in Raw264.7 Macrophages. *Mol. Med. Rep.* **2013**, *7*, 242–246. [CrossRef]
43. Liu, L.; Shang, Y.; Li, M.; Han, X.; Wang, J.; Wang, J. Curcumin Ameliorates Asthmatic Airway Inflammation by Activating Nuclear Factor-E2-Related Factor 2/Haem Oxygenase (Ho)-1 Signalling Pathway. *Clin. Exp. Pharmacol. Physiol.* **2015**, *42*, 520–529. [CrossRef]
44. Abuarqoub, H.; Green, C.J.; Foresti, R.; Motterlini, R. Curcumin Reduces Cold Storage-Induced Damage in Human Cardiac Myoblasts. *Exp. Mol. Med.* **2007**, *39*, 139–148. [CrossRef]
45. Yang, X.; Jiang, H.; Shi, Y. Upregulation of Heme Oxygenase-1 Expression by Curcumin Conferring Protection from Hydrogen Peroxide-Induced Apoptosis in H9c2 Cardiomyoblasts. *Cell Biosci.* **2017**, *7*, 20. [CrossRef]
46. Pae, H.O.; Jeong, G.S.; Jeong, S.O.; Kim, H.S.; Kim, S.A.; Kim, Y.C.; Yoo, S.J.; Kim, H.D.; Chung, H.T. Roles of Heme Oxygenase-1 In Curcumin-Induced Growth Inhibition in Rat Smooth Muscle Cells. *Exp. Mol. Med.* **2007**, *39*, 267–277. [CrossRef]
47. Olszanecki, R.; Gebska, A.; Korbut, R. The Role of Haem Oxygenase-1 in The Decrease of Endothelial Intercellular Adhesion Molecule-1 Expression by Curcumin. *Basic Clin. Pharmacol. Toxicol.* **2007**, *101*, 411–415. [CrossRef]
48. Kanitkar, M.; Bhonde, R.R. Curcumin Treatment Enhances Islet Recovery by Induction of Heat Shock Response Proteins, Hsp70 And Heme Oxygenase-1, During Cryopreservation. *Life Sci.* **2008**, *82*, 182–189. [CrossRef]
49. Abdel Aziz, M.T.; El-Asmar, M.F.; El Nadi, E.G.; Wassef, M.A.; Ahmed, H.H.; Rashed, L.A.; Obaia, E.M.; Sabry, D.; Hassouna, A.A.; Abdel Aziz, A.T. The Effect of Curcumin on Insulin Release in Rat-Isolated Pancreatic Islets. *Angiology* **2010**, *61*, 557–566. [CrossRef]
50. Wei, S.M.; Yan, Z.Z.; Zhou, J. Curcumin Attenuates Ischemia-Reperfusion Injury in Rat Testis. *Fertil. Steril.* **2009**, *91*, 271–277. [CrossRef]
51. Abd El-Fattah, A.A.; Fahim, A.T.; Sadik, N.A.H.; Ali, B.M. Resveratrol and Curcumin Ameliorate Di-(2-Ethylhexyl) Phthalate Induced Testicular Injury in Rats. *Gen. Comp. Endocrinol.* **2016**, *225*, 45–54. [CrossRef]
52. Scharstuhl, A.; Mutsaers, H.A.; Pennings, S.W.; Szarek, W.A.; Russel, F.G.; Wagener, F.A. Curcumin-Induced Fibroblast Apoptosis And In Vitro Wound Contraction Are Regulated by Antioxidants and Heme Oxygenase: Implications for Scar Formation. *J. Cell Mol. Med.* **2009**, *13*, 712–725. [CrossRef]
53. He, H.J.; Wang, G.Y.; Gao, Y.; Ling, W.H.; Yu, Z.W.; Jin, T.R. Curcumin Attenuates Nrf2 Signaling Defect, Oxidative Stress in Muscle and Glucose Intolerance in High Fat Diet-Fed Mice. *World J. Diabetes* **2012**, *3*, 94–104. [CrossRef]
54. Wu, S.Y.; Lee, Y.R.; Huang, C.C.; Li, Y.Z.; Chang, Y.S.; Yang, C.Y.; Wu, J.D.; Liu, Y.W. Curcumin-Induced Heme Oxygenase-1 Expression Plays A Negative Role for Its Anti-Cancer Effect in Bladder Cancers. *Food Chem. Toxicol.* **2012**, *50*, 3530–3536. [CrossRef]
55. Lee, W.Y.; Chen, Y.C.; Shih, C.M.; Lin, C.M.; Cheng, C.H.; Chen, K.C.; Lin, C.W. The Induction of Heme Oxygenase-1 Suppresses Heat Shock Protein 90 And the Proliferation of Human Breast Cancer Cells Through Its Byproduct Carbon Monoxide. *Toxicol. Appl. Pharmacol.* **2014**, *274*, 55–62. [CrossRef]
56. Chen, M.H.; Lee, M.Y.; Chuang, J.J.; Li, Y.Z.; Ning, S.T.; Chen, J.C.; Liu, Y.W. Curcumin Inhibits HCV Replication by Induction of Heme Oxygenase-1 And Suppression of Akt. *Int. J. Mol. Med.* **2012**, *30*, 1021–1028. [CrossRef]

57. Han, S.; Xu, J.; Guo, X.; Huang, M. Curcumin Ameliorates Severe Influenza Pneumonia via Attenuating Lung Injury and Regulating Macrophage Cytokines Production. *Clin. Exp. Pharmacol. Physiol.* **2018**, *45*, 84–93. [CrossRef]
58. Youn, G.S.; Kwon, D.J.; Ju, S.M.; Choi, S.Y.; Park, J. Curcumin Ameliorates TNF-α-Induced Icam-1 Expression and Subsequent Thp-1 Adhesiveness via The Induction of Heme Oxygenase-1 in The Hacat Cells. *BMB Rep.* **2013**, *46*, 410–415. [CrossRef]
59. El-Bassossy, H.M.; Hassan, N.; Zakaria, M.N. Heme Oxygenase-1 Alleviates Vascular Complications Associated with Metabolic Syndrome: Effect on Endothelial Dependent Relaxation and No Production. *Chem. Biol. Interact.* **2014**, *223*, 109–115. [CrossRef]
60. Scapagnini, G.; Colombrita, C.; Amadio, M.; D'agata, V.; Arcelli, E.; Sapienza, M.; Quattrone, A.; Calabrese, V. Curcumin Activates Defensive Genes and Protects Neurons Against Oxidative Stress. *Antioxid. Redox Signal.* **2006**, *8*, 395–403. [CrossRef]
61. Shin, H.J.; Lee, J.Y.; Son, E.; Lee, D.H.; Kim, H.J.; Kang, S.S.; Cho, G.J.; Choi, W.S.; Roh, G.S. Curcumin Attenuates the Kainic Acid-Induced Hippocampal Cell Death in the Mice. *Neurosci. Lett.* **2007**, *416*, 49–54. [CrossRef]
62. Park, E.; Chun, H.S. Protective Effects of Curcumin on Manganese-Induced Bv-2 Microglial Cell Death. *Biol. Pharm. Bull.* **2017**, *40*, 1275–1281. [CrossRef]
63. Yang, C.; Zhang, X.; Fan, H.; Liu, Y. Curcumin Upregulates Transcription Factor Nrf2, Ho-1 Expression and Protects Rat Brains Against Focal Ischemia. *Brain Res.* **2009**, *1282*, 133–141. [CrossRef]
64. Gonzalez-Reyes, S.; Guzman-Beltran, S.; Medina-Campos, O.N.; Pedraza-Chaverri, J. Curcumin Pretreatment Induces Nrf2 And an Antioxidant Response and Prevents Hemin-Induced Toxicity in Primary Cultures of Cerebellar Granule Neurons Of Rats. *Oxid. Med. Cell. Longev.* **2013**, *2013*, 801418. [CrossRef]
65. Cui, X.; Song, H.; Su, J. Curcumin Attenuates Hypoxic-Ischemic Brain Injury in Neonatal Rats Through Induction of Nuclear Factor Erythroid-2-Related Factor 2 And Heme Oxygenase-1. *Exp. Ther. Med.* **2017**, *14*, 1512–1518. [CrossRef]
66. Wang, Y.F.; Gu, Y.T.; Qin, G.H.; Zhong, L.; Meng, Y.N. Curcumin Ameliorates the Permeability of The Blood-Brain Barrier During Hypoxia by Upregulating Heme Oxygenase-1 Expression in Brain Microvascular Endothelial Cells. *J. Mol. Neurosci.* **2013**, *51*, 344–351. [CrossRef]
67. Eckert, G.P.; Schiborr, C.; Hagl, S.; Abdel-Kader, R.; Muller, W.E.; Rimbach, G.; Frank, J. Curcumin Prevents Mitochondrial Dysfunction in the Brain of the Senescence-Accelerated Mouse-Prone 8. *Neurochem. Int.* **2013**, *62*, 595–602. [CrossRef]
68. Zheng, K.M.; Zhang, J.; Zhang, C.L.; Zhang, Y.W.; Chen, X.C. Curcumin Inhibits Appoptosin-Induced Apoptosis via Upregulating Heme Oxygenase-1 Expression in Sh-Sy5y Cells. *Acta Pharmacol. Sin.* **2015**, *36*, 544–552. [CrossRef]
69. Cui, Q.; Li, X.; Zhu, H. Curcumin Ameliorates Dopaminergic Neuronal Oxidative Damage via Activation of The Akt/Nrf2 Pathway. *Mol. Med. Rep.* **2016**, *13*, 1381–1388. [CrossRef]
70. Jin, M.; Park, S.Y.; Shen, Q.; Lai, Y.; Ou, X.; Mao, Z.; Lin, D.; Yu, Y.; Zhang, W. Anti-Neuroinflammatory Effect of Curcumin on Pam3csk4-Stimulated Microglial Cells. *Int. J. Mol. Med.* **2018**, *41*, 521–530. [CrossRef]
71. Parada, E.; Buendia, I.; Navarro, E.; Avendano, C.; Egea, J.; Lopez, M.G. Microglial Ho-1 Induction by Curcumin Provides Antioxidant, Antineuroinflammatory, And Glioprotective Effects. *Mol. Nutr. Food Res.* **2015**, *59*, 1690–1700. [CrossRef]
72. Woo, J.M.; Shin, D.Y.; Lee, S.J.; Joe, Y.; Zheng, M.; Yim, J.H.; Callaway, Z.; Chung, H.T. Curcumin Protects Retinal Pigment Epithelial Cells Against Oxidative Stress Via Induction of Heme Oxygenase-1 Expression and Reduction of Reactive Oxygen. *Mol. Vis.* **2012**, *18*, 901–908.
73. Fetoni, A.R.; Eramo, S.L.; Paciello, F.; Rolesi, R.; Podda, M.V.; Troiani, D.; Paludetti, G. Curcuma Longa (Curcumin) Decreases In Vivo Cisplatin-Induced Ototoxicity Through Heme Oxygenase-1 Induction. *Otol. Neurotol.* **2014**, *35*, E169–E177. [CrossRef] [PubMed]
74. Yin, W.; Zhang, X.; Li, Y. Protective Effects of Curcumin in Appswe Transfected Sh-Sy5y Cells. *Neural. Regen Res.* **2012**, *7*, 405–412.
75. Mancuso, C.; Barone, E. Curcumin in Clinical Practice: Myth or Reality? *Trends Pharmacol. Sci.* **2009**, *30*, 333–334. [CrossRef] [PubMed]

76. Lao, C.D.; Ruffin, M.T.T.; Normolle, D.; Heath, D.D.; Murray, S.I.; Bailey, J.M.; Boggs, M.E.; Crowell, J.; Rock, C.L.; Brenner, D.E. Dose Escalation of a Curcuminoid Formulation. *BMC Complement. Altern Med.* **2006**, *6*, 10. [CrossRef]
77. Garcea, G.; Jones, D.J.; Singh, R.; Dennison, A.R.; Farmer, P.B.; Sharma, R.A.; Steward, W.P.; Gescher, A.J.; Berry, D.P. Detection of Curcumin and Its Metabolites in Hepatic Tissue and Portal Blood of Patients Following Oral Administration. *Br. J. Cancer* **2004**, *90*, 1011–1015. [CrossRef]
78. Baum, L.; Lam, C.W.; Cheung, S.K.; Kwok, T.; Lui, V.; Tsoh, J.; Lam, L.; Leung, V.; Hui, E.; Ng, C.; et al. Six-Month Randomized, Placebo-Controlled, Double-Blind, Pilot Clinical Trial of Curcumin in Patients with Alzheimer Disease. *J. Clin. Psychopharmacol.* **2008**, *28*, 110–113. [CrossRef] [PubMed]
79. Garcea, G.; Berry, D.P.; Jones, D.J.; Singh, R.; Dennison, A.R.; Farmer, P.B.; Sharma, R.A.; Steward, W.P.; Gescher, A.J. Consumption of The Putative Chemopreventive Agent Curcumin by Cancer Patients: Assessment of Curcumin Levels in the Colorectum And Their Pharmacodynamic Consequences. *Cancer Epidemiol. Biomarkers Prev.* **2005**, *14*, 120–125.
80. Klickovic, U.; Doberer, D.; Gouya, G.; Aschauer, S.; Weisshaar, S.; Storka, A.; Bilban, M.; Wolzt, M. Human Pharmacokinetics of High Dose Oral Curcumin and Its Effect on Heme Oxygenase-1 Expression in Healthy Male Subjects. *Biomed. Res. Int.* **2014**, *2014*, 458592. [CrossRef]
81. Jamwal, R. Bioavailable Curcumin Formulations: A Review of Pharmacokinetic Studies in Healthy Volunteers. *J. Integr. Med.* **2018**, *16*, 367–374. [CrossRef]
82. Mancuso, C. Key Factors Which Concur to the Correct Therapeutic Evaluation of Herbal Products in Free Radical-Induced Diseases. *Front. Pharmacol.* **2015**, *6*, 86. [CrossRef] [PubMed]
83. Rakotoarisoa, M.; Angelova, A. Amphiphilic Nanocarrier Systems for Curcumin Delivery in Neurodegenerative Disorders. *Medicines* **2018**, *5*, 126. [CrossRef]
84. Abdel Aziz, M.T.; El-Asmar, M.F.; El-Ibrashy, I.N.; Rezq, A.M.; Al-Malki, A.L.; Wassef, M.A.; Fouad, H.H.; Ahmed, H.H.; Taha, F.M.; Hassouna, A.A.; et al. Effect of Novel Water Soluble Curcumin Derivative On Experimental Type- 1 Diabetes Mellitus (Short Term Study). *Diabetol. Metab. Syndr.* **2012**, *4*, 30. [CrossRef]
85. Abdel Aziz, M.T.; El Asmer, M.F.; Rezq, A.; Kumosani, T.A.; Mostafa, S.; Mostafa, T.; Atta, H.; Abdel Aziz Wassef, M.; Fouad, H.H.; Rashed, L.; et al. Novel Water-Soluble Curcumin Derivative Mediating Erectile Signaling. *J. Sex. Med.* **2010**, *7*, 2714–2722. [CrossRef] [PubMed]
86. Aqil, F.; Munagala, R.; Jeyabalan, J.; Agrawal, A.K.; Gupta, R. Exosomes for The Enhanced Tissue Bioavailability and Efficacy of Curcumin. *Aaps J.* **2017**, *19*, 1691–1702. [CrossRef]
87. Wang, H.; Sui, H.; Zheng, Y.; Jiang, Y.; Shi, Y.; Liang, J.; Zhao, L. Curcumin-Primed Exosomes Potently Ameliorate Cognitive Function in Ad Mice by Inhibiting Hyperphosphorylation of The Tau Protein Through the Akt/Gsk-3beta Pathway. *Nanoscale* **2019**, *11*, 7481–7496. [CrossRef] [PubMed]
88. Burgos-Moron, E.; Calderon-Montano, J.M.; Salvador, J.; Robles, A.; Lopez-Lazaro, M. The Dark Side of Curcumin. *Int. J. Cancer* **2010**, *126*, 1771–1775. [CrossRef] [PubMed]
89. Baum, L.; Cheung, S.K.; Mok, V.C.; Lam, L.C.; Leung, V.P.; Hui, E.; Ng, C.C.; Chow, M.; Ho, P.C.; Lam, S.; et al. Curcumin Effects on Blood Lipid Profile in A 6-Month Human Study. *Pharmacol. Res.* **2007**, *56*, 509–514. [CrossRef] [PubMed]
90. Bahramsoltani, R.; Rahimi, R.; Farzaei, M.H. Pharmacokinetic Interactions of Curcuminoids With Conventional Drugs: A Review. *J. Ethnopharmacol* **2017**, *209*, 1–12. [CrossRef]
91. Turck, D.; Bresson, J.L.; Burlingame, B.; Dean, T.; Fairweather-Tait, S.; Heinonen, M.; Hirsch-Ernst, K.I.; Mangelsdorf, I.; Mcardle, H.J.; Naska, A.; et al. Curcumin and Normal Functioning of Joints: Evaluation of A Health Claim Pursuant to Article 13(5) Of Regulation (Ec) No 1924/2006. *EFSA J.* **2017**, *15*. [CrossRef]

International Journal of
Molecular Sciences

MDPI

Review

The Potential of Flavonoids for the Treatment of Neurodegenerative Diseases

Pamela Maher

Salk Institute for Biological Studies, La Jolla, CA 92037, USA; pmaher@salk.edu; Tel.: +1-858-453-4100

Received: 24 May 2019; Accepted: 18 June 2019; Published: 22 June 2019

Abstract: Neurodegenerative diseases, including Alzheimer's disease (AD), Parkinson's disease (PD), Huntington's disease (HD), and amyotrophic lateral sclerosis (ALS), currently affect more than 6 million people in the United States. Unfortunately, there are no treatments that slow or prevent disease development and progression. Regardless of the underlying cause of the disorder, age is the strongest risk factor for developing these maladies, suggesting that changes that occur in the aging brain put it at increased risk for neurodegenerative disease development. Moreover, since there are a number of different changes that occur in the aging brain, it is unlikely that targeting a single change is going to be effective for disease treatment. Thus, compounds that have multiple biological activities that can impact the various age-associated changes in the brain that contribute to neurodegenerative disease development and progression are needed. The plant-derived flavonoids have a wide range of activities that could make them particularly effective for blocking the age-associated toxicity pathways associated with neurodegenerative diseases. In this review, the evidence for beneficial effects of multiple flavonoids in models of AD, PD, HD, and ALS is presented and common mechanisms of action are identified. Overall, the preclinical data strongly support further investigation of specific flavonoids for the treatment of neurodegenerative diseases.

Keywords: oxidative stress; cognitive dysfunction; inflammation; cell death; synapse loss; protein aggregation; neurodegenerative disease

---

1. Introduction-What Is Neurodegeneration?

Before reviewing the potential beneficial effects of natural products, and in the case of this review, specifically flavonoids, on neurodegeneration, it is essential that a definition of neurodegeneration be established. Over 15 years ago, Przedborski et al. [1] published a comprehensive discussion of this topic that is still highly relevant today. They defined neurodegeneration generally as "any pathological condition primarily affecting neurons". More specifically, they characterized neurodegenerative diseases as a large, heterogeneous group of neurological disorders that affect distinct subsets of neurons in specific anatomical locations. They also noted that a number of disorders that are either not primary neuronal diseases or where neurons die of a known cause, such as hypoxia or poison, are not neurodegenerative diseases. While hundreds of neurodegenerative disorders are known, most of the attention has focused on four: Alzheimer's disease (AD), Parkinson's disease (PD), Huntington's disease (HD), and amyotrophic lateral sclerosis (ALS), although others, such as frontotemporal dementia (FTD), are as common, if not more so, than either HD or ALS [2]. In the United States, there are currently 5.8 million people with AD [3], over 700,000 with PD [4], ~30,000 with HD [5], ~16,000 with ALS [6], and 50,000–60,000 with FTD [2]. For AD, PD, and ALS, there are both genetic and sporadic forms of the disease, with the vast majority of the cases of all three being sporadic. In contrast, FTD has a stronger genetic component [2] and almost all cases of HD are dominantly inherited [5]. Regardless, for all of these diseases and irrespective of the cause, the strongest risk factor for developing any of them is increasing age. This suggests that changes that occur in the aging brain put it at increased risk for the development of a neurodegenerative disease and that the identification of those changes could provide a means to develop therapeutics that can at least slow, if not prevent, disease development and/or progression.

1.1. Aging and Age-Associated Changes in the Brain

Among the pathophysiological changes that occur in the aging brain, those that have been identified as potentially contributing to neurodegeneration include increases in oxidative stress, alterations in energy metabolism, loss of neurotrophic support, alterations in protein processing leading to the accumulation of protein aggregates, dysfunction of the neurovascular system, and immune system activation [7,8]. Given this multiplicity of changes, it is unlikely that targeting a single change will prove effective at preventing nerve cell damage and death. In addition, there is a strong possibility that the relative contributions of each of these changes will vary among individuals. Importantly, these changes interact with lifestyle, environmental, and genetic risk factors with varying degrees of penetrance. For example, although AD is defined in terms of plaque and tangle pathology, it is most frequently associated with other detrimental events, such as microvascular damage and inflammation [9]. Thus, it is likely it will be necessary to use combinations of drugs directed against different targets in order to effectively prevent these age-related changes to the brain. However, this approach is subject to a number of potential problems, including pharmacokinetic and bioavailability challenges, which in central nervous system (CNS) diseases are exacerbated by the difficulty of getting multiple compounds across the blood brain barrier and the potential for adverse drug–drug interactions. A better approach is to identify small molecules that have multiple biological activities that can impact the multiplicity of age-associated pathophysiological changes to the brain that contribute to neurodegenerative disease development and progression [10].

1.2. Approaches to Drug Discovery for Neurodegenerative Diseases

Since the 1990s, the combination of molecular and structural biology, combinatorial chemistry, and high throughput screening has dominated the drug discovery process [11]. This approach provides a rapid process for the discovery of drug candidates with high selectivity and high affinity for a specific molecular target. However, it has not produced the successes that were initially expected, especially with respect to complicated problems such as neurodegenerative diseases. Prior to the development of this target-based drug discovery approach, new drugs were discovered by evaluating chemicals against observable characteristics or phenotypes, in biological systems such as cells or animals. While this approach has fallen out of favor with the pharmaceutical industry, surprisingly a recent study showed that it still continues to be more successful than target-based approaches for the identification of first-in-class small molecule drugs [12]. It has been argued that this is because target-based discovery is based on a priori assumptions that do not take into account the complexities of biological systems or diseases [7,12].

The ideal phenotypic drug screening paradigm would employ the ultimate end user—humans—and this is how most of the natural product-based, first-in-class drugs were originally discovered. However, this is no longer an ethically viable approach. Laboratory animals, primarily disease models in mice, are currently used for preclinical testing but using them for the initial screening of drug candidates is impractical due to cost and time constraints, as well as the drive to reduce animal use in research. A reasonable alternative is to create cell-based assays that define molecular toxicity pathways relevant to age-associated neurodegeneration and select drug candidates that work in multiple assays, not just one [7]. In this way, the screening paradigms have disease relevance, reproducibility, and reasonable throughput. Many arguments can be made against the relevance of any single cellular screening assay based on the cell type or the nature of the toxic insult. Thus, to account for individual weaknesses, phenotypic screening paradigms for neurodegenerative diseases should combine multiple assays that address the different toxicities associated with the aging brain. This enables the identification of potent, disease-modifying compounds for preclinical testing in animal models of neurodegenerative diseases. In general, for screening for drug candidates against neurodegenerative diseases, these assays will utilize primary neurons, neuron-like cell lines, or microglial cell lines that are subjected to a toxic insult that has been observed to occur in the aging brain. However, the critical question still remains of what exactly should be screened.

1.3. What to Start with

One excellent source for multi-target compounds is the original pharmacopeia—plants. The earliest records describing the use of plants for medicinal purposes date back to 2600–2900 BC [13]. Still today, ~25% of all prescribed drugs are thought to be derived from plants [14]. Plants synthesize a huge array of compounds called secondary metabolites that are not required for plant growth. These compounds are derived from a limited number of basic chemical scaffolds, which are modified by specific types of substitutions. It has been suggested [14] that these compounds, as well as receptors, enzymes, and regulatory proteins, originated from a relatively small number of parental molecules, which may have co-evolved to interact with one another. Although their biological functions and structures have since diverged, structural features shared from their common past may be the reason that they interact with medically relevant targets.

1.4. Why Focus on Flavonoids

Among the huge number of plant-derived secondary metabolites, several epidemiological studies have specifically highlighted the potential beneficial role of flavonoids for the prevention of neurodegenerative diseases. The over 5000 different flavonoids can be divided into six groups (flavones, flavonols, flavanones, flavanols, anthocyanidins, and isoflavones) based on the degree of oxidation of the central C ring, the hydroxylation pattern of the rings, and the substitution at the 3 position (Figure 1). Within each group, diversity is generated by the arrangement of the hydroxyl groups combined with glycosylation or alkylation [15].

A retrospective study that looked at flavonoid intake versus disability adjusted life years (a measure of the burden that a disease has on those affected in a population) due to dementia in 23 developed countries found that total combined flavonoid intake was significantly and negatively correlated with dementia [16]. Among the flavonoid groups, only flavonol consumption showed a significant, negative correlation with dementia. Consistent with these results, a prospective cohort study [17] found that the risk ratio for dementia (risk of dementia in the high flavonoid group (dementia patients/total people in the group)/risk of dementia in the low flavonoid group (dementia patients/total people in the group)) between the highest and lowest tertiles of flavonoid intake was 0.49.

A very large epidemiological study (total of ~130,000 people followed for 20–22 years) published several years ago [18] examined whether higher intakes of total flavonoids were associated with a lower risk of PD. Five major sources of dietary flavonoids (tea, berry fruits, apples, red wine, and orange or orange juice) were examined using a food composition database and a food frequency questionnaire. In men, those with the highest quintile of flavonoid consumption had a 40% lower risk of developing PD as compared to those in the lowest quintile. However, a significant relationship between overall flavonoid consumption and PD risk was not seen in women.

Flavonoids were historically characterized on the basis of their antioxidant and free radical scavenging effects. However, more recent studies have shown that flavonoids have a wide range of activities that could make them particularly effective for blocking the age-associated toxicity pathways associated with neurodegenerative diseases [19–22].

In the following paragraphs, the results of pre-clinical, and in a few cases, clinical studies, that looked at the beneficial effects of different flavonoids in animal models of AD, PD, HD, or ALS, and when information is available, their possible modes of action, will be described. Although the goal was to be as comprehensive as possible, some studies may have been missed inadvertently. Several recent reviews also cover a subset of these topics [20–22]. There are no studies on flavonoids in models of FTD, although mouse models of the disease do exist [23].

For each disease, a brief overview is given followed by a description of the models used to study the effects of flavonoids on the disease, and then the flavonoid results based on the subclasses of flavonoids are discussed. The focus is primarily on studies employing single flavonoids and the analysis mainly utilizes primary reports.

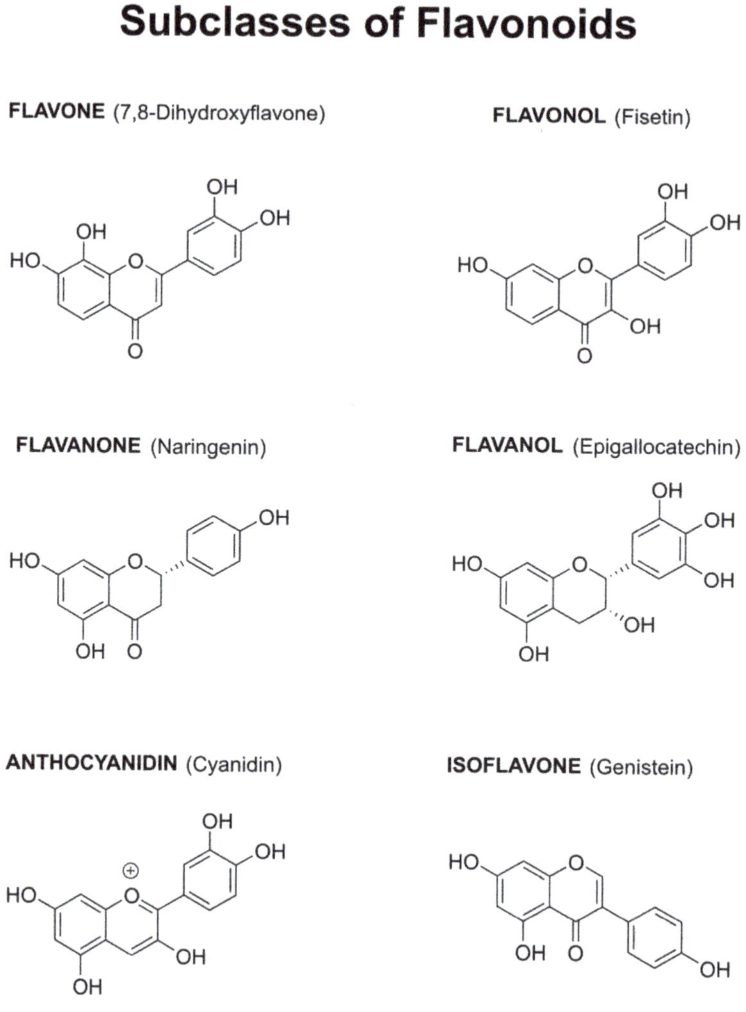

Figure 1. Structures of Representative Flavonoids from the Six Classes.

1.5. Flavonoids and Alzheimer's Disease (AD)

Alzheimer's disease is the most common type of dementia. It is characterized pathologically by the presence of both extracellular neuritic plaques containing amyloid beta (Aβ) peptide and intracellular neurofibrillary tangles containing tau [24]. Clinically, AD results in a progressive loss of cognitive ability and eventually daily function activities [25,26]. Current approved therapies are only symptomatic, providing moderate improvements in memory without altering the progression of the disease pathology [27,28]. Although a large number of clinical trials have been conducted in recent years with drug candidates designed to directly or indirectly reduce the amyloid plaque load, all of these trials have failed [29].

Three different types of models have been used to study the possible beneficial effects of flavonoids in AD: interventional, transgenic, and sporadic. For the interventional studies, Aβ is injected directly into the cerebral ventricles in the brains of the rodents (intracerebroventricular (icv) injection). There are numerous transgenic models of AD that are based on the mutations associated with the rare genetic form of the disease (familial AD or FAD). The models develop different degrees of cognitive impairment, levels of plaques and tangles, synaptic loss, gliosis, and nerve cell death depending on the type and number of mutations (1–5) [30]. Although AD drug discovery has largely focused on these FAD models, this form of the disease accounts for only ~1% of the total cases [31], and may be quite distinct from the much more prevalent, old-age-associated, sporadic form of AD. Importantly, while many therapies directed against the amyloid pathway are effective in FAD transgenic mice, to date none has translated into the clinic [29]. Since old age is by far the greatest risk factor for AD [31,32], animal models that incorporate aging into disease development may prove more useful for the development of therapies. One mouse model of aging that also develops characteristics of AD is the senescence-accelerated prone 8 (SAMP8) mouse that was developed in Japan by selective breeding of a rapidly aging phenotype. These mice exhibit a progressive, age-associated decline in brain function similar to human AD patients [33–35]. As they age, SAMP8 mice develop an early deterioration in learning and memory, as well as a number of pathophysiological alterations in the brain, including increased oxidative stress, inflammation, vascular impairment, gliosis, Aβ accumulation, and tau hyperphosphorylation. Using an integrative multiomics approach, we recently identified a number of behavioral and physiological changes that are altered with aging in these mice [36]. Although much less widely used, the senescence-accelerated OXYS rat also spontaneously develops all of the brain changes associated with AD, including structural alterations, neuronal loss, Aβ accumulation, tau hyperphosphorylation, and cognitive impairment [37].

The flavone 7,8-dihydroxyflavone (7,8-DHF) has been tested by several labs in the 5 x FAD model (multiple AD-linked mutations in the amyloid precursor protein (APP) and presenilin 1 (PS1)). Improvement in performance in the Y maze, a test for working memory, was seen following short term (10 days) intraperitoneal (ip) injection of 7,8-DHF (5 mg/kg) into 12–15 months old 5 x FAD mice [38]. Chronic oral administration of 7,8-DHF (5 mg/kg/day) from 2–6 months of age in this same model also improved memory and reduced synapse loss [39]. In contrast, in a 2 x FAD model (AD-linked mutations in APP and PS1), daily ip administration of 7,8-DHF (5 mg/kg) for 4 weeks beginning at 6 weeks of age showed no effect on learning and memory impairments in the Morris water maze [40]. However, in the first two studies [38,39], clear activation of TrkB, the proposed target of 7,8-DHF, was seen, while in the third study [40], this was not examined. Thus, the lack of effects on memory could be due to a failure to activate TrkB in this study. Further studies on this flavone and AD are clearly warranted.

The flavone apigenin has been tested in a 2 × FAD model (AD-linked mutations in APP and PS1), where oral administration (40 mg/kg/day) for 12 weeks beginning at 4 months of age resulted in improved learning and memory, reduced deposition of insoluble Aβ, a decrease in markers of oxidative stress, and an increase in the activity of the ERK-CREB pathway, an indicator of neurotrophic activity [41].

In a 1 × FAD model (AD-linked mutation in APP), a four month daily ip treatment with the citrus flavone nobiletin (10 mg/kg) improved memory and reduced the levels of both soluble and insoluble Aβ [42]. Consistent with these results, three months of daily ip injections of nobiletin (10 and 30 mg/kg) starting at 6 months of age in the 3 × FAD model (AD-linked mutations in APP, PS1, and tau) resulted in an improvement in memory on multiple tests and a reduction in soluble Aβ levels, as well as reactive oxygen species (ROS) in the mice treated with 30 mg/kg [43]. Similarly, 2 months of daily ip injections of nobiletin (10 and 50 mg/kg) starting at 4–6 months in the SAMP8 mice improved memory in multiple assays and reduced some markers of oxidative stress at both doses [44].

Daily ip injections of the flavone baicalein (10 mg/kg) beginning at 6 months of age also prevented deficits in working memory and reduced the production of Aβ in a 1 x FAD model (AD-linked mutation in APP) [45].

The flavonol fisetin has been tested in all three types of AD models (icv Aβ injection, 2 × FAD mice, SAMP8 mice), where it consistently prevented the loss of cognitive function [46–48]. Both oral administration (25 mg/kg/day) [46,47] and daily ip injections (20 mg/kg) [48] proved effective. In all of the models, fisetin was found to maintain the levels of synaptic proteins and to reduce markers of inflammation. It also reduced markers of oxidative stress and particularly lipid peroxidation and activated the ERK pathway, which is involved in both memory [49] and neurotrophic factor production and signaling [50]. However, the effects on the levels of soluble and insoluble Aβ varied between the different models, suggesting that this may not be the critical target.

Another flavonol, quercetin, was also shown to have similar benefits in multiple models of AD [51] following either oral administration (SAMP8) (25 mg/kg/day) [52] or ip injection (25 mg/kg every 2 days) (3 x FAD) [53]. Similar to fisetin, not only was quercetin able to reduce cognitive impairment but also modulated multiple targets in the brains of the treated mice, including the levels of soluble and insoluble Aβ and the activation of astrocytes or microglia, indicators of an on-going inflammatory response. Interestingly, in the study on SAMP8 mice [52], the effects of the administration of free quercetin (25 mg/kg/day) to those of nanoencapsulated quercetin particles (25 mg/kg/every 2 days) were compared. An almost 2-fold higher level of quercetin was found in the brains of the quercetin nanoparticle-treated mice, which correlated with significant effects on learning and memory, as well as astrogliosis [52]. Rutin, a glycoside of quercetin that combines quercetin with rutinose, was also found to have beneficial effects in rats injected icv with Aβ [54]. Daily ip injection of rutin (100 mg/kg) for 3 weeks after Aβ administration prevented memory loss, reduced lipid peroxidation, and increased markers of neurotrophic factor activity [54]. A recent review covered some of these same studies in more detail [20].

Similar to the results with the quercetin nanoparticles, it was found that daily ip injection of a mixture of anthocyanins (glycosylated form of anthocyanidins) from Korean black soybeans encapsulated in gold nanoparticles (10 mg/kg/day) were much more effective at reducing memory impairments, loss of synaptic proteins, and neuroinflammation in icv Aβ-injected mouse brains than the free anthocyanins [55,56].

Non-fermented teas, such as green tea, contain high levels of catechins (flavanols), including (−)-epigallocatechin gallate (EGCG), (−)-epigallocatechin, (−)-epicatechin gallate, (−)-epicatechin, and (+)-catechin. Studies on EGCG in both a 1 x FAD mouse model (50 mg/kg/day) [57] and the SAMP8 mouse (15 mg/kg/day) [58] found that long-term, oral administration improved cognitive function, reduced the levels of soluble Aβ, and prevented the decrease in some synaptic proteins. In addition to these animal studies, over 10 clinical studies have been conducted on green tea and AD [59]. These include cross-sectional and longitudinal studies where the frequency of drinking green tea and cognitive function were assessed either at a single time point or over time and interventional studies where participants were given a green tea extract and followed over time. Most, but not all, of the longitudinal and cross-sectional studies showed an inverse relationship between green tea consumption and cognitive dysfunction. Furthermore, a meta-analysis showed a dose-dependent effect of green tea consumption on cognitive impairment. However, the interventional studies had many fewer participants and the results were less consistent.

Other flavonoids that have shown benefits in animal models of AD include the flavanone glycoside hesperidin (100 mg/kg/day) [60,61] and the isoflavone puerarin (30 mg/kg/day) [62].

In summary, multiple flavonoids have shown significant benefits in three distinct models of AD (Table 1). All of the flavonoids described above improved cognitive function, and where examined, reduced markers of inflammation, oxidative stress, and synaptic dysfunction, and increased neurotrophic factor signaling. In addition, many reduced the accumulation of soluble or insoluble Aβ. Together these results support the idea that multi-target compounds, such as flavonoids that act on several different pathophysiological changes that occur in the aging brain and that are exacerbated in AD, have a strong potential for the treatment of the disease. Unfortunately, very few human studies have been performed, so whether this potential will ever be realized is not clear. In addition, the significantly enhanced effects seen with the nanoparticles of quercetin [52] and anthocyanins [55,56] strongly suggest that if flavonoids are to be used pharmacologically, then the formulation needs to be more carefully considered.

Table 1. Flavonoids that have shown efficacy in preclinical models of Alzheimer's disease (AD), Parkinson's disease (PD), Huntington's disease (HD), or amyotrophic lateral sclerosis (ALS).

	AD	PD	HD	ALS
Flavones				
7,8-DHF	X	X	X	X
Apigenin	X	X		
Baicalein	X	X		
Chrysin		X	X	
Luteolin		X		
Morin		X		
Nobiletin	X	X		
Flavonols				
Fisetin	X	X	X	X
Kaempferol			X	
Myricetin		X		
Myricitrin		X		
Quercetin	X	X	X	
Rutin	X	X	X	
Flavanols				
Catechin		X		
Epicatechin		X		
ECGC	X	X		X
Flavanones				
Hesperetin		X		
Hesperidin	X	X	X	
Naringenin		X		
Naringin		X		
Anthocyanidins				
Anthocyanins	X		X	
Isoflavones				
Genistein		X	X	
Puerarin	X	X		

1.6. Flavonoids and Parkinson's Disease (PD)

Parkinson's disease (PD) is a chronic, progressive neurodegenerative disease and the second most common neurodegenerative disease after Alzheimer's. The characteristic features of PD include resting tremor, bradykinesia (slowness of movement), rigidity, and postural instability [63]. PD is also associated with a variety of non-motor symptoms that contribute to disability. The majority of PD cases are sporadic with only about 10% of PD patients reporting a family history of the disease [64]. Age is the greatest risk factor for disease development. The pathological hallmark of PD is the degeneration of the dopaminergic (DA) neurons in the substantia nigra pars compacta (SNc) [63]. Since these neurons synapse with neurons in the striatum, their demise leads to the depletion of striatal dopamine. PD is also characterized by the presence of cytoplasmic protein aggregates, called Lewy bodies, in the remaining DA neurons of the SNc. Currently, there is no test to diagnose PD prior to the onset of motor symptoms and available treatments only improve the symptoms but do not stop disease progression. Importantly, by the time that the first symptoms appear, striatal dopamine is reduced by ~80%, and ~60% of the DA neurons of the SNc have died [65]. Thus, both better methods of diagnosis and treatments that can begin before overt disease onset are needed.

Animal models of PD generally involve treatment with a toxin, such as a pesticide or other toxic compound that has been associated with PD in vivo. The two most widely used models are 6-hydroxydopamine (6-OHDA) and 1-methyl-4-phenyl-1,2,3,6-tetrahydropyridine (MPTP) [66,67]. Both the herbicide paraquat and the pesticide rotenone have also been used to model PD. However, none of these models recapitulates all of the aspects of human PD [66,67] and most are very rapid onset, as compared to the age-dependent development of PD in human patients. Although animal models in which one or more of the genes associated with familial PD are mutated have been developed [67], most of these genetic PD models lack nigrostriatal degeneration and there is also a problem with inconsistent phenotypes between different mouse lines with the same mutation [67]. Thus, they have not been used extensively for testing of potential therapeutic compounds.

Quite a large number of different flavonoids from most of the different classes have been tested in the different rodent toxin models of PD (Table 1). Some of these results have been recently reviewed [22] and these and others are described below.

The flavone baicalein has been tested in several different models in rodents, including the MPTP model using both ip injection in rats (40 mg/kg/day) [68] and oral administration in mice (200 mg/kg/day) [69], and the rotenone model using ip injection in rats (2.5 mg/kg/day) [70]. In all cases, baicalein attenuated the loss of DA neurons, while in the mouse MPTP model and the rotenone model, it also reduced behavioral impairments and markers of oxidative stress [70]. In addition, in the MPTP models, it reduced markers of inflammation [68,69].

The flavone 7,8-DHF has also been tested in several different PD models. Oral pre- and post-treatment (12-16 mg/kg/day) in the 6-OHDA model in rats improved behavior and reduced the loss of DA neurons in the SNc [71]. Pre- and post-treatment ip injection of 7,8-DHF (5 mg/kg/day) also reduced motor function impairment and prevented DA neuron loss in the MPTP model in mice [72]. The 7,8-DHF was also able to prevent further decreases in motor function and tyrosine hydroxylase (TH) levels when it was given by ip injection (5 mg/kg/day) after MPTP treatment in a slower model of disease progression in mice [73]. Similarly, oral administration of 7,8-DHF (30 mg/kg/day) prevented the MPP+-induced progressive loss of DA neurons in a monkey model of PD [74]. The 7,8-THF was also reported to activate the TrkB receptor, thereby activating neurotrophic factor signaling pathways, and all of the rodent studies [71–73] showed a maintenance of TrkB activation by 7,8-DHF in the presence of the different toxins.

Several other flavones have also shown protective effects in PD models, including apigenin using both ip injection (10 and 20 mg/kg/day) in the rotenone model in rats [75] and oral administration (5, 10, and 20 mg/kg/day) in the MPTP model in mice [76], oral administration of chrysin (10 mg/kg/day) in the 6-OHDA model in mice [77], oral administration of luteolin (10 and 20 mg/kg/day) in the MPTP model in mice [76], oral administration of nobiletin (10 mg/kg/day only) in the MPTP model in rats [78], and daily ip injection of morin (50 mg/kg) in the MPTP model in mice [79]. All five flavones helped to preserve the DA neurons and reduced markers of inflammation, while apigenin, chrysin, and luteolin prevented toxin-mediated decreases in neurotrophic factor gene expression. Apigenin, chrysin, luteolin, and morin also reduced toxin-induced behavioral alterations.

The flavonol quercetin has also been tested in several different models, and except for a study using oral pre-administration (20 mg/kg/day) in the 6-OHDA model [80], has shown positive results. However, in a more recent study in the same model, oral administration of quercetin (50 mg/kg/day) did show a beneficial effect where it reduced the loss of striatal dopamine and the increase in markers of oxidative stress [81]. Using the rat model of rotenone toxicity [82], quercetin was found to attenuate striatal dopamine depletion in a dose-dependent manner when given by ip injection (50 and 75 mg/kg/day) for 4 days after the administration of the toxin. Quercetin also reduced rotenone-induced behavioral changes and the loss of tyrosine hydroxylase (TH) immunoreactivity in both the SN and striatum. TH immunoreactivity is a marker for the integrity of the nigrostriatal pathway. Oral administration of quercetin (100 and 200 mg/kg/day) prior to the administration of the toxin improved motor function in MPTP-treated mice in a dose-dependent manner [83], which correlated with a significant increase in striatal dopamine levels and a significant decrease in a marker of lipid peroxidation. More recently, quercetin was tested in the MitoPark transgenic mouse model of PD [84]. These mice have a conditional disruption of mitochondrial transcription factor A, specifically in DA neurons, and recapitulate several aspects of human PD, including adult onset, slow impairment of motor function, and degeneration of the nigrostriatal pathway [85]. Oral administration of quercetin (25 mg/kg/day) to these mice for 6–8 weeks beginning at 12 weeks of age moderately but significantly reduced behavioral deficits, striatal dopamine loss, and nigrostriatal degeneration. The quercetin glycoside rutin was also tested in the 6-OHDA model in rats, where daily ip injection (10 and 30 mg/kg) was shown to partially reduce motor deficits when treatment was initiated beginning 3 weeks before administration of the toxin [86]. This correlated with a moderate but significant increase in striatal dopamine levels, as well as an increase in brain GSH levels. In contrast, markers of both lipid and protein oxidation were reduced.

Although the flavonol fisetin (20 mg/kg/day) did not show positive effects in the same 6-OHDA study in rats where quercetin also failed to show a beneficial effect [80], a more recent study using MPTP-treated mice found that oral administration of fisetin (10-25 mg/kg/day) prior to treatment with the toxin dose-dependently increased striatal dopamine levels and largely prevented the loss of TH immunoreactivity in the striatum [87]. Oral administration of the flavonol kaempferol (25, 50, and 100 mg/kg/day) had similar, dose-dependent effects in the same model when started prior to treatment with MPTP [88].

Both the flavonol myricetin (2.5 μg/day) and its glycoside myricitrin (60 mg/kg/day) maintained TH-positive neurons in the 6-OHDA model in rodents when administered by ip (myricitrin) or icv (myricetin) injection [22]. Myricitrin also reduced markers of inflammation and improved motor function, while myricetin increased dopamine levels.

Several flavanols have also shown benefits in PD models. Daily ip injection of catechin (10 and 30 mg/kg) in the 6-OHDA model in rats [89], oral administration of EGCG (25 mg/kg/day) in the MPTP model in mice [90], and oral administration of epicatechin (100 mg/kg/day) in the MPTP model in rats [91] all reduced toxin-induced behavioral deficits. For both catechin and EGCG, these functional improvements correlated with a reduction in striatal dopamine loss.

Flavanones have also shown benefits in PD models, including naringenin in the MPTP model in mice [92] and the 6-OHDA model in rats (oral; 50 mg/kg/day) [80] and hesperetin in the 6-OHDA model in rats [93]. Oral administration of naringenin (25, 50, and 100 mg/kg/day) increased dopamine levels and reduced the loss of TH immunoreactivity, while also lowering markers of inflammation and oxidative stress [92]. The glycoside of naringenin, naringin (ip; 80 mg/kg/day), was tested using both pre- and post-treatment in the 6-OHDA model in rats [94]. While pre-treatment protected against the toxin-induced loss of DA neurons and prevented microglial activation, post-treatment had no beneficial effects [94]. Oral administration of hesperetin (50 mg/kg/day) reduced 6-OHDA-induced behavioral deficits and prevented the loss of DA neurons [93]. These effects correlated with a reduction in some, but not all, inflammatory markers and lower levels of indices of oxidative stress, as well as an increase in GSH levels. Similarly, oral administration of the hesperetin glycoside, hesperidin

(50 mg/kg/day), reversed behavioral deficits, reduced striatal dopamine loss, and decreased markers of oxidative stress in the brains of 6-OHDA-treated aged mice [95].

The isoflavones genistein and puerarin have also been tested in rodent PD models. Genistein administered by ip injection improved neuronal survival in both the 6-OHDA model in rats (10 mg/kg/day) and the MPTP model in mice (10 mg/kg/day) [22]. Daily ip injection of puerarin (0.12 mg/kg) reduced DA neuronal loss in the MPTP model in mice, which correlated with decreases in markers of oxidative stress and inflammation and increases in markers of neurotrophic factor signaling [62].

In summary, a wide range of flavonoids have shown significant benefits in multiple rodent toxin-based models of PD. Where examined, they reduced markers of inflammation and oxidative stress and increased markers of neurotrophic factor signaling. Together, these effects contributed to the prevention of nerve cell death and the reduction in behavioral deficits. In addition, many prevented increases in α-synuclein, a protein associated with neuronal damage and death in PD. Thus, similar to the situation with AD, the flavonoids appear to have multiple targets in the PD models, further supporting the idea that multi-target compounds are likely to provide the best treatments for this neurodegenerative disease. However, as genetic models of PD become more reproducible, it will be important to test some of the most promising flavonoids in these models as well to provide further evidence for potential clinical efficacy.

1.7. Flavonoids and Huntington's Disease (HD)

Huntington's disease is a late onset, progressive, and fatal neurodegenerative disorder characterized by movement and psychiatric disturbances, as well as cognitive impairment. There is, at present, no cure. HD is an autosomally dominant inherited disease that is caused by an unstable expansion of a trinucleotide repeat (CAG) that encodes an abnormally long polyglutamine tract in the huntingtin protein. The age at disease onset inversely correlates with the CAG repeat number. The identification of the disease-causing mutation has allowed the development of a number of cellular and animal models of HD, and these have been used to try to elucidate the mechanisms underlying disease development and progression [96–99].

Both chemical and genetic rodent models of HD have been used to test the potential preventive role of flavonoids in HD development or progression, although no single model broadly replicates both the behavioral and neuropathological changes seen in humans [100]. The chemical approach uses 3-nitropropionic acid (3-NP), which when administered systemically at low doses to rats or mice causes selective degeneration of striatal neurons—the same neurons that are lost in HD [101]. The genetic models can be divided into three groups: N-terminal transgenic animals that carry only the 5′ portion of the human *HTT* gene, which contains the CAG repeats; full length transgenic animals that carry the full length *HTT* sequence, including the CAG repeats; and knock-in models, in which CAG repeats are engineered directly into the mouse *Htt* genomic locus [100]. These genetic models differ in the time of disease onset and disease severity, with the N-terminal transgenic animals showing the most severe phenotypes.

While there have only been a limited number of studies on the potential beneficial effects of flavonoids in HD models, the published results suggest that specific flavonoids could be of potential clinical use (Table 1). Oral administration of the flavone chrysin (50 mg/kg/day) improved behavior and reduced markers of oxidative stress and cell death, and enhanced the survival of striatal neurons in the 3-NP model of HD in rats [102].

Chronic oral administration of 7,8-DHF (5 mg/kg/day) to the R6/1 N-terminal transgenic mouse model of HD delayed the development of motor and cognitive deficits, prevented the loss of striatal volume, enhanced a marker of neurotrophic factor signaling, and reduced some markers of inflammation [103]. However, the effects on lifespan, which is greatly reduced in this mouse model of HD, were not assessed.

The flavonol fisetin was tested in the R6/2 mouse model of HD, which like the R6/1 model, is a N-terminal transgenic line that has an aggressive disease phenotype and shortened lifespan [104]. Fisetin was fed to genotyped R6/2 mice and their wild type littermates in the food beginning at ~6 weeks of age (25 mg/kg/day). The mice were tested on the rotarod from ~7–13 weeks of age and survival was followed. At the time of acquisition of the animals, rotarod performance was already impaired in the R6/2 mice as compared to their wild type littermates. Motor performance in the rotarod test declined significantly more rapidly in the animals on the control diet as compared to those on the fisetin diet. Similarly, while the median life span of the R6/2 mice on the control diet was 104 days, that of fisetin-fed mice was increased by ~30% to 139 days. The in vivo mechanisms underlying the effects of fisetin were not explored in this study.

The closely related flavonol quercetin was tested by two different groups in the 3-NP model in rats. In the first study, which used oral administration [105], quercetin (25 mg/kg/day) was found to reduce motor deficits, improve mitochondrial function, and attenuate some markers of oxidative stress. Although there was some suggestion of beneficial effects on striatal neuronal survival, the results were not clear. In the second study, which used daily ip injection of quercetin (50 mg/kg) [106], it was also found to improve motor function, as well as reduce a marker of inflammation, but it did not prevent the loss of striatal neurons. The quercetin glycoside, rutin, was also tested in the 3-NP model in rats but using a different protocol for 3-NP treatment in conjunction with oral administration [107]. Similar to quercetin, rutin (25 and 50 mg/kg/day) prevented 3-NP-induced impairments in motor function and decreased markers of inflammation. It also reduced markers of oxidative stress.

In contrast to quercetin, daily ip injection of the flavonol kaempferol (25 mg/kg) in 3-NP-treated rats not only reduced motor deficits but also attenuated the loss of striatal neurons [108]. These effects correlated with a reduction in markers of oxidative stress.

The flavanone glycoside hesperidin was also tested in the 3-NP model in rats [109] using oral administration (100 mg/kg/day). Similar to the other flavonoids in this model, hesperidin reduced motor deficits, as well as markers of inflammation and oxidative stress. The isoflavone genistein was also found to be protective in the 3-NP model in rats when given by daily ip injection (10 and 20 mg/kg) [110]. Genistein also reduced motor deficits and decreased markers of oxidative stress, inflammation, and nerve cell death.

Dietary supplementation with mixed berry anthocyanins (~100 mg/kg/day) was shown to delay the loss of motor function in the R6/1 N-terminal transgenic mouse model of HD [111]. However, this effect was only seen in female HD mice. The effects on lifespan were not examined.

In summary, a number of different flavonoids have shown benefits, particularly with regard to preserving motor function, in both chemical and transgenic models of HD. However, since most, if not all, of the studies with the 3-NP model involve pretreatment with the flavonoid, it is not clear if some of the effects of the flavonoids could be due to directly inhibiting the actions of 3-NP itself rather than reducing the consequences of 3-NP treatment. Thus, it would be worth testing those flavonoids that showed promise in the 3-NP assay in a transgenic model of HD. Similar to the effects of the flavonoids in AD models, the flavonoids appear to have multiple targets in the HD models, including reducing markers of inflammation and oxidative stress.

1.8. Flavonoids and Amyotrophic Lateral Sclerosis (ALS)

ALS is a fatal neurodegenerative disease that is characterized by the loss of the motor neurons that control the voluntary movement of muscles, resulting in paralysis and death, usually within 5 years of a diagnosis [112]. Approximately 10% of ALS cases are due to heritable gene mutations but different gene mutations are increasingly being found in patients with no family history of ALS, suggesting that the genetic component is more complicated than originally thought [113]. Moreover, there are overlaps between ALS and FTD [114]. Although there are three FDA-approved drugs for ALS, they all have very modest effects on survival (https://alsnewstoday.com/approved-treatments/) (access on 21 June 2019).

The most commonly used mouse model of ALS is based on Cu/Zn superoxide dismutase 1 (*SOD1*), the first gene mutation that was shown to cause ALS [113]. SOD1 mutations are found in ~20% of patients with the familial form of ALS. The most extensively used form of these mouse models is in the SOD1-G93A transgenic mouse. Although the different SOD1 transgenic mouse lines are not identical, they all show protein aggregation, motor neuron loss, axonal denervation, progressive paralysis, and reduced lifespan [113].

Despite the absence of effective treatments for ALS and the promising results with flavonoids in other neurodegenerative diseases as described above, very few flavonoids have been tested in animal models of ALS (Table 1). In all studies, the SOD1-G93A model was used.

Three times per week ip injection of the SOD1-G93A mice with 7,8-DHF (5 mg/kg) beginning at 1 month of age reduced the age-dependent decrease in motor performance and preserved total motor neuron count and dendritic spine density on motor neurons [115]. However, effects on lifespan were not examined.

Oral administration of the flavonol fisetin (9 mg/kg) beginning at 2 months of age significantly delayed the development of motor deficits, reduced their rate of progression, and increased lifespan [116]. This correlated with a significant increase in the motor neuron count in the spinal cord. At the molecular level, fisetin increased the levels of both phospho-ERK and the antioxidant protein heme oxygenase 1. Interestingly, fisetin also increased ERK phosphorylation in a transgenic AD model [46] and in HD flies [104], suggesting that this may at least partly contribute to its beneficial effects in these different models of neurodegenerative diseases.

Several studies have shown that oral administration of the flavanol EGCG (5.8–10 mg/kg) [117,118] can also delay symptom onset and extend lifespan in the SOD1-G93A mice. Consistent with these observations, EGCG increased motor neuron survival. These effects were correlated with a decrease in multiple markers of inflammation.

In summary, while there have been few studies with flavonoids in models of ALS, the published results suggest that this is an area that warrants further exploration, especially as all of the flavonoids that have shown benefits in the transgenic ALS model have also had positive effects in other age-associated neurodegenerative diseases.

2. Summary and Outlook

A number of different flavonoids from all of the six groups have been shown to have beneficial effects in models of AD, PD, HD, and ALS (Table 1). While many flavonoids have only been tested in models of one neurodegenerative disease, others, such as fisetin and 7,8-DHF, have shown efficacy in models of all four of the diseases highlighted in this review. These results strongly support the idea that common changes associated with the aging brain underlie the development of these diseases and that compounds that can address these changes have the best chance of clinical success. These changes include increases in oxidative stress, alterations in protein processing, decreases in neurotrophic factor signaling, synaptic dysfunction, increased inflammation, and cell death, which together contribute to behavioral impairments and cognitive dysfunction (Figure 2). As discussed in this review, flavonoids have the potential to reduce or prevent all of these changes. However, it appears that more work is needed before these compounds are taken seriously as possible therapeutics for neurodegenerative disease treatment. This includes developing better approaches to administration, such as nanoparticles [52,55,56] or other types of formulations [119] that will improve their ability to get into the brain and comparing different flavonoids head to head in the same model in order to determine which ones might have the best chance of clinical success. In addition, there may be synergism between the actions of some of the flavonoids, and this possibility is worth exploring both in vitro and in vivo.

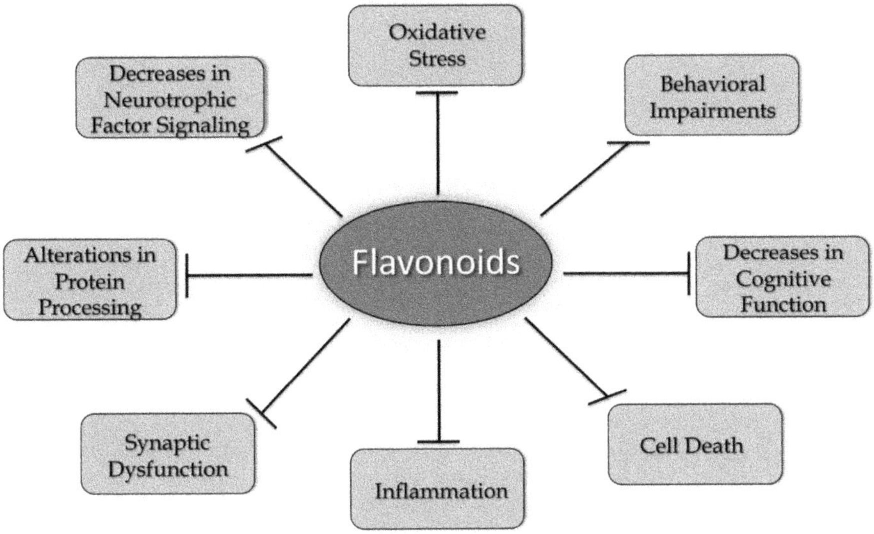

Figure 2. Flavonoids alter multiple pathways implicated in brain aging and neurodegenerative diseases. As discussed in this review, flavonoids can increase brain cell function and neuronal survival by reducing oxidative stress, activating neurotrophic factor signaling pathways, preventing alterations in protein processing, reducing synaptic dysfunction, and inhibiting inflammatory responses. Flavonoids can also enhance cognitive function and modulate behavioral impairments. Therefore, they have the potential to act as multi-factorial therapeutics for reducing the impact of neurodegenerative diseases.

Funding: This research was funded by the National Institutes of Health, grant numbers AG046153, AI104034, and NS106305.

Acknowledgments: The author would like to thank Zhibin Liang for preparing Figure 1.

Conflicts of Interest: The author declares no conflict of interest. The funders had no role in the design of the study; in the collection, analyses, or interpretation of data; in the writing of the manuscript, or in the decision to publish the results.

Abbreviations

Aβ	Amyloid beta peptide
AD	Alzheimer's disease
ALS	Amyotrophic lateral sclerosis
APP	Amyloid precursor protein
CNS	Central nervous system
CREB	cAMP response element binding protein
DA	dopaminergic
DHF	dihydroflavone
EGCG	(−)-epigallocatechin gallate
ERK	Extracellular signal regulated kinase
FAD	Familial Alzheimer's disease
FTD	Fronto-temporal dementia
GSH	glutathione
HD	Huntington's disease
ip	Intraperitoneal
icv	Intracerebroventricular
MPTP	1-methyl-4-phenyl-1,2,3,6-tetrahydropyridine
3-NP	3-nitropropionic acid

6-OHDA	6-hydroxydopamine
PD	Parkinson's disease
PS1	Presenilin 1
ROS	Reactive oxygen species
SNc	Substantia nigra pars compacta
SOD	Superoxide dismutase
TH	Tyrosine hydroxylase
TrkB	Tyrosine receptor kinase B

References

1. Przedborski, S.; Vila, M.; Jackson-Lewis, V. Neurodegeneration: What is it and where are we? *J. Clin. Investig.* **2003**, *111*, 3–10. [CrossRef] [PubMed]
2. Young, J.J.; Lavakumar, M.; Tampi, D.; Balachandran, S.; Tampi, R.R. Frontotemporal dementia: Latest evidence and clinical implications. *Ther. Adv. Psychopharmacol.* **2018**, *8*, 33–48. [CrossRef] [PubMed]
3. Available online: https://www.alz.org/alzheimers-dementia/facts-figures (accessed on 21 June 2019).
4. Available online: https://parkinson.org/Understanding-Parkinsons/Statistics (accessed on 21 June 2019).
5. Available online: https://rarediseases.org/rare-diseases/huntingtons-disease (accessed on 21 June 2019).
6. Available online: https://www.cdc.gov/mmwr/volumes/67/wr/mm6707a3.htm (accessed on 21 June 2019).
7. Prior, M.; Chiruta, C.; Currais, A.; Goldberg, J.; Ramsey, J.; Dargusch, R.; Maher, P.A.; Schubert, D. Back to the future with phenotypic screening. *ACS Chem. Neurosci.* **2014**, *5*, 503–513.
8. Bishop, N.A.; Lu, T.; Yankner, B.A. Neural mechanisms of ageing and cognitive decline. *Nature* **2010**, *464*, 529–535. [CrossRef] [PubMed]
9. Schubert, D.; Maher, P. An alternative approach to drug discovery for Alzheimer's disease dementia. *Future Med. Chem.* **2012**, *4*, 1681–1688. [CrossRef] [PubMed]
10. 1Schubert, D.; Currais, A.; Goldberg, J.; Finley, K.; Petrascheck, M.; Maher, P. Geroneuroprotectors: Effective geroprotectors for the brain. *Trends Pharmacol. Sci.* **2018**, *39*, 1004–1007. [CrossRef] [PubMed]
11. Hao, H.; Zheng, X.; Wang, G. Insights into drug discovery from natural medicines using reverse pharmacokinetics. *Trends Pharmacol. Sci.* **2014**, *35*, 168–177. [CrossRef] [PubMed]
12. Swinney, D.C. Phenotypic vs. target-based drug discovery for firrst-in-class medicines. *Clin. Pharm. Therap.* **2013**, *93*, 299–301. [CrossRef]
13. Dias, D.A.; Urban, S.; Roessner, U. A historical overview of natural products in drug discovery. *Metabolites* **2012**, *2*, 303–336. [CrossRef]
14. Yun, B.-W.; Yan, Z.; Amir, R.; Hong, S.; Jin, Y.-W.; Lee, E.-K.; Loake, G.J. Plant natural products: History, limitations and the potential of cambial meristematic cells. *Biotech. Gen. Eng. Rev.* **2013**, *28*, 47–60. [CrossRef]
15. Bakoyiannis, I.; Daskalopoulou, A.; Pergialiotis, V.; Perrea, D. Phytochemicals and cognitive health: Are flavonoids doing the trick? *Biomed. Pharmacother.* **2019**, *109*, 1488–1497. [CrossRef] [PubMed]
16. Beking, K.; Vieira, A. Flavonoid intake and disability-adjuested life years due to Alzheimer's and related dementias: A population-based study involving twenty-three developed countries. *Public Health Nutr.* **2010**, *13*, 1403–1409. [CrossRef] [PubMed]
17. Commenges, D.; Scotet, V.; Renaud, S.; Jacqmin-Gadda, H.; Barberger-Gateau, P.; Dartigues, J.F. Intake of flavonoids and risk of dementia. *Eur. J. Epidemiol.* **2000**, *16*, 357–363. [CrossRef] [PubMed]
18. Gao, X.; Cassidy, A.; Schwarzschild, M.A.; Rimm, E.B.; Ascherio, A. Habitual intake of dietary flavonoids and risk of Parkinson's disease. *Neurology* **2012**, *78*, 1138–1145. [CrossRef] [PubMed]
19. Jones, Q.R.D.; Warford, J.; Rupasinghe, H.P.V.; Robertson, G.S. Target-based selection of flavonoids for neurodegenerative disorders. *Trends Pharmacol. Sci.* **2012**, *33*, 602–610. [CrossRef] [PubMed]
20. Duarte, A.C.B.; Santana, M.G.S.; Ofrfali, G.; Parisi de Oliveira, C.T.; Priolli, D.G. Literature evidence and ARRIVE assessment on neuroprotective effects of flavonols in neurodegenerative diseases' models. *CNS Neurolog. Dis. Drug Targets* **2018**, *17*, 34–42. [CrossRef] [PubMed]
21. de Andrade Teles, R.B.; Diniz, T.C.; Costa Pinto, T.C.; de Oliviera, R.G.; Gama e Silva, M.; de Lavor, E.M.; Fernandes, A.W.C.; de Oliviera, A.P.; de Almeida Ribeiro, F.P.R.; da Silva, A.A.M.; et al. Flavonoids as therapeutic agents in Alzheimer's and Parkinson's diseases: A systematice review of preclinical evidences. *Oxid. Med. Cell. Longev.* **2018**, *2018*, 7043213. [CrossRef]

22. Kujawska, M.; Jodynis-Liebert, J. Polyphenols in Parkinson's disease: A systematic review of in vivo studies. *Nutrients* **2018**, *10*, 642. [CrossRef]
23. Ahmed, R.M.; Irish, M.; van Eersel, J.; Ittner, A.; Ke, Y.D.; Volkerling, A.; van der Hoven, J.; Tanaka, K.; Karl, T.; Kassiou, M.; et al. Mouse models of frontotemporal dementia: A comparison of phenotypes with clinical symptomatology. *Neurosc. Biobehav. Rev.* **2017**, *74*, 126–138. [CrossRef]
24. Goedert, M.; Spillantini, M.G. A century of Alzheimer's disease. *Science* **2006**, *314*, 777–781. [CrossRef]
25. McKhann, G.; Drachman, D.; Folstein, M.; Katzman, R.; Price, D.; Stadlan, E.M. Clinical diagnosis of Alzheimer's disease: Report of the NINCDS-ADRDA work group under the auspices of Department of Health and Humans Services Task Force on Alzheimer's disease. *Neurology* **1984**, *34*, 939–944. [CrossRef] [PubMed]
26. McKeith, I.; Cummings, J.L. Behavioral changes and psychological symptoms in dementia disorders. *Lancet Neurol.* **2005**, *4*, 735–742. [CrossRef]
27. Haas, C. Strategies, development and pitfalls of therapeutic options for Alzheimer's disease. *J. Alzheimer's Dis.* **2012**, *28*, 241–281. [CrossRef] [PubMed]
28. Rafil, M.S.; Aisen, P.S. Recent developments in Alzheimer's disease therapeutics. *BMC Med.* **2009**, *7*, 7.
29. Gold, M. Phase II clinical trials of anti-amyloid β antibodies: When is enough, enough? *Alzheimer's Dement.* **2017**, *3*, 402–409. [CrossRef] [PubMed]
30. Available online: https://www.alzforum.org/research-models/alzheimers-disease (accessed on 21 June 2019).
31. Swerdlow, R.H. Is aging part of Alzheimer's disease, or is Alzheimer's disease part of aging? *Neurobiol. Aging* **2007**, *28*, 1465–1480. [CrossRef] [PubMed]
32. Herrup, K. The case for rejecting the amyloid cascade hypothesis. *Nat. Neurosci.* **2015**, *18*, 794–799. [CrossRef]
33. Pallas, M. Senescence-accelerated mice P8: A tool to study brain aging and Alzheimer's disease in a mouse model. *ISRN Cell Biol.* **2012**, *2012*, 917167. [CrossRef]
34. Morley, J.E.; Armbrecht, H.J.; Farr, S.A.; Kumar, V.B. The senescence accelerated mouse (SAMP8) as a model for oxidative stress and Alzheimer's disease. *Biochim. Biophys. Acta* **2012**, *1822*, 650–656. [CrossRef]
35. Cheng, X.R.; Zhou, W.X.; Zhang, Y.X. The behavioral, pathological and therapeutic features of the senescence-accelerated mouse prone 8 strain as an Alzheimer's disease animal model. *Ageing Res. Rev.* **2014**, *13*, 13–37. [CrossRef]
36. Currais, A.; Goldberg, J.; Farrokhi, C.; Chang, M.; Prior, M.; Dargusch, R.; Daugherty, D.; Armando, A.; Quehenberger, O.; Maher, P.; et al. A comprehensive multiomics approach toward understanding the relationship between aging and dementia. *Aging* **2015**, *7*, 937–955. [CrossRef] [PubMed]
37. Stefanova, N.A.; Kozhevnikova, O.S.; Vitovtov, A.O.; Logvinov, S.V.; Rudnitskaya, E.A.; Korbolina, E.E.; Muraleva, N.A.; Kolosova, N.G. Senescence-accelerated OXYS rats: A model of age-related cognitive decline with relevance to Alzheimer disease. *Cell Cycle* **2014**, *13*, 898–909. [CrossRef] [PubMed]
38. Devi, L.; Ohno, M. 7,8 Dihydroxyflavone, a small molecule TrkB agonist, reverses memory deficits and BACE1 elevation in a mouse model of Alzheimer's disease. *Neuropsychopharmacology* **2012**, *37*, 434–444. [CrossRef] [PubMed]
39. Zhang, Z.; Liu, X.; Schroeder, J.P.; Chan, C.-B.; Song, M.; Yu, S.P.; Weinshenker, D.; Ye, K. 7,8 Dihydroxyflavone prevents synaptic loss and memory deficits in a mouse model of Alzheimer's disease. *Neuropsychopharmacology* **2014**, *39*, 638–650. [CrossRef]
40. Zhou, W.X.; Li, X.; Huang, D.; Zhou, W.X.; Li, T.; Song, W. No significant effect of 7,8 dihydroxyflavone on APP processing and Alzheimer-associated phenotypes. *Curr. Alzheimer Res.* **2015**, *12*, 47–52. [CrossRef]
41. Zhao, L.; Wang, J.-L.; Liu, R.; Li, X.-X.; Li, J.-F.; Zhang, L.-L. Neuroprotective, anti-amyloidgenic and neurotrophic effects of apigenin in an Alzheimer's disease model. *Molecules* **2013**, *18*, 9949–9965. [CrossRef] [PubMed]
42. Onozuka, H.; Nkajima, A.; Matsuzaki, K.; Shin, R.W.; Ogino, K.; Tetsu, N.; Yokosuka, A.; Sashida, Y.; Mimaki, Y.; Yamakuni, T.; et al. Nobiletin, a citrus flavonoid, improves memory impairment and Abeta pathology in a transgenic model of Alzheimer's disease. *J. Pharmacol. Exp. Ther.* **2008**, *326*, 739–744. [CrossRef] [PubMed]
43. Nakajima, A.; Aoyama, Y.; Shin, E.-J.; Nam, Y.; Kim, H.-C.; Nagai, T.; Yokosuka, A.; Mimaki, Y.; Yokoi, T.; Ohizumi, Y.; et al. Nobiletin, a citrus flavonoid, improves cognitive impairment and reduces soluble Aβ levels in a triple transgenic mouse model of Alzheimer's disease (3XTg-AD). *Behav. Brain Res.* **2015**, *289*, 69–77. [CrossRef]

44. Nakajima, A.; Aoyama, Y.; Nguyen, T.-T.L.; Shin, E.-J.; Kim, H.-C.; Yamada, S.; Nakai, T.; Nagai, T.; Yokosuka, A.; Mimaki, Y.; et al. Nobiletin, a citrus flavonoid, ameliorates cognitive impairment, oxidative burden and hyperphosphorylation of tau in senescence-accelerated mouse. *Behav. Brain Res.* **2013**, *250*, 351–360. [CrossRef]
45. Zhang, S.-Q.; Obregon, D.F.; Ehrhart, J.; Deng, J.Y.; Tian, J.; Hou, H.; Giunta, B.; Sawmiller, D.; Tan, J. Baicalein reduces b-amyloid and promotes nonamyloidgenic amyloid precursor protein processing in an Alzheimer's disease transgenic mouse model. *J. Neurosci. Res.* **2013**, *91*, 1239–1246. [CrossRef]
46. Currais, A.; Prior, M.; Dargusch, R.; Armando, A.; Ehren, J.; Schubert, D.; Quehenberger, O.; Maher, P. Modulation of p25 and inflammatory pathways by fisetin maintains cognitive function in Alzheimer's disease transgenic mice. *Aging Cell* **2014**, *13*, 379–390. [CrossRef] [PubMed]
47. Currais, A.; Farrokhi, C.; Dargusch, R.; Armando, A.; Quehenberger, O.; Schubert, D.; Maher, P. Fisetin reduces the impact of aging on behavior and physiology in the rapidly aging SAMP8 mouse. *J. Gerentol. Biol. Sci. Med. Sci.* **2018**, *73*, 299–307. [CrossRef] [PubMed]
48. Ahmad, A.; Ali, T.; Park, H.Y.; Badshah, H.; Rehman, S.U.; Kim, M.O. Neuroprotective effect of fisetin against amyloid-beta-induced cognitive/synaptic dysfunction, neuroinflammation and neurodegeneration in mice. *Mol. Neurobiol.* **2017**, *54*, 2269–2285. [CrossRef] [PubMed]
49. Sweatt, J.D. Mitogen-activated protein kinases in synaptic plasticity and memory. *Curr. Opin. Neurobiol.* **2004**, *14*, 311–317. [CrossRef] [PubMed]
50. Sagara, Y.; Vahnnasy, J.; Maher, P. Induction of PC12 cell differentiation by flavonoids is dependent upon extracellular signal-regulated kinase activation. *J. Neurochem.* **2004**, *90*, 1144–1155. [CrossRef] [PubMed]
51. Zaplatic, E.; Bule, M.; Shah, S.Z.A.; Uddin, M.S.; Niaz, K. Molecular mechanisms underlying protective role of quercetin in attenuating Alzheimer disease. *Life Sci.* **2019**, *224*, 109–119. [CrossRef] [PubMed]
52. Gayoso e Ibiapina Moreno, L.C.; Puerta, E.; Suarez-Santiago, J.E.; Santos-Magalhaes, N.S.; Ramirez, M.J.; Irache, J.M. Effect of the oral administration of nanoencapsulated quercetin on a mouse model of Alzheimer's disease. *Int. J. Pharmaceut.* **2017**, *517*, 50–57. [CrossRef]
53. Sabogal-Guaqueta, A.M.; Munoz-Manco, J.I.; Ramirez-Pineda, J.R.; Lamprea-Rodriguez, M.; Osorio, E.; Cardona-Gomez, G.P. The flavonoid quercetin ameliorates Alzheimer's disease pathology and protects cognitive and emotional function in aged triple transgenic Alzheimer's disease model mice. *Neuropsychopharmacology* **2015**, *93*, 134–145. [CrossRef]
54. Moghbelinejad, S.; Nassiri-Asi, M.; Farivar, T.N.; Abbasi, E.; Sheikhi, M.; Taghiloo, M.; Farsad, F.; Samimi, A.; Haijali, F. Rutin activates the MAPK pathway and BDNF gene expression on beta-amyloid induce neurotoxicity in rats. *Toxicol. Lett.* **2014**, *224*, 108–113. [CrossRef]
55. Ali, T.; Kim, M.J.; Rehman, S.U.; Ahmad, A.; Kim, M.O. Anthocyanin-loaded PEG-gold nanoparticles enhanced the neuroprotection of anthocyanins in $A\beta_{1-42}$ mouse model of Alzheimer's disease. *Mol. Neurobiol.* **2017**, *54*, 6490–6506. [CrossRef]
56. Kim, M.J.; Rehman, S.U.; Amin, F.U.; Kim, M.O. Enhanced neuroprotection of anthocyanin-loaded PEG-gold nanoparticles against $A\beta_{1-42}$-induced neuroinflammation and neurodegeneration via the NF-kB/JNK/GSK3β signaling pathway. *Nanomed. Nanotech. Biol. Med.* **2017**, *13*, 2533–2544. [CrossRef] [PubMed]
57. Walker, J.M.; Klakotskaia, D.; Ajit, D.; Weisman, G.A.; Wood, W.G.; Sun, G.Y.; Serfozo, P.; Simonyi, A.; Schachtman, T.R. Beneficial effects of dietary EGCG and voluntary exercise on behavior in an Alzheimer's disease mouse model. *J. Alzheimer's Dis.* **2015**, *44*, 561–572. [CrossRef] [PubMed]
58. Guo, Y.; Zhao, Y.; Nan, Y.; Wang, X.; Chen, Y.; Wang, S. (−)-epigallocatechin-3-gallate ameliorates memory impairment and rescues the abnormal synpatic protein levels in the frontal cortex and hippocampus on a mouse model of Alzheimer's disease. *Neuroreport* **2017**, *28*, 590–597. [CrossRef] [PubMed]
59. Ide, K.; Matsuoka, N.; Yamada, H.; Furushima, D.; Kawakami, K. Effects of tea catechins on Alzheimer's disease: Recent updates and perspectives. *Molecules* **2018**, *23*, 2357. [CrossRef] [PubMed]
60. Li, C.; Zug, C.; Qu, H.; Schluesener, H.; Zhang, Z. Hesperidin ameliorates behavioral impairments and neuropathology of transgenic APP/PS1 mice. *Behav. Brain Res.* **2015**, *281*, 32–42. [CrossRef] [PubMed]
61. Hajialyani, M.; Farzaei, M.H.; Echeverria, J.; Nabavi, S.M.; Uriarte, E.; Sobarzo-Sanchez, E. Hesperidin as a neuroprotective agent: A review of animal and clinical evidence. *Molecules* **2019**, *24*, 648. [CrossRef] [PubMed]
62. Zhang, S.-Q.; Wang, J.; Zhao, H.; Luo, Y. Effects of three flavonoids from an ancient Chinese medicine Radix puerariae on geriatric diseases. *Brain Circ.* **2018**, *4*, 174–184.
63. Weintraub, D.; Comella, C.L.; Horn, S. Parkinson's disease. *Am. J. Manag. Care* **2008**, *14*, S40–S69.

64. Klein, C.; Westenberger, A. Genetics of Parkinson's disease. *Cold Spring Harb. Perspect. Med.* **2012**, *2*, a008888. [CrossRef]
65. Schulz, J.B.; Falkenburger, B.H. Neuronal pathology in Parkinson's disease. *Cell Tissue Res.* **2004**, *318*, 135–147. [CrossRef]
66. Tieu, K. A guide to neurotoxic animal models of Parkinson's disease. *Cold Spring Harb. Perspect. Med.* **2011**, *1*, a009316. [CrossRef] [PubMed]
67. Blesa, J.; Przedborski, S. Parkinson's disease: Animal models and dopaminergic cell vulnerability. *Front. Neuroanat.* **2014**, *8*, 155. [CrossRef] [PubMed]
68. Hung, K.-C.; Huang, H.-J.; Wang, Y.-T.; Lin, A.M.-Y. Baicalein attenuates α-synuclein aggregation, inflammasome activation and autophagy in the MPP^+-treated nigrostriatal dopaminergic system in vivo. *J. Ethnopharmcol.* **2016**, *194*, 522–529. [CrossRef] [PubMed]
69. Cheng, Y.; He, G.; Mu, X.; Zhang, T.; Li, X.; Hu, J.; Xu, B.; Du, G. Neuroprotective effect of baicalein against MPTP neurotoxicity: Behavioral, biochemical and immunohistochemical profile. *Neurosci. Lett.* **2008**, *441*, 16–20. [CrossRef] [PubMed]
70. Zhang, X.; Du, L.; Zhang, W.; Yang, Y.; Zhou, Q.; Du, G. Therapeutic effects of baicalein on rotenone-induced Parkinson's disease through protecting mitochondrial function and biogenesis. *Sci. Rep.* **2017**, *7*, 9968. [CrossRef] [PubMed]
71. Luo, D.; Shi, Y.; Wang, J.; Lin, Q.; Sun, Y.; Ye, K.; Yan, Q.; Zhang, H. 7,8 Dihydroxyflavone protects 6-OHDA and MPTP induced dpaminergic neurons degeneration through activation of TrkB in rodents. *Neurosci. Lett.* **2016**, *620*, 43–49. [CrossRef]
72. Li, X.H.; Dai, C.F.; Chen, L.; Zhou, W.T.; Han, H.L.; Dong, Z.F. 7,8 Dihydroxyflavone ameliorates motor deficits via supressing a-synuclein expression and oxidative stress in the MPTP-indiuced mouse model of Parkinson's disease. *CNS Neurosci. Ther.* **2016**, *22*, 617–624. [CrossRef]
73. Sconce, M.D.; Churchill, M.J.; Moore, C.; Meshul, C.K. Intervention with 7,8 dihydorxyflavone blocks further striatal terminal loss and restores motor deficits in a progressive mouse model of Parkinson's disease. *Neuroscience* **2015**, *290*, 454–471. [CrossRef] [PubMed]
74. He, J.C.; Xiang, Z.; Zhu, X.; Ai, Z.; Shen, J.; Huang, T.; Liu, L.; Ji, W.; Li, T. Neuroprotective effects of 7,8, dihydroxyflavone on midbrain dopaminergic neurons in MPP+-treated monkeys. *Sci. Rep.* **2016**, *6*, 34339. [CrossRef] [PubMed]
75. Anusha, C.; Sumathi, T.; Joseph, L.D. Protective role of apigenin on rotenone induced rat model of Parkinson's disease: Suppression of neuroinflammation and oxidative stress mediated apoptosis. *Chem.-Biol. Interactions* **2017**, *269*, 67–79. [CrossRef]
76. Patil, S.P.; Jain, P.D.; Sancheti, J.S.; Ghumatkar, P.J.; Tambe, R.; Sathye, S. Neuroprotective and neurotrophic effects of apigenin and luteolin in MPTP induced parkinsonism in mice. *Neuropharmacology* **2014**, *86*, 192–202. [CrossRef] [PubMed]
77. Goes, A.T.R.; Jesse, C.R.; Antunes, M.S.; Ladd, F.V.L.; Ladd, A.A.B.L.; Luchese, C.; Paroul, N.; Boeira, S.P. Protective role of chrysin on 6-hydroxydopamine-induced neurodegeneration a mouse model of Parkinson's disease: Involvement of neuroinflammation and neurotrophins. *Chem.-Biol. Interact.* **2018**, *279*, 111–120. [CrossRef]
78. Jeong, K.H.; Jeon, M.T.; Kim, H.D.; Jung, U.J.; Jang, M.C.; Chu, J.W.; Yang, S.J.; Choi, I.Y.; Choi, M.S.; Kim, S.R. Nobiletin protects dopaminergic neurons in the 1-mehtyl-4-phenylpyridinium rat model of Parkinson's disease. *J. Med. Food* **2015**, *18*, 409–414. [CrossRef] [PubMed]
79. Lee, K.M.; Lee, Y.J.; Chun, A.H.; Kim, A.H.; Kim, J.Y.; Lee, J.Y.; Ishigami, A.; Lee, J.H. Neuroprotective and anti-inflammatory effects of morin in a murine model of Parkinson's disease. *J. Neurosci. Res.* **2016**, *94*, 865–878. [CrossRef] [PubMed]
80. Zbarsky, V.; Datla, K.P.; Parkar, S.; Rai, D.K.; Arouma, O.I.; Dexter, D.T. Neuroprotective properties of the natural phenolic antioxidants curcumin and naringenin but not quercetin and fisetin in a 6-OHDA model of Parkinson's disease. *Free Rad. Res.* **2005**, *39*, 1119–1125. [CrossRef]
81. Haleagrahara, N.; Siew, C.J.; Ponnusamy, K. Effect of quercetin and desferriooxamine on 6-hydroxydopamine (6-OHDA) induced neurotoxicity in striatum of rats. *J. Toxicol. Sci.* **2013**, *38*, 25–33. [CrossRef] [PubMed]
82. Karuppagounder, S.S.; Madathil, S.K.; Pandey, M.; Haobam, R.; Rajamma, U.; Mohanakumar, K.P. Quercetin up-regulates mitochondrial complex-I activity to protect against programmed cell death in rotenone model of Parkinson's disease in rats. *Neuroscience* **2013**, *236*, 136–148. [CrossRef]

83. Lv, C.; Hong, T.; Yang, Z.; Zhang, Y.; Wang, L.; Dong, M.; Zhao, J.; Mu, J.; Meng, Y. Effect of quercetin in the 1-methyl-4-phenyl-1,2,3,6-tetrahydropyridine-induced mouse model of Parkinson's disease. *Evid.-Based Complemen. Alternat. Med.* **2012**, *2012*, 928643. [CrossRef] [PubMed]
84. Ay, M.; Luo, J.; Langley, M.; Jin, H.; Anatharam, V.; Kanthasamy, A.; Kanthasamy, A.G. Molecular mechanisms underlying protective effects of quercetin agianst mitochondrial dysfunction and progressive dopaminergic neurodegeneration in cell culture and MitoPark transgenic mouse models of Parkinson's disease. *J. Neurochem.* **2017**, *141*, 766–782. [CrossRef]
85. Ekstrand, M.I.; Galter, D. The MitoPark Mouse-An animal model of Parkinson's diease with impaired respiratory chain function in dopamine neurons. *Parkinsonism Rel. Dis.* **2009**, *1553*, S185–S188. [CrossRef]
86. Khan, M.M.; Raza, S.S.; Javed, H.; Ahmad, A.; Khan, A.; Islam, F.; Safhi, M.M.; Islam, F. Rutin protects dopaminergic neurons from oxidative stress in an animal model of Parkinson's disease. *Neurotox. Res.* **2012**, *22*, 1–15. [CrossRef] [PubMed]
87. Maher, P. Protective effects of fisetin and other berry flavonoids in Parkinson's disease. *Food Func.* **2017**, *8*, 3033–3042. [CrossRef] [PubMed]
88. Li, S.; Pu, X.-P. Neuroprotective effect of kaempferol against 1-methyl-4-phenyl-1,2,3,6,-tetrahydropyridine-induced mouse model of Parkinson's disease. *Biol. Pharm. Bull.* **2011**, *34*, 1291–1296. [CrossRef] [PubMed]
89. Teixeira, M.D.A.; Souza, C.M.; Menezes, A.P.F.; Carmo, M.R.S.; Fonteles, A.A.; Gurgel, J.P.; Lima, F.A.V.; Viana, G.S.B.; Andrade, G.M. Catechin attenuates behavioral neurotoxicity induced by 6-OHDA in rats. *Pharmacol. Biochem. Behav.* **2013**, *110*, 1–7. [CrossRef] [PubMed]
90. Xu, Q.; Langley, M.; Kanthasamy, A.G.; Reddy, M.B. Epigallocatechin gallate has a neurorescue effect in a mouse model of Parkinson Disease. *J. Nutr.* **2017**, *147*, 1926–1931. [CrossRef] [PubMed]
91. Rubio-Osornio, M.; Gorostieta-Salas, E.; Montes, S.; Perez-Severiano, F.; Rubio, C.; Gomez, C.; Rios, C.; Guevara, J. Epicatechin reduces striatal MPP+-induce damage in rats through slight increases in SOD-Cu,Zu Activity. *Oxid. Med. Cell. Longev.* **2015**, *2015*, 276039. [CrossRef] [PubMed]
92. Mani, S.; Sekar, S.; Barathidasan, R.; Manivasagam, T.; Thenmozhi, A.J.; Sevanan, M.; Chidambaram, S.B.; Essa, M.M.; Guillemin, G.J.; Sakharkar, M.K. Naringenin decreases a-synuclein expression and neuroinflammation in MPTP-induced Parkinson's disease model in mice. *Neurotox. Res.* **2018**, *33*, 656–670. [CrossRef] [PubMed]
93. Kiasalari, Z.; Khalili, M.; Baluchnejadmojarad, T.; Roghani, M. Protective effect of oral hesperetin against unilateral striatal 6-hydorxydopamine damage in the rat. *Neurochem. Res.* **2016**, *41*, 1065–1072. [CrossRef]
94. Kim, H.D.; Jeong, K.H.; Jung, U.J.; Kim, S.R. Naringin treatment induces neuroprotective effects in a mouse model of Parkinson's disease in vivo, but not enough to restore the lesioned dopaminergic system. *J. Nutr. Biochem.* **2016**, *28*, 140–146. [CrossRef]
95. Antunes, M.S.; Goes, A.T.R.; Boeira, S.P.; Prigol, M.; Jesse, C.R. Protective effect of hesperidin in a model of Parkinson's disease induced by 6-hydroxydopmaine in aged mice. *Nutrition* **2014**, *30*, 1415–1422. [CrossRef]
96. Borrell-Pages, M.; Zala, D.; Humbert, S.; Saudou, F. Huntington's disease: From huntingtin function and dysfunction to therapeutic strategies. *Cell. Mol. Life Sci.* **2006**, *63*, 2642–2660. [CrossRef] [PubMed]
97. Ramaswamy, S.; Shannon, K.M.; Kordower, J.H. Huntington's disease: Pathological mechanisms and therapeutic strategies. *Cell Transplant.* **2007**, *16*, 301–312. [CrossRef] [PubMed]
98. Imarisio, S.; Carmichael, J.; Korolchuk, V.; Chen, C.-W.; Saiki, S.; Rose, C.; Krishna, G.; Davies, J.E.; Ttofi, E.; Underwood, B.R.; et al. Huntington's disease: From pathology and genetics to potential therapeutics. *Biochem. J.* **2008**, *412*, 191–209. [CrossRef] [PubMed]
99. Gil, J.M.; Rego, A.C. Mechanisms of neurodegeneration in Huntington's disease. *Eur. J. Neurosci.* **2008**, *27*, 2803–2820. [CrossRef] [PubMed]
100. Rangel-Barajas, C.; Rebec, G.V. Overview of Huntington's disease models: Neuropathological, molecular and behavioral differences. *Curr. Protoc. Neurosci.* **2018**, *83*, e47. [CrossRef]
101. Borlongan, C.V.; Koutouzis, T.K.; Sanberg, P.R. 3-Nitropropionic acid animal model and Huntington's disease. *Neurosci. Biobehav. Rev.* **1997**, *21*, 289–293. [CrossRef]
102. Thangarajan, S.; Ramachandran, S.; Krishnamurthy, P. Chrysin exerts neuroprotective effects against 3-nitropropionic acid induced behavioral despair-mitochondrial dysfunction and striatal apoptosis via upregulating Bcl-2 gene and downregulating Bax-Bad gnes in male wistar rats. *Biomed. Pharmacother.* **2016**, *84*, 514–525. [CrossRef]

103. Barriga, G.G.-D.; Giralt, A.; Anglada-Huguet, M.; Gaja-Capdevila, N.; Orlandi, J.G.; Soriano, J.; Canals, J.-M.; Alberch, J. 7,8 Dihydroxyflavone ameliorates cognitive and motor deficits in a Huntington's disease mouse model through specific activation of the PLCγ1 pathway. *Hum. Mol. Gen.* **2017**, *26*, 3144–3160.
104. Maher, P.; Dargusch, R.; Bodai, L.; Gerard, P.; Purcell, J.M.; Marsh, J.L. ERK activation by the polyphenols fisetin and resveratrol provides neuroprotection in multiple models of Huntington's disease. *Hum. Mol. Gen.* **2011**, *20*, 261–270. [CrossRef]
105. Sandhir, R.; Mehrotra, A. Quercetin supplementation is effective in improving mitochondrial dysfunctions induced by 3-nitropropionic acid: Implications in Huntington's disease. *Biochim. Biophys. Acta* **2013**, *1832*, 421–430. [CrossRef]
106. Chakraborty, J.; Singh, R.J.; Dutta, D.; Naskar, A.; Rajamma, U.; Mohanakumar, K.P. Quercetin improves behavioral deficiencies, restores astrocytes and microglia, and reduces serotonin metabolism in 3-nitropropionic acid-induced rat model of Huntington's disease. *CNS Neurosci. Ther.* **2014**, *20*, 10–19. [CrossRef] [PubMed]
107. Suganya, S.N.; Sumathi, T. Effect of rutin against a mitochondrial toxin, 3-nitropropionic acid induced biochemical, behavioral and histological alterations-a pilot study on Huntington's disease model in rats. *Metab. Brain Dis.* **2017**, *32*, 471–481. [CrossRef] [PubMed]
108. Lagoa, R.; Lopez-Sanchez, C.; Samhan-Arias, A.K.; Ganan, C.M.; Garcia-Martinez, V.; Gutierrez-Merino, C. Kaempferol protects against rat striatal degeneration induced by 3-nitropropionic acid. *J. Neurochem.* **2009**, *111*, 473–487. [CrossRef] [PubMed]
109. Menze, E.T.; Tadros, M.G.; Abdel-Tawab, A.M.; Khalifa, A.E. Potential neuroprotective effects of hesperidin on 3-nitropropionic acid-induced neurotoxicity in rats. *Neurotoxicology* **2012**, *33*, 1265–1275. [CrossRef] [PubMed]
110. Menze, E.T.; Esmat, A.; Tadros, M.G.; Khalifa, A.E.; Abdel-Naim, A.B. Genistein improves sensorimotor gating: Mechanisms related to its neuroprotective effects on the striatum. *Neuropharmacology* **2016**, *105*, 35–46. [CrossRef]
111. Kreilaus, F.; Spiro, A.S.; Hannan, A.J.; Garner, B.; Jenner, A.M. Therapeutic effects of anthocyanins and environmental enrichment in R6/1 Huntongton's disease mice. *J. Huntington's Dis.* **2016**, *5*, 285–296. [CrossRef] [PubMed]
112. Riancho, J.; Gil-Bea, F.J.; Santurtan, A.; Lopez de Munain, A. Amytotrophic lateral sclerosis: A complex syndrome that needs an integrated research approach. *Neural Regen. Res.* **2019**, *14*, 193–196. [CrossRef]
113. Lutz, C. Mouse models of ALS. *Brain Res.* **2018**, *1693*, 1–10. [CrossRef]
114. Ittner, L.M.; Halliday, G.M.; Kril, J.J.; Gotz, J.; Hodges, J.R.; Kiernan, M.C. FTD and ALS-translating mouse studies into clinical trials. *Nat. Rev. Neurol.* **2015**, *11*, 360–366. [CrossRef]
115. Korkmaz, O.T.; Aytan, N.; Carreras, I.; Choi, J.-K.; Kowall, N.W.; Jenkins, B.G.; Dedeoglu, A. 7,8 Dihydroxyflavone improves motor performance and enhances lower motor neuronal survival in a mouse model of amyotrophic lateral sclerosis. *Neurosci. Lett.* **2014**, *566*, 286–291. [CrossRef]
116. Wang, T.H.; Wang, S.Y.; Wang, X.D.; Jiang, H.Q.; Yang, Y.Q.; Wang, Y.; Cheng, J.L.; Zhang, C.T.; Liang, W.W.; Feng, H.L. Fisetin exerts antioxidant and neuroprotective effects in mulitple mutant hSOD1 modles of amyotrophic lateral sclerosis by activating ERK. *Neuroscience* **2018**, *379*, 152–166. [CrossRef] [PubMed]
117. Xu, Z.; Chen, S.; Li, X.; Luo, G.; Li, L.; Le, W. Neuroprotective effects of (−)-epigallocatechin-3-gallate in a transgenic mouse model of amyotrophic lateral sclerosis. *Neurochem. Res.* **2006**, *31*, 1263–1269. [CrossRef] [PubMed]
118. Koh, S.H.; Lee, S.M.; Kim, H.Y.; Lee, K.Y.; Lee, Y.J.; Kim, H.T.; Kim, J.; Kim, M.H.; Hwang, M.S.; Song, C.; et al. The effect of epigallocatechin gallate on suppressing disease progression of ALS model mice. *Neurosci. Lett.* **2006**, *395*, 103–107. [CrossRef] [PubMed]
119. Jamwal, R. Bioavailable curcumin formulations: A review of pharmacokinetic studies in healthy volunteers. *J. Integr. Med.* **2018**, *16*, 367–374. [CrossRef] [PubMed]

International Journal of
Molecular Sciences

MDPI

Review

Review of the Effect of Natural Compounds and Extracts on Neurodegeneration in Animal Models of Diabetes Mellitus

Carmen Infante-Garcia [1,2] and Monica Garcia-Alloza [1,2,*]

1 Division of Physiology, School of Medicine, Universidad de Cádiz, Edificio Andres Segovia. C/Dr. Marañon 3, 3er piso, 11002 Cádiz, Spain; carmeninfante@uca.es

2 Division of Physiology, School of Medicine, Instituto de Investigación e Innovación en Ciencias Biomedicas de la Provincia de Cadiz (INiBICA), Universidad de Cádiz, 11002 Cádiz, Spain

* Correspondence: monica.garcia@uca.es

Received: 30 April 2019; Accepted: 18 May 2019; Published: 23 May 2019

Abstract: Diabetes mellitus is a chronic metabolic disease with a high prevalence in the Western population. It is characterized by pancreas failure to produce insulin, which involves high blood glucose levels. The two main forms of diabetes are type 1 and type 2 diabetes, which correspond with >85% of the cases. Diabetes shows several associated alterations including vascular dysfunction, neuropathies as well as central complications. Brain alterations in diabetes are widely studied; however, the mechanisms implicated have not been completely elucidated. Diabetic brain shows a wide profile of micro and macrostructural changes, such as neurovascular deterioration or neuroinflammation leading to neurodegeneration and progressive cognition dysfunction. Natural compounds (single isolated compounds and/or natural extracts) have been widely assessed in metabolic disorders and many of them have also shown antioxidant, antiinflamatory and neuroprotective properties at central level. This work reviews natural compounds with brain neuroprotective activities, taking into account several therapeutic targets: Inflammation and oxidative stress, vascular damage, neuronal loss or cognitive impairment. Altogether, a wide range of natural extracts and compounds contribute to limit neurodegeneration and cognitive dysfunction under diabetic state. Therefore, they could broaden therapeutic alternatives to reduce or slow down complications associated with diabetes at central level.

Keywords: type 2 diabetes; inflammation; vascular damage; learning; memory; neuroprotection; natural extract; natural compound

1. Type 2 Diabetes Mellitus: Central Complications

Metabolic disorders include a broad range of alterations. Moreover, the terminology used to refer to many of the diseases and complications is confusing in many cases [1,2]. Among these, diabetes mellitus (DM) plays a preponderant role, due to its prevalence and societal and economical burden. In 2013 over 380 million people suffered diabetes and it is estimated that by 2035 there will be 592 million diabetic patients [3]. World Health Organization (WHO) defines DM as a chronic metabolic disease caused by inherited and/or acquired deficiency in the production of insulin by the pancreas, or by the ineffectiveness of the insulin produced. Such a deficiency results in increased concentrations of glucose in the blood, which in turn damage many of the body's systems, in particular the blood vessels and nerves [4]. The two main forms of diabetes are type 1 diabetes (T1D) and type 2 diabetes (T2D), which account for >85% of the cases [3]. T1D and T2D differentially impact populations based on age, race, ethnicity, geography and socioeconomic status [5]. T1D is the most frequent type of diabetes in children and adolescents [6]. T1D patients suffer the destruction of over

90% of β-pancreatic islets, with consequent reduction of insulin and glycaemia control. On the other hand, T2D affects adults preferentially. However, the prevalence of T2D in adolescents and young adults is dramatically increasing [7]. T2D is characterized by an initial stage of insulin resistance. To compensate hyperglycaemia, β-pancreatic cells respond by increasing insulin production and establishing a prediabetic state. When exhausted β-pancreatic cells can no longer overproduce insulin, diabetes evolves. T2D is associated to a large list of risk factors, including familiar risk, previous gestational diabetes or life styles, among others [8].

While peripheral micro and macrovascular complications associated with T2D, such as neuropathies, retinopathies or nephropathies, have been widely studied [9], only in recent years attention has been paid to central complications associated with long-term metabolic alterations [10]. The mechanisms implicated have not been completely elucidated; however, cognitive impairment, vascular dementia, Alzheimer's disease, stroke or anxiety/depression have been related to diabetes [1,11]. In this sense, the diabetic brain (with controlled or uncontrolled hyperglycemia) show brain injury with a wide profile of micro and macrostructural changes, leading to neurodegeneration, neurovascular deterioration, neuroinflammation and progressive cognition dysfunction [12–19]. However, the study of central complications associated with T2D has been probably hampered by the difficulty of the measurements [2], the lack of ideal animal models, or the fact that T2D is a complex disorder and, therefore, it is likely that multiple different, synergistic processes may interact to promote central alterations. Accordingly, the vast majority of the research are epidemiological studies in which T2D is identified as a risk factor for Alzheimer's disease or vascular dementia [20–23]. Only a few studies have captured quality data regarding metabolic and cognitive status to allow reliable diagnosis of both T2D and dementia subtype. Main limitations are due to the fact that many of the studies rely on self reported diabetes, underestimating the prevalence by up to 50%, medical records are incomplete or may even include undiagnosed diabetics as control samples [2]. Moreover, patients with diabetes are often presumed to have dementia of vascular origin. However, the main limitation might be to determine the effects of medication, since treatments for T2D may also affect brain-associated complications [2]. Hence, in order to accurately delineate the pathogenesis of cognitive impairment in people with T2D, large-scale, prospective epidemiological studies are still required [24].

2. Natural Compounds and Central Complications in DM

The wide and countless number of natural compounds from plants, animals, fungi, microorganisms and other natural resources provides a rich and a unique source in the search of new drugs [25]. The potential health risk in the indiscriminate use of natural products cannot be obviated [26]. However, plant compounds, including different natural products (single isolated compounds) and/or natural extracts (including different compounds and/or secondary metabolites), have been long analyzed and assessed in relation with different pathologies. Usually, biological activity in plants' natural extracts is mainly due to secondary metabolites. Plant secondary metabolites include two extensive categories: Nitrogen-containing compounds and those without it [27,28]. In line with these observations, several studies have shown a wide range of biological activities in these extracts, including anti-inflammatory [29,30], anti-microbial [31], anti-diabetic [18,32] or neuroprotective [27,33,34] properties, among others.

One of the most extensive group of secondary metabolites in the plant kingdom are polyphenols [35]. Structurally, they are characterized by the presence of at least one hydroxyl functional group (-HO) linked to an aromatic ring [36]. Polyphenols classification is referred to the number of phenol rings in the molecule, and the main subgroups include phenolic acids, stilbenes, flavonoids, coumarins and lignans [35]. The wider group of polyphenols in plants is represented by flavonoids, which account for over 10,000 different compounds [28,35]. As other natural compounds, flavonoids have shown several properties including antioxidant, neuroprotective [37] or anti-diabetic [38–40] effect. Another particularity of polyphenols is their role in human nutrition, which extends their utility, including not only a pharmacological, but also a nutritional perspective. This singularity of polyphenols contributes to further study of these compounds in other fields, such as human diet supplements [35,41].

As mentioned above, DM, or even prediabetes state, are associated with an increased risk to suffer neurodegenerative diseases, specially vascular dementia and Alzheimer's disease [42,43]. Therefore diabetic control may be an important and modifiable risk factor to reduce diabetes-associated neurodegeneration [44]. In this sense, while the number of articles published worldwide in relation with antidiabetic natural products is growing each year, most of them focus on metabolic control and related alterations [45]. On the other hand, studies on the effect of natural products and extracts on central complications associated with DM are more scarce. This is mainly due to the difficulty to identify individual components in complex extracts, the capability of different molecules to cross the blood brain barrier, or even discriminate the direct effect of diabetes on the pharmacokinetics, bioavailability and brain distribution of the compounds and metabolites [46]. However, given the well established complications of DM on the central nervous system, there are different targets of interest that may be covered by natural compounds, including vascular damage, neuroinflammation, neurodegeneration or cognition. Following this idea, several natural compounds and extracts have been reported to show neuroprotective effects [34,38].

2.1. Natural Compounds and DM-Related Vascular Injury

2.1.1. Vascular Damage and DM

Vascular complications are the leading cause of morbidity and mortality in diabetic patients. Vascular alterations are derived from the chronic hyperglycemic state that can affect both large and small blood vessels, characterizing diabetes macro and microangiopathy, respectively [47]. Several vascular alterations including irreversible non-enzymatic glycation of proteins, cellular redox potential alteration, increased oxidative stress or inflammatory response, as well as endothelial dysfunction or hypercoagulability contribute to vascular abnormalities associated to DM [47–49]. These underlying alterations may support the fact that diabetic patients present arterial stiffness as well as increased risk of atherosclerosis and cerebral stroke [50–52]. In line with these observations, previous studies have reported that DM patients have smaller brain volumes and white matter lesions, which have been associated to neurovascular unit dysfunction and blood brain barrier alterations. In this context T2D could cause loss of homeostasis of the cerebral microenvironment, leading to vascular damage and astrocyte alterations [53]. In addition, preclinical studies in diabetes animal models have shown exacerbated neurovascular damage, and ultrastructural abnormalities, characterized by mural endothelial cell tight and adherens junction or perycite attenuation or loss [54]. Likewise, studies in mouse models reveal brain overspread microbleeding, reproducing small vessel disease [55,56]. DM not only exacerbates neurovascular damage but also hinders the brain repair process, likely contributing to the impairment of stroke recovery [57]. In this sense, in vitro and in vivo experimental models have showed that the integrity of the blood brain barrier is affected in diabetic conditions [58–60]. Concretely, diabetes disrupts the blood brain barrier endothelium by downregulation of cell junction proteins [61–63] and upregulation of integrin expression [64,65], leading to abnormal vascular permeability [66,67]. In addition, this effect might be mediated by oxidative stress, which induces blood brain barrier disruption through osmotic damage and pericyte loss [68], ultimately leading to the leak of toxic substances and further damage to the nervous structures [69]. Interestingly, microvascular alterations seem to be present also in prediabetic animal models [70], suggesting that early hyperinsulinemia and insulin resistance are enough to induce vascular damage.

2.1.2. Natural Compounds and Extracts in Vascular Damage Associated with DM

In order to try and reverse many of these complications different natural compounds and extracts have been used in animal models. In this sense berberine, a protoberberine present in a number of medicinal plants [71], and the main active component of *Coptis chinensis French* has been used for years, and studies in patients have shown its capability to regulate glucose and lipid metabolism [72]. Moreover, at central level it has also been reported that berberine may reduce diabetes induced ectopic expression of miR-133a in endothelial cells, which is involved in endothelial dysfunction in DM. In addition, berberine may inhibit acetylcholine-induced vasorelaxation in the middle cerebral artery, guaranteeing better blood supply to the brain in streptozotozin (STZ)-treated rats, as a T1D model [73]. It has also been reported that patchouli alcohol, a natural tricyclic sesquiterpene in the traditional Chinese herb Pogostemonisherba [74], reduces ishcemia/reperfusion damage after middle cerebral artery occlusion in ob/ob mice by limiting infarct volume, protecting blood brain barrier function and decreasing inflammatory markers [74]. In line with these observations, *Mangifera indica Lin* extract, rich in natural polyphenols, reduces spontaneous central bleeding detected in db/db mice. While the actual size of the microbleeds is not affected, *Mangifera indica* extract reduces the appearance of new vascular lesions [18]. In addition, poor cerebral perfusion may contribute to cognitive impairment in diabetic state and resveratrol, a natural phenol isolated from plants like *Polygonum cuspidatum*, *Paeonia lactiflora* and *Vitis amurensis*, among others [75], may improve neurovascular coupling capacity in T2D patients [76] and reduce blood brain barrier permeability and vascular endothelial growth factor expression in the hippocampus of diabetic rats [77] (Table 1 and Figure 1).

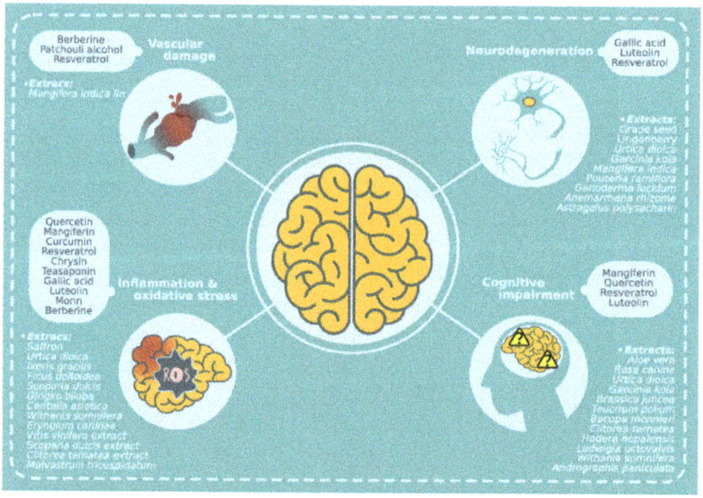

Figure 1. Central activities of natural compounds and extracts.

2.2. *Natural Compounds and Neuroinflammation Associated with DM*

2.2.1. Brain Neuroinflammation and DM

Inflammation is an immune response against several conditions including disease and infection. Acute inflammatory events are resolved efficiently and inflammation levels return to baseline in physiological conditions. However, in chronic inflammation the resolution phase is not achieved due to excessive pro-inflammatory signalling and it can provoke relevant detrimental effects [78]. Following this idea, insulin resistance and diabetes are closely associated with chronic inflammation [79]. Moreover, the finding two decades ago that proinflammatory cytokines like tumor necrosis factor-α (TNF-α), among others, are overexpressed in adipose tissue of obese mice provided a relation between obesity, diabetes and chronic inflammation [79–81].

Inflammation in the central nervous system is complexly regulated and astrocytes [82], blood inflammatory cells and even neurons seem to participate and mediate inflammation in the injured brain. However, microglia still play the most significant role at this level [83]. Microglia are a specific type of macrophage in the brain; they are held without external replenishment and they are not in contact with plasmatic proteins, which contributes to keep an immupriviledged environment in the central nervous system [84]. The classical dual role of microglia as a protective (with a typical anti-inflammatory profile) or damaging agent (with a proinflammatory response) has been recently reviewed and microglia-mediated responses seem to be more prone towards neuronal survival, regeneration [85] and overall neuroprotection [86]. The role of microglia in neurodegenerative diseases has been long studied and they also seem to be highly activated in metabolic disease models, ranging from prediabetic [87], T1D [88], T2D [55,56] models, or even diabetic mothers offspring [89]. Under diabetic conditions, hyperglycemia leads to increased mitochondrial respiration in pericytes, astrocites as well as endothelial cells [90]. This causes an increase in the production of reactive oxygen species that may consequently lead to neurovascular damage and blood brain barrier dysfunction, contributing to the inflammatory process. Increased levels of reactive oxygen species may also affect protein fuction, signaling pathways or induce upregulation of inflammatory cytokines [90]. Therefore, previous studies have shown that, in metabolic alterations, microglia mediated neuroinflammation may contribute to the neurodegenerative process by promoting the release of cytokines and chemokines including TNF-α [91,92]. In line with these ideas, studies in patients with metabolic disorders have detected a decrease in mRNA levels of the IL10-mediated anti-inflammatory defense, while iNOS-mediated inflammatory activity seems to be favored in the cortex from obese patients [93].

2.2.2. Effect of Natural Compounds on DM-Related Inflammation

Antioxidant and anti-inflammatory activities are probably the most widely explored roles of natural compounds and extracts [30,94,95]. Following this idea, many studies have previously used products of natural origin to counterbalance oxidative stress, neuroinflammation and alterations in related markers and cytokines. Even though the role of flavonoids in neuroprotection might be due to different mechanisms of action it is mediated, at least in part, by direct scavenging of free radicals as antioxidant action [35,96]. Several plants extracts constitute a relevant source of polyphenols. While in many cases they share common mechanisms and show potent anti-inflammatory and antioxidant activities, not all of them have been completely characterized. Concretely quercetin, present in many fruits and vegetables, may enhance glyoxalase pathway activity, inhibit advanced glycation end products (AGEs) formation and reduce oxidative stress [97]. Quercetin is a flavonoid present in a wide variety of plants, including *Rosa canina*, *Opuntia ficusindica* and *Allium cepa* [75]. Oral administration of quercetin to diabetic rats has shown antioxidant effects, increasing superoxide dismutase (SOD) and catalase activity, while also restoring the blood levels of vitamin C and E, which finally contribute to ameliorate the diabetes-induced in oxidative stress [98]. On the other hand, it has been described that quercetin also protects neuronal PC12 cells against high-glucose-induced oxidative stress, inflammation and apoptosis [99]. While the final underlying mechanisms involved in quercetin neuroprotective effects are not completely known, a recent study has shown that neuroprotection might mediated by phosphorylation regulation of Nrf2/ARE/glyoxalase-1 pathway in central neurons under chronic hyperglycemia, reducing AGEs and oxidative stress [38]. In line with these observations mangiferin, which is mainly present in *Mangifera indica L.* but also in Chinese herbal medicines like *Rhizoma Anemarrhenae* and *Rhizoma Belamcandae*, has anti-inflammatory [100] and antioxidant [100,101] activities. Mangiferin also enhances the function of glyoxalase-1 through activation Nrf2/ARE pathway in neurons exposed to chronic high glucose [101]. In addition, *Mangifera indica L.* extracts with a high content in mangiferin and quercetin reduce microglia activation and associated inflammation in db/db mice after long-term treatment [18].

On the other hand curcumin, a bright yellow compound isolated from the rhizome of *Curcuma longa* [75] has shown neuroprotective effects in diabetic rats reducing blood glucose, oxidative stress markers and astrocyte activation in the hippocampus [102]. A recent study has reported the potent neuroprotective effect of J147, a novel curcumin derivative developed to increase curcumin bioavailability and blood brain barrier permeability [103]. J147 reduces inflammation by decreasing TNF-α pathway activation and several other markers of neuroinflammation in mice treated with STZ [103], supporting that different curcumin extracts and derivates are potent antioxidants with the capability to limit associated central complications in diabetes. Resveratrol has a well established antioxidant activity. It reduces astrocytic activation as well as TNF-α, IL-6 transcripts the hippocampus of diabetic rats [77]. Resveratrol also normalizes malonedialdehyde and oxidezed glutathione levels in diabetic rats and it strengthens the action of antioxidants enzymes SOD and catalase [104]. *Ficus deltoidea* leaf extract also increases SOD and glutathione peroxidase values, while reducing thiobarbituric acid reactive substances [105]. Similar outcomes have been reported for saffron extracts with antidiabietic activity, which also modulate antiinflamatory pathways at central level [106]. Likewise, *Scoparia dulcis* plant extract also increases activities of plasma SOD, catalase or glutahione peroxidase or glutathione-S-transferase while reduces gluthatione in the brain from STZ diabetic male rats [107]. Similar outcomes have been described for chrysin, a flavonoid isolated from *Oroxylum indicum, Passiflora caerulea, Passiflora incarnata, Teloxys graveolens* and *Artocarpus heterophyllus* that also ameliorates oxidative stress by reducing catalase levels, SOD and glutathione in the cerebral cortex and hippocampus of diabetic rats [108].

One of the most widely studied preparations is Gingko biloba extract EGb 761, which has been described to scavenge reactive nitrogen and oxygen species, as well as peroxyl radicals [35,96,109]. A similar scavenging effect has been described for green tea extracts [35,110]. In this sense, tea extract, teasaponin, also reduces proinflammatory citokines and inflammatory signaling in the hypothalamus from mice on high fat diet [111]. For its part, *Clitorea ternatea* leaf extract, has showed protection against oxidative stress increasing SOD, total nitric oxide, catalase and glutathione levels in the brain of diabetic rats [112]. Similar antioxidant effects have been reported for grape seed extracts (*Vitis vinifera* sp.), rich in flavonoids like proanthocyanidins, showing beneficial effects on oxidative stress in the hippocampus of STZ-induced diabetes rats, to a larger extend than a classical antioxidant as viatamin E [113]. The expression of inflammatory TNF-α, and NF-κB genes are significantly reduced and other studies have also reported the role of grape seed extract in modulating AGEs/RAGE/NF-kappaB inflammatory pathway in the brain [114]. *Urtica dioica* leaves extract, rich in scopoletin, rutin, esculetin and quercetin, has also shown antioxidant and anti-inflamatory activities in the hippocampus from STZ-induced diabetic mice [115,116]. In addition, the number of astrocytes in the hippocampus from diabetic rats is reduced after treatment with *Urtica dioica* extract, supporting its anti-inflammatory role at different levels [117].

Gallic acid, is a type of phenolic acid, which is isolated from several plants including *Phaleria macrocarpa, Peltiphyllum peltatum,* and *Pistacia lentiscus.* Gallic acid may inhibit hippocampal neurodegeneration via its potent antioxidant and anti-inflammatory effects in diabetic rats [118]. Similarly, *Scoparia dulcis* extract also reduces thiobarbiyutic acid reactive substances and hydroperoxides formation in the brain from diabetic rats, supporting its role in protection against lipid peroxidation induced membrane damage [107]. Luteolin can also reduce neuroinflammation by reducing plasma and brain cytokines in a prediabetic mouse model [119]. Moreover, similar antioxidant and anti-inflammatory effects have been reported for luteolin in diabetic mice [120]. Other studies in prediabetic models have shown a protective role for *Withania somnifera*, which may reduce gliosis and microgliosis as well as expression of inflammation markers such as PPARγ, iNOS, MCP-1, TNF-α, IL-1β, and IL-6 [121]. In line with these observations, oral administration of an hexanic extract of *Eryngium carlinae* inflorescences to diabetic rats not only reduced glucose levels but also limited overall oxidation, by reducing lipid peroxidation, protein carbonylation and reactive oxigen species production, while increasing catalase activity in the brain [122]. Morin is another flavonoid isolated from *Maclura pomifera* and *Maclura tinctoria*, with similar properties [123,124]. Additionally, the flavonoid rutin has also shown antioxidant properties in the diabetic rat retina [125]. In line with these observations, berberine has been shown to reduce oxidative stress and astrogliosis in the hippocampus from diabetic rats [126]. A natural extract from *Centella asiatica,* rich in ascorbic acid, asiatic acid, oleanolic acid, stevioside, stigmasterol and α-humulene protects diabetes tissues from stress via antioxidant and anti-inflammatory mechanisms eliciting brain reduced levels of malondialdehyde, TNF-α, IFN-γ, IL-4 or IL10 [127]. Similar outcomes have been reported for *Ixeris gracilis* extract used in mice with alloxan-induced diabetes [128]. Specific assessment of mitochondrial status in

STZ-induced diabetes has also revealed the capacity of *Malvastrum tricuspidatum* extract to restore oxidative damage [129] (Tables 1 and 2, and Figure 1).

2.3. Natural Compounds and Brain Neurodegeneration in DM

2.3.1. Neurodegeneration in Diabetic Brain

A wide range of clinical [15,19,130,131] and preclinical studies [70,88,132] have shown an association of prediabetes and diabetes with brain atrophy. In this sense, magnetic resonance studies have shown that both T1D and T2D patients have reduced grey matter density and white matter lesions, as well as cortical and hippocampal atrophy [133,134]. However, it seems that brain atrophy is more severe in T2D patients, probably given that this population is older on average [135–137]. As previously pointed out, the prediabetes process seems to be enough to induce brain atrophy in patients [138] and synaptic loss is also detected in animal models when prediabetes is combined with other central complications [132]. Likewise, animal models of metabolic alterations show neuronal simplification, synaptic alterations [44], reduced neuronal density and overall brain atrophy [55,56].

Neurodegeneration in diabetes is mediated by multiple neuropathogenic factors including hyperglycemia mediated damage, but also hypoglycemic episodes, cerebrovascular alterations or insulin derregulation in the brain or among others [139]. In this sense, dysfunction of insulin/insulin receptor mediated signaling might be responsible for alterations in synaptic plasticity, cognition and memory [139,140]. Once more, oxidative stress mediated by free radicals is related with the diabetes neurodegenerative process [141], given that hyperglycemic state reduces antioxidants levels and consequently increases the production of free radicals [139]. Neurons are especially vulnerable to oxidative stress and this can induce mitochondrial oxidative damage, resulting in apoptosis and/or necrosis [142]. On the other hand, several proteins implicated in neurodegeneration, such as tau protein, which is hyperphosphorylated in diabetic mouse models, may also underlie neuronal death [70,88,143]. In overall terms, neurodegeneration is perceived as a cause of cognitive dysfunction observed in diabetes conditions.

2.3.2. Effect of Natural Compounds and Extracts on Brain Neurodegeneration Associated with DM

The majority of the studies on natural compounds and extracts have focused on their antioxidant and anti-inflammatory activities. However, neurodegeneration is a multifactorial pathogenic process and it is feasible than various, concomitant underlying mechanisms are responsible for their final neuroprotective effect. In this sense, polyphenols are able to modulate the activity of multiple involved targets, which contribute their pleiotropic effects (anti-inflamatory, antioxidant or inmunomodulatory) [144], and, indeed, phenolic compounds have shown their neuroprotective role in vitro, in animal models and in clinical studies [145–148]. In line with these observations, flavonoids are not only implicated in scavenging of free radicals and reducing oxidative stress [35,96], but they can also modulate brain signaling cascades implicated in neuronal apoptosis, alter the expression of specific genes and modify mitochondrial activity [149].

Mangifera indica extract has shown its capability to limit brain atrophy in db/db mice. Cortex and hippocampus are largely preserved after long-term administration [18]. Interestingly, oral treatment with *Mangifera indica* also reduces tau hyperphosphorylation, an early marker of neuronal damage, and it also preserves compromised neuronal population in this model [18]. In line with these observations quercetin has also been shown to protect neuronal PC12 cells against high-glucose-induced oxidative stress, inflammation and apoptosis [99], as described for gallic acid in diabetic mice [118]. Curcumin protects against structural alterations of the hippocampus associated with diabetes, by reducing disorganization of small pyramidal cells in CA1, cellular loss in the pyramidal cells of CA3 and degenerated granule cells in the dentate gyrus [102]. In addition, curcumin derivate, J147, has been shown to upregulate nervous system development functions in diabetic mice. Moreover, functions related with neuron growth, such as proliferation, axon growth and long-term potentiation are the most significantly changed [103]. Luteolin also shows neuroprotective activity by increasing the levels of brain-derived neurotrophic factor, the action of synapsin I and postsynaptic density protein 95 in the cortex and hippocampus from mice on high fat diet [119]. Likewise, resveratrol has also been widely assessed and chronic treatment improves neuronal injury, not only through attenuation of oxidative stress and neuroinflammation, but also by reducing synaptic loss and increasing synaptic plasticity markers SYN and GAP-43 [150], as well as by inhibiting hippocampal apoptosis through the Bcl-2, Bax and caspase-3 signaling pathways in STZ-induced diabetic rats [151]. Gallic acid may inhibit hippocampal neurodegeneration in diabetic mice not only through its potent antioxidant and anti-inflammatory activities, but also due to its anti-apoptotic properties [118].

Other mechanisms of action have been presented for different compounds and extracts, many of which have focused on the hippocampus, a key area in learning and memory. In this sense *Astragalus Polysacharin* extract may upregulate phosphorylation levels of N-methyl-D-aspartate receptor, calcium/calmodulin-dependent protein kinase II and cAMP response element-binding protein, as well as reduce the number of dead cells in the CA1 region of the hippocampus from STZ-treated diabetic rats [152]. On the other hand, antioxidants present in bilberry fruits, rich in anthocyanins, influence the morphology of and possibly exhibit beneficial and neuroprotective effects on hippocampal neurons during diabetes [153]. *Pouteria ramiflora* extract administration to STZ-treated rats exerts hippocampal neuroprotection by restoring myosin-Va expression and the nuclear diameters of pyramidal neurons of the CA3 and the polymorphic cells of the hilus [154]. In a T1D rat model, *Garcinia kola* seeds limit neuronal loss in regions involved in cognitive and motor functions, including the motor cortex, the medial septal nucleus an cerebellar Purkinje /granular cell layers [155]. *Urtica dioica* leaves extract also seems to exert it neuroprotective activities by modulating different pathways. It downregulates iNOS, while it upregulates BDNF, TrKB, cyclin D1, Bcl2, autophagy5 and autophagy7 mRNA expression and reduces TNF-α expression in diffrent hippocampal regions. In addition, an overall reduction of neuronal damage and DNA fragmentacion has been observed in the hippocampus from diabetic mice [156]. Other studies have also shown that *Urtica dioica* extract may limit granule cell loss of the dentate gyrus from young diabetic rats. While the positive effect is not observed when the extract is used preventively, it seems to ameliorate hippocampus cell loss when used as a treatment [157]. Similar outcomes have been observed after ginger extract administration, in combination with insulin, to male diabetic rats, showing changes in the expression of cyclin D1 gene and reducing apoptosis in hippocamapal cells [158]. Apart from its well established antioxidant activity, grape seed extract reduces caspases 3 and 9 expression in the hippocampus, ameliorating apoptosis in diabetic rats [113]. Another way of maintaining hippocampus integrity has been observed with an aqueous extract of *Anemarrhena rhizome*, capable of increasing cell proliferation and neurpeptide Y expression in the dentate gyrus from diabetic rats [159]. Lingonberry extract also exerts neuroprotective activity in diabetic rats by reducing oxidative stress, but also by restoring the density of purinergic receptors in the cortex [160]. In addition, in T2D mice with cerebral ischemic injury, chronic treatment with a water-soluble extract from the culture medium of *Ganoderma lucidum* mycelia reduced neuronal cell death and vacuolation in the ischemic penumbra, with reduced number of TUNEL, cleaved caspase-3 cells and the expression of receptor-interaging protein kinase 3 mRNA and protein, confering resistance to apoptosis and necroptosis [161] (Tables 1 and 2, and Figure 1).

2.4. Natural Compounds and Cognitive Impairment in DM

2.4.1. Cognitive Dysfunction Associated with Diabetes

Substantial epidemiological evidence supports that cognitive dysfunction is a common complication of diabetes [162–164]. It has been estimated that 20–70% of patients with DM show cognitive decline, and 60% present at higher risk of dementia [11,12]. Following this idea, it is noteworthy that even prediabetic adults shown accelerated cognitive decline, associated with smaller total brain tissue volume [131]. Different stages of cognitive dysfunction have been associated with diabetes, depending on affected cognitive features, age or prognosis, andprobably with different underlying mechanisms [165–167]. Previous studies in patients have reported a wide range of diabetes-associated cognitive decrements ranging from subtle changes in cognitive function (that might give rise to cognitive complaints, but should not affect activities of daily life) and mild cognitive impairment, to severe forms like dementia [162,168]. Several factors, including vascular injury, insulin resistance, inflammation and depression, are potential risk factors for cognitive dysfunction in diabetic patients [168–170]. These data are also supported by studies in animal models, where severe cognitive impairment is observed in diabetic animal models that are also dependent on the model under study, the age and evolution of the disease [70,171].

2.4.2. Effect of Natural Compounds and Extracts on Cognitive Impairment Associated with DM

As previously discussed, the mechanisms of action of natural compounds and extracts remain largely elusive, and it is feasible that a combination of different positive effects, including antioxidant, anti-inflammatory, vascular protection, antiapoptotic or proregenerative activities are responsible for observed beneficial effects in DM associated cognitive alterations. Concretely, mangiferin has been shown to counterbalance learning and memory impairments in diabetic rats, treated with STZ, when assessed in the Morris water maze [172]. Similarly, db/db mice on long-term *Mangifera indica* extract, with a high content of mangiferin, significantly improve their performance in the Morris water maze [18]. Moreover, episodic memory alterations are also ameliorated in a very demanding version of the new object discrimination test, and "what", "where" and "when" paradigms are significantly improved [18]. Quercetin also ameliorates STZ-induced spatial learning and memory impairment in the Morris water maze [173,174], reducing the time spent in target quadrant in the test trial and increasing escape latendcy in the elevated plus maze. Similar results have been reported when chrysin [108] or *Andrographis paniculata* extract [175] are administered to STZ-treated rats. Similar outcomes have been reported when *Hedera nepalensis* extract is administered to STZ-aluminium trichloride rat model [176]. Likewise, grape seed proanthocyanidin extract [177], kola nut extract [178] or *Garcinia kola* seeds [155] also improve cognitive impairment in diabetic rat models. *Andrographis paniculata* extract, enriched in andrographolide, improves cognitive function in STZ-treated rats and the effect seems to be mediated by reducing oxidative stress and acetylcholinesterase activity [175]. Similar underlying mechanisms have been described for *Clitorea ternatea* leaves extract, which also improve spatial working memory, spatial reference memory, and spatial working-reference in the Y maze, the Morris water maze and radial arm maze, respectively, in diabetic rats [112]. In addition, studies with *Brassica juncea* extract [179] or resveratrol [150] have reported positive effects on learning and memory in diabetic rats. Equally, hydroalcoholic extract of *Teucrium polium* also limits cognitive impairment in the passive avoidance test while reducing oxidative stress markers in diabetic rats [180]. In addition, cognitive impairment is ameliorated in mice models after administration of *Rosa canina* hydro-alcoholic extract [181] or *Ludwigia octovalvis* extract [182]. Other studies on diabetic mice have shown that *Flos puerariae* extract also improves cognitive impairment after STZ administration, by reducing oxidative stress and restoring cholinergic activity (enhancing cholinacetyltransferase and alleviating acetylcholinesterase activities) in the the cortex from STZ-treated mice [183], and similar outcomes have been reported with *Withania somnifera* and *Aloe vera* extracts [184]. *Bacopa monnieri* [185] and *Urtica dioica* [115,186] extracts restore memory deficits in different diabetic mouse models. Additionaly, cognitive impairment in early metabolic alterations, such as prediabetic mice on a high fat diet, improve in the Morris water maze and the step-through task after luteolin [119] or *Ludwigia octovalvis* extract administation [182] (Tables 1 and 2, and Figure 1).

Conclusions: Altogether, natural components and extracts show antioxidant and anti-inflammatory activities at central level, as well as a relevant capacity to reduce vascular damage, contributing altogether to limit neurodegeneration and cognitive derived alterations. Therefore, while the ultimate underlying mechanisms remain largely unknown, they could contribute to expand therapeutic options to treat or reduce central complications associated with DM.

Table 1. Natural compounds and extracts with activity at central level associated with metabolic disorders.

Natural Compound	Action	Plant Source	References
Berberine	Regulation of glucose and lipid metabolism. Reduction of diabetes induced ectopic expression of miR-133a involved in endothelial dysfunction associated with DM. Inhibition of acetylcholine-induced vasorelaxation in the middle cerebral artery → better blood supply to the brain in STZ-treated rats. Reduction of oxidative stress and astrogliosis in the hippocampus from diabetic rats.	*Coptis chinensis* French and others.	[72,73,126]
Patchouli alcohol	Reduction of ischemia/reperfusion damage after middle cerebral artery occlusion in ob/ob mice by limiting infarct volume, protecting blood brain barrier function and decreasing inflammatory markers.	Pogostemonisherba	[74]
Resveratrol	Improvement of neurovascular coupling capacity in T2D patients. Reduction of blood brain barrier permeability and VEGF expression in the hippocampus of diabetic rats. Restriction of astrocytic activation as well as TNF-α, IL-6 transcripts the hippocampus of diabetic rats. Normalization of malonedialdehyde and oxidezed glutathione levels in diabetic rats and strengthening of the action of antioxidants enzymes SOD and catalase. Improvement of neuronal injury by attenuation of oxidative stress and neuroinflammation, and by reducing synaptic loss and increasing synaptic plasticity markers SYN and GAP-43, as well as by inhibiting hippocampal apoptosis through the Bcl-2, Bax and caspase-3 signaling pathways in STZ-induced diabetic rats. Protection against learning and memory alterations in diabetic rats.	*Polygonum cuspidatum*, *Paeonia lactiflora* and *Vitis amurensis*, among others	[75–77,104,150,151]
Quercetin	Enhancement glyoxalase pathway activity, inhibition of AGEs formation and reduction of oxidative stress. Increase of SOD and catalase activities, restoring blood levels of vitamin C and E and ameliorating diabetes-induced oxidative stress. Protection of neuronal PC12 cells against high-glucose-induced oxidative stress, inflammation and apoptosis. Improvement in learning and spatial memory in the Morris water maze.	*Rosa canina*, *Opuntia ficusindica* and *Allium cepa*	[38,75,97–99,173,174]
Mangiferin	Improvement of the function of glyoxalase-1 through activationNrf2/ARE pathway in neurons exposed to chronic high glucose. Protections against learning and memory impairments in diabetic rats, treated with STZ.	*Mangifera indica Lin*, *Rhizoma Anemarrhenae* and *Rhizoma Belamcandae* among others	[100,101,172]
Curcumin	Neuroprotective effects in diabetic rats reducing blood glucose, oxidative stress markers and astrocyte activation in hippocampus. Protection against structural alterations of the hippocampus associated with diabetes.	*Curcuma longa*	[75,102]

Table 1. *Cont.*

Natural Compound	Action	Plant Source	References
J147 curcumin derivative	Increase of curcumin bioavailability and blood brain barrier permeability. Reduction of inflammation by decreasing TNF-α pathway activation and several other markers of neuroinflammation in mice treated with STZ. Upregulation of nervous system development functions in diabetic mice including functions related with neuron growth, proliferation, axon growth and long-term potentiation.	*Curcumin derivate*	[103]
Chrysin	Amelioration of oxidative stress by reducing catalase levels, SOD, and glutathione in the cerebral cortex and hippocampus from diabetic rats. Improvement in spatial memory and learning abilities in Morris water maze test.	*Oroxylum indicum*, *Passiflora caerulea*, *Passiflora incarnata*, *Teloxys graveolens* and *Artocarpus heterophyllus*	[108]
Teasaponin	Reduction of proinflammatory citokines and inflammatory signaling in the hypothalamus from mice on high fat diet.	*Camellia sinensis*	[111]
Gallic acid	Inhibition of hippocampal neurodegeneration via its potent antioxidant and anti-inflammatory effects in diabetic rats as well as its anti-apoptotic properties.	*Phaleria macrocarpa*, *Peltiphyllum peltatum*, and *Pistacia lentiscus*	[118]
Luteolin	Neuroinflammation amelioration by reducingplasma and brain cytokines levels in a prediabetic mice. Antioxidant and anti-inflammatory effects in diabetic mice. Neuroprotection by increasing the levels of brain-derived neurotrophic factor, the action of synapsin I and postsynaptic density protein 95 in the cortex and hippocampus from mice on high fat diet. Protection against cognitive impairment in early metabolic alterations, such as prediabetic mice on a high fat diet, improvements in the Morris water maze and the step-through task.	*Salvia officinalis*, *Artemisa annua*, and others	[119,120]
Morin	Inhibition of oxidative stress and inflammation in the brain of STZ-induced diabetic rats. Neuroprotection via attenuation of ROS induced oxidative damage and neuroinflammation in experimental diabetic neuropathy.	*Maclura pomifera* and *Maclura tinctoria*	[123,124]
Rutin	Antioxidant properties in the diabetic rat retina.	*Urtica dioica* and others	[125]

Table 2. Natural extracts with activity at central level associated with metabolic disorders.

Natural Extract	Action	References
Mangifera indica Lin. extract	Reduction of spontaneous central bleeding db/db mice Restriction of microglia activation and associated inflammation in db/db mice after long-term treatment. Limitation of brain atrophy and reduction of tau hyperphosphorylation in db/db mice. Protections against learning and memory impairments in db/db mice in the Morris water maze and new object discrimination tests.	[18]
Ficus deltoidea leaf extract	Increased SOD and glutathione peroxidase values and reduction of thiobarbituric acid reactive substances.	[105]
Scoparia dulcis extract	Increase of plasma SOD, catalase or glutahione peroxidase or glutathione-S-transferase activities and reduction of gluthatione in the brain from STZ diabetic male rats. Reduction of thiobarbituric acid reactive substances and hydroperoxides formation in the brain from diabetic rats	[107]
Gingko biloba extract EGb 761	Scavenging reactive nitrogen and oxygen species, as well as peroxyl radicals.	[35,96,109]
Green tea extracts	Scavenging reactive nitrogen and oxygen species, as well as peroxyl radicals.	[35,110]
Clitorea ternatea leaf extract	Protection against oxidative stress increasing SOD, total nitric oxide, catalase and glutathione levels in the brain of diabetic rats. Improvement of spatial working memory, spatial reference memory, and spatial working-reference memory in the Y maze, the Morris water maze and radial arm maze in diabetic rats.	[112]
Grape seed extract	Beneficial effects on oxidative stress in the hippocampus of STZ-induced diabetes rats. Reduction in expression of inflammatory TNF-α, and NF-κB genes and modulation of AGEs/RAGE/NF-kappaB inflammatory pathway in the brain. Reduction of caspases 3 and 9 expression in the hippocampus, ameliorating apoptosis in diabetic rats. Improvement of cognitive impairment in diabetic rat models.	[113,114,177]
Urtica dioica leaves extract	Antioxidant and anti-inflamatory activities in hippocampus from STZ-induced diabetes in mice. Reduction in the number of astrocytes in the hippocampus from diabetic rats. Protection against memory deficits in different diabetic mouse models. Neuroprotective activities by iNOS downregulation, while it upregulates BDNF, TrKB, cyclin D1, Bcl2, autophagy5 and autophagy7 mRNA expression and reduces TNF-α expression in the hippocampus. Reduction of neuronal damage and DNA fragmentation. Limitation of granule cell loss of the dentate gyrus from young diabetic rats.	[115–117,156,157,186]
Withania somnifera leaf powder	Reduction of gliosis and microgliosis as well as expression of inflammation markers such as PPARγ, iNOS, MCP-1, TNF-α, IL-1β, and IL-6. Improvement of cognitive impairment STZ-treated mice, by reducing oxidative stress.	[121,184]
Extract of *Eryngium carlinae* inflorescences	Reduction of glucose levels by reducing lipid peroxidation, protein carbonylation and reactive oxigen species production, while increasing catalase activity in the brain of diabetic rats.	[122]
Centella asiatica extract	Protection of diabetes tissues from stress via antioxidant and anti-inflammatory mechanisms by brain reduced levels of malondialdehyde, TNF-α, IFN-γ, IL-4 or IL10.	[127]
Ixeris gracilis extract	Antidiabetic, antioxidant, and TNF- α lowering properties in alloxan-induced diabetic mice.	[128]

Table 2. *Cont.*

Natural Extract	Action	References
Malvastrum tricuspidatum extract	Restoration oxidative damage of mitochondrial status in STZ-induced diabetes.	[129]
Astragalus Polysacharin extract	Upregulation of phosphorylation levels of N-methyl-D-aspartate receptor, calcium/calmodulin-dependent protein kinase II and cAMP response element-binding protein, as well as reduction of the number of dead cells in the CA1 region of the hippocampus from STZ-treated diabetic rats.	[152]
Pouteria ramiflora extract	Hippocampal neuroprotection by restoring myosin-Va expression and the nuclear diameters of pyramidal neurons of the CA3 and the polymorphic cells of the hilus in STZ-treated rats.	[154]
Garcinia kola seeds	Reduced neuronal loss in regions involved in cognitive and motor functions, including the motor cortex, the medial septal nucleus a cerebellar Purkinje /granular cell layers in a T1D rat model. Improvement of cognitive abilities in diabetic rat models	[155]
Anemarrhena rhizome aqueous extract	Maintenance of hippocampus integrity by increasing cell proliferation and neurpeptide Y expression in the dentate gyrus from diabetic rats.	[159]
Lingonberry extract	Neuroprotective activity in diabetic rats by reducing oxidative stress and by restoring the density of purinergic receptors in the cortex.	[160]
Ganoderma lucidum mycelia extract	Increased resistance to apoptosis and necroptosis in T2D mice with cerebral ischemic injury.	[161]
Andrographis paniculata extract	Improvement of cognitive function in STZ-treated rats by reducing oxidative stress and acetylcholinesterase activity.	[175]
Hedera nepalensis extract	Improvement of cognitive abilities in STZ-aluminium trichloride rat model.	[176]
Kola nut extract	Protection against cognitive dysfunction in diabetic rat models.	[178]
Brassica juncea extract	Positive effects on learning and memory in diabetic rats.	[179]
hydroalcoholic extract of *Teucrium polium*	Limitation of cognitive impairment in the passive avoidance test and reduction of oxidative stress markers in diabetic rats.	[180]
Rosa canina hydro-alcoholic extract	Amelioration of cognitive impairment in mouse models after treatment.	[181]
Ludwigia octovalvis extract	Improvement of glycemic control and memory performance in mice fed with high fat diet. Protection against cognitive impairment in diabetic mice.	[182]
Flos Puerariae extract	Improvement of cognitive impairment after STZ administration, by reducing oxidative stress and restoring cholinergic activity (enhancing cholinacetyltransferase and alleviating acetylcholinesterase activities) in the cortex.	[183]
Aloe vera extract	Protection against cognitive impairment after STZ administration in mice, by reducing oxidative stress.	[184]
Bacopa monnieri extracts (CDRI-08)	Enhancement of spatial memory in T1D and T2D mice and reduction of oxidative stress.	[185]

Author Contributions: C.I.-G. concept and design, drafting the manuscript. M.G.-A. concept and design, drafting and critical revision of manuscript for intellectual content.

Funding: M.G.-A.: Programa Estatal de I+D+I orientada a los Retos de la Sociedad (BFU 2016-75038-R) Ministerio de Economía y Competitividad, financed by Agencia Estatal de Investigación (AEI) and Fondo Europeo de Desarrollo Regional (FEDER). Programa Explora Ciencia. Ministerio de Ciencia, Innovación y Universidades (BFU2017-91910-EXP). Subvención para la financiación de la investigación y la innovación biomédica y en Ciencias de la Salud en el marco de la iniciativa territorial integrada 2014–2020 para la provincia de Cádiz. Consejeria de Salud. Junta de Andalucia. Union Europea, financed by the Fondo de Desarrollo Regional (FEDER) (PI-0008-2017).

Conflicts of Interest: Authors declare no conflict of interest

Abbreviations

AGEs	Advanced glycation end products
DM	Diabetes mellitus
SOD	Superoxide dismutase
STZ	Streptozotocin
TNF-α	Tumor necrosis factor α
T1D	Type 1 diabetes
T2D	Type 2 diabetes
WHO	World Health Organization

References

1. Cornier, M.A.; Dabelea, D.; Hernandez, T.L.; Lindstrom, R.; Steig, A.J.; Stob, N.R.; Van Pelt, R.E.; Wang, H.; Eckel, R.H. The metabolic syndrome. *Endocr. Rev.* **2008**, *29*, 777–822. [CrossRef] [PubMed]
2. Craft, S. The role of metabolic disorders in Alzheimer disease and vascular dementia: two roads converged. *Arch Neurol.* **2009**, *66*, 300–305. [CrossRef]
3. Forouhi, N.G.; Wareham, N.J. Epidemiology of diabetes. *Medicine* **2014**, *42*, 698–702. [CrossRef] [PubMed]
4. World Health Organization. Diabetes. Available online: https://www.who.int/diabetes/en/ (accessed on 30 April 2019).
5. Skyler, J.S.; Bakris, G.L.; Bonifacio, E.; Darsow, T.; Eckel, R.H.; Groop, L.; Groop, P.H.; Handelsman, Y.; Insel, R.A.; Mathieu, C.; et al. Differentiation of Diabetes by Pathophysiology, Natural History, and Prognosis. *Diabetes* **2017**, *66*, 241–255. [CrossRef]
6. Craig, M.E.; Jefferies, C.; Dabelea, D.; Balde, N.; Seth, A.; Donaghue, K.C. ISPAD Clinical Practice Consensus Guidelines Definition, epidemiology, and classification of diabetes in children and adolescents. *Pediatr. Diabetes* **2014**, *15*, 4–17. [CrossRef]
7. Lascar, N.; Brown, J.; Pattison, H.; Barnett, A.H.; Bailey, C.J.; Bellary, S. Type 2 diabetes in adolescents and young adults. *Lancet Diabetes Endocrinol.* **2018**, *6*, 69–80. [CrossRef]
8. Martin-Timon, I.; Sevillano-Collantes, C.; Segura-Galindo, A.; Del Canizo-Gomez, F.J. Type 2 diabetes and cardiovascular disease: Have all risk factors the same strength? *World J. Diabetes* **2014**, *5*, 444–470. [CrossRef]
9. Rosenson, R.S.; Fioretto, P.; Dodson, P.M. Does microvascular disease predict macrovascular events in type 2 diabetes? *Atherosclerosis* **2011**, *218*, 13–18. [CrossRef]
10. Craft, S. Alzheimer disease: Insulin resistance and AD–extending the translational path. *Nat. Rev. Neurol.* **2012**, *8*, 360–362. [CrossRef]
11. Strachan, M.W.; Reynolds, R.M.; Frier, B.M.; Mitchell, R.J.; Price, J.F. The role of metabolic derangements and glucocorticoid excess in the aetiology of cognitive impairment in type 2 diabetes. Implications for future therapeutic strategies. *Diabetes Obesity Metab.* **2009**, *11*, 407–414. [CrossRef] [PubMed]
12. Hamed, S.A. Brain injury with diabetes mellitus: evidence, mechanisms and treatment implications. *Expert Rev. Clin. Pharmacol.* **2017**, *10*, 409–428.
13. Kodl, C.T.; Franc, D.T.; Rao, J.P.; Anderson, F.S.; Thomas, W.; Mueller, B.A.; Lim, K.O.; Seaquist, E.R. Diffusion tensor imaging identifies deficits in white matter microstructure in subjects with type 1 diabetes that correlate with reduced neurocognitive function. *Diabetes* **2008**, *57*, 3083–3089. [CrossRef]
14. Ryan, C.M.; Geckle, M.O.; Orchard, T.J. Cognitive efficiency declines over time in adults with Type 1 diabetes: effects of micro- and macrovascular complications. *Diabetologia* **2003**, *46*, 940–948. [CrossRef]

15. Moran, C.; Beare, R.; Phan, T.G.; Bruce, D.G.; Callisaya, M.L.; Srikanth, V.; Alzheimer's Disease Neuroimaging Initiative (ADNI). Type 2 diabetes mellitus and biomarkers of neurodegeneration. *Neurology* **2015**, *85*, 1123–1130. [CrossRef]
16. Fishel, M.A.; Watson, G.S.; Montine, T.J.; Wang, Q.; Green, P.S.; Kulstad, J.J.; Cook, D.G.; Peskind, E.R.; Baker, L.D.; Goldgaber, D.; et al. Hyperinsulinemia provokes synchronous increases in central inflammation and beta-amyloid in normal adults. *Arch. Neurol.* **2005**, *62*, 1539–1544. [CrossRef]
17. Wang, T.; Fu, F.; Han, B.; Zhang, L.; Zhang, X. Danshensu ameliorates the cognitive decline in streptozotocin-induced diabetic mice by attenuating advanced glycation end product-mediated neuroinflammation. *J. Neuroimmunol.* **2012**, *245*, 79–86. [CrossRef]
18. Infante-Garcia, C.; Jose Ramos-Rodriguez, J.; Marin-Zambrana, Y.; Teresa Fernandez-Ponce, M.; Casas, L.; Mantell, C.; Garcia-Alloza, M. Mango leaf extract improves central pathology and cognitive impairment in a type 2 diabetes mouse model. *Brain Pathol.* **2017**, *27*, 499–507. [CrossRef]
19. Moran, C.; Beare, R.; Wang, W.; Callisaya, M.; Srikanth, V.; Alzheimer's Disease Neuroimaging Initiative (ADNI). Type 2 diabetes mellitus, brain atrophy, and cognitive decline. *Neurology* **2019**, *92*, e823–e830. [CrossRef]
20. Luchsinger, J.A.; Reitz, C.; Honig, L.S.; Tang, M.X.; Shea, S.; Mayeux, R. Aggregation of vascular risk factors and risk of incident Alzheimer disease. *Neurology* **2005**, *65*, 545–551. [CrossRef]
21. Luchsinger, J.A.; Tang, M.X.; Shea, S.; Mayeux, R. Hyperinsulinemia and risk of Alzheimer disease. *Neurology* **2004**, *63*, 1187–1192. [CrossRef]
22. Matsuzaki, T.; Sasaki, K.; Tanizaki, Y.; Hata, J.; Fujimi, K.; Matsui, Y.; Sekita, A.; Suzuki, S.O.; Kanba, S.; Kiyohara, Y.; et al. Insulin resistance is associated with the pathology of Alzheimer disease: the Hisayama study. *Neurology* **2010**, *75*, 764–770. [CrossRef]
23. Schrijvers, E.M.; Witteman, J.C.; Sijbrands, E.J.; Hofman, A.; Koudstaal, P.J.; Breteler, M.M. Insulin metabolism and the risk of Alzheimer disease: the Rotterdam Study. *Neurology* **2010**, *75*, 1982–1987. [CrossRef]
24. Strachan, M.W.; Reynolds, R.M.; Frier, B.M.; Mitchell, R.J.; Price, J.F. The relationship between type 2 diabetes and dementia. *Br. Med. Bull.* **2008**, *88*, 131–146. [CrossRef]
25. Newman, D.J.; Cragg, G.M. Natural Products as Sources of New Drugs from 1981 to 2014. *J. Nat. Prod.* **2016**, *79*, 629–661. [CrossRef]
26. Flores-Jimenez, N.G.; Rojas-Lemus, M.; Fortoul, T.I.; Zepeda-Rodriguez, A.; Lopez-Camacho, P.Y.; Anacleto-Santos, J.; Malagon-Gutierrez, F.; Basurto-Islas, G.; Rivera-Fernandez, N. Histopathological alterations in mice under sub-acute treatment with Hintonia latiflora methanolic stem bark extract. *Histol. Histopathol.* **2018**, *33*, 1299–1309.
27. Spagnuolo, C.; Napolitano, M.; Tedesco, I.; Moccia, S.; Milito, A.; Russo, G.L. Neuroprotective Role of Natural Polyphenols. *Curr. Top. Med. Chem.* **2016**, *16*, 1943–1950. [CrossRef]
28. Cheynier, V.; Comte, G.; Davies, K.M.; Lattanzio, V.; Martens, S. Plant phenolics: recent advances on their biosynthesis, genetics, and ecophysiology. *Plant Physiol. Biochem.* **2013**, *72*, 1–20. [CrossRef]
29. Sevastre-Berghian, A.C.; Toma, V.A.; Sevastre, B.; Hanganu, D.; Vlase, L.; Benedec, D.; Oniga, I.; Baldea, I.; Olteanu, D.; Moldovan, R.; et al. Characterization and biological effects of Hypericum extracts on experimentally-induced - anxiety, oxidative stress and inflammation in rats. *J. Physiol. Pharmacol.* **2018**, *6*, 9.
30. Spagnuolo, C.; Moccia, S.; Russo, G.L. Anti-inflammatory effects of flavonoids in neurodegenerative disorders. *Eur. J. Med. Chem.* **2018**, *153*, 105–115. [CrossRef]
31. Lima, M.C.; Paiva de Sousa, C.; Fernandez-Prada, C.; Harel, J.; Dubreuil, J.D.; de Souza, E.L. A review of the current evidence of fruit phenolic compounds as potential antimicrobials against pathogenic bacteria. *Microb. Pathog.* **2019**, *130*, 259–270. [CrossRef]
32. Christman, L.M.; Dean, L.L.; Allen, J.C.; Godinez, S.F.; Toomer, O.T. Peanut skin phenolic extract attenuates hyperglycemic responses in vivo and in vitro. *PloS ONE* **2019**, *14*, e0214591. [CrossRef]
33. Pohl, F.; Kong Thoo Lin, P. The Potential Use of Plant Natural Products and Plant Extracts with Antioxidant Properties for the Prevention/Treatment of Neurodegenerative Diseases: In Vitro, In Vivo and Clinical Trials. *Molecules* **2018**, *23*, 3283. [CrossRef]
34. Infante-Garcia, C.; Ramos-Rodriguez, J.J.; Delgado-Olmos, I.; Gamero-Carrasco, C.; Fernandez-Ponce, M.T.; Casas, L.; Mantell, C.; Garcia-Alloza, M. Long-Term Mangiferin Extract Treatment Improves Central Pathology and Cognitive Deficits in APP/PS1 Mice. *Mol. Neurobiol.* **2017**, *54*, 4696–4704. [CrossRef]

35. Figueira, I.; Menezes, R.; Macedo, D.; Costa, I.; Dos Santos, C.N. Polyphenols Beyond Barriers: A Glimpse into the Brain. *Curr. Neuropharmacol.* **2017**, *15*, 562–594. [CrossRef]
36. Tsao, R. Chemistry and biochemistry of dietary polyphenols. *Nutrients* **2010**, *2*, 1231–1246. [CrossRef]
37. Garcia-Alloza, M.; Dodwell, S.A.; Meyer-Luehmann, M.; Hyman, B.T.; Bacskai, B.J. Plaque-derived oxidative stress mediates distorted neurite trajectories in the Alzheimer mouse model. *J. Neuropathol. Exp. Neurol.* **2006**, *65*, 1082–1089. [CrossRef]
38. Liu, Y.W.; Liu, X.L.; Kong, L.; Zhang, M.Y.; Chen, Y.J.; Zhu, X.; Hao, Y.C. Neuroprotection of quercetin on central neurons against chronic high glucose through enhancement of Nrf2/ARE/glyoxalase-1 pathway mediated by phosphorylation regulation. *Biomed. Pharmacother.* **2019**, *109*, 2145–2154. [CrossRef]
39. Fu, Q.Y.; Li, Q.S.; Lin, X.M.; Qiao, R.Y.; Yang, R.; Li, X.M.; Dong, Z.B.; Xiang, L.P.; Zheng, X.Q.; Lu, J.L.; et al. Antidiabetic Effects of Tea. *Molecules* **2017**, *22*, 849. [CrossRef]
40. Dominguez Avila, J.A.; Rodrigo Garcia, J.; Gonzalez Aguilar, G.A.; de la Rosa, L.A. The Antidiabetic Mechanisms of Polyphenols Related to Increased Glucagon-Like Peptide-1 (GLP1) and Insulin Signaling. *Molecules* **2017**, *22*, 903. [CrossRef]
41. Serna-Thome, G.; Castro-Eguiluz, D.; Fuchs-Tarlovsky, V.; Sanchez-Lopez, M.; Delgado-Olivares, L.; Coronel-Martinez, J.; Molina-Trinidad, E.M.; de la Torre, M.; Cetina-Perez, L. Use of Functional Foods and Oral Supplements as Adjuvants in Cancer Treatment. *Rev. Inves. Clin.* **2018**, *70*, 136–146. [CrossRef]
42. Biessels, G.J.; Staekenborg, S.; Brunner, E.; Brayne, C.; Scheltens, P. Risk of dementia in diabetes mellitus: A systematic review. *Lancet Neurol.* **2006**, *5*, 64–74. [CrossRef]
43. Crane, P.K.; Walker, R.; Hubbard, R.A.; Li, G.; Nathan, D.M.; Zheng, H.; Haneuse, S.; Craft, S.; Montine, T.J.; Kahn, S.E.; et al. Glucose levels and risk of dementia. *N. Engl. J. Med.* **2013**, *369*, 540–548. [CrossRef] [PubMed]
44. Infante-Garcia, C.; Ramos-Rodriguez, J.J.; Hierro-Bujalance, C.; Ortegon, E.; Pickett, E.; Jackson, R.; Hernandez-Pacho, F.; Spires-Jones, T.; Garcia-Alloza, M. Antidiabetic Polypill Improves Central Pathology and Cognitive Impairment in a Mixed Model of Alzheimer's Disease and Type 2 Diabetes. *Mol. Neurobiol.* **2018**, *55*, 6130–6144. [CrossRef] [PubMed]
45. Munhoz, A.C.M.; Frode, T.S. Isolated Compounds from Natural Products with Potential Antidiabetic Activity - A Systematic Review. *Curr. Diabetes Rev.* **2018**, *14*, 36–106. [CrossRef]
46. Chen, T.Y.; Ferruzzi, M.G.; Wu, Q.L.; Simon, J.E.; Talcott, S.T.; Wang, J.; Ho, L.; Todd, G.; Cooper, B.; Pasinetti, G.M.; et al. Influence of diabetes on plasma pharmacokinetics and brain bioavailability of grape polyphenols and their phase II metabolites in the Zucker diabetic fatty rat. *Mol. Nutr. Food Res.* **2017**, *61*, 1700111. [CrossRef]
47. Domingueti, C.P.; Dusse, L.M.; Carvalho, M.; de Sousa, L.P.; Gomes, K.B.; Fernandes, A.P. Diabetes mellitus: The linkage between oxidative stress, inflammation, hypercoagulability and vascular complications. *J. Diabetes Complicat.* **2016**, *30*, 738–745. [CrossRef]
48. Goldberg, R.B. Cytokine and cytokine-like inflammation markers, endothelial dysfunction, and imbalanced coagulation in development of diabetes and its complications. *J. Clin. Endocrinol. Metab.* **2009**, *94*, 3171–3182. [CrossRef]
49. Wautier, J.L.; Guillausseau, P.J. Diabetes, advanced glycation endproducts and vascular disease. *Vasc. Med.* **1998**, *3*, 131–137. [CrossRef]
50. Reddy, G.K. AGE-related cross-linking of collagen is associated with aortic wall matrix stiffness in the pathogenesis of drug-induced diabetes in rats. *Microvasc. Res.* **2004**, *68*, 132–142. [CrossRef]
51. Idris, I.; Thomson, G.A.; Sharma, J.C. Diabetes mellitus and stroke. *Int. J. Clin. Pract.* **2006**, *60*, 48–56. [CrossRef]
52. Callahan, A.; Amarenco, P.; Goldstein, L.B.; Sillesen, H.; Messig, M.; Samsa, G.P.; Altafullah, I.; Ledbetter, L.Y.; MacLeod, M.J.; Scott, R.; et al. Risk of stroke and cardiovascular events after ischemic stroke or transient ischemic attack in patients with type 2 diabetes or metabolic syndrome: secondary analysis of the Stroke Prevention by Aggressive Reduction in Cholesterol Levels (SPARCL) trial. *Arch. Neurol.* **2011**, *68*, 1245–1251. [CrossRef] [PubMed]
53. Mogi, M.; Horiuchi, M. Neurovascular coupling in cognitive impairment associated with diabetes mellitus. *Circ. J.* **2011**, *75*, 1042–1048. [CrossRef] [PubMed]

54. Hayden, M.R.; Grant, D.G.; Aroor, A.R.; DeMarco, V.G. Empagliflozin Ameliorates Type 2 Diabetes-Induced Ultrastructural Remodeling of the Neurovascular Unit and Neuroglia in the Female db/db Mouse. *Brain Sci.* **2019**, *9*, 57. [CrossRef]
55. Infante-Garcia, C.; Ramos-Rodriguez, J.J.; Galindo-Gonzalez, L.; Garcia-Alloza, M. Long-term central pathology and cognitive impairment are exacerbated in a mixed model of Alzheimer's disease and type 2 diabetes. *Psychoneuroendocrinology* **2016**, *65*, 15–25. [CrossRef] [PubMed]
56. Ramos-Rodriguez, J.J.; Jimenez-Palomares, M.; Murillo-Carretero, M.I.; Infante-Garcia, C.; Berrocoso, E.; Hernandez-Pacho, F.; Lechuga-Sancho, A.M.; Cozar-Castellano, I.; Garcia-Alloza, M. Central vascular disease and exacerbated pathology in a mixed model of type 2 diabetes and Alzheimer's disease. *Psychoneuroendocrinology* **2015**, *62*, 69–79. [CrossRef] [PubMed]
57. Zhang, L.; Chopp, M.; Zhang, Y.; Xiong, Y.; Li, C.; Sadry, N.; Rhaleb, I.; Lu, M.; Zhang, Z.G. Diabetes Mellitus Impairs Cognitive Function in Middle-Aged Rats and Neurological Recovery in Middle-Aged Rats After Stroke. *Stroke* **2016**, *47*, 2112–2118. [CrossRef]
58. Pasquier, F.; Boulogne, A.; Leys, D.; Fontaine, P. Diabetes mellitus and dementia. *Diabetes Metab.* **2006**, *32*, 403–414. [CrossRef]
59. Wang, S.; Cao, C.; Chen, Z.; Bankaitis, V.; Tzima, E.; Sheibani, N.; Burridge, K. Pericytes regulate vascular basement membrane remodeling and govern neutrophil extravasation during inflammation. *PloS ONE* **2012**, *7*, e45499. [CrossRef]
60. Bogush, M.; Heldt, N.A.; Persidsky, Y. Blood Brain Barrier Injury in Diabetes: Unrecognized Effects on Brain and Cognition. *J. Neuroimmune Pharmacol.* **2017**, *12*, 593–601. [CrossRef]
61. Manasson, J.; Tien, T.; Moore, C.; Kumar, N.M.; Roy, S. High glucose-induced downregulation of connexin 30.2 promotes retinal vascular lesions: implications for diabetic retinopathy. *Investig. Ophthalmol. Vis. Sci.* **2013**, *54*, 2361–2366. [CrossRef]
62. Sajja, R.K.; Prasad, S.; Cucullo, L. Impact of altered glycaemia on blood-brain barrier endothelium: an in vitro study using the hCMEC/D3 cell line. *Fluids Barriers CNS.* **2014**, *11*, 8. [CrossRef]
63. Li, B.; Li, Y.; Liu, K.; Wang, X.; Qi, J.; Wang, B.; Wang, Y. High glucose decreases claudins-5 and -11 in cardiac microvascular endothelial cells: Antagonistic effects of tongxinluo. *Endocr. Res.* **2017**, *42*, 15–21. [CrossRef]
64. Maile, L.A.; Gollahon, K.; Wai, C.; Dunbar, P.; Busby, W.; Clemmons, D. Blocking alphaVbeta3 integrin ligand occupancy inhibits the progression of albuminuria in diabetic rats. *J. Diabetes Res.* **2014**, *2014*, 421827. [CrossRef]
65. Park, S.W.; Yun, J.H.; Kim, J.H.; Kim, K.W.; Cho, C.H.; Kim, J.H. Angiopoietin 2 induces pericyte apoptosis via alpha3beta1 integrin signaling in diabetic retinopathy. *Diabetes* **2014**, *63*, 3057–3068. [CrossRef]
66. Lee, Y.J.; Jung, S.H.; Kim, S.H.; Kim, M.S.; Lee, S.; Hwang, J.; Kim, S.Y.; Kim, Y.M.; Ha, K.S. Essential Role of Transglutaminase 2 in Vascular Endothelial Growth Factor-Induced Vascular Leakage in the Retina of Diabetic Mice. *Diabetes* **2016**, *65*, 2414–2428. [CrossRef]
67. Abu El-Asrar, A.M.; Mohammad, G.; Nawaz, M.I.; Abdelsaid, M.; Siddiquei, M.M.; Alam, K.; Van den Eynde, K.; De Hertogh, G.; Opdenakker, G.; Al-Shabrawey, M.; et al. The Chemokine Platelet Factor-4 Variant (PF-4var)/CXCL4L1 Inhibits Diabetes-Induced Blood-Retinal Barrier Breakdown. *Investig. Ophthalmol. Vis. Sci.* **2015**, *56*, 1956–1964. [CrossRef]
68. Price, T.O.; Eranki, V.; Banks, W.A.; Ercal, N.; Shah, G.N. Topiramate treatment protects blood-brain barrier pericytes from hyperglycemia-induced oxidative damage in diabetic mice. *Endocrinology* **2012**, *153*, 362–372. [CrossRef]
69. Takechi, R.; Lam, V.; Brook, E.; Giles, C.; Fimognari, N.; Mooranian, A.; Al-Salami, H.; Coulson, S.H.; Nesbit, M.; Mamo, J.C.L. Blood-Brain Barrier Dysfunction Precedes Cognitive Decline and Neurodegeneration in Diabetic Insulin Resistant Mouse Model: An Implication for Causal Link. *Front. Aging Neurosci.* **2017**, *9*, 399. [CrossRef]
70. Ramos-Rodriguez, J.J.; Ortiz, O.; Jimenez-Palomares, M.; Kay, K.R.; Berrocoso, E.; Murillo-Carretero, M.I.; Perdomo, G.; Spires-Jones, T.; Cozar-Castellano, I.; Lechuga-Sancho, A.M.; et al. Differential central pathology and cognitive impairment in pre-diabetic and diabetic mice. *Psychoneuroendocrinology* **2013**, *38*, 2462–2475. [CrossRef]
71. Jin, Y.; Khadka, D.B.; Cho, W.J. Pharmacological effects of berberine and its derivatives: A patent update. *Expert Opin. Ther. Pat.* **2016**, *26*, 229–243. [CrossRef]

72. Yin, J.; Xing, H.; Ye, J. Efficacy of berberine in patients with type 2 diabetes mellitus. *Metabolism* **2008**, *57*, 712–717. [CrossRef]
73. Yin, S.; Bai, W.; Li, P.; Jian, X.; Shan, T.; Tang, Z.; Jing, X.; Ping, S.; Li, Q.; Miao, Z.; et al. Berberine suppresses the ectopic expression of miR-133a in endothelial cells to improve vascular dementia in diabetic rats. *Clin. Exp. Hypertens.* **2018**, 1–9. [CrossRef]
74. Wei, L.L.; Chen, Y.; Yu, Q.Y.; Wang, Y.; Liu, G. Patchouli alcohol protects against ischemia/reperfusion-induced brain injury via inhibiting neuroinflammation in normal and obese mice. *Brain Res.* **2018**, *1682*, 61–70. [CrossRef]
75. Patel, S.S.; Udayabanu, M. Effect of natural products on diabetes associated neurological disorders. *Rev. Neurosci.* **2017**, *28*, 271–293. [CrossRef]
76. Wong, R.H.; Raederstorff, D.; Howe, P.R. Acute Resveratrol Consumption Improves Neurovascular Coupling Capacity in Adults with Type 2 Diabetes Mellitus. *Nutrients* **2016**, *8*, 425. [CrossRef]
77. Jing, Y.H.; Chen, K.H.; Kuo, P.C.; Pao, C.C.; Chen, J.K. Neurodegeneration in streptozotocin-induced diabetic rats is attenuated by treatment with resveratrol. *Neuroendocrinology* **2013**, *98*, 116–127. [CrossRef]
78. Newcombe, E.A.; Camats-Perna, J.; Silva, M.L.; Valmas, N.; Huat, T.J.; Medeiros, R. Inflammation: The link between comorbidities, genetics, and Alzheimer's disease. *J. Neuroinflammation* **2018**, *15*, 276. [CrossRef]
79. Hotamisligil, G.S. Inflammation and metabolic disorders. *Nature* **2006**, *444*, 860–867. [CrossRef]
80. Hotamisligil, G.S.; Shargill, N.S.; Spiegelman, B.M. Adipose expression of tumor necrosis factor-alpha: Direct role in obesity-linked insulin resistance. *Science* **1993**, *259*, 87–91. [CrossRef]
81. Wellen, K.E.; Hotamisligil, G.S. Inflammation, stress, and diabetes. *J. Clin. Investig.* **2005**, *115*, 1111–1119. [CrossRef]
82. Colombo, E.; Farina, C. Astrocytes: Key Regulators of Neuroinflammation. *Trends Immunol.* **2016**, *37*, 608–620. [CrossRef]
83. Jeong, H.K.; Ji, K.; Min, K.; Joe, E.H. Brain inflammation and microglia: Facts and misconceptions. *Exp. Neurobiol.* **2013**, *22*, 59–67. [CrossRef] [PubMed]
84. Ransohoff, R.M.; Engelhardt, B. The anatomical and cellular basis of immune surveillance in the central nervous system. *Nat. Rev. Immunol.* **2012**, *12*, 623–635. [CrossRef]
85. Ferreira, R.; Bernardino, L. Dual role of microglia in health and disease: pushing the balance toward repair. *Front Cell Neurosci.* **2015**, *9*, 51. [CrossRef]
86. Chen, Z.; Trapp, B.D. Microglia and neuroprotection. *J. Neurochem.* **2016**, *136*, 10–17. [CrossRef]
87. Ramos-Rodriguez, J.J.; Ortiz-Barajas, O.; Gamero-Carrasco, C.; de la Rosa, P.R.; Infante-Garcia, C.; Zopeque-Garcia, N.; Lechuga-Sancho, A.M.; Garcia-Alloza, M. Prediabetes-induced vascular alterations exacerbate central pathology in APPswe/PS1dE9 mice. *Psychoneuroendocrinology* **2014**, *48*, 123–135. [CrossRef]
88. Ramos-Rodriguez, J.J.; Infante-Garcia, C.; Galindo-Gonzalez, L.; Garcia-Molina, Y.; Lechuga-Sancho, A.; Garcia-Alloza, M. Increased Spontaneous Central Bleeding and Cognition Impairment in APP/PS1 Mice with Poorly Controlled Diabetes Mellitus. *Mol. Neurobiol.* **2016**, *53*, 2685–2697. [CrossRef]
89. Ramos-Rodriguez, J.J.; Sanchez-Sotano, D.; Doblas-Marquez, A.; Infante-Garcia, C.; Lubian-Lopez, S.; Garcia-Alloza, M. Intranasal insulin reverts central pathology and cognitive impairment in diabetic mother offspring. *Mol. Neurodegener.* **2017**, *12*, 57. [CrossRef]
90. Van Dyken, P.; Lacoste, B. Impact of Metabolic Syndrome on Neuroinflammation and the Blood-Brain Barrier. *Front. Neurosci.* **2018**, *12*, 930. [CrossRef]
91. Hwang, I.K.; Choi, J.H.; Nam, S.M.; Park, O.K.; Yoo, D.Y.; Kim, W.; Yi, S.S.; Won, M.H.; Seong, J.K.; Yoon, Y.S. Activation of microglia and induction of pro-inflammatory cytokines in the hippocampus of type 2 diabetic rats. *Neurol. Res.* **2014**, *36*, 824–832. [CrossRef]
92. Ibrahim, A.S.; El-Shishtawy, M.M.; Pena, A., Jr.; Liou, G.I. Genistein attenuates retinal inflammation associated with diabetes by targeting of microglial activation. *Mol. Vis.* **2010**, *16*, 2033–2042.
93. Lauridsen, J.K.; Olesen, R.H.; Vendelbo, J.; Hyde, T.M.; Kleinman, J.E.; Bibby, B.M.; Brock, B.; Rungby, J.; Larsen, A. High BMI levels associate with reduced mRNA expression of IL10 and increased mRNA expression of iNOS (NOS2) in human frontal cortex. *Transl. Psychiatry* **2017**, *7*, e1044. [CrossRef]
94. Chen, W.; Jia, Z.; Pan, M.-H.; Babu, P.V.A. Natural Products for the Prevention of Oxidative Stress-Related Diseases: Mechanisms and Strategies. *Oxidative Med. Cell. Longev.* **2016**, *2016*, 1–2. [CrossRef]
95. Jia, Z.; Babu, P.V.A.; Chen, W.; Sun, X. Natural Products Targeting on Oxidative Stress and Inflammation: Mechanisms, Therapies, and Safety Assessment. *Oxidative Med. Cell. Longev.* **2018**, *2018*, 1–3. [CrossRef]

96. Maitra, I.; Marcocci, L.; Droy-Lefaix, M.T.; Packer, L. Peroxyl radical scavenging activity of Ginkgo biloba extract EGb 761. *Biochem. Pharmacol.* **1995**, *49*, 1649–1655. [CrossRef]
97. Frandsen, J.R.; Narayanasamy, P. Neuroprotection through flavonoid: Enhancement of the glyoxalase pathway. *Redox Biol.* **2018**, *14*, 465–473. [CrossRef]
98. Mahesh, T.; Menon, V.P. Quercetin allievates oxidative stress in streptozotocin-induced diabetic rats. *Phytother. Res.* **2004**, *18*, 123–127. [CrossRef]
99. Bournival, J.; Francoeur, M.A.; Renaud, J.; Martinoli, M.G. Quercetin and sesamin protect neuronal PC12 cells from high-glucose-induced oxidation, nitrosative stress, and apoptosis. *Rejuvenation Res.* **2012**, *15*, 322–333. [CrossRef]
100. Marquez, L.; Garcia-Bueno, B.; Madrigal, J.L.; Leza, J.C. Mangiferin decreases inflammation and oxidative damage in rat brain after stress. *Eur. J. Nutr.* **2012**, *51*, 729–739. [CrossRef]
101. Liu, Y.W.; Cheng, Y.Q.; Liu, X.L.; Hao, Y.C.; Li, Y.; Zhu, X.; Zhang, F.; Yin, X.X. Mangiferin Upregulates Glyoxalase 1 Through Activation of Nrf2/ARE Signaling in Central Neurons Cultured with High Glucose. *Mol. Neurobiol.* **2017**, *54*, 4060–4070. [CrossRef]
102. Faheem, N.M.; El Askary, A. Neuroprotective role of curcumin on the hippocampus against the structural and serological alterations of streptozotocin-induced diabetes in Sprague Dawely rats. *Iran. J. Basic Med. Sci.* **2017**, *20*, 690–699.
103. Daugherty, D.J.; Marquez, A.; Calcutt, N.A.; Schubert, D. A novel curcumin derivative for the treatment of diabetic neuropathy. *Neuropharmacology* **2018**, *129*, 26–35. [CrossRef]
104. Sadi, G.; Konat, D. Resveratrol regulates oxidative biomarkers and antioxidant enzymes in the brain of streptozotocin-induced diabetic rats. *Pharm. Biol.* **2016**, *54*, 1156–1163. [CrossRef]
105. Nurdiana, S.; Goh, Y.M.; Hafandi, A.; Dom, S.M.; Nur Syimal'ain, A.; Noor Syaffinaz, N.M.; Ebrahimi, M. Improvement of spatial learning and memory, cortical gyrification patterns and brain oxidative stress markers in diabetic rats treated with Ficus deltoidea leaf extract and vitexin. *J. Tradit. Complement. Med.* **2018**, *8*, 190–202. [CrossRef]
106. Samarghandian, S.; Azimi-Nezhad, M.; Samini, F. Ameliorative Effect of Saffron Aqueous Extract on Hyperglycemia, Hyperlipidemia, and Oxidative Stress on Diabetic Encephalopathy in Streptozotocin Induced Experimental Diabetes Mellitus. *BioMed Int.* **2014**, *2014*, 1–12. [CrossRef]
107. Pari, L.; Latha, M. Protective role of Scoparia dulcis plant extract on brain antioxidant status and lipidperoxidation in STZ diabetic male Wistar rats. *BMC Complement Altern Med.* **2004**, *4*, 16. [CrossRef]
108. Li, R.; Zang, A.; Zhang, L.; Zhang, H.; Zhao, L.; Qi, Z.; Wang, H. Chrysin ameliorates diabetes-associated cognitive deficits in Wistar rats. *Neurol. Sci.* **2014**, *35*, 1527–1532. [CrossRef]
109. Marcocci, L.; Packer, L.; Droy-Lefaix, M.T.; Sekaki, A.; Gardes-Albert, M. Antioxidant action of Ginkgo biloba extract EGb. *Methods Enzymol.* **1994**, *234*, 462–475.
110. Choi, H.R.; Choi, J.S.; Han, Y.N.; Bae, S.J.; Chung, H.Y. Peroxynitrite scavenging activity of herb extracts. *Phytother. Res.* **2002**, *16*, 364–367. [CrossRef]
111. Yu, Y.; Wu, Y.; Szabo, A.; Wu, Z.; Wang, H.; Li, D.; Huang, X.F. Teasaponin reduces inflammation and central leptin resistance in diet-induced obese male mice. *Endocrinology* **2013**, *154*, 3130–3140. [CrossRef]
112. Talpate, K.A.; Bhosale, U.A.; Zambare, M.R.; Somani, R.S. Neuroprotective and nootropic activity of Clitorea ternatea Linn.(Fabaceae) leaves on diabetes induced cognitive decline in experimental animals. *J. Pharm. Bioallied Sci.* **2014**, *6*, 48–55.
113. Yonguc, G.N.; Dodurga, Y.; Adiguzel, E.; Gundogdu, G.; Kucukatay, V.; Ozbal, S.; Yilmaz, I.; Cankurt, U.; Yilmaz, Y.; Akdogan, I. Grape seed extract has superior beneficial effects than vitamin E on oxidative stress and apoptosis in the hippocampus of streptozotocin induced diabetic rats. *Gene* **2015**, *555*, 119–126. [CrossRef]
114. Lu, M.; Xu, L.; Li, B.; Zhang, W.; Zhang, C.; Feng, H.; Cui, X.; Gao, H. Protective effects of grape seed proanthocyanidin extracts on cerebral cortex of streptozotocin-induced diabetic rats through modulating AGEs/RAGE/NF-kappaB pathway. *J. Nutr. Sci. Vitaminol.* **2010**, *56*, 87–97. [CrossRef]
115. Patel, S.S.; Gupta, S.; Udayabanu, M. Urtica dioica modulates hippocampal insulin signaling and recognition memory deficit in streptozotocin induced diabetic mice. *Metab. Brain Dis.* **2016**, *31*, 601–611. [CrossRef]
116. Patel, S.S.; Parashar, A.; Udayabanu, M. Urtica dioica leaves modulates muscarinic cholinergic system in the hippocampus of streptozotocin-induced diabetic mice. *Metab. Brain Dis.* **2015**, *30*, 803–811. [CrossRef]

117. Jahanshahi, M.; Golalipour, M.J.; Afshar, M. The effect of Urtica dioica extract on the number of astrocytes in the dentate gyrus of diabetic rats. *Folia Morphol.* **2009**, *68*, 93–97.
118. Abdel-Moneim, A.; Yousef, A.I.; Abd El-Twab, S.M.; Abdel Reheim, E.S.; Ashour, M.B. Gallic acid and p-coumaric acid attenuate type 2 diabetes-induced neurodegeneration in rats. *Metab. Brain Dis.* **2017**, *32*, 1279–1286. [CrossRef]
119. Liu, Y.; Fu, X.; Lan, N.; Li, S.; Zhang, J.; Wang, S.; Li, C.; Shang, Y.; Huang, T.; Zhang, L. Luteolin protects against high fat diet-induced cognitive deficits in obesity mice. *Behav. Brain Res.* **2014**, *267*, 178–188. [CrossRef]
120. Liu, Y.; Tian, X.; Gou, L.; Sun, L.; Ling, X.; Yin, X. Luteolin attenuates diabetes-associated cognitive decline in rats. *Brain Res. Bull.* **2013**, *94*, 23–29. [CrossRef]
121. Kaur, T.; Kaur, G. Withania somnifera as a potential candidate to ameliorate high fat diet-induced anxiety and neuroinflammation. *J. Neuroinflammation* **2017**, *14*, 201. [CrossRef]
122. Pena-Montes, D.J.; Huerta-Cervantes, M.; Rios-Silva, M.; Trujillo, X.; Huerta, M.; Noriega-Cisneros, R.; Salgado-Garciglia, R.; Saavedra-Molina, A. Protective Effect of the Hexanic Extract of Eryngium carlinae Inflorescences In Vitro, in Yeast, and in Streptozotocin-Induced Diabetic Male Rats. *Antioxidants* **2019**, *8*, 73. [CrossRef]
123. Ola, M.S.; Aleisa, A.M.; Al-Rejaie, S.S.; Abuohashish, H.M.; Parmar, M.Y.; Alhomida, A.S.; Ahmed, M.M. Flavonoid, morin inhibits oxidative stress, inflammation and enhances neurotrophic support in the brain of streptozotocin-induced diabetic rats. *Neurol. Sci.* **2014**, *35*, 1003–1008. [CrossRef]
124. Bachewal, P.; Gundu, C.; Yerra, V.G.; Kalvala, A.K.; Areti, A.; Kumar, A. Morin exerts neuroprotection via attenuation of ROS induced oxidative damage and neuroinflammation in experimental diabetic neuropathy. *BioFactors* **2018**, *44*, 109–122. [CrossRef]
125. Ola, M.S.; Ahmed, M.M.; Ahmad, R.; Abuohashish, H.M.; Al-Rejaie, S.S.; Alhomida, A.S. Neuroprotective Effects of Rutin in Streptozotocin-Induced Diabetic Rat Retina. *J. Mol. Neurosci.* **2015**, *56*, 440–448. [CrossRef]
126. Moghaddam, H.K.; Baluchnejadmojarad, T.; Roghani, M.; Khaksari, M.; Norouzi, P.; Ahooie, M.; Mahboobi, F. Berberine ameliorate oxidative stress and astrogliosis in the hippocampus of STZ-induced diabetic rats. *Mol. Neurobiol.* **2014**, *49*, 820–826. [CrossRef]
127. Masola, B.; Oguntibeju, O.O.; Oyenihi, A.B. Centella asiatica ameliorates diabetes-induced stress in rat tissues via influences on antioxidants and inflammatory cytokines. *Biomed. Pharmacother.* **2018**, *101*, 447–457. [CrossRef]
128. Syiem, D.; Warjri, P. Antidiabetic, antioxidant, and TNF-alpha lowering properties of extract of the traditionally used plant Ixeris gracilis in alloxan-induced diabetic mice. *Pharmaceutical Biol.* **2015**, *53*, 494–502. [CrossRef]
129. Solanki, I.; Parihar, P.; Shetty, R.; Parihar, M.S. Synaptosomal and mitochondrial oxidative damage followed by behavioral impairments in streptozotocin induced diabetes mellitus: restoration by Malvastrum tricuspidatum. *Cell Mol. Biol.* **2017**, *63*, 94–101. [CrossRef]
130. van Harten, B.; de Leeuw, F.E.; Weinstein, H.C.; Scheltens, P.; Biessels, G.J. Brain imaging in patients with diabetes: a systematic review. *Diabetes Care* **2006**, *29*, 2539–2548. [CrossRef]
131. Marseglia, A.; Fratiglioni, L.; Kalpouzos, G.; Wang, R.; Backman, L.; Xu, W. Prediabetes and diabetes accelerate cognitive decline and predict microvascular lesions: A population-based cohort study. *Alzheimer's Dement.* **2019**, *15*, 25–33. [CrossRef]
132. Ramos-Rodriguez, J.J.; Spires-Jones, T.; Pooler, A.M.; Lechuga-Sancho, A.M.; Bacskai, B.J.; Garcia-Alloza, M. Progressive Neuronal Pathology and Synaptic Loss Induced by Prediabetes and Type 2 Diabetes in a Mouse Model of Alzheimer's Disease. *Mol. Neurobiol.* **2017**, *54*, 3428–3438. [CrossRef]
133. Moran, C.; Tapp, R.J.; Hughes, A.D.; Magnussen, C.G.; Blizzard, L.; Phan, T.G.; Beare, R.; Witt, N.; Venn, A.; Munch, G.; et al. The Association of Type 2 Diabetes Mellitus with Cerebral Gray Matter Volume Is Independent of Retinal Vascular Architecture and Retinopathy. *J. Diabetes Res.* **2016**, *2016*, 6328953. [CrossRef] [PubMed]
134. Bednarik, P.; Moheet, A.A.; Grohn, H.; Kumar, A.F.; Eberly, L.E.; Seaquist, E.R.; Mangia, S. Type 1 Diabetes and Impaired Awareness of Hypoglycemia Are Associated with Reduced Brain Gray Matter Volumes. *Front. Neurosci.* **2017**, *11*, 529. [CrossRef] [PubMed]
135. McCrimmon, R.J.; Ryan, C.M.; Frier, B.M. Diabetes and cognitive dysfunction. *Lancet* **2012**, *379*, 2291–2299. [CrossRef]

136. Kumar, A.; Haroon, E.; Darwin, C.; Pham, D.; Ajilore, O.; Rodriguez, G.; Mintz, J. Gray matter prefrontal changes in type 2 diabetes detected using MRI. *J. Magn. Reson. Imaging: Jmri.* **2008**, *27*, 14–19. [CrossRef]
137. de Bresser, J.; Tiehuis, A.M.; van den Berg, E.; Reijmer, Y.D.; Jongen, C.; Kappelle, L.J.; Mali, W.P.; Viergever, M.A.; Biessels, G.J.; Utrecht Diabetic Encephalopathy Study Group. Progression of cerebral atrophy and white matter hyperintensities in patients with type 2 diabetes. *Diabetes Care* **2010**, *33*, 1309–1314. [CrossRef]
138. Convit, A.; Wolf, O.T.; Tarshish, C.; de Leon, M.J. Reduced glucose tolerance is associated with poor memory performance and hippocampal atrophy among normal elderly. *Proc. Natl. Acad. Sci. USA* **2003**, *100*, 2019–2022. [CrossRef]
139. Muriach, M.; Flores-Bellver, M.; Romero, F.J.; Barcia, J.M. Diabetes and the brain: oxidative stress, inflammation, and autophagy. *Oxidative Med. Cell. Longev.* **2014**, *2014*, 102158. [CrossRef]
140. Zhao, W.Q.; Alkon, D.L. Role of insulin and insulin receptor in learning and memory. *Mol. Cell. Endocrinol.* **2001**, *177*, 125–134. [CrossRef]
141. Beckman, K.B.; Ames, B.N. The free radical theory of aging matures. *Physiol. Rev.* **1998**, *78*, 547–581. [CrossRef]
142. Merad-Boudia, M.; Nicole, A.; Santiard-Baron, D.; Saille, C.; Ceballos-Picot, I. Mitochondrial impairment as an early event in the process of apoptosis induced by glutathione depletion in neuronal cells: relevance to Parkinson's disease. *Biochem. Pharmacol.* **1998**, *56*, 645–655. [CrossRef]
143. Bharadwaj, P.; Wijesekara, N.; Liyanapathirana, M.; Newsholme, P.; Ittner, L.; Fraser, P.; Verdile, G. The Link between Type 2 Diabetes and Neurodegeneration: Roles for Amyloid-beta, Amylin, and Tau Proteins. *J. Alzheimer's Dis.* **2017**, *59*, 421–432. [CrossRef]
144. Kimura, Y.; Ito, H.; Ohnishi, R.; Hatano, T. Inhibitory effects of polyphenols on human cytochrome P450 3A4 and 2C9 activity. *Food Chem. Toxicol.* **2010**, *48*, 429–435. [CrossRef] [PubMed]
145. Espargaro, A.; Ginex, T.; Vadell, M.D.; Busquets, M.A.; Estelrich, J.; Munoz-Torrero, D.; Luque, F.J.; Sabate, R. Combined in Vitro Cell-Based/in Silico Screening of Naturally Occurring Flavonoids and Phenolic Compounds as Potential Anti-Alzheimer Drugs. *J. Nat. Products.* **2017**, *80*, 278–289. [CrossRef]
146. Cittadini, M.C.; Repossi, G.; Albrecht, C.; Di Paola Naranjo, R.; Miranda, A.R.; de Pascual-Teresa, S.; Soria, E.A. Effects of bioavailable phenolic compounds from Ilex paraguariensis on the brain of mice with lung adenocarcinoma. *Phytother. Res.* **2019**, *33*, 1142–1149. [CrossRef] [PubMed]
147. Chan, E.W.L.; Yeo, E.T.Y.; Wong, K.W.L.; See, M.L.; Wong, K.Y.; Gan, S.Y. Piper sarmentosum Roxb. Root Extracts Confer Neuroprotection by Attenuating Beta Amyloid-Induced Pro-Inflammatory Cytokines Released from Microglial Cells. *Curr. Alzheimer Res.* **2019**, *16*, 251–260. [CrossRef] [PubMed]
148. Kean, R.J.; Lamport, D.J.; Dodd, G.F.; Freeman, J.E.; Williams, C.M.; Ellis, J.A.; Butler, L.T.; Spencer, J.P. Chronic consumption of flavanone-rich orange juice is associated with cognitive benefits: an 8-wk, randomized, double-blind, placebo-controlled trial in healthy older adults. *Am. J. Clin. Nutrition.* **2015**, *101*, 506–514. [CrossRef] [PubMed]
149. Vauzour, D. Dietary polyphenols as modulators of brain functions: biological actions and molecular mechanisms underpinning their beneficial effects. *Oxidative Med. Cell. Longev.* **2012**, *2012*, 914273. [CrossRef] [PubMed]
150. Tian, X.; Liu, Y.; Ren, G.; Yin, L.; Liang, X.; Geng, T.; Dang, H.; An, R. Resveratrol limits diabetes-associated cognitive decline in rats by preventing oxidative stress and inflammation and modulating hippocampal structural synaptic plasticity. *Brain Res.* **2016**, *1650*, 1–9. [CrossRef]
151. Tian, Z.; Wang, J.; Xu, M.; Wang, Y.; Zhang, M.; Zhou, Y. Resveratrol Improves Cognitive Impairment by Regulating Apoptosis and Synaptic Plasticity in Streptozotocin-Induced Diabetic Rats. *Cell. Physiol. Biochem.* **2016**, *40*, 1670–1677. [CrossRef]
152. Zhang, G.; Fang, H.; Li, Y.; Xu, J.; Zhang, D.; Sun, Y.; Zhou, L.; Zhang, H. Neuroprotective Effect of Astragalus Polysacharin on Streptozotocin (STZ)-Induced Diabetic Rats. *Med. Sci. Monit.* **2019**, *25*, 135–141. [CrossRef]
153. Matysek, M.; Mozel, S.; Szalak, R.; Zacharko-Siembida, A.; Obszanska, K.; Arciszewski, M.B. Effect of feeding with bilberry fruit on the expression pattern of alphaCaMKII in hippocampal neurons in normal and diabetic rats. *Polish J. Vet. Sci.* **2017**, *20*, 313–319. [CrossRef]
154. da Costa, A.V.; Calabria, L.K.; Furtado, F.B.; de Gouveia, N.M.; Oliveira, R.J.; de Oliveira, V.N.; Beletti, M.E.; Espindola, F.S. Neuroprotective effects of Pouteria ramiflora (Mart.) Radlk (Sapotaceae) extract on the brains of rats with streptozotocin-induced diabetes. *Metab. Brain Dis.* **2013**, *28*, 411–419. [CrossRef]

155. Seke Etet, P.F.; Farahna, M.; Satti, G.M.H.; Bushara, Y.M.; El-Tahir, A.; Hamza, M.A.; Osman, S.Y.; Dibia, A.C.; Vecchio, L. Garcinia kola seeds may prevent cognitive and motor dysfunctions in a type 1 diabetes mellitus rat model partly by mitigating neuroinflammation. *J. Complement Integr. Med.* **2017**, *14*. [CrossRef]
156. Patel, S.S.; Ray, R.S.; Sharma, A.; Mehta, V.; Katyal, A.; Udayabanu, M. Antidepressant and anxiolytic like effects of Urtica dioica leaves in streptozotocin induced diabetic mice. *Metab. Brain* **2018**, *33*, 1281–1292. [CrossRef]
157. Fazeli, S.A.; Gharravi, A.M.; Ghafari, S.; Jahanshahi, M.; Golalipour, M.J. The granule cell density of the dentate gyrus following administration of Urtica dioica extract to young diabetic rats. *Folia Morphol.* **2008**, *67*, 196–204.
158. Molahosseini, A.; Taghavi, M.M.; Taghipour, Z.; Shabanizadeh, A.; Fatehi, F.; Kazemi Arababadi, M.; Eftekhar Vaghefe, S.H. The effect of the ginger on the apoptosis of hippochampal cells according to the expression of BAX and Cyclin D1 genes and histological characteristics of brain in streptozotocin male diabetic rats. *Cell. Mol. Biol.* **2016**, *62*, 1–5.
159. Shin, M.S.; Kim, S.K.; Kim, Y.S.; Kim, S.E.; Ko, I.G.; Kim, C.J.; Kim, Y.M.; Kim, B.K.; Kim, T.S. Aqueous extract of Anemarrhena rhizome increases cell proliferation and neuropeptide Y expression in the hippocampal dentate gyrus on streptozotocin-induced diabetic rats. *Fitoterapia* **2008**, *79*, 323–327. [CrossRef]
160. Reichert, K.P.; Schetinger, M.R.C.; Gutierres, J.M.; Pelinson, L.P.; Stefanello, N.; Dalenogare, D.P.; Baldissarelli, J.; Lopes, T.F.; Morsch, V.M. Lingonberry Extract Provides Neuroprotection by Regulating the Purinergic System and Reducing Oxidative Stress in Diabetic Rats. *Mol. Nutr. Food Res.* **2018**, *62*, e1800050. [CrossRef]
161. Xuan, M.; Okazaki, M.; Iwata, N.; Asano, S.; Kamiuchi, S.; Matsuzaki, H.; Sakamoto, T.; Miyano, Y.; Iizuka, H.; Hibino, Y. Chronic Treatment with a Water-Soluble Extract from the Culture Medium of Ganoderma lucidum Mycelia Prevents Apoptosis and Necroptosis in Hypoxia/Ischemia-Induced Injury of Type 2 Diabetic Mouse Brain. *Evid Based Complement Alternat. Med.* **2015**, *2015*, 865986. [CrossRef]
162. Koekkoek, P.S.; Kappelle, L.J.; van den Berg, E.; Rutten, G.E.; Biessels, G.J. Cognitive function in patients with diabetes mellitus: guidance for daily care. *Lancet Neurol.* **2015**, *14*, 329–340. [CrossRef]
163. Gudala, K.; Bansal, D.; Schifano, F.; Bhansali, A. Diabetes mellitus and risk of dementia: A meta-analysis of prospective observational studies. *J. Diabetes Investig.* **2013**, *4*, 640–650. [CrossRef]
164. Zhang, J.; Chen, C.; Hua, S.; Liao, H.; Wang, M.; Xiong, Y.; Cao, F. An updated meta-analysis of cohort studies: Diabetes and risk of Alzheimer's disease. *Diabetes Res. Clin. Pract.* **2017**, *124*, 41–47. [CrossRef]
165. Gaudieri, P.A.; Chen, R.; Greer, T.F.; Holmes, C.S. Cognitive function in children with type 1 diabetes: A meta-analysis. *Diabetes Care* **2008**, *31*, 1892–1897. [CrossRef]
166. Hughes, T.M.; Ryan, C.M.; Aizenstein, H.J.; Nunley, K.; Gianaros, P.J.; Miller, R.; Costacou, T.; Strotmeyer, E.S.; Orchard, T.J.; Rosano, C. Frontal gray matter atrophy in middle aged adults with type 1 diabetes is independent of cardiovascular risk factors and diabetes complications. *J. Diabetes Its Complicat.* **2013**, *27*, 558–564. [CrossRef]
167. Ferguson, S.C.; Blane, A.; Wardlaw, J.; Frier, B.M.; Perros, P.; McCrimmon, R.J.; Deary, I.J. Influence of an early-onset age of type 1 diabetes on cerebral structure and cognitive function. *Diabetes Care* **2005**, *28*, 1431–1437. [CrossRef]
168. Biessels, G.J.; Despa, F. Cognitive decline and dementia in diabetes mellitus: mechanisms and clinical implications. *Nat. Rev. Endocrinol.* **2018**, *14*, 591–604. [CrossRef]
169. Feinkohl, I.; Price, J.F.; Strachan, M.W.; Frier, B.M. The impact of diabetes on cognitive decline: Potential vascular, metabolic, and psychosocial risk factors. *Alzheimer's Res. Ther.* **2015**, *7*, 46. [CrossRef]
170. Geijselaers, S.L.C.; Sep, S.J.S.; Stehouwer, C.D.A.; Biessels, G.J. Glucose regulation, cognition, and brain MRI in type 2 diabetes: a systematic review. *Lancet Diabetes Endocrinol.* **2015**, *3*, 75–89. [CrossRef]
171. Jeon, B.T.; Heo, R.W.; Jeong, E.A.; Yi, C.O.; Lee, J.Y.; Kim, K.E.; Kim, H.; Roh, G.S. Effects of caloric restriction on O-GlcNAcylation, Ca(2+) signaling, and learning impairment in the hippocampus of ob/ob mice. *Neurobiol. Aging* **2016**, *44*, 127–137. [CrossRef]
172. Liu, Y.W.; Zhu, X.; Yang, Q.Q.; Lu, Q.; Wang, J.Y.; Li, H.P.; Wei, Y.Q.; Yin, J.L.; Yin, X.X. Suppression of methylglyoxal hyperactivity by mangiferin can prevent diabetes-associated cognitive decline in rats. *Psychopharmacology* **2013**, *228*, 585–594. [CrossRef]

173. Bhutada, P.; Mundhada, Y.; Bansod, K.; Bhutada, C.; Tawari, S.; Dixit, P.; Mundhada, D. Ameliorative effect of quercetin on memory dysfunction in streptozotocin-induced diabetic rats. *Neurobiol. Learn. Mem.* **2010**, *94*, 293–302. [CrossRef]
174. Maciel, R.M.; Carvalho, F.B.; Olabiyi, A.A.; Schmatz, R.; Gutierres, J.M.; Stefanello, N.; Zanini, D.; Rosa, M.M.; Andrade, C.M.; Rubin, M.A.; et al. Neuroprotective effects of quercetin on memory and anxiogenic-like behavior in diabetic rats: Role of ectonucleotidases and acetylcholinesterase activities. *Biomed. Pharmacother.* **2016**, *84*, 559–568. [CrossRef]
175. Thakur, A.K.; Rai, G.; Chatterjee, S.S.; Kumar, V. Beneficial effects of an Andrographis paniculata extract and andrographolide on cognitive functions in streptozotocin-induced diabetic rats. *Pharm. Biol.* **2016**, *54*, 1528–1538. [CrossRef]
176. Hashmi, W.J.; Ismail, H.; Mehmood, F.; Mirza, B. Neuroprotective, antidiabetic and antioxidant effect of Hedera nepalensis and lupeol against STZ + AlCl3 induced rats model. *Daru* **2018**, *26*, 179–190. [CrossRef]
177. Sanna, R.S.; Muthangi, S.; Devi, S.A. Grape seed proanthocyanidin extract and insulin prevents cognitive decline in type 1 diabetic rat by impacting Bcl-2 and Bax in the prefrontal cortex. *Metab. Brain Dis.* **2019**, *34*, 103–117. [CrossRef]
178. Imam-Fulani, A.O.; Sanusi, K.O.; Owoyele, B.V. Effects of acetone extract of Cola nitida on brain sodium-potassium adenosine triphosphatase activity and spatial memory in healthy and streptozotocin-induced diabetic female Wistar rats. *J. Basic Clin. Physiol. Pharmacol.* **2018**, *29*, 411–416. [CrossRef]
179. Thakur, A.K.; Chatterjee, S.S.; Kumar, V. Beneficial effects of Brassica juncea on cognitive functions in rats. *Pharm. Biol.* **2013**, *51*, 1304–1310. [CrossRef]
180. Mousavi, S.M.; Niazmand, S.; Hosseini, M.; Hassanzadeh, Z.; Sadeghnia, H.R.; Vafaee, F.; Keshavarzi, Z. Beneficial Effects of Teucrium polium and Metformin on Diabetes-Induced Memory Impairments and Brain Tissue Oxidative Damage in Rats. *Int. J. Alzheimers Dis.* **2015**, *2015*, 493729.
181. Farajpour, R.; Sadigh-Eteghad, S.; Ahmadian, N.; Farzipour, M.; Mahmoudi, J.; Majdi, A. Chronic Administration of Rosa canina Hydro-Alcoholic Extract Attenuates Depressive-Like Behavior and Recognition Memory Impairment in Diabetic Mice: A Possible Role of Oxidative Stress. *Med Princ Pract.* **2017**, *26*, 245–250. [CrossRef]
182. Lin, W.S.; Lo, J.H.; Yang, J.H.; Wang, H.W.; Fan, S.Z.; Yen, J.H.; Wang, P.Y. Ludwigia octovalvis extract improves glycemic control and memory performance in diabetic mice. *J. Ethnopharmacol.* **2017**, *207*, 211–219. [CrossRef]
183. Liu, Z.H.; Chen, H.G.; Wu, P.F.; Yao, Q.; Cheng, H.K.; Yu, W.; Liu, C. Flos Puerariae Extract Ameliorates Cognitive Impairment in Streptozotocin-Induced Diabetic Mice. *Evid Based Complement Alternat Med.* **2015**, *2015*, 873243. [CrossRef]
184. Parihar, M.S.; Chaudhary, M.; Shetty, R.; Hemnani, T. Susceptibility of hippocampus and cerebral cortex to oxidative damage in streptozotocin treated mice: prevention by extracts of Withania somnifera and Aloe vera. *J. Clin. Neurosci.* **2004**, *11*, 397–402. [CrossRef]
185. Pandey, S.P.; Singh, H.K.; Prasad, S. Alterations in Hippocampal Oxidative Stress, Expression of AMPA Receptor GluR2 Subunit and Associated Spatial Memory Loss by Bacopa monnieri Extract (CDRI-08) in Streptozotocin-Induced Diabetes Mellitus Type 2 Mice. *PLoS ONE* **2015**, *10*, e0131862. [CrossRef]
186. Patel, S.S.; Udayabanu, M. Urtica dioica extract attenuates depressive like behavior and associative memory dysfunction in dexamethasone induced diabetic mice. *Metab. Brain Dis.* **2014**, *29*, 121–130. [CrossRef]

International Journal of
Molecular Sciences

MDPI

Review

Natural Compounds for Alzheimer's Disease Therapy: A Systematic Review of Preclinical and Clinical Studies

Stephanie Andrade, Maria João Ramalho, Joana Angélica Loureiro * and Maria do Carmo Pereira *

LEPABE, Department of Chemical Engineering, Faculty of Engineering of the University of Porto, 4200-465 Porto, Portugal; stephanie@fe.up.pt (S.A.); mjramalho@fe.up.pt (M.J.R.)
* Correspondence: joana.loureiro@fe.up.pt (J.A.L.); mcsp@fe.up.pt (M.d.C.P.)

Received: 16 April 2019; Accepted: 7 May 2019; Published: 10 May 2019

Abstract: Alzheimer's Disease (AD) is a neurodegenerative disorder related with the increase of age and it is the main cause of dementia in the world. AD affects cognitive functions, such as memory, with an intensity that leads to several functional losses. The continuous increase of AD incidence demands for an urgent development of effective therapeutic strategies. Despite the extensive research on this disease, only a few drugs able to delay the progression of the disease are currently available. In the last years, several compounds with pharmacological activities isolated from plants, animals and microorganisms, revealed to have beneficial effects for the treatment of AD, targeting different pathological mechanisms. Thus, a wide range of natural compounds may play a relevant role in the prevention of AD and have proven to be efficient in different preclinical and clinical studies. This work aims to review the natural compounds that until this date were described as having significant benefits for this neurological disease, focusing on studies that present clinical trials.

Keywords: neurodegenerative disease; bioactive compound; natural extract; β-amyloid peptide; tau protein; clinical trial; human studies; animal studies; in vitro studies

1. Introduction

Neurodegenerative diseases induce alterations in the central nervous system with psychological and physiological negative effects [1]. Alzheimer's disease (AD) is known as a neurodegenerative disorder with major importance and the principal cause of dementia among the elderly [2,3]. Microscopically, intraneuronal neurofibrillary tangles (NFTs) and extracellular senile plaques (or amyloid plaques) characterize the AD. While senile plaques are constituted by extracellular deposits of β-amyloid (Aβ) peptide, the hyperphosphorylation and abnormal deposition of tau protein compose the NFTs [4].

Aβ derives from the amyloid precursor protein (APP), proteolytic cleavage of amyloid precursor protein (APP), an integral membrane protein that possesses the general properties of a cell surface receptor [5], by the consecutive action of β- and γ-secretases (amyloidogenic pathway). However, this amyloidogenic pathway can be stopped by the competition of α-secretase with γ-secretase (non-amyloidogenic pathway) [6]. The amyloid cascade hypothesis (ACH) suggests that the imbalance between the Aβ generation and its clearance causes the dysfunction and consequently cell death. Aβ polymerizes in a variety of structurally different forms including oligomeric, protofibrillar, and fibrils, forming the senile plaques [7]. Several findings suggest that oligomers play an important role in the ACH [8]. Nowadays, it is proved that Aβ oligomers, including protofibrils and prefibrils, are more toxic than fibrils [9]. Tau protein is also related with the ACH. First, tau monomers aggregate and form oligomers that aggregate into a β-sheet conformation, forming NFTs [10]. NFTs accumulate inside the neurons, resulting in their death. The ACH suggests that toxic concentrations of Aβ cause changes in tau protein and subsequent formation of NFTs, leading to synaptic and neuronal loss [11].

Though a direct relationship between the degree of AD and the amount of Aβ aggregates and tau levels have been established, numerous other mechanisms of neurodegeneration have been suggested, such as neuroinflammation [12], oxidative stress [13], genetic [14] and environmental factors [15]. So, there is an urgent need to develop efficient therapies that target the various pathogenic mechanisms associated with AD. Based on these mechanisms, different therapeutic molecules can act through different pathways [16–18]. However, the currently available medications only control the symptoms in an early stage of the disease [11].

Therefore, it is fundamental to seek for new strategies for AD therapy [19–22]. Natural compounds were the first molecules used as therapeutic agents [23]. Nowadays, the study of these natural compounds revealed that they present neuroprotective effects, arousing an increasing interest in the scientific community and in the pharmaceutical industry [24,25]. A diversity of natural compounds from different origins was described to be suitable to prevent and attenuate several pathologies, including neurological diseases, such as AD [26–28]. Several in vitro and in vivo studies have proven the therapeutic potential of natural compounds, however, just a small percentage has reached the clinical trials stage [29]. Since several causes are related with this disease, the preventive properties of the natural compounds can be associated with several mechanisms as shown in Figure 1 [6,30–34].

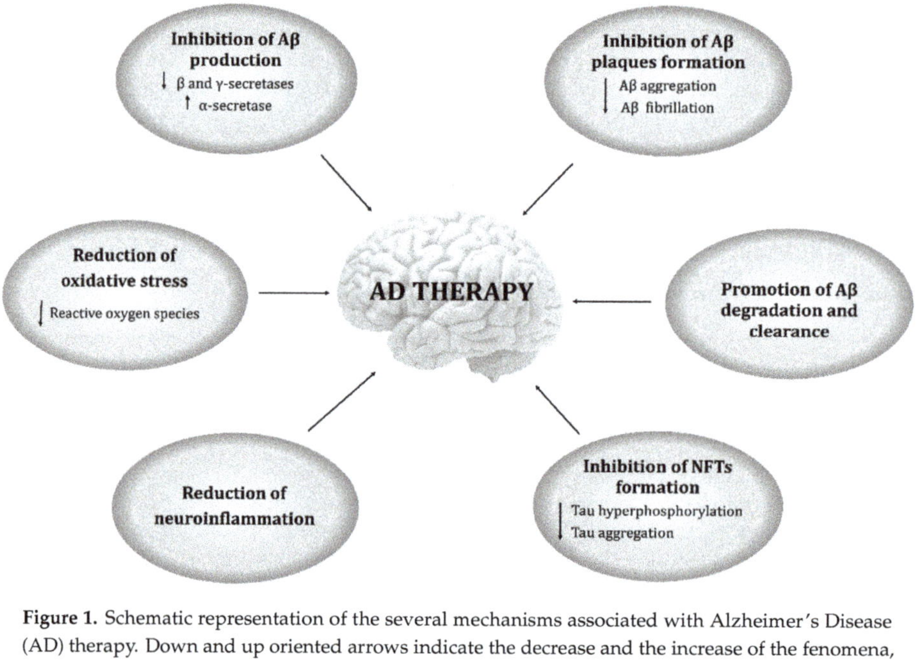

Figure 1. Schematic representation of the several mechanisms associated with Alzheimer's Disease (AD) therapy. Down and up oriented arrows indicate the decrease and the increase of the fenomena, respectively.

In this review, the natural compounds already in clinical trials phase are described and the reported results are presented and discussed. Other natural compounds with known potentially beneficial effects in AD in a preclinical development stage with in vitro and in vivo studies are also described. For preclinical studies, only the most recent reported works are cited. The systematic literature search was conducted using PubMed, Science direct, Google Scholar, Scopus and Web of Science as online databases until April 2019. Only papers written in English were considered with unlimited publication date.

2. Natural Compounds in Clinical Trials and Their Effects on AD

Natural compounds are an emerging approach for AD therapy. For the assessment of their therapeutic efficiency and potential side effects, human trials have been performed in the last years. The first natural product studied in a clinical trial was nicotine in 1992. However, no clinical trials were performed in the last two decades for this molecule. During the 90s, several other compounds were studied in clinical trials for AD therapy, such as vitamins. These molecules are still being tested in human trials up until this date. In the last years, other natural compounds are gaining interest by the scientific community and have achieved the clinical trials phase, such as bryostatin, which effects started to be evaluated in humans in 2017. A detailed report of these findings is described below. The natural compounds were divided into two groups: bioactive compounds and natural extracts, and they are summarized in Tables 1 and 2, respectively. Here, a bioactive compound refers to a therapeutic molecule while a natural extract is the mixture of several molecules. The compounds are listed from the ones with more participants and longer duration.

Table 1. Bioactive compounds in clinical trials for AD therapy.

Bioactive Compound	Condition of Participants	Number of Subjects	Duration	Outcomes	Ref.
Vitamin D	Mild cognitive impairment	8	8 weeks	Reduction of Aβ level	[35]
	Mild cognitive impairment and early AD	48	20 months	Reduction of Aβ level; Improvement of cognitive functions	[36]
Vitamin D and memantine	Moderate AD	43	24 weeks	Improvement of cognitive functions	[37]
Vitamin D and antioxidants	Mild to moderate AD	78	16 weeks	Reduction of oxidative stress	[38]
Vitamin E and vitamin C	AD	20	1 month	Reduction of oxidative stress	[39]
Vitamin E and selegiline	Moderate AD	341	2 years	Delay of AD progression	[40]
Vitamin E and donepezil	Mild cognitive impairment	769	5 years	No effectiveness in delaying AD progression	[41]
Vitamin E and memantine	Mild to moderate AD	613	5 years	Delay of AD progression	[42]
Vitamin E and selenium	Healthy patients	3786	13 years	No prevention of dementia	[43]
Docosahexaenoic acid (DHA) and eicosapentaenoic acid	AD	204	12 months	Safe and well tolerated; No effectiveness in delaying cognitive decline	[44]
DHA	AD	295	18 months	No effectiveness in delaying cognitive decline	[45]
	Cognitive impairments	485	24 weeks	Improvement of cognitive functions	[46]
	Mild cognitive impairment	36	1 year	Safe and well tolerated; Improvement of memory	[47]
Homotaurine	Mild to moderate AD	1052	78 weeks	Improvement of cognitive functions	[48,49]
		58	3 months	No harmful effects on vital signs; Side effects	[50]
		10	4 weeks	Improvement of the central cholinergic transmission	[51]
Huperzine A	AD	103	8 weeks	Safe and well tolerated; Improvement of memory and behaviour	[52]
		60	60 days	Safe and well tolerated; Reduction of oxidative stress	[53]
	Mild to moderate AD	177	16 weeks	Safe and well tolerated; Improvement of cognitive functions	[54]
Bryostatin	AD	9	46 weeks	Safe and well tolerated: Improvement of cognitive functions	[55]
		150	12 weeks	Improvement of cognitive functions	[56]
Melatonin	AD	150	12 weeks	Improvement of memory	[57]
		14	22 to 35 months	Improvement of cognitive functions	[58]
	Mild cognitive impairment	50	9 to 18 months	Improvement of cognitive functions	[59]
	Mild to moderate AD	80	24 weeks	Safe; Improvement of cognitive functions	[60]

Table 1. *Cont.*

Bioactive Compound	Condition of Participants	Number of Subjects	Duration	Outcomes	Ref.
Resveratrol	Mild to moderate AD	119	52 weeks	Side effects; No effectiveness in reducing biomarkers levels	[61]
		39	1 year	Safe and well tolerated; No effectiveness in treat AD	[62]
Nicotine	AD	70	2 weeks	Improvement of perceptual and visual attentional deficits	[63]
		6	9 weeks	Safe; Improvement of learning	[64]
		8	10 weeks	Improvement of attentional performance	[65]
Curcumin	AD	34	6 months	Safe and well tolerated	[66]

Table 2. Natural extracts and other natural products in clinical trials for AD therapy.

Natural Extracts and Other Products	Condition of Participants	Number of Subjects	Duration	Outcomes	Ref.
Ginkgo biloba	Mild to moderate dementia	410	24 weeks	Safe; Improvement of neuropsychiatric symptoms	[67,68]
		410	24 weeks	Improvement of cognitive and functional functions	[69]
	AD or vascular dementia	404	24 weeks	Improvement of cognitive functions and functional abilities; Improvement of neuropsychiatric symptoms	[70]
	Mild cognitive impairment	160	24 weeks	Safe and well tolerated; Improvement of cognitive functions	[71]
Saffron	Mild to moderate AD	46	16 weeks	Safe; Improvement of cognitive functions and memory	[72]
Lemon balm	Mild to moderate AD	40	4 months	Improvement of cognition function and agitation	[73]
Green tea	Severe AD	30	2 months	Improvement of cognitive functions	[74]
Papaya	AD	20	6 months	Reduction of oxidative stress	[75]
Sage	Mild to moderate AD	20	4 months	Improvement of cognitive functions; No side effects except agitation	[76]
Coconut	AD	44	21 days	Improvement of cognitive functions	[77]
Apple	Moderate to severe AD	21	1 month	No improvement of cognitive functions; Improvement behavioural and psychotic symptoms; Reduction of anxiety, agitation and delusion	[78]
Blueberry	Early memory failures	9	12 weeks	Improvement of learning; Reduction of depressive symptoms	[79]
Colostrinin	AD	n. d.	15 weeks	Improvement of cognitive and daily functions	[80]

n. d.—The information was not provided by the authors.

2.1. Bioactive Compounds

Vitamins have been described as therapeutic compounds for AD. Among them, vitamin C, E and D have aroused great interest. Vitamin C (Figure 2A) is found in several vegetables and fruits, mostly citrus fruits. In vivo studies reported that vitamin C prevented the neuroinflammation [81] and the brain oxidative damage due to its potent antioxidant activity [82]. Also, it was observed in an AD mouse model that Vitamin C reduced the Aβ oligomers formation and tau phosphorylation, improving the behavioral decline. The reduction of Aβ levels [83] and Aβ plaque burden [84] was also observed in vivo.

On the other hand, vitamin E, which is present in several fruits and vegetables (Figure 2B), also showed in vivo antioxidant and anti-inflammatory effects [85]. Other in vivo study revealed that vitamin E reduced the Aβ levels [86].

Other vitamin with reported beneficial effects for AD, is vitamin D. Adding to several benefits of vitamin D [87], its therapeutic effect in AD has also been studied in last years. Although the major source of vitamin D is sunlight exposure (vitamin D_3, Figure 2C) [88], around 20% can be obtained from food, including fatty fish and fish-liver oils (vitamin D_2, Figure 2D) [89]. In vivo studies revealed that vitamin D is an anti-inflammatory compound [90] with the ability to inhibit the activity of β and γ-secretases, reducing the Aβ production and amyloid plaques and to increase the Aβ degradation [91]. As result, an improvement on learning and memory performance was verified in AD rats [92,93]. Also, low plasma Aβ is linked to the incidence of AD.

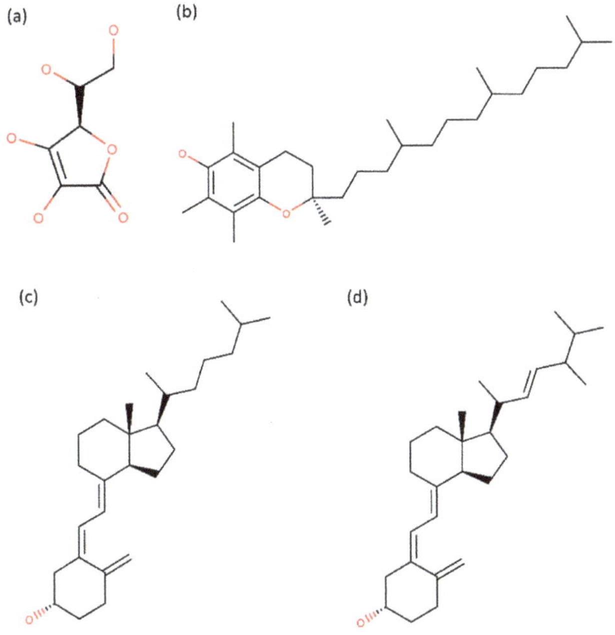

Figure 2. Chemical structures of: (**a**) vitamin C, (**b**) vitamin E, (**c**) vitamin D_3 and (**d**) vitamin D_2.

Clinical trials revealed that vitamin D increased plasma Aβ in mild cognitive impairment patients, suggesting a reduction in Aβ levels in the brain. In fact, Miller et al. (2016) studied the effect of vitamin D supplementation on the plasma levels of Aβ in eight patients over 60 years old in a pilot study. Patients were randomly divided in two groups, treatment and placebo groups. Patients from the treatment group were administered with 50,000 IU per week for eight weeks. The obtained results showed that vitamin D intake increased plasma Aβ levels, suggesting a decrease in Aβ brain levels [35].

SanMartin et al. (2017) evaluated the role of vitamin D in the Aβ clearance from the brain. Patients with mild cognitive impairment and very early AD (n = 47) were orally supplemented with vitamin D at 50,000 IU once a week for six weeks, followed by 1500–2000 IU daily for 18 months. The obtained results showed that lymphocyte susceptibility to death, Aβ plasma levels and cognitive status improved after six months of vitamin D supplementation in cognitive impairment patients, but not in very early AD patients. Thus, supplementation with vitamin D proved to be beneficial in cognitive impairment patients. The lack of effects in very early AD patients suggest that vitamin D intake is not able to delay the progression of the disease in a more advanced stage [36].

Co-therapy with vitamin D and other molecules for AD therapy has also been explored in clinical trials. In fact, Annweiler et al. (2012) conducted a double-blind, placebo-controlled pilot trial with 43 white patients over 60 years with moderate AD symptoms [37]. The main goal of this trial was to evaluate the combination of neuroprotective effects of memantine and vitamin D in preventing neuronal loss and cognitive decline. Memantine was selected because is one of the most prescribed drugs for AD therapy [94]. Patients were randomly divided in three groups, being administered with memantine plus vitamin D (n = 8), or memantine alone (n = 18), or vitamin D alone (n = 17). Patients were administered with drugs for 24 weeks. Memantine was administered orally at 5 mg per week for the first four weeks and then 20 mg per day for the rest of the trial. Patients received a drinking solution of vitamin D at 100,000 IU every four weeks. After the study, patients co-treated with memantine and vitamin D showed better cognitive performance than patients treated with vitamin D or memantine alone [37].

Co-supplementation with vitamin D and other natural compounds was also studied in clinical trials. In fact, Galasko et al. (2012) conducted a double-blind, placebo-controlled clinical trial to evaluate what antioxidant supplementation affected the levels of AD's histopathological marks, such as Aβ peptide and tau protein [38]. Patients with mild to moderate AD (n = 78) received placebo or daily supplement containing 800 IU of vitamin E, 500 mg of vitamin D, 900 mg of α-lipoic acid and 400 mg of coenzyme Q for 16 weeks. The attained results showed that the co-supplementation did not affect amyloid or tau levels, but a reduction on levels of an oxidative stress biomarker, the cerebrospinal fluid F2-isoprostane, was verified.

Also, co-supplementation with multivitamins was evaluated in clinical trials. In fact, Kontush et al. (2001) evaluated the efficiency of supplementation with both vitamin E and vitamin C to decrease oxidation of lipoproteins in AD patients [39]. Lipid oxidation is related with AD progression. Twenty patients with AD were randomly divided in two groups. The first group received a daily supplement for one month of 400 IU vitamin E alone, and the second group received a daily combination of 400 IU vitamin E and 1000 mg of vitamin C. The obtained results proved that combined supplementation was more efficient in maintaining active doses of vitamins in the plasma and decreasing lipid oxidation.

Co-therapy of different drugs with vitamin E was also studied in clinical trials. Sano et al. (1997) evaluated the effects of vitamin E and selegiline co-administration [40]. Selegiline is a monoamine oxidase inhibitor, that prevents dopamine degradation [95]. For that, a double-blind, placebo-controlled clinical trial was conducted with 341 patients with moderate AD's symptoms for two years. The patients were randomly divided in four groups, a placebo group, one receiving vitamin E, one receiving selegiline, and another one receiving both drugs. Vitamin E was daily administered at a dose of 2000 IU per day, and 10 mg of selegiline daily. Co-therapy proved to efficiently slow the progression of the disease [40].

The combined effect of donepezil and vitamin E was also studied. Donepezil is a drug used for AD therapy to control the symptoms. To compare the effects of this drug with vitamin E on the outcome effects on patients with mild cognitive impairment, a double-blind, placebo-controlled clinical trial was conducted by Petersen et al. (2005) [41]. Patients over the age of 55 (n = 769) were randomly divided in three groups, placebo, vitamin E alone or donepezil alone. The daily dose of vitamin E was 1000 IU, and after six weeks the dose was increased to 2000 IU, for five years. Vitamin E proved to not be able to delay the disease progression.

Dysken et al. (2014) studied the combination effects of vitamin E and memantine [42]. For that, a double-blind, placebo-controlled clinical trial was conducted with 613 patients with mild to moderate AD's symptoms for five years. The patients were randomly divided in three groups, one receiving vitamin E, one receiving memantine, and another one receiving both vitamin E and memantine. The used doses for vitamin E were 2000 IU per day, and 20 mg of memantine daily. Treatment with vitamin E alone proved to be more efficient in slowing disease cognitive decline comparatively with the placebo group. However, no differences were verified for co-therapy comparatively with treatment with memantine alone.

Kryscio et al. (2017) intended to assess if vitamin E and selenium intake could prevent dementia in healthy men over 60 [43]. Although no evidence exists to support the use of selenium in the treatment of AD, some works suggest that this product has a preventive potential [96]. A double-blind, placebo-controlled clinical trial involving 3786 male patients was conducted for 13 years. The participants were randomly divided into four groups. The first group received vitamin E, to the second only selenium was administered, the third group received a combination of vitamin E and selenium, and the fourth received placebo. The conclusions of this trials were that neither of the supplementation regimen proved to be able in preventing dementia [43].

Docosahexaenoic acid (DHA) is a polyunsaturated fatty acid from marine fish and algae [97] and its structural formula is presented in Figure 3. DHA demonstrated to have an antioxidant activity reducing the lipid peroxide and reactive oxygen species (ROS) levels in the brain of AD rats, improving the learning [98]. In addition, in vivo experiments showed that DHA reduces the Aβ levels, Aβ accumulation and plaque burden [99]. Some in vitro experiments demonstrated that DHA decreases the β- and γ-secretase activity and increases the α-secretase activity [100]. An in vitro study suggests that DHA reduced soluble Aβ oligomers levels and inhibited the formation and polymerization of Aβ fibrils [101]. Furthermore, DHA stimulated the Aβ degradation [102] and disaggregation of preformed Aβ fibrils in vitro [103].

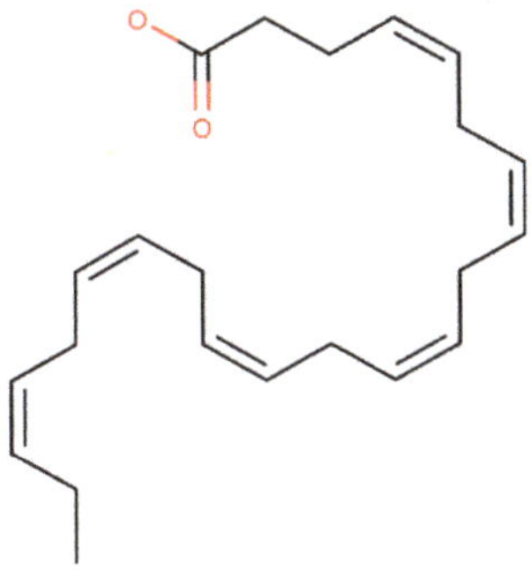

Figure 3. Chemical structure of docosahexaenoic acid (DHA).

The effects of supplementation with DHA in AD patients were studied in different clinical trials. In fact, Freund-Levi et al. (2006) conducted a double-blind, placebo-controlled clinical trial with 204 AD patients [44]. The main goal of this study was to evaluate the efficacy of dietary co-supplementation of DHA with other fatty acid, the eicosapentaenoic acid, on the cognitive functions of patients with mild to moderate AD. The patients were randomly divided in two groups, treatment and placebo. Patients on treatment group received a daily dose of 1.7 g of DHA and 0.6 g of eicosapentaenoic acid for six months. After this period, all patients received fatty acid co-supplementation for six more months. Despite the treatment being safe and well tolerated, the supplementation with these fatty acids did not delay the rate of cognitive decline of the patients.

Quinn et al. (2010) conducted a double-blind, placebo-controlled clinical trial to evaluate the efficacy of supplementation with DHA on the cognitive and functional decline in AD patients [45]. A daily dose of 2 g of DHA or placebo was administered to 295 patients for 18 months. The extent of brain atrophy was measured, and the results showed that DHA did not alter the patients' condition. The attained results also proved that administration of DHA did not slow the rate of cognitive and functional decline.

The same group conducted a double-blind, placebo-controlled, clinical study in the same year to evaluate the ability of DHA to improve the cognitive functions of 485 participants with age-related cognitive decline [46]. The subjects were randomly assigned to a daily oral administration of 900 mg of DHA orally or placebo for 24 weeks. The attained results proved that supplementation with DHA improved cognitive health, since the participants showed enhanced learning and memory functions.

Lee et al. (2013) studied the effects of DHA administration using fish oil on the cognitive function in patients over 60 diagnosed with mild cognitive impairment [47]. The participants ($n = 36$) were randomly divided in two groups, placebo and treatment group. The treatment group was orally administered with 430 g of DHA three times a day, for one year. No significant side effects were verified, suggesting the potential of DHA to improve memory functions. However, studies with more patients and longer intervention periods, are necessary to define the optimal dosage.

Homotaurine, also known as tramiprosate, is an aminosulfonate metabolite extracted from marine red alga *Grateloupia livida* and its structural formula is presented in Figure 4 [104]. In in vitro experiments, homotaurine proved to efficiently inhibit the Aβ aggregation [105] and reduce the Aβ plaque formation. This compound was also able to reduce the Aβ levels in vivo [106]. Additionally, the compound stabilized Aβ monomers and inhibited the Aβ oligomers formation in vitro [107].

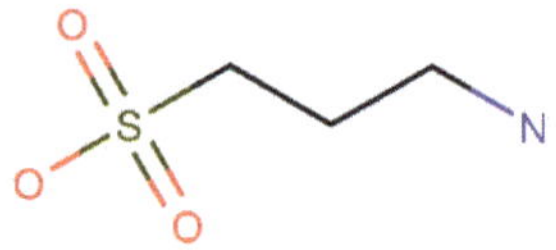

Figure 4. Chemical structure of homotaurine.

Aisen et al. (2011) conducted a phase III double-blind, placebo-controlled trial with 1052 patients with mild to moderate AD symptoms to evaluate the effect of homotaurine in slowing AD progression [48,49]. This compound was the first inhibitor of Aβ aggregation that has reached a phase III clinical trial. The participants were randomly divided in three groups. The first group was the placebo group, and the other two groups received daily treatment with homotaurine at dose of 100 and 150 mg for 78 weeks, respectively. The authors proved that homotaurine administration had beneficial effect on cognition [108,109]

The safety and tolerability of this compound administered to 58 patients with mild to moderate AD symptoms, were studied previously in a phase II clinical trial conducted by the same group [50]. Patients received placebo, 100 or 150 mg of homotaurine for three months. No harmful effects on vital signs were verified and the most frequent side effects were nausea, vomiting, and diarrhoea.

Martorana et al. (2014) conducted a study with 10 patients with mild cognitive impairment with ages between 59 and 74 [51]. The participants were administered daily with 100 mg of homotaurine for four weeks. The obtained results showed that homotaurine improved the central cholinergic transmission.

Huperzine A is isolated from *Huperzia serrata* (Thunb.) Trevis. (Lycopodiaceae) and its structural formula is presented in Figure 5. This compound demonstrated to have antioxidant properties. Huperzine A was able to reduce ROS and lipid peroxidation in an AD rat model [110]. Also, this product presents the in vitro ability to increase the α-secretase activity, significantly decreasing the Aβ levels, suggesting a blocking action in the Aβ production [111].

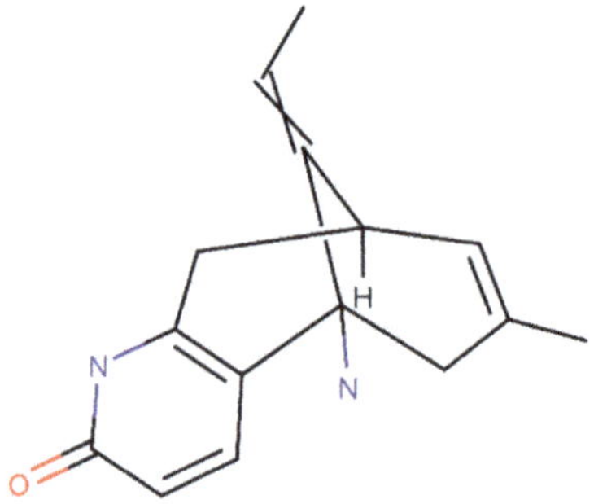

Figure 5. Chemical structure of huperzine A.

Xu et al. (1995) evaluated the efficacy and safety of huperzine A in AD patients. Four tablets of huperzine A (200 μg) or placebo were administered orally to 103 patients, twice a day, for eight weeks [52]. The results showed that the administration of huperzine A improved the memory and behaviour of AD patients. Also, the obtained results for the compound were better than for placebo. Huperzine A did not induce side effects.

To further compare the efficacy and safety of huperzine A administered into capsules and tablets in AD patients, the same group conducted a new trial four years later [53]. In this study, 200 μg of huperzine A or placebo into capsules and tablets were administered twice a day to 60 patients, for 60 days. Both groups revealed a reduction in ROS levels in the plasma and erythrocytes of AD patients, without side effects besides nausea. This trial suggests that huperzine A in capsules and tablets is safe to be used in AD patients.

Later, Rafii et al. (2011) studied the safety and efficacy of two concentrations of huperzine A, 200 and 400 μg twice a day, in patients with mild to moderate AD in a phase II clinical trial [54]. Placebo or huperzine A was administered to 177 patients for 16 weeks. The results demonstrated that at 400 μg/day huperzine A was not efficient, not being able to treat AD. However, at the concentration of 800 μg/day, the compound improved the cognition of AD patients. Huperzine A was safe at both studied doses.

Bryostatin is a macrolide lactone extracted from *bryozoan Bugula neritina* [112]. The structural formula of the compound is presented in Figure 6. An in vivo study showed that bryostatin reduced the Aβ production by the stimulation of α-secretase activity, reducing the mortality of AD mice model [113]. Also, bryostatin revealed to enhance the learning and memory in AD mice model [114].

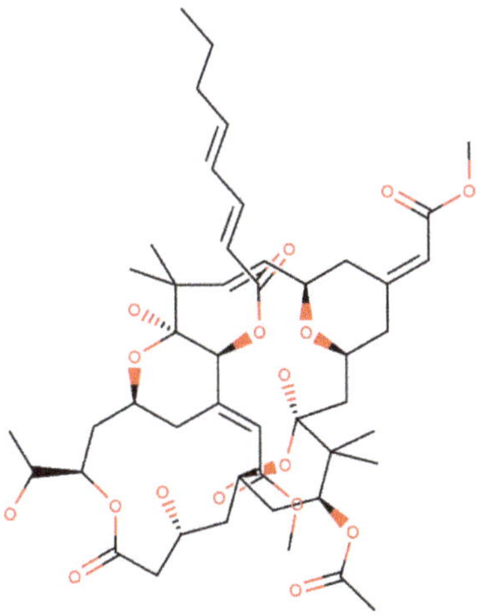

Figure 6. Chemical structure of bryostatin.

Recently, Nelson et al. (2017) evaluated the safety, tolerability and effects on cognitive function of bryostatin on AD patients in a phase II clinical trial [55]. A single dose of bryostatin at 25 $\mu g/m^2$ was administered to six patients, while three patients received placebo. Bryostatin proved to improve cognitive functions and to be safe and well tolerated.

Another phase II clinical trial was performed with the same goals [56]. Farlow et al. (2018) administered 20 or 40 μg of bryostatin or placebo to 150 AD patients, for 12 weeks. This study confirmed the safety of both doses of bryostatin. Also, an improvement of cognitive functions was observed using doses of 20 μg of bryostatin.

Melatonin is collected from animals, plants, fungi and bacteria and its structural formula is presented in Figure 7. This compound demonstrated to have antioxidant properties due to its ability to decrease ROS in vivo [115]. In addition, an in vivo study reported the beneficial effects on neuroinflammation [116]. Further, an in vitro study proved the ability to inhibit the β-sheet conformation and, consequently, Aβ fibrils [117], decreasing the Aβ levels in the brain of AD rat model [118]. Another in vitro study proved that melatonin inhibits β- and γ-secretase activity and enhances the α-secretase activity, blocking the Aβ monomers production [119].

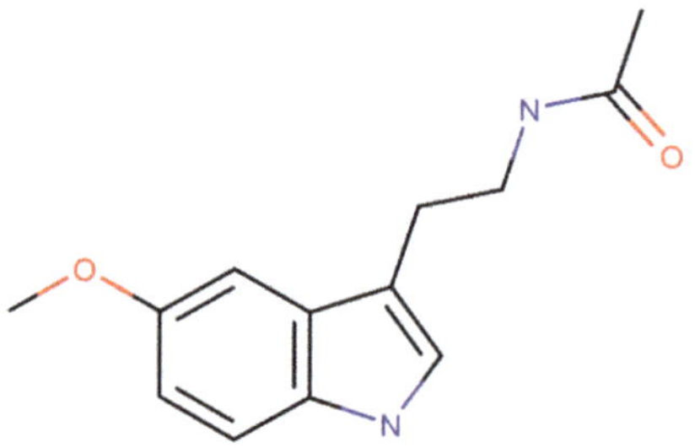

Figure 7. Chemical structure of melatonin.

Brusco et al. (1998) evaluated the efficacy of melatonin in monozygotic twins with AD, with similar cognitive and neuropsychologic impairments [57]. Only one of the twins orally received daily 6 mg of melatonin for 36 months. The results suggest that melatonin improved the memory of the treated patient. Also, the clinical evaluation revealed that the twin that did not receive the treatment presented a more advanced state of the disease.

Later, the same group studied the effect of melatonin in cognitive dysfunctions of 14 AD patients [58]. The patients received 9 mg of melatonin daily for 22 to 35 months. The results showed an improvement in cognitive functions, after the treatment.

The same results were obtained by Furio et al. (2007) that performed a clinical trial with 50 outpatients diagnosed with mild cognitive impairment, where half of patients received 3 to 9 mg of melatonin for 9 to 18 months [59].

Wade et al. (2014) investigated the ability of 2 mg of melatonin to improve the cognitive functions of patients with mild to moderate AD [60]. Melatonin or placebo was administered to 80 patients for 24 weeks. Placebo was also administered two weeks before and after melatonin treatment. The results revealed an improvement in cognitive functions of AD patients treated with melatonin, comparing to placebo. Also, treatment was safe for both groups. Thus, these clinical trials suggested that melatonin administration can be a suitable therapeutic strategy for the treatment of AD.

Resveratrol is a naturally occurring non-flavonoid polyphenol present in grapes (*Vitis vinifera* L. (Vitaceae)) and red wine and its structural formula is presented in Figure 8 [120]. In vitro experiments demonstrated that resveratrol induces the inhibition of studies proved a reduction of Aβ fibrils formation [121] and induced the in vitro Aβ disaggregation by an intracellular proteasomal action [108]. In vitro results showed that resveratrol has the ability to reshape toxic aggregates into a non-toxic aggregate type [109]. As result, resveratrol decreased the Aβ levels [122] and plaque levels in brain of AD rats [123]. In addition, in vivo evidence suggests that resveratrol has anti-inflammatory [122] and antioxidant effects [124]. Also, an in vitro study showed that resveratrol prevents the tau hyperphosphorylation [125].

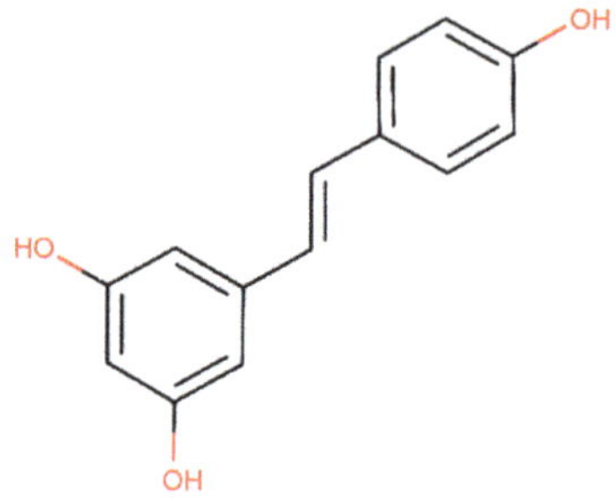

Figure 8. Chemical structure of resveratrol.

Turner et al. (2015) performed a phase 2 clinical trial for 52 weeks in mild to moderate AD patients. The group studied the safety, tolerability and the ability of resveratrol to reduce the biomarkers of the disease (Aβ and tau). Here, 119 individuals were orally administered once a day with placebo or 500 mg of resveratrol, with an increase of 500 mg each 13 weeks. Although this study suggests that resveratrol can cross the blood-brain barrier (BBB), the results were not satisfactory. Besides inducing some side effects like nausea, diarrhea, and weight loss, the brain volume and biomarkers levels were lower in the placebo group than resveratrol group [61].

Recently, Zhu et al. (2018) evaluated the safety, tolerability and efficacy of a mixture containing 5 mg of resveratrol, 5 g dextrose and 5 g of malate. Fifteen mL of the mixture or placebo were orally administered twice a day to 39 patients with mild to moderate AD for one year. The administration was done together with an 8 oz glass of commercial grape juice. The results revealed that the preparation was safe and well tolerated. However, no evidence was observed concerning the efficacy of the product for AD therapy [62].

Nicotine is extracted from the tobacco plant leaves (*Nicotiana tabacum* L., Solanaceae) and its structural formula is presented Figure 9. Nicotine presents the ability to delay the amyloidogenesis by inhibiting the β-sheet structures in vitro [126], decreasing in vivo β-secretase expression [127] and inhibiting in vivo Aβ aggregation [128]. An in vitro study revealed that nicotine inhibits the Aβ fibrils formation and their length, and disaggregate Aβ fibrils [129], causing an in vivo decrease of Aβ [127] and plaque amounts [128]. In addition, an in vitro study suggested valuable effects of nicotine due to their antioxidant properties [130]. Also, the decrease of APP containing Aβ peptide observed in in vivo experiments can be the reason to the diminution of Aβ and amyloid plaque levels [131].

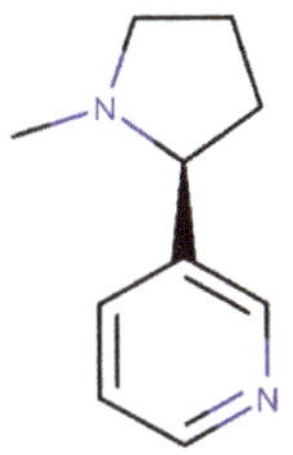

Figure 9. Chemical structure of nicotine.

Jones et al. (1992) studied the effect of nicotine on AD patients [63]. Three acute doses of nicotine (0.4, 0.6 and 0.8 mg) were subcutaneously administered to 22 AD patients and 48 controls. The results revealed that nicotine improved the perceptual and visual attentional deficits observed in AD patients.

The effect of nicotine on behaviour, cognition, and physiology of six AD patients was evaluated in a pilot study proposed by Wilson et al. (1995) [64]. Placebo, nicotine and washout were sequentially administered for seven, eight and seven days, respectively. After nicotine administration, an improvement in learning was observed, which persisted with washout. Memory, behaviour and cognition were not affected. Also, the safety of nicotine was proved.

The clinical and neuropsychological effects of nicotine was evaluated in eight AD patients by White et al. (1999) [65]. Transdermal nicotine was administered for two periods of four weeks, separated by two weeks of washout. A nicotine patch was used daily for 16 h with the following doses: 5 mg/day in the first week, 10 mg/day in the second and third week, and finally, 5 mg/day in the fourth week. The results suggest that nicotine significantly improved the attentional performance. However, the limited sample of the study does not allow conclusive results.

Curcumin is an active component founded in the root of *Curcuma longa* L. (Zingiberaceae) and its structural formula is presented in Figure 10. This compound presents the in vivo ability to prevent the Aβ aggregation and disaggregate preformed Aβ fibrils [132,133]. Also, curcumin presents in vitro and in vivo anti-inflammatory and antioxidant beneficial effects, respectively [134,135]. Also, in vitro experiments showed that curcumin decreases β and γ-secretase levels [133,136,137]. As result, the spatial learning of AD rat model was improved, as well as the memory impairment [133].

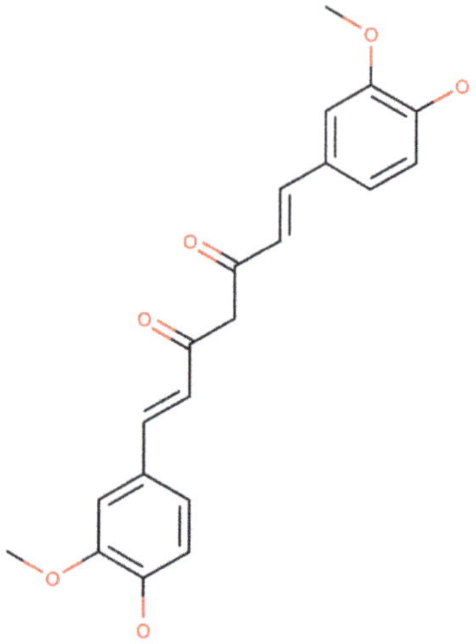

Figure 10. Chemical structure of curcumin.

Baum et al. (2008) performed a clinical trial to study the safety of curcumin on AD patients [66]. For six months, the authors administered 1 g, 4 g of curcumin or placebo in 34 AD patients. The results proved that curcumin did not produce side effects in AD patients, but the authors revealed the necessity of additional trials to confirm the efficacy of curcumin in AD treatment.

2.2. Natural Extracts and Other Natural Products

Ginkgo biloba (*Ginkgo biloba* L., Ginkgoaceae) has been studied as therapeutic drug for AD and other neurological diseases therapy. In vitro evidence revealed that ginkgo biloba extract can prevent Aβ aggregation, decrease Aβ fibrillogenesis and destabilize preformed fibril [138]. Substantial in vivo experimental evidence indicates that ginkgo biloba has antioxidant [139] and anti-inflammatory properties, ameliorating the cognitive and memory impairment in an AD rat model [140]. In vivo studies showed that ginkgo biloba favors the non-amyloidogenic via of APP by increasing α-secretase activity, inhibiting the Aβ production [141,142].

Several clinical trials have been carried out in the last 10 years to test the viability of the compound in treating patients with dementia. Bachinskaya et al. (2011) examined the effect of gingko biloba extract EGb 761® on neuropsychiatric symptoms of dementia [67,68]. Outpatients with mild to moderate dementia (AD with or without cerebrovascular disease or vascular dementia) ($n = 410$) were considered in this study. Patients received 240 mg of extract or placebo once daily for 24 weeks. The treatment with gingko biloba was safe and improved the neuropsychiatric symptoms, which include apathy, irritability, depression, among others.

Also, with the same conditions, Herrschaft et al. (2012) revealed that the treatment with gingko biloba improved the cognition and the life quality of patients [69].

Ihl et al. (2012) performed a similar 24-week randomised controlled trial involving 404 outpatients [70]. Patients were diagnosed with AD ($n = 333$) or vascular dementia ($n = 71$). In addition to confirming the improvement of neuropsychiatric symptoms observed in the previous trial, the extract improved the cognitive functions and functional abilities of patients.

Gavrilova et al. (2014) also conducted a clinical trial to study the effects of gingko biloba in neuropsychiatric symptoms and cognition in 160 patients with mild cognitive impairment [71].

The patients received 240 mg of EGb 761® or placebo for 24 weeks. The trial proved that the extract improved the neuropsychiatric symptoms and cognitive functions of patients. Also, the extract was safe and well tolerated. Taking together, the last clinical trials proved that a 240 mg daily dose of ginkgo biloba extract is safe in the treatment of dementia.

Saffron (*Crocus sativus* L., Iridaceae) is a stem-less herb with antioxidant [143] and anti-inflammatory activities in vivo [144]. This product inhibited the in vitro Aβ aggregation and fibrillogenesis [145].

Akhondzadeh et al. (2010) evaluated the efficacy of 30 mg saffron in the treatment of mild to moderate AD [72]. Saffron or placebo were orally administered daily for 16 weeks, to 46 patients. The phase II study showed that the administration of saffron improved the cognition and memory of AD patients. Also, no side effects differences were observed with saffron or placebo administration. Thus, saffron seems to be safe in the treatment.

Lemon balm (*Melissa officinalis* L., Lamiaceae) from the mint family that is native to Europe with antioxidant activity in vitro [146]. In vivo studies proved the ability of lemon balm extract to improve the memory of an AD model, probably due to the inhibition of β-secretase activity [147]. To assess the efficacy and safety of *Melissa officinalis* extract on patients with mild to moderate AD, Akhondzadeh et al. (2013) administered to 40 patients 60 drops of extract or placebo, for four months [73]. The results proved that *Melissa officinalis* extract ameliorated the cognition and agitation of AD patients.

Green tea (*Camellia sinensis* (L.) Kuntze, Theaceae) from steaming and drying of leaves of the *Camellia sinensis* plant proved to be a rich source of antioxidants in in vivo studies [148]. In addition, the green tea prevented the spatial learning and memory destruction in an AD mice model by decreasing Aβ oligomers levels [149] and hyperphosphorylated tau protein [150].

Recently, Arab et al. (2016) developed a clinical trial with 30 patients to study the antioxidant activity of green tea in patients with severe AD and its ability to improve cognitive functions [74]. Patients received daily 2 g of green tea through the ingestion of pills, for two months. The results showed an improvement on cognitive functions, confirming the effects of the antioxidant activity of green tea.

Papaya (*Carica papaya* L., Caricaceae) is a fruit often used in medicine that has amino acids, β-carotene, oligosaccharides and vitamins, with benefits in AD.

A clinical trial performed by Barbagallo et al. (2015) studied the antioxidant activity of fermented papaya powder extract in AD patients [75]. AD patients (n = 20) received 4.5 g of extract daily for six months, while the 12 controls did not receive any treatment. The results showed that the supplementation with fermented papaya powder reduced the ROS generation and nitric oxide production in AD patients, with no significant changes in controls. Thus, the papaya can be used as antioxidant in the AD therapy.

Sage (*Salvia officinalis* L., Lamiaceae) is a medicinal plant with a long-standing reputation in European medical herbalism due to its anti-inflammatory and antioxidant properties observed in vivo [151].

Akhondzadeh et al. (2008) developed a clinical trial to evaluate the efficacy and safety of *Salvia officinalis* extract in the treatment of patients with mild to moderate AD [76]. Patients received daily 60 drops of sage extract or placebo for four months. The results showed that sage extract improved cognitive functions. Also, after the treatment, any group revealed side effects except agitation, that seems to be more pronounced in placebo group. This study proved that sage can be useful in the therapy of mild to moderate AD.

Coconut (*Cocos nucifera* L., Arecaceae) demonstrated to be able to reduce the Aβ deposition and aggregation and the oxidative stress in a transgenic *Caenorhabditis elegans* AD model [152]. Coconut oil also enhanced the memory of rats [153]. Also, in vitro studies demonstrated that the coconut oil reduced de APP expression, decreasing the Aβ secretion [154] and protected neuronal cells against Aβ-induced neurotoxicity.

Ortí et al. (2018) performed a clinical trial with 44 AD patients [77]. Half of individuals received daily 40 mL of coconut oil, distributed by the breakfast (20 mL) and lunch (20 mL), for 21 days. Before and after the oil administration, cognitive function was evaluated. The trial revealed that the patients treated with coconut oil demonstrated an improvement of cognitive functions.

Apple (*Malus domestica* Borkh., Rosaceae) showed to be a promising approach to prevent AD. In vivo evidence demonstrated that the apple extract prevents the oxidative stress and reduces the Aβ levels, improving the memory of AD rats [155]. Besides, in vivo studies demonstrated that apple juice is able to reduce γ-secretase expression, which leads to the reduction of Aβ production [156].

Remington et al. (2010) performed an open-label pilot clinical trial with 21 patients with moderate to severe AD [78]. The authors administered two 4-oz of apple juice daily for one month. Although the results suggest that there was no modification in the degree of dementia, a significant improvement in behavioural and psychotic symptoms was observed, with reduction of anxiety, agitation, and delusion. This study suggests that the supplementation with apple juice can attenuate the AD-related decline.

Blueberry (*Vaccinium myrtillus* L., Ericaceae) is a fruit composed by several polyphenols named anthocyanins, with antioxidant [157] and anti-inflammatory properties in vivo [158]. In vitro works suggested that blueberries increase the Aβ clearance [159] and inhibit the Aβ aggregation, decreasing the amount of toxic species [157]. As a result, an improvement in cognitive functions and motor performance was observed in an AD mouse model [160].

Krikorian et al. (2010) evaluated the effects of daily administration of wild blueberry juice in a group of nine elderly subjects with early memory failures [79]. The daily consumption of blueberry juice was proportional with body weight, varying between 6 and 9 mL/kg. After 12 weeks of treatment, an improvement in learning was observed as well as a reduction of depressive symptoms. The study suggests that the blueberry supplementation can confer neuroprotection.

Colostrinin, a milk form produced by mammary glands [161], presents in vitro antioxidant and anti-inflammatory activities, and inhibits the Aβ fibrils formation and disassembles Aβ aggregates [162]. Also, the ability of colostrinin to inhibit tau phosphorylation and eliminate Aβ was proved in vitro [163].

The effect of colostrinin on AD patients was studied in a clinical trial conducted by Szaniszlo et al. (2009) [80]. Patients over 50 received 100 μg of colostrinin or placebo for 15 weeks. The results showed an enhancement in cognitive and daily function of AD patients treated with colostrinin. Thus, this compound can be a suitable approach for AD therapy.

3. Preclinical In Vivo Studies of Natural Compounds and Their Effects on AD

Besides the natural compounds that have been studied in clinical trials, several other products have proved to have a potential beneficial effect in AD therapy in a preclinical stage, namely in in vivo studies. The preclinical phase involving in vivo studies is conducted to assess if the new compounds are safe and effective, before they can proceed to the clinical trials phase. A detailed report of animal studies results is described below. The natural compounds were divided into two groups: bioactive compounds and natural extracts and organized by the number of mechanisms associated with AD therapy, from the highest to the least.

3.1. Bioactive Compounds

Epigallocatechin gallate (EGCG) is a polyphenol found in green tea with several neuroprotective effects in AD. In vivo evidence suggests that EGCG decreased β- and γ-secretase actions and enhanced the α-secretase activity, leading to the decrease of Aβ levels improving the memory [164]. Besides that, EGCG inhibited the in vitro Aβ aggregation [165] and the in vivo Aβ oligomerization [166]. Moreover, EGCG inhibited the in vitro tau aggregation [167] and increased the in vivo clearance of phosphorylated tau [168]. Lastly, EGCG has been reported in in vivo experiments to demonstrate antioxidant [169] and anti-inflammatory actions [170].

Retinoic acid is a terpenoid and a metabolite of vitamin A. In vitro studies revealed that retinoic acid inhibited Aβ fibrils formation and their extension and destabilized Aβ fibrils [171]. In vitro evidence demonstrated that retinoic acid decreases the Aβ levels by inhibiting β- [172] and γ-secretase [173] and increasing α-secretase activity [172]. An in vivo study reported the ability of retinoic acid reducing brain Aβ deposition, APP phosphorylation and tau phosphorylation. This work also proved the anti-inflammatory activity of this compound, improving the learning and memory of AD mice model [174].

Caffeine is perhaps the most consumed psychoactive compound. It is present in the coffee bean, but it can be also found in some teas, cocoa drinks, candy bars, among other herbs. In vivo studies suggest that caffeine reduced the β-secretase and γ-secretase levels, decreasing the Aβ production [175]. An in vitro study showed that the inhibition of the β-sheets conformation can be related with the ability of caffeine to reduce Aβ levels [176]. Also, it was observed in vivo that this natural product promotes Aβ clearance [177]. In vivo evidence suggested that caffeine have anti-inflammatory and antioxidant properties [178]. In vivo studies demonstrated that the improvement observed in the memory could result from hippocampal tau phosphorylation reduction [179].

Baicalein is a naturally occurring flavonoid from the roots of *Scutellaria baicalensis* Georgi (Lamiaceae). In vitro studies suggested that baicalein inhibits the ROS production, reducing the oxidative stress [180]. In vitro results proved that baicalein inhibits Aβ fibrillation and oligomerisation and disaggregates Aβ fibrils [181]. In vivo studies proved that baicalein is able to increase the α-secretase and decrease the β-secretase activities, reducing the Aβ production [182,183]. Also, the tau phosphorylation in AD model mice was prevented and the cognitive function improved [183].

Berberine is an isoquinoline alkaloid found in rhizoma coptidis, an herb frequently used in Chinese herbal medicine. In vivo evidence suggests that berberine inhibited the β-secretase expression, reducing the Aβ production. Also, berberine stimulated the Aβ clearance and inhibited the Aβ plaque deposition and hyperphosphorylation of APP and tau [184]. Berberine has been also described as having in vivo anti-inflammatory and antioxidative activities [185].

Kaempferol is a polyphenolic flavonoid found in fruits, vegetables and herbs. In vivo studies proved its antioxidant effect, improving the learning and memory of a transgenic drosophila AD model [186]. Also, in vitro evidence showed that kaempferol has anti-inflammatory activity [187], inhibits Aβ aggregation [188] and destabilizes Aβ fibrils [189]. Also, another in vitro study proved that kaempferol inhibits the β-secretase activity [190].

Quercetin is a flavonol, naturally occurring polyphenolic compounds present in fruits, vegetables and herbs. In vivo studies showed that quercetin improved the memory and cognitive impairments of an AD model and reduced the oxidative stress [191]. Moreover, in vitro evidence suggested that quercetin prevents the Aβ aggregation [192], inhibits the Aβ fibrils formation and destabilizes Aβ fibrils [193], decreasing the Aβ levels in brain of AD model mice [194]. Additionally, this compound was reported in in vivo studies as inhibitor of β-secretase and taupathy [195].

Fisetin is a flavonoid extracted from *Rhus succedanea* L. (Anacardiaceae) and also found in some fruits and vegetables. Fisetin proved to inhibit Aβ aggregation in vivo [196] and fibril formation in vitro [188], reducing the in vivo Aβ accumulation [197]. Also, an in vivo experiment described fisetin as a β-secretase inhibitor and anti-inflammatory product [197]. Additionally, fisetin promotes the in vitro degradation of phosphorylated tau [198] and reduced the in vivo tau hyperphosphorylation [197].

Oleuropein is a polyphenol present in extra virgin olive oil with antioxidant [199] and anti-inflammatory properties in vivo [200]. The Aβ levels and amyloid plaque load were reduced in vivo, resulting in an amelioration of cognitive functions [201]. Also, the compound inhibited the Aβ aggregation in vivo [200], favouring the formation of non-toxic aggregates in vitro [202]. Additionally, in vitro evidence suggested that oleuropein decreased the Aβ oligomers levels through the promotion of α-secretase activity [203]. Lastly, oleuropein was described as tau aggregation inhibitor in vitro [204].

Tannic acid is a polyphenol found in herbs and fruits. An in vivo experiment showed that tannic acid is a natural inhibitor of β-secretase with anti-inflammatory properties, preventing the cognitive impairment of AD mice [205]. One in vitro study affirmed that tannic acid inhibits Aβ formation associated with less amyloidogenic APP proteolysis, inhibits Aβ fibrils formation as their extension and still destabilizes Aβ fibrils [206]. Another in vitro study demonstrated that tannic acid inhibits the tau aggregation [207].

Crocin is a carotenoid mainly found in the stigma of saffron flower. In vitro experiments showed that crocin inhibits the Aβ fibril formation [208] through the inhibition of the Aβ fibrillogenesis [145]. Also, in vitro evidence suggests that crocin reduces the number of fibrils as well as their length [208]. An in vitro study confirmed that crocin can also disrupt Aβ aggregates [209]. Also, the therapeutic effects of crocin can be linked to its antioxidant [210] and anti-inflammatory activities [211] observed in in vivo studies.

Epicatechin represents one of the antioxidants from the flavonoids family. High amounts of this compound can be found in cocoa beans, green tea and grapes. In vivo data showed that epicatechin has antioxidant [212] and anti-inflammatory activities [213]. Further, in vitro studies suggest that epicatechin is an inhibitor of β-secretase [214]. As result, epicatechin decreased the Aβ levels in an AD mice model [212]. Also, epicatechin has the in vitro ability to inhibit tau aggregation [215] and fibril formation changing the secondary structure [216].

Gallic acid is a phenolic acid present in fruits, vegetables and herbs. Gallic acid proved to have antioxidant [217] and anti-inflammatory activities, improving the learning and memory in vivo [218]. Also, gallic acid can reduce the in vitro Aβ aggregation by the inhibition of conformational transition to β-sheet [219]. An in vivo experiment observed a reduction in Aβ levels after gallic acid administration due to the increase of α-secretase action, promoting the non-amyloidogenic route and consequently the decreases the Aβ oligomerization [220].

Ferulic acid is a phenolic compound naturally present in numerous fruits and vegetables. In vivo results revealed that ferulic acid is an antioxidant [221] and anti-inflammatory compound [222]. Also, it can reduce the in vivo Aβ production by reducing the β-secretase activity [222]. The decrease of β-sheets structures was also observed in an in vitro experiment, inhibiting the Aβ aggregation [223]. Additionally, ferulic acid decreased the Aβ deposition and improved the cognitive performance of an AD mouse model [224]. Also, ferulic acid decreased the Aβ fibrils levels in vitro [225].

Rutin is a bioflavonoid extracted from some vegetables and fruits. This product is a glycoside of the flavonoid quercetin with antioxidant and anti-inflammatory properties in vivo [226]. The same in vivo study showed that this compound inhibited the Aβ aggregation [226]. Also, rutin decreased the Aβ fibrils formation in vitro [193]. This can be due to its ability to inhibit the β-secretase activity in vitro [193]. Also, rutin disaggregated Aβ fibrils in vitro [193].

Salvianolic acid B is a phenylpropanol founded in the *Salvia miltiorrhiza* Bunge (Lamiaceae) root. In vivo experiments showed a strong antioxidant and anti-inflammatory activities, improving the memory and learning of an AD mouse model [227]. Also, salvianolic acid B inhibited the Aβ aggregation and disaggregated preformed Aβ fibrils in vitro [228]. Another in vitro work suggested that salvianolic acid B inhibits the β-secretase which leads to the inhibition of Aβ production [229].

Myricetin is a flavonoid extracted from several fruits, vegetables and herbs. In vitro proofs showed that myricetin prevents Aβ aggregation and consequent fibrillation [189,230] due to its capacity to inhibit β-secretase and increase the α-secretase activity [231]. Also, myricetin blocked the structural changes on Aβ in vitro, inducing a reduction in Aβ levels [231]. Also, the disaggregation of Aβ fibrils was observed in vitro [189]. As result, an in vivo study showed that myricetin enhanced the learning and memory impairments in an AD rat model [232].

Naringenin is a natural compound present in citrus fruits and tomatoes. It is the major flavanone constituent found in *Citrus junos* Siebold ex Tanaka, Rutaceae. An in vitro study revealed that naringenin inhibited the APP and β-secretase activity and reduced the levels of phosphorylated tau [233]. As result, brain levels of Aβ were reduced in vivo [234]. In vivo evidence also proved the antioxidant [235] and anti-inflammatory activities of the compound, improving motor coordination, learning and memory of AD rats [236].

Luteolin, a polyphenol flavonoid found in fruits, vegetables and herbs, exhibits potent anti-inflammatory activity in vitro [237] and antioxidant activity against induced-oxidative stress in a in vivo AD model [238], ameliorating the spatial learning and memory impairment [239]. An in vitro study also proved that this compound is a potent inhibitor of β-secretase [240]. Another in vitro study demonstrated that luteolin is able to reduce tau hyperphosphorylation [241].

Asiatic acid is a pentacyclic triterpene found in plants. Asiatic acid demonstrates an ability to inhibit the β-secretase and increase the α-secretase activity in vitro. Also, it demonstrates an ability to activate Aβ clearance [242], which explains the substantial reduction in Aβ levels in AD mice [243]. Numerous in vivo works suggest that asiatic acid has antioxidant properties, clearing free radicals and decreasing lipid peroxidation, improving the learning and memory [244].

Puerarin is an isoflavanone glycoside isolated from *Pueraria lobata* (Willd.) Ohwi (Leguminosae) used to treat some diseases. In vivo studies found that puerarin inhibited the tau phosphorylation and reduced Aβ levels, ameliorating the spatial learning and memory in an AD mice model [245]. The beneficial effects of puerarin were suggested in in vivo experiments to be connected to its ability to reduce the oxidative stress [246] and neuroinflammation [247].

Oleocanthal is one of the main active components of extra virgin olive oil. In vitro evidence suggests that this compound changes the structure of tau protein, inhibiting its aggregation [248] and fibrillization [249]. In vivo results proved that oleocanthal enhances the Aβ clearance, reducing the amyloid load. Also, the anti-inflammatory activity of the compound was verified [250].

Viniferin (trans ε-viniferin) is a polyphenol present in a variety of vines, including *Vitis vinifera* L., Vitaceae. In vitro evidence proved the anti-inflammatory [251] and antioxidant [252] activities of the compound. Also, viniferin disaggregated Aβ [251] and inhibited the Aβ aggregation, reducing the fibril formation [253].

Scyllo-inositol, also known as scyllo-cyclohexanehexol, is one of the stereoisomers of inositol, found in dogwood *Cornus florida* L. (Cornaceae) and coconut palm *Cocos nucifera* L. (Arecaceae). An in vivo study showed that this compound decreases the Aβ levels and inhibits the Aβ aggregation, improving the memory of AD rat model [254]. In vitro evidence demonstrated that scyllo-inositol induces structural modifications in Aβ, stabilizes Aβ oligomers and inhibits fibril formation [255].

Honokiol is a poly-phenolic product found in *Magnolia officinalis* Rehder & E.H.Wilson, Magnoliaceae. In vivo evidence suggested that honokiol is an antioxidant [256] and anti-inflammatory compound [257]. In vivo studies revealed that honokiol inhibits the β-secretase activity, reducing the Aβ production and senile plaque deposition. Also, the Aβ degradation was enhanced by honokiol [257]. As result, honokiol decreased Aβ-induced hippocampal neuronal apoptosis, improving learning and memory of AD mice model [256].

Apigenin is a flavonoid found in plants, fruits and vegetables. Numerous in vitro and in vivo works showed its anti-inflammatory [258] and antioxidant [259] properties, respectively. An in vivo experiment proved that apigenin changes APP processing by the β-secretase inhibition preventing the Aβ deposition and consequently, improving the memory impairments [259].

Caffeic acid is a phenolic acid present in food, beverages and Chinese herbal medicines with antioxidant and anti-inflammatory properties in vivo. This compound improved the learning of AD rat models [260]. In vitro studies showed that caffeic acid reduced the tau phosphorylation and protected the PC12 cells against Aβ-induced toxicity [261].

β-carotene belongs to the carotenoid family. One in vitro study reported that β-carotene has an anti-aggregation activity and destabilizes Aβ [171]. Another in vivo study demonstrated the β-carotene has the ability to reduce oxidative stress, by reducing the ROS production [262].

Rosmarinic acid is a phenolic carboxylic acid found in rosemary, lemon balm and peppermint, among others. An in vivo study proved that this compound has antioxidant properties, protecting an AD mouse model against memory deficits [263]. Also, rosmarinic acid inhibited the tau hyperphosphorylation [264] and fibrillization in vitro [265].

Nordihydroguaiaretic acid (NDGA) is a compound found in *Larrea divaricata* Cav. (Zygophyllaceae) with in vivo antioxidant properties [266]. An in vitro study reported that NDGA inhibits the Aβ fibrils formation, reducing the number of fibrils and small amorphous aggregates. Additionally, this compound disrupts Aβ fibrils [267].

Osthole is a coumarin isolated from *Cnidium monnieri* (L.) Cusson (Apiaceae). An in vivo study showed that this compound significantly enhanced the memory of an AD rat model, that can be linked to its antioxidant activity [268] and with a reduction of Aβ levels found in the brain. This reduction can be due to the inhibition of β-secretase in vitro [269]. Also, in vitro evidence suggests that this product decreases the phosphorylated tau levels [270].

Ellagic acid is a polyphenol extracted from *Punica granatum* L. (Lythraceae). An in vitro study proved that this compound inhibits of β-secretase activity preventing neurotoxicity [271]. Ellagic acid has antioxidant and anti-inflammatory properties, that improve learning and memory injuries in AD rat model [272].

Glycine betaine is an organic osmolyte, which could be isolated from vegetables and marine products. In vivo evidence revealed that glycine betaine reduces tau hyperphosphorylation and Aβ production, improving memory deficits [273]. Also, glycine betaine inhibited the β-secretase activity and activated the α-secretase activity in vitro, thereby inhibiting the Aβ production [274].

Hydroxytyrosol is a phenolic compound extracted from the olive leaf and oil. In vivo studies demonstrated that it is a compound with antioxidant and anti-inflammatory properties [275]. Also, hydroxytyrol showed to reduce the levels of Aβ plaques in an AD mice model [276].

L-theanine is an amino acid present in green tea. An in vivo work showed that L-theanine decreased the oxidative stress and the Aβ levels [277]. Also, this natural product proved to inhibit tau hyperphosphorylation in vitro [278].

13-Desmethyl spirolide C is a marine compound belonging to the cyclic imine group produced by the dinoflagellate *Alexandrium ostenfeldii* and accumulate in shellfish. An in vitro study revealed that 13-desmethyl spirolide C is a spirolide that can reduce intracellular Aβ accumulation and hyperphosphorylated tau levels [279]. The reduction of intracellular Aβ levels was also observed in an in vivo study [280].

Gossypin is a flavonoid found in *Hibiscus vitifolius* L. (Malvaceae) and has been reported in in vivo experiments to exhibit anti-inflammatory [281] and antioxidant actions [282].

Gypenosides are triterpenoid saponins extracted from *Gynostemma pentaphyllum* (Thunb.) Makino (Cucurbitaceae) and they are reported in an in vivo study to be products with antioxidant and anti-inflammatory activities, improving the cognitive impairment [283].

1,2,3,4,6-Penta-*O*-galloyl-β-D-glucopyranose (PGG) is a polyphenol and the main constituent of the *Paeonia x suffruticosa Andrews* (Paeoniaceae) root, a tree peony native to China and used in traditional medicine practices. In vivo experiments proved that PGG inhibits the Aβ oligomerization, which prevents Aβ fibril formation, resulting in the decrease of Aβ levels and improvement of memory. PGG is also able to promote the destabilization of Aβ fibrils [284].

Enoxaparin is a low molecular weight heparin present in the intestinal mucosa of pigs. Enoxaparin reduced the Aβ load through the decreasing of β-secretase activity [285]. Also, enoxaparin has anti-inflammatory activity in vivo [286], improving the cognition of an AD mice model [287].

Morin, a natural flavonoid mainly found in *Maclura pomifera* (Raf.) C. K. Schneid. (Moraceae), *Maclura tinctoria* (L.) D. Don ex Steud. (Moraceae) and leaves of *Psidium guajava* L. (Myrtaceae), promoted the inhibition of β-secretase activity in vitro [190]. Besides, morin is able to reduce tau hyperphosphorylation in vivo [288].

Naringin is a flavonoid present in citrus fruits, namely in grapefruit. In vivo studies suggested that the antioxidant and anti-inflammatory activities of this compound improved the learning and memory of AD rats [289].

Vanillic acid is a phenolic acid extracted from the plant *Angelica sinensis* (Oliv.) Diels (apiaceae) with antioxidant and anti-inflammatory activities in vivo. As a result, an improvement in learning and memory of AD rats was observed [290].

Punicalagin is an ellagitannin found in the fruit peel of pomegranate (*Punica granatum* L. (Lythraceae)). In vivo studies suggest that punicalagin has potential as a nutritional preventive strategy in AD due to its anti-inflammatory activity. This natural product favors the anti-amylogenic route through the inhibition of β-secretase, reducing Aβ levels [291].

Piperine is a nitrogenous alkaloid found in fruits of the family *piperaceae*, including in *piper nigrum* L. and *piper longum* L. This compound has been used in traditional medicine to cure several diseases. In vivo trials reported that the reduction of lipid peroxidation can be linked with the neuroprotective effects of this compound [292], resulting in a significant improvement in memory of AD rat model [293].

Rhodosin is a flavonol extracted from the root of *Sedum roseum* (L.) Scop. (Crassulaceae) that improved the learning and memory injuries in an AD rat model due to its antioxidant activity [294].

3.2. Natural Extracts and Other Natural Products

Garlic (*Allium sativum* L., Amaryllidaceae) is frequently used in culinary and medicine. Several studies showed that the administration of aged garlic extract significantly improves the memory deficit by several pathways. In vitro studies demonstrated that aged garlic extract has antioxidant properties [295], inhibits Aβ fibril formation through the inhibition of Aβ aggregation [296] and it is able to defibrillate Aβ fibrils [296]. In addition, in vivo evidence showed that aged garlic extract has anti-inflammatory properties [297], increases the α-secretase activity and inhibits tau hyperphosphorylation [298].

Cinnamon (*Cinnamomum verum* J. Presl., Lauraceae) is one of the most used spices and has been traditionally applied in the treatment of some diseases and their symptoms. Cinnamon extract is found to inhibit in vitro tau aggregation and promote the disassembly of tau filaments [215]. Other in vitro studies suggested that the potential therapeutic effect of cinnamon against AD can also be due to its anti-inflammatory activity [299]. In vivo evidence showed that cinnamon extract has antioxidant activity [300], prevents Aβ oligomerization [301], reducing the Aβ level and correcting the cognitive impairment of transgenic mice [300].

Olive (*Olea europaea* L., *Oleaceae*) is the source of olive oil, one of the most important ingredients in the Mediterranean diet. In vivo studies showed that extra virgin olive oil ameliorated behavioural impairments. Also, the oil reduced the Aβ and phosphorylated tau levels [302]. This decrease can be due to the increase of Aβ clearance and APP modulation [303]. In vivo studies also proved its antioxidant activity, protecting against Aβ-induced cytotoxicity [304].

Walnut (*Juglans regia* L., Juglandaceae) is a dried fruit composed by fatty acids, vitamins, alpha tocopherol, and polyphenols, in particular ellagic acid. An in vitro study showed that walnut extract inhibited the Aβ fibril formation through the inhibition of Aβ fibrillation, and also defibrillated Aβ fibrils [305]. Additionally, in vivo studies demonstrated that walnut extract reduced the oxidative stress and neuroinflammation induced by Aβ in an AD mice model [306].

Grapes (*Vitis vinifera* L., Vitaceae) are composed by several polyphenols including catechin, epicatechin, epigallocatechin and epicatechin gallate. In vivo studies have revealed that grape seed extract increases the memory performance and reduces ROS production, thereby protecting the central nervous system [307]. An in vitro work revealed that grape seed extract blocks the Aβ fibril formation [308] through the inhibition of Aβ aggregation [309]. Therefore, the amount of amyloid plaques in the brain of AD mice was reduced. Besides, grape seed extract can attenuate the neuroinflammation in vivo [310]. In vivo works proved that the grape skin extract has antioxidant property [311] and inhibits the in vitro Aβ fibril formation [121,312].

Pomegranate (*Punica granatum* L., Lythraceae) is a fruit with a variety of antioxidant polyphenols. Pomegranate juice reduced the Aβ levels and amyloid plaques in an AD mouse model, improving spatial learning and cognitive performance [313]. Further in vivo analysis revealed that these results could be the product of the inhibition of γ-secretase activity [314]. In addition, in vivo studies demonstrated that pomegranate has anti-inflammatory [315] and antioxidant activities [316].

Skullcap (*Scutellaria baicalensis* Georgi, Lamiaceae) is a native American plant commonly used in traditional Chinese medicine. An in vivo study found that skullcap was able to protect hippocampal neurons against Aβ-induced damage through the attenuation of oxidative stress and neuroinflammation [317].

Strawberry (*Fragaria x ananassa* (Weston) Duchesne, Rosaceae) is known to contain high phenolic contents. In vivo studies showed that strawberries have anti-inflammatory [318] and antioxidant activities, protecting against oxidative stress [319].

Moringa (*Moringa oleifera* Lam., Moringaceae), an Asian and African plant, presents several nutrients, including β-carotene, vitamin C and E and phenols, including quercetin and kaempferol. In vivo studies showed that this plant improved the memory and learning due to its antioxidant activity [320].

4. Preclinical In Vitro Studies of Natural Compounds and Their Effects on AD

Besides the aforementioned natural compounds studied in human and animal studies, several other products have gained an increasing interest in scientific community for AD therapy. In fact, different compounds were tested in vitro and showed promising results. Some compounds proved to be efficient in preventing the formation of Aβ aggregates and disassembling Aβ fibrils, such as the case of tetracycline [321], methyl caffeate [322], retinol [171] and gou teng [323]. Also, other products demonstrated to be able to promote Aβ clearance, including withanolide A [242] and retinal [171].

The reduction of Aβ levels can occur through changes in the structure of Aβ aggregates induced by natural compounds such as piceatannol [324]. This product is also able to decrease Aβ levels through the activation of α-secretase. Withanolide A also promotes α-secretase expression and simultaneously inhibits β-secretase activity [242]. Other products proved to be inhibitors of β-secretase activity such as bastadin 9 [325], dictyodendrin [326], epicatechin gallate [327], gracilin [328], ianthellidone F [329], lamellarin O [329], neocoylin [330], tasiamide B [331], topsentinol K trisulfate [332] and xestosaprol [333].

Besides these mechanisms, natural compounds can prevent AD progression by other mechanisms. For example, yessotoxin [334], gambierol [335], gracilin [328], gymnodimine [336], palinurin [337] and schisandrone [338] reduced tau hyperphosphorylation. In addition, some compounds revealed to be able to suppress the oxidative stress by the scavenging of ROS and inflammatory response induced by Aβ, such as schisandrone [294], piceatannol [339], gracilin [340], sophocarpidine [294] and tetrahydroaplysulphurin-1 [340].

Despite the verified good outcomes, the study of some of these compounds was abandoned. For example, tetracycline was studied in 2001 but no more studies were reported for this compound. Also, for epicatechin gallate no studies were reported since 2003, and for retinal and retinol since 2004.

5. Discussion

Several bioactive compounds and natural extracts that were described herein to treat and prevent AD were revised and discussed. Until this date, most of the studied natural compounds are mainly derived from vegetable sources, with just a few molecules isolated from animals and marine organisms. Since AD is a multifactorial disorder, different therapeutic mechanisms were associated with these natural compounds.

The approval process for a new compound to become clinically available is an extremely lengthy process, and it is divided into different phases. Before tests on humans, new compounds must be evaluated in preclinical studies. Several natural compounds proved to be promising for AD therapy in in vitro and in vivo studies, as discussed in this work. However, due to physiological differences between tested animals and humans, clinical trials are still necessary to validate the safety and efficacy of these compounds. Clinical studies are of outmost importance for the development of new therapeutic compounds, drugs and devices. Human studies allow to assess safety, tolerance and effective therapeutic doses for treating diseases. Some of the performed clinical trials described in this review did not show significant improvement in the delay or treatment of the symptoms. However, even if the trials do not exhibit positive outcomes, the obtained results can be still used to guide the scientists in the right path for drug discovery. Also, some of the conducted clinical trials with natural compounds for AD therapy, showed no conclusive results due to the limited size of samples. However, several compounds proved to be safe in human studies and were allowed to proceed to subsequent phases. To this date, homotaurine is the only compound that reached phase III of clinical trials for AD therapy.

Despite only a few natural products having been studied in clinical trials, numerous compounds proved to have beneficial properties in preclinical studies, as shown in Figure 11. Based on the works mentioned in this review, 21% of natural compounds achieved the clinical trials phase. However, it needs to be taken into account that since these types of products are commonly consumed in the daily life, it is easier to reach the phase I of clinical trials as they are supposed to be safe for humans. Unfortunately, not all these natural products demonstrated significant effects in the AD treatment. However, they could be used for AD prevention. In the next few years, it is expected that the number of natural compounds being studied in clinical trials for the prevention and treatment of AD will significantly increase. Since the enrichment of several food and beverages is a recent trend, fortification strategies using natural products could be a promising approach for AD prevention. In fact, some groups have studied the combination of different natural compounds. In 2009, a group started clinical trials for a beverage with supplementation of a mixture of natural compounds to be consumed by AD patients [341]. This supplement, commercially called Souvenaid®, demonstrates beneficial effects in the patients. This product is already commercially available in some countries being partially financially supported by the public health care systems.

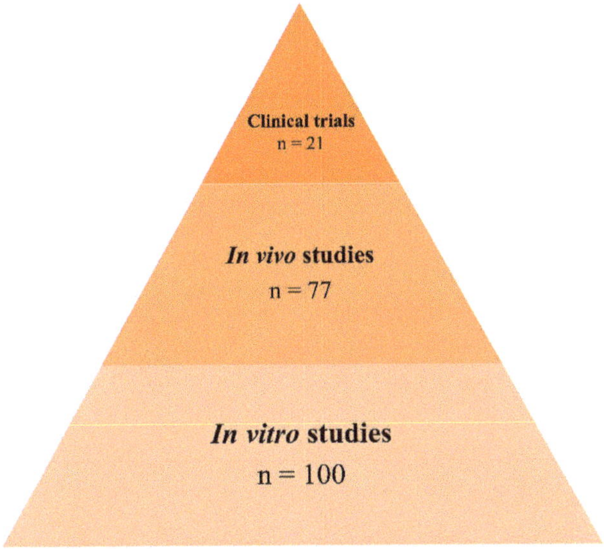

Figure 11. Number of natural products studied in different development phases.

Still, the neuroprotective effects of natural compounds depend of their ability to cross BBB. The low bioavailability of drugs and the difficulty to cross the BBB remains the major obstacles for the development of new therapies [342]. Drug delivery systems (DDS) targeting the brain seem to be a promising strategy to increase the bioavailability of compounds and the transport across the BBB [343]. DDS can protect the natural compounds from biological degradation and transport the molecules to the brain by masking their limiting physicochemical properties [344]. Thus, low doses of natural compounds are slowly released in the brain, increasing the efficiency of the therapeutic effects.

Among the studied natural compounds, only a small percentage have been encapsulated in DDS for brain targeting. Only the encapsulation of curcumin [345–348], epigallocatechin gallate [349,350], grape extracts [312], huperzine A [351], piperine [352], quercetin [353] and resveratrol [312] in functionalized DDS was reported in the literature. Therefore, some of these compounds seem to be the most promising for the AD treatment. One interesting approach could be the co-encapsulation in the same DDS of more than one natural compound with different therapeutic mechanisms, obtaining a synergistic effect. In the future, in addition to being necessary further studies to understand how natural compounds exert their therapeutic effects on AD, further experiments to target the drugs to the brain need to be performed.

6. Conclusions

AD is a disabling disorder with a major negative impact on our current society. At this moment, no drugs have been developed to prevent or treat AD. The existing molecules only aim to control the symptoms. With the increase of average life expectancy, it is fundamental to discover and develop new molecules able to prevent and treat AD. Several natural products have proven to be promising for AD therapy in clinical and preclinical studies. Clinical trials have shown that several compounds appear to be effective for AD therapy, whereas others have failed in human trials. Natural compounds in earlier phases of research need further studies to uncover their therapeutic potential for AD.

Author Contributions: S.A. did the bibliographic research and wrote the manuscript. M.J.R., J.A.L. and M.d.C.P. contributed with the paper organization, writing and discussion.

Funding: This work was financially supported by: project UID/EQU/00511/2019—Laboratory for Process Engineering, Environment, Biotechnology and Energy—LEPABE funded by national funds through FCT/MCTES (PIDDAC); Project POCI-01-0145-FEDER-006939, funded by FEDER funds through COMPETE2020—Programa Operacional Competitividade e Internacionalização (POCI) and by national funds (PIDDAC) through FCT/MCTES; Project "LEPABE-2-ECO-INNOVATION"—NORTE-01-0145-FEDER-000005, funded by Norte Portugal Regional Operational Program (NORTE 2020), under PORTUGAL 2020 Partnership Agreement, through the European Regional Development Fund (ERDF), and FCT doctoral grant—SFRH/BD/129312/2017.

Conflicts of Interest: The authors declare no conflict of interest. The funders had no role in the design of the study; in the collection, analyses, or interpretation of data; in the writing of the manuscript, or in the decision to publish the results.

Abbreviations

Aβ	β-amyloid
ACH	Amyloid cascade hypothesis
AD	Alzheimer's disease
APP	Amyloid precursor protein
BBB	Blood-brain barrier
DDS	Drug delivery systems
DHA	Docosahexanoic acid
EGCG	Epigallocatechin gallate
NDGA	Nordihydroguaiaretic acid
NFTs	Neurofibrillary tangles
PGG	1,2,3,4,6-Penta-*O*-galloyl-β-D-glucopyranose
ROS	Reactive oxygen species

References

1. Reitz, C.; Mayeux, R. Alzheimer disease: Epidemiology, diagnostic criteria, risk factors and biomarkers. *Biochem. Pharmacol.* **2014**, *88*, 640–651. [CrossRef]
2. Kumar, A.; Singh, A.; Ekavali. A review on Alzheimer's disease pathophysiology and its management: An update. *Pharmacol. Rep.* **2015**, *67*, 195–203. [CrossRef]

3. Solomon, A.; Mangialasche, F.; Richard, E.; Andrieu, S.; Bennett, D.A.; Breteler, M.; Fratiglioni, L.; Hooshmand, B.; Khachaturian, A.S.; Schneider, L.S.; et al. Advances in the prevention of Alzheimer's disease and dementia. *J. Int. Med.* **2014**, *275*, 229–250. [CrossRef]
4. Selkoe, D.J.; Hardy, J. The amyloid hypothesis of Alzheimer's disease at 25 years. *EMBO Mol. Med.* **2016**, *8*, 595–608. [CrossRef]
5. Dawkins, E.; Small, D.H. Insights into the physiological function of the β-amyloid precursor protein: Beyond Alzheimer's disease. *J. Neurochem.* **2014**, *129*, 756–769. [CrossRef] [PubMed]
6. Zhang, Z.; Song, M.; Liu, X.; Su Kang, S.; Duong, D.M.; Seyfried, N.T.; Cao, X.; Cheng, L.; Sun, Y.E.; Ping Yu, S.; et al. Delta-secretase cleaves amyloid precursor protein and regulates the pathogenesis in Alzheimer's disease. *Nat. Commun.* **2015**, *6*, 8762. [CrossRef]
7. Mohamed, T.; Shakeri, A.; Rao, P.P.N. Amyloid cascade in Alzheimer's disease: Recent advances in medicinal chemistry. *Eur. J. Med. Chem.* **2016**, *113*, 258–272. [CrossRef] [PubMed]
8. Sengupta, U.; Nilson, A.N.; Kayed, R. The Role of Amyloid-β Oligomers in Toxicity, Propagation, and Immunotherapy. *EBioMedicine* **2016**, *6*, 42–49. [CrossRef] [PubMed]
9. Verma, M.; Vats, A.; Taneja, V. Toxic species in amyloid disorders: Oligomers or mature fibrils. *Ann. Indian Acad. Neurol.* **2015**, *18*, 138–145.
10. Swerdlow, R.H.; Burns, J.M.; Khan, S.M. The Alzheimer's disease mitochondrial cascade hypothesis: Progress and perspectives. *Biochim. Biophys. Acta (BBA) Mol. Basis Dis.* **2014**, *1842*, 1219–1231. [CrossRef]
11. Barage, S.H.; Sonawane, K.D. Amyloid cascade hypothesis: Pathogenesis and therapeutic strategies in Alzheimer's disease. *Neuropeptides* **2015**, *52*, 1–18. [CrossRef] [PubMed]
12. Heneka, M.T.; Carson, M.J.; Khoury, J.E.; Landreth, G.E.; Brosseron, F.; Feinstein, D.L.; Jacobs, A.H.; Wyss-Coray, T.; Vitorica, J.; Ransohoff, R.M.; et al. Neuroinflammation in Alzheimer's disease. *Lancet Neurol.* **2015**, *14*, 388–405. [CrossRef]
13. Huang, W.J.; Zhang, X.; Chen, W.W. Role of oxidative stress in Alzheimer's disease. *Biomed. Rep.* **2016**, *4*, 519–522. [CrossRef] [PubMed]
14. Karch, C.M.; Cruchaga, C.; Goate, Alison M. Alzheimer's Disease Genetics: From the Bench to the Clinic. *Neuron* **2014**, *83*, 11–26. [CrossRef]
15. Chin-Chan, M.; Navarro-Yepes, J.; Quintanilla-Vega, B. Environmental pollutants as risk factors for neurodegenerative disorders: Alzheimer and Parkinson diseases. *Front. Cell. Neurosci.* **2015**, *9*, 124. [CrossRef]
16. Karch, C.M.; Goate, A.M. Alzheimer's Disease Risk Genes and Mechanisms of Disease Pathogenesis. *Biol. Psychiatry* **2015**, *77*, 43–51. [CrossRef] [PubMed]
17. Barbara, G.; Joana, A.L.; Manuel, A.N.C.; Maria do Carmo, P. The Potential Effect of Fluorinated Compounds in the Treatment of Alzheimer's Disease. *Curr. Pharm. Des.* **2015**, *21*, 5725–5735.
18. Loureiro, J.A.; Crespo, R.; Börner, H.; Martins, P.M.; Rocha, F.A.; Coelho, M.; Pereira, M.C.; Rocha, S. Fluorinated beta-sheet breaker peptides. *J. Mater. Chem. B* **2014**, *2*, 2259–2264. [CrossRef]
19. Folch, J.; Petrov, D.; Ettcheto, M.; Abad, S.; Sánchez-López, E.; García, M.L.; Olloquequi, J.; Beas-Zarate, C.; Auladell, C.; Camins, A. Current Research Therapeutic Strategies for Alzheimer's Disease Treatment. *Neural Plast.* **2016**, *2016*, 15. [CrossRef] [PubMed]
20. Loureiro, J.A.; Gomes, B.; Fricker, G.; Coelho, M.A.N.; Rocha, S.; Pereira, M.C. Cellular uptake of PLGA nanoparticles targeted with anti-amyloid and anti-transferrin receptor antibodies for Alzheimer's disease treatment. *Colloids Surf. B Biointerfaces* **2016**, *145*, 8–13. [CrossRef] [PubMed]
21. Loureiro, J.A.; Gomes, B.; Fricker, G.; Cardoso, I.; Ribeiro, C.A.; Gaiteiro, C.; Coelho, M.A.N.; Pereira, M.D.C.; Rocha, S. Dual ligand immunoliposomes for drug delivery to the brain. *Colloids Surf. B Biointerfaces* **2015**, *134*, 213–219. [CrossRef]
22. Loureiro, J.A.; Gomes, B.; Coelho, M.A.; do Carmo Pereira, M.; Rocha, S. Targeting nanoparticles across the blood-brain barrier with monoclonal antibodies. *Nanomedicine (Lond.)* **2014**, *9*, 709–722. [CrossRef]
23. Silva, T.; Reis, J.; Teixeira, J.; Borges, F. Alzheimer's disease, enzyme targets and drug discovery struggles: From natural products to drug prototypes. *Ageing Res. Rev.* **2014**, *15*, 116–145. [CrossRef]
24. Dey, A.; Bhattacharya, R.; Mukherjee, A.; Pandey, D.K. Natural products against Alzheimer's disease: Pharmaco-therapeutics and biotechnological interventions. *Biotechnol. Adv.* **2017**, *35*, 178–216. [CrossRef]
25. David, B.; Wolfender, J.-L.; Dias, D.A. The pharmaceutical industry and natural products: Historical status and new trends. *Phytochem. Rev.* **2015**, *14*, 299–315. [CrossRef]

26. Asha, H. A Review: Natural Compounds as Anti-Alzheimer´s Disease Agents. *Curr. Nutr. Food Sci.* **2017**, *13*, 247–254.
27. Rasool, M.; Malik, A.; Qureshi, M.S.; Manan, A.; Pushparaj, P.N.; Asif, M.; Qazi, M.H.; Qazi, A.M.; Kamal, M.A.; Gan, S.H.; et al. Recent Updates in the Treatment of Neurodegenerative Disorders Using Natural Compounds. *Evid. Based Complement. Altern. Med.* **2014**, *2014*, 979730. [CrossRef]
28. Andrade, S.; Ramalho, M.J.; Loureiro, J.A.; Pereira, M.C. Interaction of natural compounds with biomembrane models: A biophysical approach for the Alzheimer's disease therapy. *Colloids Surf. B Biointerfaces* **2019**, *180*, 83–92. [CrossRef]
29. Butler, M.S.; Robertson, A.A.; Cooper, M.A. Natural product and natural product derived drugs in clinical trials. *Nat. Prod. Rep.* **2014**, *31*, 1612–1661. [CrossRef]
30. Doig, A.J.; Derreumaux, P. Inhibition of protein aggregation and amyloid formation by small molecules. *Curr. Opin. Struct. Biol.* **2015**, *30*, 50–56. [CrossRef]
31. Baranello, R.J.; Bharani, K.L.; Padmaraju, V.; Chopra, N.K.; Lahiri, D.H.; Greig, N.A.; Pappolla, M.; Sambamurti, K. Amyloid-Beta Protein Clearance and Degradation (ABCD) Pathways and their Role in Alzheimer's Disease. *Curr. Alzheimer Res.* **2015**, *12*, 32–46. [CrossRef]
32. Wischik, C.M.; Harrington, C.R.; Storey, J.M.D. Tau-aggregation inhibitor therapy for Alzheimer's disease. *Biochem. Pharmacol.* **2014**, *88*, 529–539. [CrossRef]
33. Morales, I.; Guzmán-Martínez, L.; Cerda-Troncoso, C.; Farías, G.A.; Maccioni, R.B. Neuroinflammation in the pathogenesis of Alzheimer's disease. A rational framework for the search of novel therapeutic approaches. *Front. Cell. Neurosci.* **2014**, *8*, 112. [CrossRef]
34. Ahmed, T.; Javed, S.; Javed, S.; Tariq, A.; Šamec, D.; Tejada, S.; Nabavi, S.F.; Braidy, N.; Nabavi, S.M. Resveratrol and Alzheimer's Disease: Mechanistic Insights. *Mol. Neurobiol.* **2017**, *54*, 2622–2635. [CrossRef]
35. Miller, B.J.; Whisner, C.M.; Johnston, C.S. Vitamin D Supplementation Appears to Increase Plasma Aβ 40 in Vitamin D Insufficient Older Adults: A Pilot Randomized Controlled Trial. *J. Alzheimer's Dis.* **2016**, *52*, 843–847. [CrossRef]
36. SanMartín, C.D.; Henriquez, M.; Chacon, C.; Ponce, D.P.; Salech, F.; Rogers, N.K.; Behrens, M.I. Vitamin D Increases Aβ140 Plasma Levels and Protects Lymphocytes from Oxidative Death in Mild Cognitive Impairment Patients. *Curr. Alzheimer Res.* **2018**, *15*, 561–569. [CrossRef]
37. Annweiler, C.; Herrmann, F.R.; Fantino, B.; Brugg, B.; Beauchet, O. Effectiveness of the Combination of Memantine Plus Vitamin D on Cognition in Patients With Alzheimer Disease: A Pre-Post Pilot Study. *Cogn. Behav. Neurol.* **2012**, *25*, 121–127. [CrossRef]
38. Galasko, D.R.; Peskind, E.; Clark, C.M.; et al. Antioxidants for alzheimer disease: A randomized clinical trial with cerebrospinal fluid biomarker measures. *Arch. Neurol.* **2012**, *69*, 836–841. [CrossRef]
39. Kontush, A.; Mann, U.; Arlt, S.; Ujeyl, A.; Lührs, C.; Müller-Thomsen, T.; Beisiegel, U. Influence of vitamin E and C supplementation on lipoprotein oxidation in patients with Alzheimer's disease. *Free Radic. Biol. Med.* **2001**, *31*, 345–354. [CrossRef]
40. Sano, M.; Ernesto, C.; Thomas, R.G.; Klauber, M.R.; Schafer, K.; Grundman, M.; Woodbury, P.; Growdon, J.; Cotman, C.W.; Pfeiffer, E.; et al. A Controlled Trial of Selegiline, Alpha-Tocopherol, or Both as Treatment for Alzheimer's Disease. *N. Engl. J. Med.* **1997**, *336*, 1216–1222. [CrossRef]
41. Petersen, R.C.; Thomas, R.G.; Grundman, M.; Bennett, D.; Doody, R.; Ferris, S.; Galasko, D.; Jin, S.; Kaye, J.; Levey, A.; et al. Vitamin E and Donepezil for the Treatment of Mild Cognitive Impairment. *N. Engl. J. Med.* **2005**, *352*, 2379–2388. [CrossRef]
42. Dysken, M.W.; Sano, M.; Asthana, S.; et al. Effect of vitamin e and memantine on functional decline in alzheimer disease: The team-ad va cooperative randomized trial. *JAMA* **2014**, *311*, 33–44. [CrossRef] [PubMed]
43. Kryscio, R.J.; Abner, E.L.; Caban-Holt, A.; Lovell, M.; Goodman, P.; Darke, A.K.; Yee, M.; Crowley, J.; Schmitt, F.A. Association of Antioxidant Supplement Use and Dementia in the Prevention of Alzheimer's Disease by Vitamin E and Selenium Trial (PREADViSE). *JAMA Neurol.* **2017**, *74*, 567–573. [CrossRef]
44. Freund-Levi, Y.; Eriksdotter-Jönhagen, M.; Cederholm, T.; Basun, H.; Faxén-Irving, G.; Garlind, A.; Vedin, I.; Vessby, B.; Wahlund, L.-O.; Palmblad, J. ω-3 Fatty Acid Treatment in 174 Patients With Mild to Moderate Alzheimer Disease: OmegAD Study: A Randomized Double-blind Trial. *Arch. Neurol.* **2006**, *63*, 1402–1408. [CrossRef] [PubMed]

45. Quinn, J.F.; Raman, R.; Thomas, R.G.; Yurko-Mauro, K.; Nelson, E.B.; Van Dyck, C.; Galvin, J.E.; Emond, J.; Jack, C.R.; Weiner, M.; et al. Docosahexaenoic Acid Supplementation and Cognitive Decline in Alzheimer Disease: A Randomized Trial. *JAMA* **2010**, *304*, 1903–1911. [CrossRef]
46. Yurko-Mauro, K.; McCarthy, D.; Rom, D.; Nelson, E.B.; Ryan, A.S.; Blackwell, A.; Salem, N.; Stedman, M. Beneficial effects of docosahexaenoic acid on cognition in age-related cognitive decline. *Alzheimer's Dement.* **2010**, *6*, 456–464. [CrossRef] [PubMed]
47. Lee, L.K.; Shahar, S.; Chin, A.-V.; Yusoff, N.A.M. Docosahexaenoic acid-concentrated fish oil supplementation in subjects with mild cognitive impairment (MCI): A 12-month randomised, double-blind, placebo-controlled trial. *Psychopharmacology* **2013**, *225*, 605–612. [CrossRef] [PubMed]
48. Gauthier, S.; Aisen, P.S.; Ferris, S.H.; Saumier, D.; Duong, A.; Haine, D.; Garceau, D.; Suhy, J.; Oh, J.; Lau, W.; et al. Effect of tramiprosate in patients with mild-to-moderate alzheimer's disease: Exploratory analyses of the MRI sub-group of the alphase study. *JNHA J. Nutr. Health Aging* **2009**, *13*, 550–557. [CrossRef]
49. Aisen, P.S.; Gauthier, S.; Ferris, S.H.; Saumier, D.; Haine, D.; Garceau, D.; Duong, A.; Suhy, J.; Oh, J.; Lau, W.C.; et al. Tramiprosate in mild-to-moderate Alzheimer's disease—A randomized, double-blind, placebo-controlled, multi-centre study (the Alphase Study). *Arch. Med. Sci. AMS* **2011**, *7*, 102–111. [CrossRef] [PubMed]
50. Aisen, P.S.; Saumier, D.; Briand, R.; Laurin, J.; Gervais, F.; Tremblay, P.; Garceau, D. A Phase II study targeting amyloid-β with 3APS in mild-to-moderate Alzheimer disease. *Neurology* **2006**, *67*, 1757–1763. [CrossRef]
51. Martorana, A.; Di Lorenzo, F.; Manenti, G.; Semprini, R.; Koch, G. Homotaurine Induces Measurable Changes of Short Latency Afferent Inhibition in a Group of Mild Cognitive Impairment Individuals. *Front. Aging Neurosci.* **2014**, *6*, 254. [CrossRef]
52. Xu, S.S.; Gao, Z.X.; Weng, Z.; Du, Z.M.; Xu, W.A.; Yang, J.S.; Zhang, M.L.; Tong, Z.H.; Fang, Y.S.; Chai, X.S. Efficacy of tablet huperzine-A on memory, cognition, and behavior in Alzheimer's disease. *Zhongguo Yao Li Xue Bao* **1995**, *16*, 391–395.
53. Xu, S.S.; Cai, Z.Y.; Qu, Z.W.; Yang, R.M.; Cai, Y.L.; Wang, G.Q.; Su, X.Q.; Zhong, X.S.; Cheng, R.Y.; Xu, W.A.; et al. Huperzine-A in capsules and tablets for treating patients with Alzheimer disease. *Zhongguo Yao Li Xue Bao* **1999**, *20*, 486–490.
54. Rafii, M.S.; Walsh, S.; Little, J.T.; Behan, K.; Reynolds, B.; Ward, C.; Jin, S.; Thomas, R.; Aisen, P.S. A phase II trial of huperzine A in mild to moderate Alzheimer disease. *Neurology* **2011**, *76*, 1389–1394. [CrossRef]
55. Nelson, T.J.; Sun, M.K.; Lim, C.; Sen, A.; Khan, T.; Chirila, F.V.; Alkon, D.L. Bryostatin Effects on Cognitive Function and PKCvarepsilon in Alzheimer's Disease Phase IIa and Expanded Access Trials. *J. Alzheimers Dis.* **2017**, *58*, 521–535. [CrossRef]
56. Farlow, M.R.; Thompson, R.E.; Wei, L.J.; Tuchman, A.J.; Grenier, E.; Crockford, D.; Wilke, S.; Benison, J.; Alkon, D.L. A Randomized, Double-Blind, Placebo-Controlled, Phase II Study Assessing Safety, Tolerability, and Efficacy of Bryostatin in the Treatment of Moderately Severe to Severe Alzheimer's Disease. *J. Alzheimers Dis.* **2018**. [CrossRef]
57. Brusco, L.I.; Márquez, M.; Cardinali, D.P. Monozygotic twins with Alzheimer's disease treated with melatonin: Case report. *J. Pineal Res.* **1998**, *25*, 260–263. [CrossRef]
58. Brusco, L.I.; Marquez, M.; Cardinali, D.P. Melatonin treatment stabilizes chronobiologic and cognitive symptoms in Alzheimer's disease. *Neuro Endocrinol. Lett.* **2000**, *21*, 39–42.
59. Furio, A.M.; Brusco, L.I.; Cardinali, D.P. Possible therapeutic value of melatonin in mild cognitive impairment: A retrospective study. *J. Pineal Res.* **2007**, *43*, 404–409. [CrossRef]
60. Wade, A.G.; Farmer, M.; Harari, G.; Fund, N.; Laudon, M.; Nir, T.; Frydman-Marom, A.; Zisapel, N. Add-on prolonged-release melatonin for cognitive function and sleep in mild to moderate Alzheimer's disease: A 6-month, randomized, placebo-controlled, multicenter trial. *Clin. Interv. Aging* **2014**, *9*, 947–961.
61. Turner, R.S.; Thomas, R.G.; Craft, S.; van Dyck, C.H.; Mintzer, J.; Reynolds, B.A.; Brewer, J.B.; Rissman, R.A.; Raman, R.; Aisen, P.S.; et al. A randomized, double-blind, placebo-controlled trial of resveratrol for Alzheimer disease. *Neurology* **2015**, *85*, 1383–1391. [CrossRef]
62. Zhu, C.W.; Grossman, H.; Neugroschl, J.; Parker, S.; Burden, A.; Luo, X.; Sano, M. A randomized, double-blind, placebo-controlled trial of resveratrol with glucose and malate (RGM) to slow the progression of Alzheimer's disease: A pilot study. *Alzheimer's Dement. Transl. Res. Clin. Interv.* **2018**, *4*, 609–616. [CrossRef]

63. Jones, G.M.; Sahakian, B.J.; Levy, R.; Warburton, D.M.; Gray, J.A. Effects of acute subcutaneous nicotine on attention, information processing and short-term memory in Alzheimer's disease. *Psychopharmacol* **1992**, *108*, 485–494. [CrossRef]
64. Wilson, A.L.; Langley, L.K.; Monley, J.; Bauer, T.; Rottunda, S.; McFalls, E.; Kovera, C.; McCarten, J.R. Nicotine patches in Alzheimer's disease: Pilot study on learning, memory, and safety. *Pharmacol. Biochem. Behav.* **1995**, *51*, 509–514. [CrossRef]
65. White, H.K.; Levin, E.D. Four-week nicotine skin patch treatment effects on cognitive performance in Alzheimer's disease. *Psychopharmacology (Berl)* **1999**, *143*, 158–165. [CrossRef]
66. Baum, L.; Lam, C.W.K.; Cheung, S.K.-K.; Kwok, T.; Lui, V.; Tsoh, J.; Lam, L.; Leung, V.; Hui, E.; Ng, C.; et al. Six-Month Randomized, Placebo-Controlled, Double-Blind, Pilot Clinical Trial of Curcumin in Patients With Alzheimer Disease. *J. Clin. Psychopharmacol.* **2008**, *28*, 110–113. [CrossRef]
67. Bachinskaya, N.; Hoerr, R.; Ihl, R. Alleviating neuropsychiatric symptoms in dementia: The effects of Ginkgo biloba extract EGb 761. Findings from a randomized controlled trial. *Neuropsychiatr. Dis. Treat.* **2011**, *7*, 209–215.
68. Ihl, R.; Bachinskaya, N.; Korczyn, A.D.; Vakhapova, V.; Tribanek, M.; Hoerr, R.; Napryeyenko, O.; Group, G.S. Efficacy and safety of a once-daily formulation of Ginkgo biloba extract EGb 761 in dementia with neuropsychiatric features: A randomized controlled trial. *Int. J. Geriatr. Psychiatry* **2011**, *26*, 1186–1194. [CrossRef]
69. Herrschaft, H.; Nacu, A.; Likhachev, S.; Sholomov, I.; Hoerr, R.; Schlaefke, S. Ginkgo biloba extract EGb 761®in dementia with neuropsychiatric features: A randomised, placebo-controlled trial to confirm the efficacy and safety of a daily dose of 240 mg. *J. Psychiatr. Res.* **2012**, *46*, 716–723. [CrossRef]
70. Ihl, R.; Tribanek, M.; Bachinskaya, N.; for the, G.S.G. Efficacy and Tolerability of a Once Daily Formulation of Ginkgo biloba Extract EGb 761®in Alzheimer's Disease and Vascular Dementia: Results from a Randomised Controlled Trial. *Pharmacopsychiatry* **2012**, *45*, 41–46. [CrossRef]
71. Gavrilova, S.I.; Preuss, U.W.; Wong, J.W.M.; Hoerr, R.; Kaschel, R.; Bachinskaya, N. Efficacy and safety of Ginkgo biloba extract EGb 761®in mild cognitive impairment with neuropsychiatric symptoms: A randomized, placebo-controlled, double-blind, multi-center trial. *Int. J. Geriatr. Psychiatry* **2014**, *29*, 1087–1095. [CrossRef] [PubMed]
72. Akhondzadeh, S.; Shafiee Sabet, M.; Harirchian, M.H.; Togha, M.; Cheraghmakani, H.; Razeghi, S.; Hejazi, S.S.; Yousefi, M.H.; Alimardani, R.; Jamshidi, A.; et al. A 22-week, multicenter, randomized, double-blind controlled trial of Crocus sativus in the treatment of mild-to-moderate Alzheimer's disease. *Psychopharmacology* **2010**, *207*, 637–643. [CrossRef] [PubMed]
73. Akhondzadeh, S.; Noroozian, M.; Mohammadi, M.; Ohadinia, S.; Jamshidi, A.; Khani, M. Melissa officinalis extract in the treatment of patients with mild to moderate Alzheimer's disease: A double blind, randomised, placebo controlled trial. *J. Neurol. Neurosurg. Psychiatry* **2003**, *74*, 863–866. [CrossRef] [PubMed]
74. Arab, H.; Mahjoub, S.; Hajian-Tilaki, K.; Moghadasi, M. The effect of green tea consumption on oxidative stress markers and cognitive function in patients with Alzheimer's disease: A prospective intervention study. *CASP J. Int. Med.* **2016**, *7*, 188–194.
75. Barbagallo, M.; Marotta, F.; Dominguez, L.J. Oxidative stress in patients with Alzheimer's disease: Effect of extracts of fermented papaya powder. *Mediat. Inflamm.* **2015**. [CrossRef] [PubMed]
76. Scholey, A.B.; Tildesley, N.T.; Ballard, C.G.; Wesnes, K.A.; Tasker, A.; Perry, E.K.; Kennedy, D.O. An extract of Salvia (sage) with anticholinesterase properties improves memory and attention in healthy older volunteers. *Psychopharmacology (Berl)* **2008**, *198*, 127–139. [CrossRef]
77. De la Rubia Orti, J.E.; Garcia-Pardo, M.P.; Drehmer, E.; Sancho Cantus, D.; Julian Rochina, M.; Aguilar, M.A.; Hu Yang, I. Improvement of Main Cognitive Functions in Patients with Alzheimer's Disease after Treatment with Coconut Oil Enriched Mediterranean Diet: A Pilot Study. *J. Alzheimers Dis.* **2018**, *65*, 577–587. [CrossRef]
78. Remington, R.; Chan, A.; Lepore, A.; Kotlya, E.; Shea, T.B. Apple juice improved behavioral but not cognitive symptoms in moderate-to-late stage Alzheimer's disease in an open-label pilot study. *Am. J. Alzheimers Dis. Dement.* **2010**, *25*, 367–371. [CrossRef]
79. Krikorian, R.; Shidler, M.D.; Nash, T.A.; Kalt, W.; Vinqvist-Tymchuk, M.R.; Shukitt-Hale, B.; Joseph, J.A. Blueberry supplementation improves memory in older adults. *J. Agric. Food Chem.* **2010**, *58*, 3996–4000. [CrossRef]

80. Szaniszlo, P.; German, P.; Hajas, G.; Saenz, D.N.; Kruzel, M.; Boldogh, I. New insights into clinical trial for Colostrinin in Alzheimer's disease. *J. Nutr. Health Aging* **2009**, *13*, 235–241. [CrossRef]
81. Ahmad, A.A.; Shah, S.; Badshah, H.J.; Kim, M.; Ali, T.H.; Yoon, G.H.; Kim, T.B.; Abid, N.; Ur Rehman, S.; Khan, S.O.; et al. Neuroprotection by vitamin C against ethanol -induced neuroinflammation associated neurodegeneration in developing rat brain. *CNS Neurol. Disord. Drug Targets* **2016**, *15*, 360–370. [CrossRef]
82. Sil, S.; Ghosh, T.; Gupta, P.; Ghosh, R.; Kabir, S.N.; Roy, A. Dual Role of Vitamin C on the Neuroinflammation Mediated Neurodegeneration and Memory Impairments in Colchicine Induced Rat Model of Alzheimer Disease. *J. Mol. Neurosci.* **2016**, *60*, 421–435. [CrossRef]
83. Murakami, K.; Murata, N.; Ozawa, Y.; Kinoshita, N.; Irie, K.; Shirasawa, T.; Shimizu, T. Vitamin C restores behavioral deficits and amyloid-beta oligomerization without affecting plaque formation in a mouse model of Alzheimer's disease. *J. Alzheimers Dis.* **2011**, *26*, 7–18. [CrossRef]
84. Kook, S.Y.; Lee, K.M.; Kim, Y.; Cha, M.Y.; Kang, S.; Baik, S.H.; Lee, H.; Park, R.; Mook-Jung, I. High-dose of vitamin C supplementation reduces amyloid plaque burden and ameliorates pathological changes in the brain of 5XFAD mice. *Cell Death Dis.* **2014**, *5*, e1083. [CrossRef]
85. Rosales-Corral, S.; Tan, D.X.; Reiter, R.J.; Valdivia-Velazquez, M.; Martinez-Barboza, G.; Acosta-Martinez, J.P.; Ortiz, G.G. Orally administered melatonin reduces oxidative stress and proinflammatory cytokines induced by amyloid-beta peptide in rat brain: A comparative, in vivo study versus vitamin C and E. *J. Pineal Res.* **2003**, *35*, 80–84. [CrossRef]
86. Sung, S.; Yao, Y.; Uryu, K.; Yang, H.; Lee, V.M.; Trojanowski, J.Q.; Pratico, D. Early vitamin E supplementation in young but not aged mice reduces Abeta levels and amyloid deposition in a transgenic model of Alzheimer's disease. *FASEB J.* **2004**, *18*, 323–325.
87. Ramalho, M.J.; Loureiro, J.A.; Gomes, B.; Frasco, M.F.; Coelho, M.A.N.; Pereira, M.C. PLGA nanoparticles for calcitriol delivery. In Proceedings of the 2015 IEEE 4th Portuguese Meeting on Bioengineering, ENBENG, Porto, Portugal, 26–28 Februray 2015.
88. Ramalho, M.J.; Loureiro, J.A.; Gomes, B.; Frasco, M.F.; Coelho, M.A.N.; Pereira, M.C. PLGA nanoparticles as a platform for vitamin D-based cancer therapy. *Beilstein J. Nanotechnol.* **2015**, *6*, 1306–1318. [CrossRef]
89. Ramalho, M.J.; Coelho, M.A.N.; Pereira, M.C. Nanoparticles for Delivery of Vitamin D: Challenges and Opportunities. In *A Critical Evaluation of Vitamin D—Clinical Overview*; Gowder, S., Ed.; InTech: Rijeka, Croatia, 2017; p. 11.
90. Yu, J.; Gattoni-Celli, M.; Zhu, H.; Bhat, N.R.; Sambamurti, K.; Gattoni-Celli, S.; Kindy, M.S. Vitamin D3-enriched diet correlates with a decrease of amyloid plaques in the brain of AbetaPP transgenic mice. *J. Alzheimers Dis.* **2011**, *25*, 295–307. [CrossRef]
91. Grimm, M.; Thiel, A.; Lauer, A.; Winkler, J.; Lehmann, J.; Regner, L.; Nelke, C.; Janitschke, D.; Benoist, C.; Streidenberger, O.; et al. Vitamin D and Its Analogues Decrease Amyloid-β (Aβ) Formation and Increase Aβ-Degradation. *Int. J. Mol. Sci.* **2017**, *18*, 2764. [CrossRef]
92. Landel, V.; Millet, P.; Baranger, K.; Loriod, B.; Féron, F. Vitamin D interacts with Esr1 and Igf1 to regulate molecular pathways relevant to Alzheimer's disease. *Mol. Neurodegener.* **2016**, *11*, 22. [CrossRef]
93. Morello, M.; Landel, V.; Lacassagne, E.; Baranger, K.; Annweiler, C.; Féron, F.; Millet, P. Vitamin D Improves Neurogenesis and Cognition in a Mouse Model of Alzheimer's Disease. *Mol. Neurobiol.* **2018**, *55*, 6463–6479. [CrossRef]
94. Kishi, T.; Matsunaga, S.; Oya, K.; Nomura, I.; Ikuta, T.; Iwata, N. Memantine for Alzheimer's disease: An updated systematic review and meta-analysis. *J. Alzheimer's Dis.* **2017**, *60*, 401–425. [CrossRef]
95. Youdim, M.B.H.; Edmondson, D.; Tipton, K.F. The therapeutic potential of monoamine oxidase inhibitors. *Nat. Rev. Neurosci.* **2006**, *7*, 295. [CrossRef]
96. Varikasuvu, S.R.; Prasad, V.S.; Kothapalli, J.; Manne, M. Brain Selenium in Alzheimer's Disease (BRAIN SEAD Study): A Systematic Review and Meta-Analysis. *Biol. Trace Elem. Res.* **2018**. [CrossRef]
97. Hixson, S.M.; Sharma, B.; Kainz, M.J.; Wacker, A.; Arts, M.T. Production, distribution, and abundance of long-chain omega-3 polyunsaturated fatty acids: A fundamental dichotomy between freshwater and terrestrial ecosystems. *Environ. Rev.* **2015**, *23*, 414–424. [CrossRef]
98. Hashimoto, M.; Hossain, S.; Katakura, M.; Al Mamun, A.; Shido, O. The binding of Aβ1-42 to lipid rafts of RBC is enhanced by dietary docosahexaenoic acid in rats: Implicates to Alzheimer's disease. *Biochim. Et Biophys. Acta Biomembr.* **2015**, *1848*, 1402–1409. [CrossRef]

99. Lim, G.P.; Calon, F.; Morihara, T.; Yang, F.; Teter, B.; Ubeda, O.; Salem, N., Jr.; Frautschy, S.A.; Cole, G.M. A diet enriched with the omega-3 fatty acid docosahexaenoic acid reduces amyloid burden in an aged Alzheimer mouse model. *J. Neurosci.* **2005**, *25*, 3032–3040. [CrossRef]
100. Grimm, M.O.; Kuchenbecker, J.; Grosgen, S.; Burg, V.K.; Hundsdorfer, B.; Rothhaar, T.L.; Friess, P.; de Wilde, M.C.; Broersen, L.M.; Penke, B.; et al. Docosahexaenoic acid reduces amyloid beta production via multiple pleiotropic mechanisms. *J. Biol. Chem.* **2011**, *286*, 14028–14039. [CrossRef]
101. Hossain, S.; Hashimoto, M.; Katakura, M.; Miwa, K.; Shimada, T.; Shido, O. Mechanism of docosahexaenoic acid-induced inhibition of in vitro Aβ1-42 fibrillation and Aβ1-42-induced toxicity in SH-S5Y5 cells. *J. Neurochem.* **2009**, *111*, 568–579. [CrossRef]
102. Grimm, M.O.W.; Mett, J.; Stahlmann, C.P.; Haupenthal, V.J.; Blümel, T.; Stötzel, H.; Grimm, H.S.; Hartmann, T. Eicosapentaenoic acid and docosahexaenoic acid increase the degradation of amyloid-β by affecting insulin-degrading enzyme1. *Biochem. Cell Biol.* **2016**, *94*, 534–542. [CrossRef]
103. Hashimoto, M.; Shahdat, H.M.; Yamashita, S.; Katakura, M.; Tanabe, Y.; Fujiwara, H.; Gamoh, S.; Miyazawa, T.; Arai, H.; Shimada, T.; et al. Docosahexaenoic acid disrupts in vitro amyloid β1-40 fibrillation and concomitantly inhibits amyloid levels in cerebral cortex of Alzheimer's disease model rats. *J. Neurochem.* **2008**, *107*, 1634–1646. [CrossRef]
104. Mehdinia, A.; Rostami, S.; Dadkhah, S.; Fumani, N.S. Simultaneous screening of homotaurine and taurine in marine macro-algae using liquid chromatography–fluorescence detection. *J. Iran. Chem. Soc.* **2017**, *14*, 2135–2142. [CrossRef]
105. Santa-Maria, I.; Hernández, F.; Del Rio, J.; Moreno, F.J.; Avila, J. Tramiprosate, a drug of potential interest for the treatment of Alzheimer's disease, promotes an abnormal aggregation of tau. *Mol. Neurodegener.* **2007**, *2*, 17. [CrossRef]
106. Gervais, F.; Paquette, J.; Morissette, C.; Krzywkowski, P.; Yu, M.; Azzi, M.; Lacombe, D.; Kong, X.; Aman, A.; Laurin, J.; et al. Targeting soluble Abeta peptide with Tramiprosate for the treatment of brain amyloidosis. *Neurobiol. Aging* **2007**, *28*, 537–547. [CrossRef]
107. Kocis, P.; Tolar, M.; Yu, J.; Sinko, W.; Ray, S.; Blennow, K.; Fillit, H.; Hey, J.A. Elucidating the Aβ42 Anti-Aggregation Mechanism of Action of Tramiprosate in Alzheimer's Disease: Integrating Molecular Analytical Methods, Pharmacokinetic and Clinical Data. *Cns. Drugs* **2017**, *31*, 495–509. [CrossRef]
108. Marambaud, P.; Zhao, H.; Davies, P. Resveratrol promotes clearance of Alzheimer's disease amyloid-beta peptides. *J. Biol. Chem.* **2005**, *280*, 37377–37382. [CrossRef]
109. Ladiwala, A.R.A.; Lin, J.C.; Bale, S.S.; Marcelino-Cruz, A.M.; Bhattacharya, M.; Dordick, J.S.; Tessier, P.M. Resveratrol selectively remodels soluble oligomers and fibrils of amyloid Aβ into off-pathway conformers. *J. Biol. Chem.* **2010**, *285*, 24228–24237. [CrossRef]
110. Xiao, X.Q.; Zhang, H.Y.; Tang, X.C. Huperzine A attenuates amyloid beta-peptide fragment 25–35-induced apoptosis in rat cortical neurons via inhibiting reactive oxygen species formation and caspase-3 activation. *J. Neurosci. Res.* **2002**, *67*, 30–36. [CrossRef]
111. Peng, Y.; Jiang, L.; Lee, D.Y.; Schachter, S.C.; Ma, Z.; Lemere, C.A. Effects of huperzine A on amyloid precursor protein processing and beta-amyloid generation in human embryonic kidney 293 APP Swedish mutant cells. *J. Neurosci. Res.* **2006**, *84*, 903–911. [CrossRef]
112. Tian, X.R.; Tang, H.F.; Tian, X.L.; Hu, J.J.; Huang, L.L.; Gustafson, K.R. Review of bioactive secondary metabolites from marine bryozoans in the progress of new drugs discovery. *Future Med. Chem.* **2018**, *10*, 1497–1514. [CrossRef]
113. Etcheberrigaray, R.; Tan, M.; Dewachter, I.; Kuiperi, C.; Van der Auwera, I.; Wera, S.; Qiao, L.; Bank, B.; Nelson, T.J.; Kozikowski, A.P.; et al. Therapeutic effects of PKC activators in Alzheimer's disease transgenic mice. *Proc. Natl. Acad. Sci. USA* **2004**, *101*, 11141–11146. [CrossRef]
114. Schrott, L.M.; Jackson, K.; Yi, P.; Dietz, F.; Johnson, G.S.; Basting, T.F.; Purdum, G.; Tyler, T.; Rios, J.D.; Castor, T.P.; et al. Acute Oral Bryostatin-1 Administration Improves Learning Deficits in the APP/PS1 Transgenic Mouse Model of Alzheimer's Disease. *Curr. Alzheimer Res.* **2015**, *12*, 22–31. [CrossRef]
115. Shah, S.A.; Khan, M.; Jo, M.-H.; Jo, M.G.; Amin, F.U.; Kim, M.O. Melatonin Stimulates the SIRT1/Nrf2 Signaling Pathway Counteracting Lipopolysaccharide (LPS)-Induced Oxidative Stress to Rescue Postnatal Rat Brain. *CNS Neurosci. Ther.* **2017**, *23*, 33–44. [CrossRef]

116. Ali, T.; Badshah, H.; Kim, T.H.; Kim, M.O. Melatonin attenuates D-galactose-induced memory impairment, neuroinflammation and neurodegeneration via RAGE/NF-KB/JNK signaling pathway in aging mouse model. *J. Pineal Res.* **2015**, *58*, 71–85. [CrossRef]
117. Pappolla, M.; Bozner, P.; Soto, C.; Shao, H.; Robakis, N.K.; Zagorski, M.; Frangione, B.; Ghiso, J. Inhibition of Alzheimer beta-fibrillogenesis by melatonin. *J. Biol. Chem.* **1998**, *273*, 7185–7188. [CrossRef]
118. Rudnitskaya, E.A.; Muraleva, N.A.; Maksimova, K.Y.; Kiseleva, E.; Kolosova, N.G.; Stefanova, N.A. Melatonin Attenuates Memory Impairment, Amyloid-beta Accumulation, and Neurodegeneration in a Rat Model of Sporadic Alzheimer's Disease. *J. Alzheimers Dis.* **2015**, *47*, 103–116. [CrossRef]
119. Panmanee, J.; Nopparat, C.; Chavanich, N.; Shukla, M.; Mukda, S.; Song, W.; Vincent, B.; Govitrapong, P. Melatonin regulates the transcription of betaAPP-cleaving secretases mediated through melatonin receptors in human neuroblastoma SH-SY5Y cells. *J. Pineal Res.* **2015**, *59*, 308–320. [CrossRef]
120. Andrade, S.; Ramalho, M.J.; Pereira, M.D.C.; Loureiro, J.A. Resveratrol Brain Delivery for Neurological Disorders Prevention and Treatment. *Front. Pharmacol.* **2018**, *9*, 1261. [CrossRef]
121. Andrade, S.; Loureiro, J.A.; Coelho, M.A.N.; Pereira, M.D.C. Interaction studies of amyloid beta-peptide with the natural compound resveratrol. In Proceedings of the 2015 IEEE 4th Portuguese Meeting on Bioengineering (ENBENG), Porto, Portugal, 26–28 Februray 2015; pp. 1–3.
122. Zhao, H.F.; Li, N.; Wang, Q.; Cheng, X.J.; Li, X.M.; Liu, T.T. Resveratrol decreases the insoluble Aβ1-42 level in hippocampus and protects the integrity of the blood-brain barrier in AD rats. *Neuroscience* **2015**, *310*, 641–649. [CrossRef]
123. Karuppagounder, S.S.; Pinto, J.T.; Xu, H.; Chen, H.L.; Beal, M.F.; Gibson, G.E. Dietary supplementation with resveratrol reduces plaque pathology in a transgenic model of Alzheimer's disease. *Neurochem. Int.* **2009**, *54*, 111–118. [CrossRef]
124. Ma, X.R.; Sun, Z.K.; Liu, Y.R.; Jia, Y.J.; Zhang, B.A.; Zhang, J.W. Resveratrol improves cognition and reduces oxidative stress in rats with vascular dementia. *Neural Regen. Res.* **2013**, *8*, 2050–2059.
125. He, X.; Li, Z.; Rizak, J.D.; Wu, S.; Wang, Z.; He, R.; Su, M.; Qin, D.; Wang, J.; Hu, X. Resveratrol attenuates formaldehyde induced hyperphosphorylation of tau protein and cytotoxicity in N2a cells. *Front. Neurosci.* **2017**, *10*, 598. [CrossRef]
126. Salomon, A.R.; Marcinowski, K.J.; Friedland, R.P.; Zagorski, M.G. Nicotine inhibits amyloid formation by the beta-peptide. *Biochemistry* **1996**, *35*, 13568–13578. [CrossRef]
127. Srivareerat, M.; Tran, T.T.; Salim, S.; Aleisa, A.M.; Alkadhi, K.A. Chronic nicotine restores normal Aβ levels and prevents short-term memory and E-LTP impairment in Aβ rat model of Alzheimer's disease. *Neurobiol. Aging* **2011**, *32*, 834–844. [CrossRef]
128. Nordberg, A.; Hellstrom-Lindahl, E.; Lee, M.; Johnson, M.; Mousavi, M.; Hall, R.; Perry, E.; Bednar, I.; Court, J. Chronic nicotine treatment reduces beta-amyloidosis in the brain of a mouse model of Alzheimer's disease (APPsw). *J. Neurochem.* **2002**, *81*, 655–658. [CrossRef]
129. Ono, K.; Hasegawa, K.; Yamada, M.; Naiki, H. Nicotine breaks down preformed Alzheimer's beta-amyloid fibrils in vitro. *Biol. Psychiatry* **2002**, *52*, 880–886. [CrossRef]
130. Navarro, E.; Buendia, I.; Parada, E.; León, R.; Jansen-Duerr, P.; Pircher, H.; Egea, J.; Lopez, M.G. Alpha7 nicotinic receptor activation protects against oxidative stress via heme-oxygenase I induction. *Biochem. Pharmacol.* **2015**, *97*, 473–481. [CrossRef]
131. Lahiri, D.K.; Utsuki, T.; Chen, D.; Farlow, M.R.; Shoaib, M.; Ingram, D.K.; Greig, N.H. Nicotine reduces the secretion of Alzheimer's β-amyloid precursor protein containing β-amyloid peptide in the rat without altering synaptic proteins. *Ann. N. Y. Acad. Sci.* **2002**, *965*, 364–372. [CrossRef]
132. Yang, F.; Lim, G.P.; Begum, A.N.; Ubeda, O.J.; Simmons, M.R.; Ambegaokar, S.S.; Chen, P.P.; Kayed, R.; Glabe, C.G.; Frautschy, S.A.; et al. Curcumin inhibits formation of amyloid beta oligomers and fibrils, binds plaques, and reduces amyloid in vivo. *J. Biol. Chem.* **2005**, *280*, 5892–5901. [CrossRef]
133. Wang, P.; Su, C.; Li, R.; Wang, H.; Ren, Y.; Sun, H.; Yang, J.; Sun, J.; Shi, J.; Tian, J.; et al. Mechanisms and effects of curcumin on spatial learning and memory improvement in APPswe/PS1dE9 mice. *J. Neurosci. Res.* **2014**, *92*, 218–231. [CrossRef]
134. Reddy, P.H.; Manczak, M.; Yin, X.; Grady, M.C.; Mitchell, A.; Kandimalla, R.; Kuruva, C.S. Protective effects of a natural product, curcumin, against amyloid beta induced mitochondrial and synaptic toxicities in Alzheimer's disease. *J. Investig. Med.* **2016**, *64*, 1220–1234.

135. Liu, Z.-J.; Li, Z.-H.; Liu, L.; Tang, W.-X.; Wang, Y.; Dong, M.-R.; Xiao, C. Curcumin Attenuates Beta-Amyloid-Induced Neuroinflammation via Activation of Peroxisome Proliferator-Activated Receptor-Gamma Function in a Rat Model of Alzheimer's Disease. *Front. Pharmacol.* **2016**, *7*, 261. [CrossRef]
136. Lin, R.; Chen, X.; Li, W.; Han, Y.; Liu, P.; Pi, R. Exposure to metal ions regulates mRNA levels of APP and BACE1 in PC12 cells: Blockage by curcumin. *Neurosci. Lett.* **2008**, *440*, 344–347. [CrossRef]
137. Xiong, Z.; Hongmei, Z.; Lu, S.; Yu, L. Curcumin mediates presenilin-1 activity to reduce beta-amyloid production in a model of Alzheimer's Disease. *Pharmacol. Rep. PR* **2011**, *63*, 1101–1108. [CrossRef]
138. Xie, H.; Wang, J.R.; Yau, L.F.; Liu, Y.; Liu, L.; Han, Q.B.; Zhao, Z.; Jiang, Z.H. Catechins and procyanidins of Ginkgo biloba show potent activities towards the inhibition of beta-amyloid peptide aggregation and destabilization of preformed fibrils. *Molecules* **2014**, *19*, 5119–5134. [CrossRef]
139. Belviranlı, M.; Okudan, N. The effects of Ginkgo biloba extract on cognitive functions in aged female rats: The role of oxidative stress and brain-derived neurotrophic factor. *Behav. Brain Res.* **2015**, *278*, 453–461. [CrossRef]
140. Zhang, L.-D.; Ma, L.; Zhang, L.; Dai, J.-G.; Chang, L.-G.; Huang, P.-L.; Tian, X.-Q. Hyperbaric Oxygen and Ginkgo Biloba Extract Ameliorate Cognitive and Memory Impairment via Nuclear Factor Kappa-B Pathway in Rat Model of Alzheimer's Disease. *Chin. Med. J.* **2015**, *128*, 3088–3093. [CrossRef]
141. Colciaghi, F.; Borroni, B.; Zimmermann, M.; Bellone, C.; Longhi, A.; Padovani, A.; Cattabeni, F.; Christen, Y.; Di Luca, M. Amyloid precursor protein metabolism is regulated toward alpha-secretase pathway by Ginkgo biloba extracts. *Neurobiol. Dis.* **2004**, *16*, 454–460. [CrossRef]
142. Yao, Z.X.; Han, Z.; Drieu, K.; Papadopoulos, V. Ginkgo biloba extract (Egb 761) inhibits β-amyloid production by lowering free cholesterol levels. *J. Nutr. Biochem.* **2004**, *15*, 749–756. [CrossRef]
143. Samarghandian, S.; Azimi-Nezhad, M.; Samini, F.; Farkhondeh, T. The Role of Saffron in Attenuating Age-related Oxidative Damage in Rat Hippocampus. *Recent Pat. Food Nutr. Agric.* **2016**, *8*, 183–189. [CrossRef]
144. Moallem, S.A.; Hariri, A.T.; Mahmoudi, M.; Hosseinzadeh, H. Effect of aqueous extract of *Crocus sativus* L. (saffron) stigma against subacute effect of diazinon on specific biomarkers in rats. *Toxicol. Ind. Health* **2012**, *30*, 141–146. [CrossRef] [PubMed]
145. Papandreou, M.A.; Kanakis, C.D.; Polissiou, M.G.; Efthimiopoulos, S.; Cordopatis, P.; Margarity, M.; Lamari, F.N. Inhibitory activity on amyloid-beta aggregation and antioxidant properties of Crocus sativus stigmas extract and its crocin constituents. *J. Agric. Food Chem.* **2006**, *54*, 8762–8768. [CrossRef] [PubMed]
146. Dastmalchi, K.; Damien Dorman, H.J.; Oinonen, P.P.; Darwis, Y.; Laakso, I.; Hiltunen, R. Chemical composition and in vitro antioxidative activity of a lemon balm (*Melissa officinalis* L.) extract. *LWT Food Sci. Technol.* **2008**, *41*, 391–400. [CrossRef]
147. Ozarowski, M.; Mikolajczak, P.L.; Piasecka, A.; Kachlicki, P.; Kujawski, R.; Bogacz, A.; Bartkowiak-Wieczorek, J.; Szulc, M.; Kaminska, E.; Kujawska, M.; et al. Influence of the Melissa officinalis Leaf Extract on Long-Term Memory in Scopolamine Animal Model with Assessment of Mechanism of Action. *Evid. Based Complement. Altern. Med.* **2016**, *2016*, 9729818. [CrossRef] [PubMed]
148. Schimidt, H.L.; Garcia, A.; Martins, A.; Mello-Carpes, P.B.; Carpes, F.P. Green tea supplementation produces better neuroprotective effects than red and black tea in Alzheimer-like rat model. *Food Res. Int.* **2017**, *100*, 442–448. [CrossRef] [PubMed]
149. Li, Q.; Zhao, H.F.; Zhang, Z.F.; Liu, Z.G.; Pei, X.R.; Wang, J.B.; Li, Y. Long-term green tea catechin administration prevents spatial learning and memory impairment in senescence-accelerated mouse prone-8 mice by decreasing Abeta1-42 oligomers and upregulating synaptic plasticity-related proteins in the hippocampus. *Neuroscience* **2009**, *163*, 741–749. [CrossRef] [PubMed]
150. Li, H.; Wu, X.; Wu, Q.; Gong, D.; Shi, M.; Guan, L.; Zhang, J.; Liu, J.; Yuan, B.; Han, G.; et al. Green tea polyphenols protect against okadaic acid-induced acute learning and memory impairments in rats. *Nutrition* **2014**, *30*, 337–342. [CrossRef] [PubMed]
151. Kerem, K.U.; Cengiz, U.M.; Neslihan, T.; Derya, U.; Emine, C.; Emre, E. The Anti-Inflammatory and Antioxidant Effects of Salvia officinalis on Lipopolysaccharide-Induced Inflammation in Rats. *J. Med. Food* **2017**, *20*, 1193–1200.
152. Manalo, R.V.; Silvestre, M.A.; Barbosa, A.L.A.; Medina, P.M. Coconut (Cocos nucifera) Ethanolic Leaf Extract Reduces Amyloid-β (1-42) Aggregation and Paralysis Prevalence in Transgenic Caenorhabditis elegans Independently of Free Radical Scavenging and Acetylcholinesterase Inhibition. *Biomedicines* **2017**, *5*, 17. [CrossRef]

153. Rahim, N.S.; Lim, S.M.; Mani, V.; Abdul Majeed, A.B.; Ramasamy, K. Enhanced memory in Wistar rats by virgin coconut oil is associated with increased antioxidative, cholinergic activities and reduced oxidative stress. *Pharm. Biol.* **2017**, *55*, 825–832. [CrossRef]
154. Bansal, A.; Kirschner, M.; Zu, L.; Cai, D.; Zhang, L. Coconut Oil Decreases Expression of Amyloid Precursor Protein (APP) and Secretion of Amyloid Peptides throughInhibitionof ADP-ribosylation factor 1 (ARF1). *Brain Res.* **2018**. [CrossRef]
155. Cheng, D.; Xi, Y.; Cao, J.; Cao, D.; Ma, Y.; Jiang, W. Protective effect of apple (Ralls) polyphenol extract against aluminum-induced cognitive impairment and oxidative damage in rat. *NeuroToxicology* **2014**, *45*, 111–120. [CrossRef]
156. Chan, A.; Shea, T.B. Dietary Supplementation with Apple Juice Decreases Endogenous Amyloid-beta Levels in Murine Brain. *J. Alzheimers Dis.* **2009**, *16*, 167–171. [CrossRef]
157. Fuentealba, J.; Dibarrart, A.J.; Fuentes-Fuentes, M.C.; Saez-Orellana, F.; Quiñones, K.; Guzmán, L.; Perez, C.; Becerra, J.; Aguayo, L.G. Synaptic failure and adenosine triphosphate imbalance induced by amyloid-β aggregates are prevented by blueberry-enriched polyphenols extract. *J. Neurosci. Res.* **2011**, *89*, 1499–1508. [CrossRef]
158. Shukitt-Hale, B.; Lau, F.C.; Carey, A.N.; Galli, R.L.; Spangler, E.L.; Ingram, D.K.; Joseph, J.A. Blueberry polyphenols attenuate kainic acid-induced decrements in cognition and alter inflammatory gene expression in rat hippocampus. *Nutr. Neurosci.* **2008**, *11*, 172–182. [CrossRef]
159. Zhu, Y.; Bickford, P.C.; Sanberg, P.; Giunta, B.; Tan, J. Blueberry opposes β-amyloid peptide-induced microglial activation via inhibition of p44/42 mitogen-activation protein kinase. *Rejuvenation Res.* **2008**, *11*, 891–901. [CrossRef]
160. Tan, L.; Yang, H.; Pang, W.; Li, H.; Liu, W.; Sun, S.; Song, N.; Zhang, W.; Jiang, Y. Investigation on the Role of BDNF in the Benefits of Blueberry Extracts for the Improvement of Learning and Memory in Alzheimer's Disease Mouse Model. *J. Alzheimers Dis.* **2017**, *56*, 629–640. [CrossRef]
161. Kadakkuzha, B.M.; Liu, X.-A.; Swarnkar, S.; Chen, Y. Chapter 18—Genomic and Proteomic Mechanisms and Models in Toxicity and Safety Evaluation of Nutraceuticals. In *Nutraceuticals*; Gupta, R.C., Ed.; Academic Press: Boston, MA, USA, 2016; pp. 227–237.
162. Schuster, D.; Rajendran, A.; Hui, S.W.; Nicotera, T.; Srikrishnan, T.; Kruzel, M.L. Protective effect of colostrinin on neuroblastoma cell survival is due to reduced aggregation of β-amyloid. *Neuropeptides* **2005**, *39*, 419–426. [CrossRef]
163. Stewart, M.G. Colostrinin™: A naturally occurring compound derived from mammalian colostrum with efficacy in treatment of neurodegenerative diseases, including Alzheimer's. *Expert Opin. Pharmacother.* **2008**, *9*, 2553–2559. [CrossRef]
164. Walker, J.M.; Klakotskaia, D.; Ajit, D.; Weisman, G.A.; Wood, W.G.; Sun, G.Y.; Serfozo, P.; Simonyi, A.; Schachtman, T.R. Beneficial effects of dietary EGCG and voluntary exercise on behavior in an Alzheimer's disease mouse model. *J. Alzheimers Dis.* **2015**, *44*, 561–572. [CrossRef]
165. Harvey, B.S.; Musgrave, I.F.; Ohlsson, K.S.; Fransson, A.; Smid, S.D. The green tea polyphenol (-)-epigallocatechin-3-gallate inhibits amyloid-β evoked fibril formation and neuronal cell death in vitro. *Food Chem.* **2011**, *129*, 1729–1736. [CrossRef]
166. Abbas, S.; Wink, M. Epigallocatechin gallate inhibits beta amyloid oligomerization in Caenorhabditis elegans and affects the daf-2/insulin-like signaling pathway. *Phytomedicine Int. J. Phytother. Phytopharm.* **2010**, *17*, 902–909. [CrossRef]
167. Wobst, H.J.; Sharma, A.; Diamond, M.I.; Wanker, E.E.; Bieschke, J. The green tea polyphenol (-)-epigallocatechin gallate prevents the aggregation of tau protein into toxic oligomers at substoichiometric ratios. *FEBS Lett.* **2015**, *589*, 77–83. [CrossRef]
168. Chesser, A.S.; Ganeshan, V.; Yang, J.; Johnson, G.V.W. Epigallocatechin-3-gallate enhances clearance of phosphorylated tau in primary neurons. *Nutr. Neurosci.* **2016**, *19*, 21–31. [CrossRef]
169. Biasibetti, R.; Tramontina, A.C.; Costa, A.P.; Dutra, M.F.; Quincozes-Santos, A.; Nardin, P.; Bernardi, C.L.; Wartchow, K.M.; Lunardi, P.S.; Gonçalves, C.A. Green tea (-)epigallocatechin-3-gallate reverses oxidative stress and reduces acetylcholinesterase activity in a streptozotocin-induced model of dementia. *Behav. Brain Res.* **2013**, *236*, 186–193. [CrossRef]

170. Lee, Y.J.; Choi, D.Y.; Yun, Y.P.; Han, S.B.; Oh, K.W.; Hong, J.T. Epigallocatechin-3-gallate prevents systemic inflammation-induced memory deficiency and amyloidogenesis via its anti-neuroinflammatory properties. *J. Nutr. Biochem.* **2013**, *24*, 298–310. [CrossRef]
171. Ono, K.; Yoshiike, Y.; Takashima, A.; Hasegawa, K.; Naiki, H.; Yamada, M. Vitamin A exhibits potent antiamyloidogenic and fibril-destabilizing effects in vitro. *Exp. Neurol.* **2004**, *189*, 380–392. [CrossRef]
172. Koryakina, A.; Aeberhard, J.; Kiefer, S.; Hamburger, M.; Kuenzi, P. Regulation of secretases by all-trans-retinoic acid. *FEBS J.* **2009**, *276*, 2645–2655. [CrossRef]
173. Kapoor, A.; Wang, B.J.; Hsu, W.M.; Chang, M.Y.; Liang, S.M.; Liao, Y.F. Retinoic acid-elicited RARalpha/RXRalpha signaling attenuates Abeta production by directly inhibiting gamma-secretase-mediated cleavage of amyloid precursor protein. *ACS Chem. Neurosci.* **2013**, *4*, 1093–1100. [CrossRef]
174. Ding, Y.; Qiao, A.; Wang, Z.; Goodwin, J.S.; Lee, E.-S.; Block, M.L.; Allsbrook, M.; McDonald, M.P.; Fan, G.-H. Retinoic Acid Attenuates β-Amyloid Deposition and Rescues Memory Deficits in an Alzheimer's Disease Transgenic Mouse Model. *J. Neurosci.* **2008**, *28*, 11622–11634. [CrossRef]
175. Arendash, G.W.; Mori, T.; Cao, C.; Mamcarz, M.; Runfeldt, M.; Dickson, A.; Rezai-Zadeh, K.; Tane, J.; Citron, B.A.; Lin, X.; et al. Caffeine reverses cognitive impairment and decreases brain amyloid-beta levels in aged Alzheimer's disease mice. *J. Alzheimers Dis.* **2009**, *17*, 661–680. [CrossRef]
176. Sharma, B.; Kalita, S.; Paul, A.; Mandal, B.; Paul, S. The role of caffeine as an inhibitor in the aggregation of amyloid forming peptides: A unified molecular dynamics simulation and experimental study. *RSC Adv.* **2016**, *6*, 78548–78558. [CrossRef]
177. Qosa, H.; Abuznait, A.H.; Hill, R.A.; Kaddoumi, A. Enhanced brain amyloid-beta clearance by rifampicin and caffeine as a possible protective mechanism against Alzheimer's disease. *J. Alzheimers Dis.* **2012**, *31*, 151–165. [CrossRef]
178. Ullah, F.; Ali, T.; Ullah, N.; Kim, M.O. Caffeine prevents d-galactose-induced cognitive deficits, oxidative stress, neuroinflammation and neurodegeneration in the adult rat brain. *Neurochem. Int.* **2015**, *90*, 114–124. [CrossRef]
179. Laurent, C.; Eddarkaoui, S.; Derisbourg, M.; Leboucher, A.; Demeyer, D.; Carrier, S.; Schneider, M.; Hamdane, M.; Muller, C.E.; Buee, L.; et al. Beneficial effects of caffeine in a transgenic model of Alzheimer's disease-like tau pathology. *Neurobiol. Aging* **2014**, *35*, 2079–2090. [CrossRef]
180. Bend, J.R.; Xia, X.Y.; Chen, D.; Awaysheh, A.; Lo, A.; Rieder, M.J.; Jane Rylett, R. Attenuation of oxidative stress in HEK 293 cells by the TCM constituents schisanhenol, baicalein, resveratrol or crocetin and two defined mixtures. *J. Pharm. Pharm. Sci.* **2015**, *18*, 661–682. [CrossRef]
181. Lu, J.H.; Ardah, M.T.; Durairajan, S.S.; Liu, L.F.; Xie, L.X.; Fong, W.F.; Hasan, M.Y.; Huang, J.D.; El-Agnaf, O.M.; Li, M. Baicalein inhibits formation of alpha-synuclein oligomers within living cells and prevents Abeta peptide fibrillation and oligomerisation. *Chembiochem. A Eur. J. Chem. Biol.* **2011**, *12*, 615–624. [CrossRef]
182. Zhang, S.Q.; Obregon, D.; Ehrhart, J.; Deng, J.; Tian, J.; Hou, H.; Giunta, B.; Sawmiller, D.; Tan, J. Baicalein reduces beta-amyloid and promotes nonamyloidogenic amyloid precursor protein processing in an Alzheimer's disease transgenic mouse model. *J. Neurosci. Res.* **2013**, *91*, 1239–1246. [CrossRef]
183. Manca, M.L.; Marongiu, F.; Castangia, I.; Catalán-Latorre, A.; Caddeo, C.; Bacchetta, G.; Ennas, G.; Zaru, M.; Fadda, A.M.; Manconi, M. Protective effect of grape extract phospholipid vesicles against oxidative stress skin damages. *Ind. Crop. Prod.* **2016**, *83*, 561–567. [CrossRef]
184. Huang, M.; Jiang, X.; Liang, Y.; Liu, Q.; Chen, S.; Guo, Y. Berberine improves cognitive impairment by promoting autophagic clearance and inhibiting production of β-amyloid in APP/tau/PS1 mouse model of Alzheimer's disease. *Exp. Gerontol.* **2017**, *91*, 25–33. [CrossRef]
185. He, W.; Wang, C.; Chen, Y.; He, Y.; Cai, Z. Berberine attenuates cognitive impairment and ameliorates tau hyperphosphorylation by limiting the self-perpetuating pathogenic cycle between NF-κB signaling, oxidative stress and neuroinflammation. *Pharmacol. Rep.* **2017**, *69*, 1341–1348. [CrossRef] [PubMed]
186. Beg, T.; Jyoti, S.; Naz, F.; Ali, F.; Ali, S.K.; Reyad, A.M.; Siddique, Y.H. Protective Effect of Kaempferol on the Transgenic Drosophila Model of Alzheimer's Disease. *CNS Neurol. Disord. Drug Targets* **2018**, *17*, 421–429. [CrossRef] [PubMed]
187. Garcia-Mediavilla, V.; Crespo, I.; Collado, P.S.; Esteller, A.; Sanchez-Campos, S.; Tunon, M.J.; Gonzalez-Gallego, J. The anti-inflammatory flavones quercetin and kaempferol cause inhibition of inducible nitric oxide synthase, cyclooxygenase-2 and reactive C-protein, and down-regulation of the nuclear factor kappaB pathway in Chang Liver cells. *Eur. J. Pharmacol.* **2007**, *557*, 221–229. [CrossRef]

188. Akaishi, T.; Morimoto, T.; Shibao, M.; Watanabe, S.; Sakai-Kato, K.; Utsunomiya-Tate, N.; Abe, K. Structural requirements for the flavonoid fisetin in inhibiting fibril formation of amyloid beta protein. *Neurosci. Lett.* **2008**, *444*, 280–285. [CrossRef] [PubMed]
189. Ono, K.; Yoshiike, Y.; Takashima, A.; Hasegawa, K.; Naiki, H.; Yamada, M. Potent anti-amyloidogenic and fibril-destabilizing effects of polyphenols in vitro: Implications for the prevention and therapeutics of Alzheimer's disease. *J. Neurochem.* **2003**, *87*, 172–181. [CrossRef] [PubMed]
190. Shimmyo, Y.; Kihara, T.; Akaike, A.; Niidome, T.; Sugimoto, H. Flavonols and flavones as BACE-1 inhibitors: Structure-activity relationship in cell-free, cell-based and in silico studies reveal novel pharmacophore features. *Biochim. Biophys. Acta* **2008**, *1780*, 819–825. [CrossRef] [PubMed]
191. Kim, J.H.; Lee, J.; Lee, S.; Cho, E.J. Quercetin and quercetin-3-β-d-glucoside improve cognitive and memory function in Alzheimer's disease mouse. *Appl. Biol. Chem.* **2016**, *59*, 721–728. [CrossRef]
192. Regitz, C.; Dussling, L.M.; Wenzel, U. Amyloid-beta (Abeta(1)(-)(4)(2))-induced paralysis in Caenorhabditis elegans is inhibited by the polyphenol quercetin through activation of protein degradation pathways. *Mol. Nutr. Food Res.* **2014**, *58*, 1931–1940. [CrossRef]
193. Jimenez-Aliaga, K.; Bermejo-Bescos, P.; Benedi, J.; Martin-Aragon, S. Quercetin and rutin exhibit antiamyloidogenic and fibril-disaggregating effects in vitro and potent antioxidant activity in APPswe cells. *Life Sci.* **2011**, *89*, 939–945. [CrossRef]
194. Zhang, X.; Hu, J.; Zhong, L.; Wang, N.; Yang, L.; Liu, C.C.; Li, H.; Wang, X.; Zhou, Y.; Zhang, Y.; et al. Quercetin stabilizes apolipoprotein e and reduces brain Aβ levels in amyloid model mice. *Neuropharmacology* **2016**, *108*, 179–192. [CrossRef]
195. Sabogal-Guáqueta, A.M.; Muñoz-Manco, J.I.; Ramírez-Pineda, J.R.; Lamprea-Rodriguez, M.; Osorio, E.; Cardona-Gómez, G.P. The flavonoid quercetin ameliorates Alzheimer's disease pathology and protects cognitive and emotional function in aged triple transgenic Alzheimer's disease model mice. *Neuropharmacology* **2015**, *93*, 134–145. [CrossRef] [PubMed]
196. Prakash, D.; Sudhandiran, G. Dietary flavonoid fisetin regulates aluminium chloride-induced neuronal apoptosis in cortex and hippocampus of mice brain. *J. Nutr. Biochem.* **2015**, *26*, 1527–1539. [CrossRef] [PubMed]
197. Ahmad, A.; Ali, T.; Park, H.Y.; Badshah, H.; Rehman, S.U.; Kim, M.O. Neuroprotective Effect of Fisetin Against Amyloid-Beta-Induced Cognitive/Synaptic Dysfunction, Neuroinflammation, and Neurodegeneration in Adult Mice. *Mol. Neurobiol.* **2017**, *54*, 2269–2285. [CrossRef] [PubMed]
198. Kim, S.; Choi, K.J.; Cho, S.J.; Yun, S.M.; Jeon, J.P.; Koh, Y.H.; Song, J.; Johnson, G.V.W.; Jo, C. Fisetin stimulates autophagic degradation of phosphorylated tau via the activation of TFEB and Nrf2 transcription factors. *Sci. Rep.* **2016**, *6*, 24933. [CrossRef] [PubMed]
199. Pourkhodadad, S.; Alirezaei, M.; Moghaddasi, M.; Ahmadvand, H.; Karami, M.; Delfan, B.; Khanipour, Z. Neuroprotective effects of oleuropein against cognitive dysfunction induced by colchicine in hippocampal CA1 area in rats. *J. Physiol. Sci.* **2016**, *66*, 397–405. [CrossRef] [PubMed]
200. Luccarini, I.; Ed Dami, T.; Grossi, C.; Rigacci, S.; Stefani, M.; Casamenti, F. Oleuropein aglycone counteracts Aβ42 toxicity in the rat brain. *Neurosci. Lett.* **2014**, *558*, 67–72. [CrossRef]
201. Pantano, D.; Luccarini, I.; Nardiello, P.; Servili, M.; Stefani, M.; Casamenti, F. Oleuropein aglycone and polyphenols from olive mill waste water ameliorate cognitive deficits and neuropathology. *Br. J. Clin. Pharmacol.* **2017**, *83*, 54–62. [CrossRef]
202. Stefania, R.; Valentina, G.; Monica, B.; Daniela, N.; Annalisa, R.; Andrea, B.; Massimo, S. Aβ(1-42) Aggregates into Non-Toxic Amyloid Assemblies in the Presence of the Natural Polyphenol Oleuropein Aglycon. *Curr. Alzheimer Res.* **2011**, *8*, 841–852.
203. Kostomoiri, M.; Fragkouli, A.; Sagnou, M.; Skaltsounis, L.A.; Pelecanou, M.; Tsilibary, E.C.; Tzinia, A.K. Oleuropein, an Anti-oxidant Polyphenol Constituent of Olive Promotes α-Secretase Cleavage of the Amyloid Precursor Protein (AβPP). *Cell. Mol. Neurobiol.* **2013**, *33*, 147–154. [CrossRef]
204. Daccache, A.; Lion, C.; Sibille, N.; Gerard, M.; Slomianny, C.; Lippens, G.; Cotelle, P. Oleuropein and derivatives from olives as Tau aggregation inhibitors. *Neurochem. Int.* **2011**, *58*, 700–707. [CrossRef]
205. Mori, T.; Rezai-Zadeh, K.; Koyama, N.; Arendash, G.W.; Yamaguchi, H.; Kakuda, N.; Horikoshi-Sakuraba, Y.; Tan, J.; Town, T. Tannic acid is a natural beta-secretase inhibitor that prevents cognitive impairment and mitigates Alzheimer-like pathology in transgenic mice. *J. Biol. Chem.* **2012**, *287*, 6912–6927. [CrossRef]

206. Ono, K.; Hasegawa, K.; Naiki, H.; Yamada, M. Anti-amyloidogenic activity of tannic acid and its activity to destabilize Alzheimer's β-amyloid fibrils in vitro. *Biochim. Biophys. Acta (BBA) Mol. Basis Dis.* **2004**, *1690*, 193–202. [CrossRef]
207. Yao, J.; Gao, X.; Sun, W.; Yao, T.; Shi, S.; Ji, L. Molecular hairpin: A possible model for inhibition of tau aggregation by tannic acid. *Biochemistry* **2013**, *52*, 1893–1902. [CrossRef]
208. Ghahghaei, A.; Bathaie, S.Z.; Bahraminejad, E. Mechanisms of the Effects of Crocin on Aggregation and Deposition of Aβ1–40 Fibrils in Alzheimer's Disease. *Int. J. Pept. Res.* **2012**, *18*, 347–351. [CrossRef]
209. Ghahghaei, A.; Bathaie, S.Z.; Kheirkhah, H.; Bahraminejad, E. The protective effect of crocin on the amyloid fibril formation of Abeta42 peptide in vitro. *Cell. Mol. Biol. Lett.* **2013**, *18*, 328–339. [CrossRef]
210. Naghizadeh, B.; Mansouri, M.T.; Ghorbanzadeh, B. Protective effects of crocin against streptozotocin-induced oxidative damage in rat striatum. *Acta Med. Iran* **2014**, *52*, 101–105.
211. Heidari, S.; Mehri, S.; Hosseinzadeh, H. Memory enhancement and protective effects of crocin against D-galactose aging model in the hippocampus of Wistar rats. *Iran. J. Basic Med. Sci.* **2017**, *20*, 1250–1259.
212. Zhang, Z.; Wu, H.; Huang, H. Epicatechin Plus Treadmill Exercise are Neuroprotective Against Moderate-stage Amyloid Precursor Protein/Presenilin 1 Mice. *Pharmacogn. Mag.* **2016**, *12*, S139–S146.
213. Zeng, Y.Q.; Wang, Y.J.; Zhou, X.F. Effects of (-)epicatechin on the pathology of APP/PS1 transgenic mice. *Front. Neurol.* **2014**, *5*, 69. [CrossRef]
214. Cox, C.J.; Choudhry, F.; Peacey, E.; Perkinton, M.S.; Richardson, J.C.; Howlett, D.R.; Lichtenthaler, S.F.; Francis, P.T.; Williams, R.J. Dietary (-)-epicatechin as a potent inhibitor of betagamma-secretase amyloid precursor protein processing. *Neurobiol. Aging* **2015**, *36*, 178–187. [CrossRef]
215. George, R.C.; Lew, J.; Graves, D.J. Interaction of cinnamaldehyde and epicatechin with tau: Implications of beneficial effects in modulating alzheimer's disease pathogenesis. *J. Alzheimer's Dis.* **2013**, *36*, 21–40. [CrossRef]
216. Carbonaro, M.; Di Venere, A.; Filabozzi, A.; Maselli, P.; Minicozzi, V.; Morante, S.; Nicolai, E.; Nucara, A.; Placidi, E.; Stellato, F. Role of dietary antioxidant (−)-epicatechin in the development of β-lactoglobulin fibrils. *Biochim. Biophys. Acta (BBA) Proteins Proteom.* **2016**, *1864*, 766–772. [CrossRef] [PubMed]
217. Hajipour, S.; Sarkaki, A.; Farbood, Y.; Eidi, A.; Mortazavi, P.; Valizadeh, Z. Effect of gallic acid on dementia type of Alzheimer disease in rats: Electrophysiological and histological studies. *Basic Clin. Neurosci.* **2016**, *7*, 97–106. [CrossRef]
218. Kim, M.J.; Seong, A.R.; Yoo, J.Y.; Jin, C.H.; Lee, Y.H.; Kim, Y.J.; Lee, J.; Jun, W.J.; Yoon, H.G. Gallic acid, a histone acetyltransferase inhibitor, suppresses beta-amyloid neurotoxicity by inhibiting microglial-mediated neuroinflammation. *Mol. Nutr. Food Res.* **2011**, *55*, 1798–1808. [CrossRef]
219. Jayamani, J.; Shanmugam, G. Gallic acid, one of the components in many plant tissues, is a potential inhibitor for insulin amyloid fibril formation. *Eur. J. Med. Chem.* **2014**, *85*, 352–358. [CrossRef]
220. Valizadeh, Z.; Eidi, A.; Sarkaki, A.; Farbood, Y.; Mortazavi, P. Dementia type of Alzheimer's disease due to beta-amyloid was improved by Gallic acid in rats. *HealthMED* **2012**, *6*, 3648–3656.
221. Tsai, F.-S.; Wu, L.-Y.; Yang, S.-E.; Cheng, H.-Y.; Tsai, C.-C.; Wu, C.-R.; Lin, L.-W. Ferulic Acid Reverses the Cognitive Dysfunction Caused by Amyloid β Peptide 1-40 Through Anti-Oxidant Activity and Cholinergic Activation in Rats. *Am. J. Chin. Med.* **2015**, *43*, 319–335. [CrossRef] [PubMed]
222. Mori, T.; Koyama, N.; Guillot-Sestier, M.V.; Tan, J.; Town, T. Ferulic acid is a nutraceutical beta-secretase modulator that improves behavioral impairment and alzheimer-like pathology in transgenic mice. *PLoS ONE* **2013**, *8*, e55774. [CrossRef] [PubMed]
223. Zhang, Y.; Cui, L.; Zhang, Y.; Cao, H.; Wang, Y.; Teng, T.; Ma, G.; Li, Y.; Li, K. Ferulic acid inhibits the transition of amyloid-β42 monomers to oligomers but accelerates the transition from oligomers to fibrils. *J. Alzheimer's Dis.* **2013**, *37*, 19–28.
224. Yan, J.J.; Jung, J.S.; Kim, T.K.; Hasan, M.A.; Hong, C.W.; Nam, J.S.; Song, D.K. Protective effects of ferulic acid in amyloid precursor protein plus presenilin-1 transgenic mouse model of Alzheimer disease. *Biol. Pharm. Bull.* **2013**, *36*, 140–143. [CrossRef]
225. Ono, K.; Hirohata, M.; Yamada, M. Ferulic acid destabilizes preformed beta-amyloid fibrils in vitro. *Biochem. Biophys. Res. Commun.* **2005**, *336*, 444–449. [CrossRef] [PubMed]
226. Xu, P.-X.; Wang, S.-W.; Yu, X.-L.; Su, Y.-J.; Wang, T.; Zhou, W.-W.; Zhang, H.; Wang, Y.-J.; Liu, R.-T. Rutin improves spatial memory in Alzheimer's disease transgenic mice by reducing Aβ oligomer level and attenuating oxidative stress and neuroinflammation. *Behav. Brain Res.* **2014**, *264*, 173–180. [CrossRef]

227. Lee, Y.W.; Kim, D.H.; Jeon, S.J.; Park, S.J.; Kim, J.M.; Jung, J.M.; Lee, H.E.; Bae, S.G.; Oh, H.K.; Ho Son, K.H.; et al. Neuroprotective effects of salvianolic acid B on an Aβ25–35 peptide-induced mouse model of Alzheimer's disease. *Eur. J. Pharmacol.* **2013**, *704*, 70–77. [CrossRef] [PubMed]
228. Durairajan, S.S.; Yuan, Q.; Xie, L.; Chan, W.S.; Kum, W.F.; Koo, I.; Liu, C.; Song, Y.; Huang, J.D.; Klein, W.L.; et al. Salvianolic acid B inhibits Abeta fibril formation and disaggregates preformed fibrils and protects against Abeta-induced cytotoxicty. *Neurochem. Int.* **2008**, *52*, 741–750. [CrossRef] [PubMed]
229. Tang, Y.; Huang, D.; Zhang, M.H.; Zhang, W.S.; Tang, Y.X.; Shi, Z.X.; Deng, L.; Zhou, D.H.; Lu, X.Y. Salvianolic acid B inhibits Aβ generation by modulating BACE1 activity in SH-SY5Y-APPsw cells. *Nutrients* **2016**, *8*, 333. [CrossRef] [PubMed]
230. Fiori, J.; Naldi, M.; Bartolini, M.; Andrisano, V. Disclosure of a fundamental clue for the elucidation of the myricetin mechanism of action as amyloid aggregation inhibitor by mass spectrometry. *Electrophoresis* **2012**, *33*, 3380–3386. [CrossRef] [PubMed]
231. Shimmyo, Y.; Kihara, T.; Akaike, A.; Niidome, T.; Sugimoto, H. Multifunction of myricetin on A beta: Neuroprotection via a conformational change of A beta and reduction of A beta via the interference of secretases. *J. Neurosci. Res.* **2008**, *86*, 368–377. [CrossRef] [PubMed]
232. Ramezani, M.; Darbandi, N.; Khodagholi, F.; Hashemi, A. Myricetin protects hippocampal CA3 pyramidal neurons and improves learning and memory impairments in rats with Alzheimer's disease. *Neural Regen Res.* **2016**, *11*, 1976–1980. [CrossRef] [PubMed]
233. Md, S.; Gan, S.Y.; Haw, Y.H.; Ho, C.L.; Wong, S.; Choudhury, H. In vitro neuroprotective effects of naringenin nanoemulsion against β-amyloid toxicity through the regulation of amyloidogenesis and tau phosphorylation. *Int. J. Biol. Macromol.* **2018**, *118*, 1211–1219. [CrossRef] [PubMed]
234. Yang, W.; Ma, J.; Liu, Z.; Lu, Y.; Hu, B.; Yu, H. Effect of naringenin on brain insulin signaling and cognitive functions in ICV-STZ induced dementia model of rats. *Neurol. Sci.* **2014**, *35*, 741–751. [CrossRef]
235. Ghofrani, S.; Joghataei, M.-T.; Mohseni, S.; Baluchnejadmojarad, T.; Bagheri, M.; Khamse, S.; Roghani, M. Naringenin improves learning and memory in an Alzheimer's disease rat model: Insights into the underlying mechanisms. *Eur. J. Pharmacol.* **2015**, *764*, 195–201. [CrossRef]
236. Wu, L.-H.; Lin, C.; Lin, H.-Y.; Liu, Y.-S.; Wu, C.Y.-J.; Tsai, C.-F.; Chang, P.-C.; Yeh, W.-L.; Lu, D.-Y. Naringenin Suppresses Neuroinflammatory Responses Through Inducing Suppressor of Cytokine Signaling 3 Expression. *Mol. Neurobiol.* **2016**, *53*, 1080–1091. [CrossRef]
237. Zhang, J.X.; Xing, J.G.; Wang, L.L.; Jiang, H.L.; Guo, S.L.; Liu, R. Luteolin inhibits fibrillary β-amyloid1-40 -induced inflammation in a human blood-brain barrier mode by suppressing the p38 MAPK-mediated NF-εκB signaling pathways. *Molecules* **2017**, 22. [CrossRef]
238. Liu, R.; Gao, M.; Qiang, G.F.; Zhang, T.T.; Lan, X.; Ying, J.; Du, G.H. The anti-amnesic effects of luteolin against amyloid β25–35 peptide-induced toxicity in mice involve the protection of neurovascular unit. *Neuroscience* **2009**, *162*, 1232–1243. [CrossRef]
239. Wang, H.; Wang, H.; Cheng, H.; Che, Z. Ameliorating effect of luteolin on memory impairment in an Alzheimer's disease model. *Mol. Med. Rep.* **2016**, *13*, 4215–4220. [CrossRef]
240. Zheng, N.; Yuan, P.; Li, C.; Wu, J.; Huang, J. Luteolin Reduces BACE1 Expression through NF-κB and Estrogen Receptor Mediated Pathways in HEK293 and SH-SY5Y Cells. *J. Alzheimer's Dis.* **2015**, *45*, 659–671. [CrossRef]
241. Sawmiller, D.; Li, S.; Shahaduzzaman, M.; Smith, A.J.; Obregon, D.; Giunta, B.; Borlongan, C.V.; Sanberg, P.R.; Tan, J. Luteolin Reduces Alzheimer's Disease Pathologies Induced by Traumatic Brain Injury. *Int. J. Mol. Sci.* **2014**, *15*, 895–904. [CrossRef]
242. Patil, S.P.; Maki, S.; Khedkar, S.A.; Rigby, A.C.; Chan, C. Withanolide A and Asiatic Acid Modulate Multiple Targets Associated with Amyloid-beta Precursor Protein Processing and Amyloid-beta Protein Clearance. *J. Nat. Prod.* **2010**, *73*, 1196–1202. [CrossRef]
243. Dhanasekaran, M.; Holcomb, L.A.; Hitt, A.R.; Tharakan, B.; Porter, J.W.; Young, K.A.; Manyam, B.V. Centella asiatica extract selectively decreases amyloid β levels in hippocampus of Alzheimer's disease animal model. *Phytother. Res.* **2009**, *23*, 14–19. [CrossRef]
244. Veerendra Kumar, M.H.; Gupta, Y.K. Effect of Centella asiatica on cognition and oxidative stress in an intracerebroventricular streptozotocin model of Alzheimer's disease in rats. *Clin. Exp. Pharmacol. Physiol.* **2003**, *30*, 336–342. [CrossRef]

245. Mei, Z.R.; Tan, X.P.; Liu, S.Z.; Huang, H.H. Puerarin alleviates cognitive impairment and tau hyperphosphorylation in APP/PS1 transgenic mice. *Zhongguo Zhongyao Zazhi* **2016**, *41*, 3285–3289.
246. Zhao, S.S.; Yang, W.N.; Jin, H.; Ma, K.G.; Feng, G.F. Puerarin attenuates learning and memory impairments and inhibits oxidative stress in STZ-induced SAD mice. *NeuroToxicol* **2015**, *51*, 166–171. [CrossRef]
247. Mahdy, H.M.; Mohamed, M.R.; Emam, M.A.; Karim, A.M.; Abdel-Naim, A.B.; Khalifa, A.E. The anti-apoptotic and anti-inflammatory properties of puerarin attenuate 3-nitropropionic-acid induced neurotoxicity in rats. *Can. J. Physiol. Pharmacol.* **2014**, *92*, 252–258. [CrossRef]
248. Monti, M.C.; Margarucci, L.; Riccio, R.; Casapullo, A. Modulation of Tau Protein Fibrillization by Oleocanthal. *J. Nat. Prod.* **2012**, *75*, 1584–1588. [CrossRef]
249. Li, W.; Sperry, J.B.; Crowe, A.; Trojanowski, J.Q.; Smith, A.B., III; Lee, V.M.-Y. Inhibition of tau fibrillization by oleocanthal via reaction with the amino groups of tau. *J. Neurochem.* **2009**, *110*, 1339–1351. [CrossRef]
250. Qosa, H.; Batarseh, Y.S.; Mohyeldin, M.M.; El Sayed, K.A.; Keller, J.N.; Kaddoumi, A. Oleocanthal enhances amyloid-β clearance from the brains of TgSwDI mice and in vitro across a human blood-brain barrier model. *ACS Chem. Neurosci.* **2015**, *6*, 1849–1859. [CrossRef]
251. Vion, E.; Page, G.; Bourdeaud, E.; Paccalin, M.; Guillard, J.; Rioux Bilan, A. Trans ε-viniferin is an amyloid-β disaggregating and anti-inflammatory drug in a mouse primary cellular model of Alzheimer's disease. *Mol. Cell. Neurosci.* **2018**, *88*, 1–6. [CrossRef]
252. Jeong, H.Y.; Kim, J.Y.; Lee, H.K.; Ha, D.T.; Song, K.-S.; Bae, K.; Seong, Y.H. Leaf and stem of Vitis amurensis and its active components protect against amyloid β protein (25–35)-induced neurotoxicity. *Arch. Pharmacal Res.* **2010**, *33*, 1655–1664. [CrossRef]
253. Rivière, C.; Papastamoulis, Y.; Fortin, P.-Y.; Delchier, N.; Andriamanarivo, S.; Waffo-Teguo, P.; Kapche, G.D.W.F.; Amira-Guebalia, H.; Delaunay, J.-C.; Mérillon, J.-M.; et al. New stilbene dimers against amyloid fibril formation. *Bioorg. Med. Chem. Lett.* **2010**, *20*, 3441–3443. [CrossRef]
254. McLaurin, J.; Kierstead, M.E.; Brown, M.E.; Hawkes, C.A.; Lambermon, M.H.; Phinney, A.L.; Darabie, A.A.; Cousins, J.E.; French, J.E.; Lan, M.F.; et al. Cyclohexanehexol inhibitors of Abeta aggregation prevent and reverse Alzheimer phenotype in a mouse model. *Nat. Med.* **2006**, *12*, 801–808. [CrossRef]
255. McLaurin, J.; Golomb, R.; Jurewicz, A.; Antel, J.P.; Fraser, P.E. Inositol stereoisomers stabilize an oligomeric aggregate of Alzheimer amyloid beta peptide and inhibit abeta -induced toxicity. *J. Biol. Chem.* **2000**, *275*, 18495–18502. [CrossRef]
256. Wang, M.; Li, Y.; Ni, C.; Song, G. Honokiol Attenuates Oligomeric Amyloid $β_{1-42}$-Induced Alzheimer's Disease in Mice Through Attenuating Mitochondrial Apoptosis and Inhibiting the Nuclear Factor Kappa-B Signaling Pathway. *Cell. Physiol. Biochem.* **2017**, *43*, 69–81. [CrossRef]
257. Wang, D.; Dong, X.; Wang, C. Honokiol Ameliorates Amyloidosis and Neuroinflammation and Improves Cognitive Impairment in Alzheimer's Disease Transgenic Mice. *J. Pharmacol. Exp. Ther.* **2018**, *366*, 470–478. [CrossRef]
258. Balez, R.; Steiner, N.; Engel, M.; Munoz, S.S.; Lum, J.S.; Wu, Y.; Wang, D.; Vallotton, P.; Sachdev, P.; O'Connor, M.; et al. Neuroprotective effects of apigenin against inflammation, neuronal excitability and apoptosis in an induced pluripotent stem cell model of Alzheimer's disease. *Sci. Rep.* **2016**, *6*, 31450. [CrossRef]
259. Zhao, L.; Wang, J.-L.; Liu, R.; Li, X.-X.; Li, J.-F.; Zhang, L. Neuroprotective, Anti-Amyloidogenic and Neurotrophic Effects of Apigenin in an Alzheimer's Disease Mouse Model. *Molecules* **2013**, *18*, 9949. [CrossRef]
260. Wang, Y.; Wang, Y.; Li, J.; Hua, L.; Han, B.; Zhang, Y.; Yang, X.; Zeng, Z.; Bai, H.; Yin, H.; et al. Effects of caffeic acid on learning deficits in a model of Alzheimer's disease. *Int. J. Mol. Med.* **2016**, *38*, 869–875. [CrossRef]
261. Sul, D.; Kim, H.-S.; Lee, D.; Joo, S.S.; Hwang, K.W.; Park, S.-Y. Protective effect of caffeic acid against beta-amyloid-induced neurotoxicity by the inhibition of calcium influx and tau phosphorylation. *Life Sci.* **2009**, *84*, 257–262. [CrossRef]
262. Burton, G.; Ingold, K. beta-Carotene: An unusual type of lipid antioxidant. *Science* **1984**, *224*, 569–573. [CrossRef]
263. Alkam, T.; Nitta, A.; Mizoguchi, H.; Itoh, A.; Nabeshima, T. A natural scavenger of peroxynitrites, rosmarinic acid, protects against impairment of memory induced by Abeta(25-35). *Behav. Brain Res.* **2007**, *180*, 139–145. [CrossRef]

264. Iuvone, T.; De Filippis, D.; Esposito, G.; D'Amico, A.; Izzo, A.A. The spice sage and its active ingredient rosmarinic acid protect PC12 cells from amyloid-beta peptide-induced neurotoxicity. *J. Pharmacol. Exp. Ther.* **2006**, *317*, 1143–1149. [CrossRef]
265. Cornejo, A.; Aguilar Sandoval, F.; Caballero, L.; Machuca, L.; Muñoz, P.; Caballero, J.; Perry, G.; Ardiles, A.; Areche, C.; Melo, F. Rosmarinic acid prevents fibrillization and diminishes vibrational modes associated to β sheet in tau protein linked to Alzheimer's disease. *J. Enzym. Inhib. Med. Chem.* **2017**, *32*, 945–953. [CrossRef]
266. Siddique, Y.H.; Ali, F. Protective effect of nordihydroguaiaretic acid (NDGA) on the transgenic Drosophila model of Alzheimer's disease. *Chem. Biol. Interact.* **2017**, *269*, 59–66. [CrossRef]
267. Ono, K.; Hasegawa, K.; Yoshiike, Y.; Takashima, A.; Yamada, M.; Naiki, H. Nordihydroguaiaretic acid potently breaks down pre-formed Alzheimer's beta-amyloid fibrils in vitro. *J. Neurochem.* **2002**, *81*, 434–440. [CrossRef]
268. Dong, X.; Zhang, D.; Zhang, L.; Li, W.; Meng, X. Osthole improves synaptic plasticity in the hippocampus and cognitive function of Alzheimer's disease rats via regulating glutamate. *Neural Regen Res.* **2012**, *7*, 2325–2332.
269. Jiao, Y.; Kong, L.; Yao, Y.; Li, S.; Tao, Z.; Yan, Y.; Yang, J. Osthole decreases beta amyloid levels through up-regulation of miR-107 in Alzheimer's disease. *Neuropharmacology* **2016**, *108*, 332–344. [CrossRef]
270. Yao, Y.; Wang, Y.; Kong, L.; Chen, Y.; Yang, J. Osthole decreases tau protein phosphorylation via PI3K/AKT/GSK-3β signaling pathway in Alzheimer's disease. *Life Sci.* **2019**, *217*, 16–24. [CrossRef]
271. Kwak, H.M.; Jeon, S.Y.; Sohng, B.H.; Kim, J.G.; Lee, J.M.; Lee, K.B.; Jeong, H.H.; Hur, J.M.; Kang, Y.H.; Song, K.S. beta-Secretase (BACE1) inhibitors from pomegranate (*Punica granatum*) husk. *Arch. Pharmacal Res.* **2005**, *28*, 1328–1332. [CrossRef]
272. Kiasalari, Z.; Heydarifard, R.; Khalili, M.; Afshin-Majd, S.; Baluchnejadmojarad, T.; Zahedi, E.; Sanaierad, A.; Roghani, M. Ellagic acid ameliorates learning and memory deficits in a rat model of Alzheimer's disease: An exploration of underlying mechanisms. *Psychopharmacology* **2017**. [CrossRef]
273. Chai, G.S.; Jiang, X.; Ni, Z.F.; Ma, Z.W.; Xie, A.J.; Cheng, X.S.; Wang, Q.; Wang, J.Z.; Liu, G.P. Betaine attenuates Alzheimer-like pathological changes and memory deficits induced by homocysteine. *J. Neurochem.* **2013**, *124*, 388–396. [CrossRef]
274. Liu, X.P.; Qian, X.; Xie, Y.; Qi, Y.; Peng, M.F.; Zhan, B.C.; Lou, Z.Q. Betaine suppressed Abeta generation by altering amyloid precursor protein processing. *Neurol. Sci.* **2014**, *35*, 1009–1013.
275. Peng, Y.; Hou, C.; Yang, Z.; Li, C.; Jia, L.; Liu, J.; Tang, Y.; Shi, L.; Li, Y.; Long, J.; et al. Hydroxytyrosol mildly improve cognitive function independent of APP processing in APP/PS1 mice. *Mol. Nutr. Food Res.* **2016**, *60*, 2331–2342. [CrossRef]
276. Nardiello, P.; Pantano, D.; Lapucci, A.; Stefani, M.; Casamenti, F. Diet Supplementation with Hydroxytyrosol Ameliorates Brain Pathology and Restores Cognitive Functions in a Mouse Model of Amyloid-beta Deposition. *J. Alzheimers Dis.* **2018**, *63*, 1161–1172. [CrossRef]
277. Kim, T.I.; Lee, Y.K.; Park, S.G.; Choi, I.S.; Ban, J.O.; Park, H.K.; Nam, S.Y.; Yun, Y.W.; Han, S.B.; Oh, K.W.; et al. l-Theanine, an amino acid in green tea, attenuates beta-amyloid-induced cognitive dysfunction and neurotoxicity: Reduction in oxidative damage and inactivation of ERK/p38 kinase and NF-kappaB pathways. *Free Radic. Biol. Med.* **2009**, *47*, 1601–1610. [CrossRef]
278. Ben, P.; Zhang, Z.; Zhu, Y.; Xiong, A.; Gao, Y.; Mu, J.; Yin, Z.; Luo, L. l-Theanine attenuates cadmium-induced neurotoxicity through the inhibition of oxidative damage and tau hyperphosphorylation. *NeuroToxicology* **2016**, *57*, 95–103. [CrossRef] [PubMed]
279. Alonso, E.; Vale, C.; Vieytes, M.R.; Laferla, F.M.; Gimenez-Llort, L.; Botana, L.M. 13-Desmethyl spirolide-C is neuroprotective and reduces intracellular Abeta and hyperphosphorylated tau in vitro. *Neurochem. Int.* **2011**, *59*, 1056–1065. [CrossRef] [PubMed]
280. Alonso, E.; Otero, P.; Vale, C.; Alfonso, A.; Antelo, A.; Giménez-Llort, L.; Chabaud, L.; Guillou, C.; Botana, L.M. Benefit of 13-desmethyl spirolide C treatment in triple transgenic mouse model of Alzheimer disease: Beta-amyloid and neuronal markers improvement. *Curr. Alzheimer Res.* **2013**, *10*, 279–289. [CrossRef]
281. Ferrandiz, M.L.; Alcaraz, M.J. Anti-inflammatory activity and inhibition of arachidonic acid metabolism by flavonoids. *Agents Actions* **1991**, *32*, 283–288. [CrossRef] [PubMed]
282. Thamizhiniyan, V.; Vijayaraghavan, K.; Subramanian, S.P. Gossypin, a flavonol glucoside protects pancreatic beta-cells from glucotoxicity in streptozotocin-induced experimental diabetes in rats. *Biomed. Prev. Nutr.* **2012**, *2*, 239–245. [CrossRef]

283. Zhang, G.L.; Deng, J.P.; Wang, B.H.; Zhao, Z.W.; Li, J.; Gao, L.; Liu, B.L.; Xong, J.R.; Guo, X.D.; Yan, Z.Q.; et al. Gypenosides improve cognitive impairment induced by chronic cerebral hypoperfusion in rats by suppressing oxidative stress and astrocytic activation. *Behav. Pharmacol.* **2011**, *22*, 633–644. [CrossRef] [PubMed]
284. Fujiwara, H.; Tabuchi, M.; Yamaguchi, T.; Iwasaki, K.; Furukawa, K.; Sekiguchi, K.; Ikarashi, Y.; Kudo, Y.; Higuchi, M.; Saido, T.C.; et al. A traditional medicinal herb Paeonia suffruticosa and its active constituent 1,2,3,4,6-penta-*O*-galloyl-beta-D-glucopyranose have potent anti-aggregation effects on Alzheimer's amyloid beta proteins in vitro and in vivo. *J. Neurochem.* **2009**, *109*, 1648–1657. [CrossRef] [PubMed]
285. Cui, H.; King, A.E.; Jacobson, G.A.; Small, D.H. Peripheral treatment with enoxaparin exacerbates amyloid plaque pathology in Tg2576 mice. *J. Neurosci. Res.* **2017**, *95*, 992–999. [CrossRef]
286. Bergamaschini, L.; Rossi, E.; Storini, C.; Pizzimenti, S.; Distaso, M.; Perego, C.; De Luigi, A.; Vergani, C.; De Simoni, M.G. Peripheral treatment with enoxaparin, a low molecular weight heparin, reduces plaques and beta-amyloid accumulation in a mouse model of Alzheimer's disease. *J. Neurosci.* **2004**, *24*, 4181–4186. [CrossRef]
287. Timmer, N.M.; van Dijk, L.; der Zee, C.E.E.M.V.; Kiliaan, A.; de Waal, R.M.W.; Verbeek, M.M. Enoxaparin treatment administered at both early and late stages of amyloid β deposition improves cognition of APPswe/PS1dE9 mice with differential effects on brain Aβ levels. *Neurobiol. Dis.* **2010**, *40*, 340–347. [CrossRef]
288. Gong, E.J.; Park, H.R.; Kim, M.E.; Piao, S.; Lee, E.; Jo, D.G.; Chung, H.Y.; Ha, N.C.; Mattson, M.P.; Lee, J. Morin attenuates tau hyperphosphorylation by inhibiting GSK3beta. *Neurobiol. Dis.* **2011**, *44*, 223–230. [CrossRef]
289. Sachdeva, A.K.; Kuhad, A.; Chopra, K. Naringin ameliorates memory deficits in experimental paradigm of Alzheimer's disease by attenuating mitochondrial dysfunction. *Pharmacol. Biochem. Behav.* **2014**, *127*, 101–110. [CrossRef]
290. Ulamin, F.; Alishah, S.; Ok Kim, M. Vanillic Acid Attenuates Aβ1-42-Induced Oxidative Stress and Cognitive Impairment in Mice. *Sci. Rep.* **2017**, *7*, 40753.
291. Kim, Y.E.; Hwang, C.J.; Lee, H.P.; Kim, C.S.; Son, D.J.; Ham, Y.W.; Hellström, M.; Han, S.B.; Kim, H.S.; Park, E.K.; et al. Inhibitory effect of punicalagin on lipopolysaccharide-induced neuroinflammation, oxidative stress and memory impairment via inhibition of nuclear factor-kappaB. *Neuropharmacology* **2017**, *117*, 21–32. [CrossRef]
292. Chonpathompikunlert, P.; Wattanathorn, J.; Muchimapura, S. Piperine, the main alkaloid of Thai black pepper, protects against neurodegeneration and cognitive impairment in animal model of cognitive deficit like condition of Alzheimer's disease. *Food Chem. Toxicol.* **2010**, *48*, 798–802. [CrossRef]
293. Shrivastava, P.; Vaibhav, K.; Tabassum, R.; Khan, A.; Ishrat, T.; Khan, M.M.; Ahmad, A.; Islam, F.; Safhi, M.M.; Islam, F. Anti-apoptotic and Anti-inflammatory effect of Piperine on 6-OHDA induced Parkinson's Rat model. *J. Nutr. Biochem.* **2013**, *24*, 680–687. [CrossRef]
294. Gao, J.; Inagaki, Y.; Li, X.; Kokudo, N.; Tang, W. Research progress on natural products from traditional Chinese medicine in treatment of Alzheimer's disease. *Drug Discov. Ther.* **2013**, *7*, 46–57.
295. Jeong, J.H.; Jeong, H.R.; Jo, Y.N.; Kim, H.J.; Shin, J.H.; Heo, H.J. Ameliorating effects of aged garlic extracts against Aβ-induced neurotoxicity and cognitive impairment. *BMC Complement. Altern. Med.* **2013**, *13*, 268. [CrossRef]
296. Gupta, V.B.; Indi, S.S.; Rao, K.S. Garlic extract exhibits antiamyloidogenic activity on amyloid-beta fibrillogenesis: Relevance to Alzheimer's disease. *Phytother. Res. PTR* **2009**, *23*, 111–115. [CrossRef]
297. Nillert, N.; Pannangrong, W.; Welbat, J.U.; Chaijaroonkhanarak, W.; Sripanidkulchai, K.; Sripanidkulchai, B. Neuroprotective effects of aged garlic extract on cognitive dysfunction and neuroinflammation induced by β-amyloid in rats. *Nutrients* **2017**, *9*, 24. [CrossRef]
298. Chauhan, N.B. Effect of aged garlic extract on APP processing and tau phosphorylation in Alzheimer's transgenic model Tg2576. *J. Ethnopharmacol.* **2006**, *108*, 385–394. [CrossRef]
299. Ho, S.C.; Chang, K.S.; Chang, P.W. Inhibition of neuroinflammation by cinnamon and its main components. *Food Chem.* **2013**, *138*, 2275–2282. [CrossRef]
300. Modi, K.K.; Roy, A.; Brahmachari, S.; Rangasamy, S.B.; Pahan, K. Cinnamon and Its Metabolite Sodium Benzoate Attenuate the Activation of p21rac and Protect Memory and Learning in an Animal Model of Alzheimer's Disease. *PLoS ONE* **2015**, *10*, e0130398. [CrossRef]
301. Frydman-Marom, A.; Levin, A.; Farfara, D.; Benromano, T.; Scherzer-Attali, R.; Peled, S.; Vassar, R.; Segal, D.; Gazit, E.; Frenkel, D.; et al. Orally administrated cinnamon extract reduces beta-amyloid oligomerization and corrects cognitive impairment in Alzheimer's disease animal models. *PLoS ONE* **2011**, *6*, e16564. [CrossRef]

302. Lauretti, E.; Iuliano, L.; Praticò, D. Extra-virgin olive oil ameliorates cognition and neuropathology of the 3xTg mice: Role of autophagy. *Ann. Clin. Transl. Neurol.* **2017**, *4*, 564–574. [CrossRef]
303. Qosa, H.; Mohamed, L.A.; Batarseh, Y.S.; Alqahtani, S.; Ibrahim, B.; LeVine, H.; Keller, J.N.; Kaddoumi, A. Extra-virgin olive oil attenuates amyloid-β and tau pathologies in the brains of TgSwDI mice. *J. Nutr. Biochem.* **2015**, *26*, 1479–1490. [CrossRef]
304. Amel, N.; Wafa, T.; Samia, D.; Yousra, B.; Issam, C.; Cheraif, I.; Attia, N.; Mohamed, H. Extra virgin olive oil modulates brain docosahexaenoic acid level and oxidative damage caused by 2,4-Dichlorophenoxyacetic acid in rats. *J. Food Sci. Technol.* **2016**, *53*, 1454–1464. [CrossRef]
305. Chauhan, N.; Wang, K.C.; Wegiel, J.; Malik, M.N. Walnut extract inhibits the fibrillization of amyloid beta-protein, and also defibrillizes its preformed fibrils. *Curr. Alzheimer Res.* **2004**, *1*, 183–188. [CrossRef]
306. Zou, J.; Cai, P.-S.; Xiong, C.-M.; Ruan, J.-L. Neuroprotective effect of peptides extracted from walnut (Juglans Sigilata Dode) proteins on Aβ25-35-induced memory impairment in mice. *J. Huazhong Univ. Sci. Technol. Med. Sci.* **2016**, *36*, 21–30. [CrossRef]
307. Balu, M.; Sangeetha, P.; Murali, G.; Panneerselvam, C. Age-related oxidative protein damages in central nervous system of rats: Modulatory role of grape seed extract. *Int. J. Dev. Neurosci.* **2005**, *23*, 501–507. [CrossRef]
308. Ono, K.; Condron, M.M.; Ho, L.; Wang, J.; Zhao, W.; Pasinetti, G.M.; Teplow, D.B. Effects of grape seed-derived polyphenols on amyloid beta-protein self-assembly and cytotoxicity. *J. Biol. Chem.* **2008**, *283*, 32176–32187. [CrossRef]
309. Wang, J.; Ho, L.; Zhao, W.; Ono, K.; Rosensweig, C.; Chen, L.; Humala, N.; Teplow, D.B.; Pasinetti, G.M. Grape-derived polyphenolics prevent Abeta oligomerization and attenuate cognitive deterioration in a mouse model of Alzheimer's disease. *J. Neurosci.* **2008**, *28*, 6388–6392. [CrossRef]
310. Wang, Y.J.; Thomas, P.; Zhong, J.H.; Bi, F.F.; Kosaraju, S.; Pollard, A.; Fenech, M.; Zhou, X.F. Consumption of grape seed extract prevents amyloid-beta deposition and attenuates inflammation in brain of an Alzheimer's disease mouse. *Neurotox Res.* **2009**, *15*, 3–14. [CrossRef]
311. Pervin, M.; Hasnat, M.A.; Lee, Y.M.; Kim, D.H.; Jo, J.E.; Lim, B.O. Antioxidant activity and acetylcholinesterase inhibition of grape skin anthocyanin (GSA). *Molecules* **2014**, *19*, 9403–9418. [CrossRef]
312. Loureiro, J.; Andrade, S.; Duarte, A.; Neves, A.; Queiroz, J.; Nunes, C.; Sevin, E.; Fenart, L.; Gosselet, F.; Coelho, M.; et al. Resveratrol and Grape Extract-loaded Solid Lipid Nanoparticles for the Treatment of Alzheimer's Disease. *Molecules* **2017**, *22*, 277. [CrossRef]
313. Hartman, R.E.; Shah, A.; Fagan, A.M.; Schwetye, K.E.; Parsadanian, M.; Schulman, R.N.; Finn, M.B.; Holtzman, D.M. Pomegranate juice decreases amyloid load and improves behavior in a mouse model of Alzheimer's disease. *Neurobiol. Dis.* **2006**, *24*, 506–515. [CrossRef]
314. Ahmed, A.H.; Subaiea, G.M.; Eid, A.; Li, L.; Seeram, N.P.; Zawia, N.H. Pomegranate extract modulates processing of amyloid-beta precursor protein in an aged Alzheimer's disease animal model. *Curr. Alzheimer Res.* **2014**, *11*, 834–843. [CrossRef]
315. Essa, M.M.; Subash, S.; Akbar, M.; Al-Adawi, S.; Guillemin, G.J. Long-term dietary supplementation of pomegranates, figs and dates alleviate neuroinflammation in a transgenic mouse model of Alzheimer's disease. *PLoS ONE* **2015**, *10*, e0120964. [CrossRef]
316. Subash, S.; Essa, M.M.; Al-Asmi, A.; Al-Adawi, S.; Vaishnav, R.; Braidy, N.; Manivasagam, T.; Guillemin, G.J. Pomegranate from Oman Alleviates the Brain Oxidative Damage in Transgenic Mouse Model of Alzheimer's disease. *J. Tradit. Complement. Med.* **2014**, *4*, 232–238. [CrossRef]
317. Jeong, K.; Shin, Y.C.; Park, S.; Park, J.S.; Kim, N.; Um, J.Y.; Go, H.; Sun, S.; Lee, S.; Park, W.; et al. Ethanol extract of Scutellaria baicalensis Georgi prevents oxidative damage and neuroinflammation and memorial impairments in artificial senescense mice. *J. Biomed. Sci.* **2011**, *18*, 14. [CrossRef]
318. Ebenezer, P.J.; Wilson, C.B.; Wilson, L.D.; Nair, A.R.; J, F. The Anti-Inflammatory Effects of Blueberries in an Animal Model of Post-Traumatic Stress Disorder (PTSD). *PLoS ONE* **2016**, *11*, e0160923. [CrossRef]
319. Ma, H.; Johnson, S.L.; Liu, W.; DaSilva, N.A.; Meschwitz, S.; Dain, J.A.; Seeram, N.P. Evaluation of Polyphenol Anthocyanin-Enriched Extracts of Blackberry, Black Raspberry, Blueberry, Cranberry, Red Raspberry, and Strawberry for Free Radical Scavenging, Reactive Carbonyl Species Trapping, Anti-Glycation, Anti-beta-Amyloid Aggregation, and Microglial Neuroprotective Effects. *Int. J. Mol. Sci.* **2018**, *19*, 461.
320. Sutalangka, C.; Wattanathorn, J.; Muchimapura, S.; Thukham-mee, W. Moringa oleifera mitigates memory impairment and neurodegeneration in animal model of age-related dementia. *Oxid. Med. Cell. Longev.* **2013**, *2013*, 695936. [CrossRef]

321. Forloni, G.; Colombo, L.; Girola, L.; Tagliavini, F.; Salmona, M. Anti-amyloidogenic activity of tetracyclines: Studies in vitro. *FEBS Lett.* **2001**, *487*, 404–407. [CrossRef]
322. Airoldi, C.; Sironi, E.; Dias, C.; Marcelo, F.; Martins, A.; Rauter, A.P.; Nicotra, F.; Jimenez-Barbero, J. Natural compounds against Alzheimer's disease: Molecular recognition of Abeta1-42 peptide by Salvia sclareoides extract and its major component, rosmarinic acid, as investigated by NMR. *Chem. Asian J.* **2013**, *8*, 596–602. [CrossRef]
323. Fujiwara, H.; Iwasaki, K.; Furukawa, K.; Seki, T.; He, M.; Maruyama, M.; Tomita, N.; Kudo, Y.; Higuchi, M.; Saido, T.C.; et al. Uncaria rhynchophylla, a Chinese medicinal herb, has potent antiaggregation effects on Alzheimer's beta-amyloid proteins. *J. Neurosci. Res.* **2006**, *84*, 427–433. [CrossRef]
324. Luo, Q.; Lin, T.; Zhang, C.Y.; Zhu, T.; Wang, L.; Ji, Z.; Jia, B.; Ge, T.; Peng, D.; Chen, W. A novel glyceryl monoolein-bearing cubosomes for gambogenic acid: Preparation, cytotoxicity and intracellular uptake. *Int. J. Pharm.* **2015**, *493*, 30–39. [CrossRef]
325. Williams, P.; Sorribas, A.; Liang, Z. New methods to explore marine resources for Alzheimer's therapeutics. *Curr. Alzheimer Res.* **2010**, *7*, 210–213. [CrossRef]
326. Zhang, H.; Conte, M.M.; Khalil, Z.; Huang, X.-C.; Capon, R.J. New dictyodendrins as BACE inhibitors from a southern Australian marine sponge, Ianthella sp. *RSC Adv.* **2012**, *2*, 4209–4214. [CrossRef]
327. Jeon, S.Y.; Bae, K.; Seong, Y.H.; Song, K.S. Green tea catechins as a BACE1 (beta-secretase) inhibitor. *Bioorg. Med. Chem. Lett.* **2003**, *13*, 3905–3908. [CrossRef]
328. Leiros, M.; Alonso, E.; Rateb, M.E.; Houssen, W.E.; Ebel, R.; Jaspars, M.; Alfonso, A.; Botana, L.M. Gracilins: Spongionella-derived promising compounds for Alzheimer disease. *Neuropharmacology* **2015**, *93*, 285–293. [CrossRef]
329. Zhang, H.; Conte, M.M.; Huang, X.C.; Khalil, Z.; Capon, R.J. A search for BACE inhibitors reveals new biosynthetically related pyrrolidones, furanones and pyrroles from a southern Australian marine sponge, Ianthella sp. *Org. Biomol. Chem.* **2012**, *10*, 2656–2663. [CrossRef]
330. Williams, P.; Sorribas, A.; Howes, M.J. Natural products as a source of Alzheimer's drug leads. *Nat. Prod. Rep.* **2011**, *28*, 48–77. [CrossRef]
331. Liu, J.; Chen, W.; Xu, Y.; Ren, S.; Zhang, W.; Li, Y. Design, synthesis and biological evaluation of tasiamide B derivatives as BACE1 inhibitors. *Bioorg. Med. Chem.* **2015**, *23*, 1963–1974. [CrossRef]
332. Dai, J.; Sorribas, A.; Yoshida, W.Y.; Kelly, M.; Williams, P.G. Topsentinols, 24-isopropyl steroids from the marine sponge Topsentia sp. *J. Nat. Prod.* **2010**, *73*, 1597–1600. [CrossRef]
333. Dai, J.; Sorribas, A.; Yoshida, W.Y.; Kelly, M.; Williams, P.G. Xestosaprols from the Indonesian marine sponge Xestospongia sp. *J. Nat. Prod.* **2010**, *73*, 1188–1191. [CrossRef]
334. Alonso, E.; Vale, C.; Vieytes, M.R.; Botana, L.M. Translocation of PKC by yessotoxin in an in vitro model of Alzheimer's disease with improvement of tau and beta-amyloid pathology. *ACS Chem. Neurosci.* **2013**, *4*, 1062–1070. [CrossRef]
335. Alonso, E.; Fuwa, H.; Vale, C.; Suga, Y.; Goto, T.; Konno, Y.; Sasaki, M.; LaFerla, F.M.; Vieytes, M.R.; Gimenez-Llort, L.; et al. Design and synthesis of skeletal analogues of gambierol: Attenuation of amyloid-beta and tau pathology with voltage-gated potassium channel and *N*-methyl-D-aspartate receptor implications. *J. Am. Chem. Soc.* **2012**, *134*, 7467–7479. [CrossRef]
336. Alonso, E.; Vale, C.; Vieytes, M.R.; Laferla, F.M.; Gimenez-Llort, L.; Botana, L.M. The cholinergic antagonist gymnodimine improves Abeta and tau neuropathology in an in vitro model of Alzheimer disease. *Cell. Physiol. Biochem. Int. J. Exp. Cell. Physiol. Biochem. Pharmacol.* **2011**, *27*, 783–794. [CrossRef]
337. Bidon-Chanal, A.; Fuertes, A.; Alonso, D.; Perez, D.I.; Martinez, A.; Luque, F.J.; Medina, M. Evidence for a new binding mode to GSK-3: Allosteric regulation by the marine compound palinurin. *Eur. J. Med. Chem.* **2013**, *60*, 479–489. [CrossRef]
338. Li, G.D.; Yan, W.H.; Xing, Y. Effects of schisandrone on Tau protein hyperphosphorylation in differentiation of neural stem cells from APP transgenic mice. *J. Clin. Rehabil. Tissue Eng. Res.* **2009**, *13*, 4490–4494.
339. Choi, B.; Kim, S.; Jang, B.G.; Kim, M.J. Piceatannol, a natural analogue of resveratrol, effectively reduces beta-amyloid levels via activation of alpha-secretase and matrix metalloproteinase-9. *J. Funct. Foods* **2016**, *23*, 124–134. [CrossRef]
340. Leiros, M.; Sanchez, J.A.; Alonso, E.; Rateb, M.E.; Houssen, W.E.; Ebel, R.; Jaspars, M.; Alfonso, A.; Botana, L.M. Spongionella secondary metabolites protect mitochondrial function in cortical neurons against oxidative stress. *Mar. Drugs* **2014**, *12*, 700–718. [CrossRef]

341. Soininen, H.; Solomon, A.; Visser, P.J.; Hendrix, S.B.; Blennow, K.; Kivipelto, M.; Hartmann, T.; Hallikainen, I.; Hallikainen, M.; Helisalmi, S.; et al. 24-month intervention with a specific multinutrient in people with prodromal Alzheimer's disease (LipiDiDiet): A randomised, double-blind, controlled trial. *Lancet Neurol.* **2017**, *16*, 965–975. [CrossRef]
342. Ramalho, M.J.; Andrade, S.; Coelho, M.Á.N.; Loureiro, J.A.; Pereira, M.C. Biophysical interaction of temozolomide and its active metabolite with biomembrane models: The relevance of drug-membrane interaction for Glioblastoma Multiforme therapy. *Eur. J. Pharm. Biopharm.* **2019**, *136*, 156–163. [CrossRef]
343. Ramalho, M.J.; Sevin, E.; Gosselet, F.; Lima, J.; Coelho, M.A.N.; Loureiro, J.A.; Pereira, M.C. Receptor-mediated PLGA nanoparticles for glioblastoma multiforme treatment. *Int. J. Pharm.* **2018**, *545*, 84–92. [CrossRef]
344. Ramalho, M.J.; Pereira, M.C. Preparation and Characterization of Polymeric Nanoparticles: An Interdisciplinary Experiment. *J. Chem. Educ.* **2016**, *93*, 1446–1451. [CrossRef]
345. Jia, T.; Sun, Z.; Lu, Y.; Gao, J.; Zou, H.; Xie, F.; Zhang, G.; Xu, H.; Sun, D.; Yu, Y.; et al. A dual brain-targeting curcumin-loaded polymersomes ameliorated cognitive dysfunction in intrahippocampal amyloid-beta1-42-injected mice. *Int. J. Nanomed.* **2016**, *11*, 3765–3775.
346. Djiokeng Paka, G.; Doggui, S.; Zaghmi, A.; Safar, R.; Dao, L.; Reisch, A.; Klymchenko, A.; Roullin, V.G.; Joubert, O.; Ramassamy, C. Neuronal Uptake and Neuroprotective Properties of Curcumin-Loaded Nanoparticles on SK-N-SH Cell Line: Role of Poly(lactide-co-glycolide) Polymeric Matrix Composition. *Mol. Pharm.* **2016**, *13*, 391–403. [CrossRef]
347. Mourtas, S.; Lazar, A.N.; Markoutsa, E.; Duyckaerts, C.; Antimisiaris, S.G. Multifunctional nanoliposomes with curcumin–lipid derivative and brain targeting functionality with potential applications for Alzheimer disease. *Eur. J. Med. Chem.* **2014**, *80*, 175–183. [CrossRef]
348. Hoppe, J.B.; Coradini, K.; Frozza, R.L.; Oliveira, C.M.; Meneghetti, A.B.; Bernardi, A.; Pires, E.S.; Beck, R.C.R.; Salbego, C.G. Free and nanoencapsulated curcumin suppress β-amyloid-induced cognitive impairments in rats: Involvement of BDNF and Akt/GSK-3β signaling pathway. *Neurobiol. Learn. Mem.* **2013**, *106*, 134–144. [CrossRef]
349. Zhang, J.; Zhou, X.; Yu, Q.; Yang, L.; Sun, D.; Zhou, Y.; Liu, J. Epigallocatechin-3-gallate (EGCG)-Stabilized Selenium Nanoparticles Coated with Tet-1 Peptide To Reduce Amyloid-β Aggregation and Cytotoxicity. *ACS Appl. Mater. Interfaces* **2014**, *6*, 8475–8487. [CrossRef]
350. Debnath, K.; Shekhar, S.; Kumar, V.; Jana, N.R.; Jana, N.R. Efficient Inhibition of Protein Aggregation, Disintegration of Aggregates, and Lowering of Cytotoxicity by Green Tea Polyphenol-Based Self-Assembled Polymer Nanoparticles. *ACS Appl. Mater. Interfaces* **2016**, *8*, 20309–20318. [CrossRef]
351. Meng, Q.; Wang, A.; Hua, H.; Jiang, Y.; Wang, Y.; Mu, H.; Wu, Z.; Sun, K. Intranasal delivery of Huperzine A to the brain using lactoferrin-conjugated *N*-trimethylated chitosan surface-modified PLGA nanoparticles for treatment of Alzheimer's disease. *Int. J. Nanomed.* **2018**, *13*, 705–718. [CrossRef]
352. Elnaggar, Y.S.; Etman, S.M.; Abdelmonsif, D.A.; Abdallah, O.Y. Novel piperine-loaded Tween-integrated monoolein cubosomes as brain-targeted oral nanomedicine in Alzheimer's disease: Pharmaceutical, biological, and toxicological studies. *Int. J. Nanomed.* **2015**, *10*, 5459–5473. [CrossRef]
353. Kuo, Y.-C.; Tsao, C.-W. Neuroprotection against apoptosis of SK-N-MC cells using RMP-7- and lactoferrin-grafted liposomes carrying quercetin. *Int. J. Nanomed.* **2017**, *12*, 2857–2869. [CrossRef]

MDPI
St. Alban-Anlage 66
4052 Basel
Switzerland
Tel. +41 61 683 77 34
Fax +41 61 302 89 18
www.mdpi.com

International Journal of Molecular Sciences Editorial Office
E-mail: ijms@mdpi.com
www.mdpi.com/journal/ijms

www.ingramcontent.com/pod-product-compliance
Lightning Source LLC
LaVergne TN
LVHW070220100526
838202LV00015B/2066
* 9 7 8 3 0 3 9 3 6 2 1 6 5 *